and Metabolism

Robert P. Wagner

Department of Zoology
University of Texas

Herschel K. Mitchell

Division of Biology
California Institute of Technology

John Wiley and Sons, Inc., *New York* · *London* · *Sydney*

Library of Congress Catalog Card Number: 63-18630
Printed in the United States of America

Genetics and Metabolism

Genetics

Second Edition

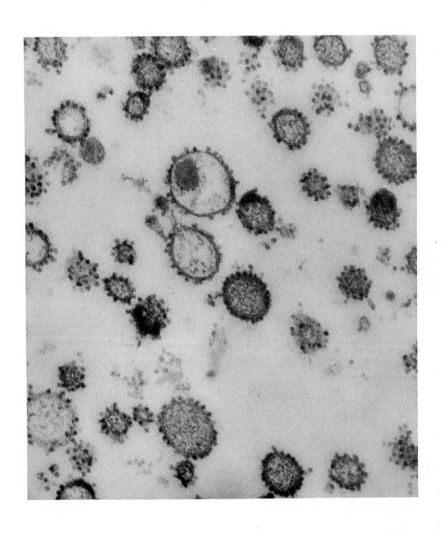

. . . inheritance is the recurrence,
in successive generations, of like
forms of metabolism.

E. B. Wilson, 1896

Preface

The purpose of this second edition remains the same as the first—to bring together a variety of facts derived from the diverse disciplines which contribute to our understanding of the biochemical basis of inheritance. The interim period between the two editions has produced many changes by introducing not only new facts but new concepts and outlooks. Chief among these concepts has been the rise of "molecular genetics," or the attempt to put inheritance on a molecular basis. There is nothing new about this idea, of course, for such attempts were made starting in the last century; but now it is possible to state unequivocally that the basis of inheritance resides in the ability of certain molecules to be reproduced. Not only are these molecules reproduced in their own kind, but they are also able by virtue of their structures to control the synthesis of the rest of the cell and, indirectly, its functioning. This has caused a major revolution not only in genetics but in biochemistry for it has led to more intensive studies in the structure and biosynthesis of proteins and nucleic acids at many different levels of integration, and to a more sophisticated attitude toward the nature of the genetic material.

Despite the major advances made in the last few years, however, there still remains much to be known. Major riddles remain to be solved, and for this reason we attempt here not only to cover the material that is of immediate interest but to draw attention to problems that are "unfashionable" but still important. Furthermore, we have attempted to approach our subject on as broad a front as possible. As a result we have retained in this edition a considerable body of material that is not considered "biochemical genetics" per se by most modern workers, but which we believe is important for any student of the subject to know if he is to have a well-rounded picture of the general field.

We wish to acknowledge with thanks the invaluable assistance of Miss Nina Steher, Mrs. Esther Eakin, Mrs. Annamarie Mitchell, Miss Kathleen Mitchell, and Mrs. Margaret Wagner in the preparation of the manuscript. We also give thanks to the many investigators who provided us with original photographs and figures. To our many colleagues around the world who have made suggestions and offered constructive criticism, we give our thanks with gratitude.

ROBERT P. WAGNER
HERSCHEL K. MITCHELL

Austin, Texas
Pasadena, California
January 1964

Contents

ix

1 Introduction

This book deals almost totally with how genes act and the effects of their actions, rather than with how they are inherited. Before taking up the main discussion, however, it is first necessary to make some attempt to define the term *gene*. This is, as will be seen, a rather difficult task to undertake, because the term has changed in meaning over the years, resulting in a certain amount of confusion regarding its definition.

Perhaps the best way to approach the definition of an important concept in science, such as the gene, is to trace its origins in history. In so doing we will attempt to focus attention on certain ancillary facts which will be of value in understanding the succeeding chapters.

Origin and Development of the Gene Concept

In order to insure the continued existence of their species, all organisms must be able to survive as individuals until they reproduce their own kind. This statement is so obvious it seems trivial. But from it flows practically all of biology. Intrinsic in the statement is the fact that all organisms have a certain form and function which they transmit to their progeny. The ancient Greek thinkers addressed themselves to the explanation of this fact, and arrived at a variety of explanations, none of which was satisfactory, but all of which had a powerful influence on future thought.

Early Theory. As a general rule, most of the ancients thought in terms of spontaneous generation, especially for the lower forms of life which were believed to arise from unorganized masses or slimes. It was recognized that some of the higher forms came from eggs, but no one seemed to have any idea about how eggs originated. Aristotle (384–322 B.C.) considered the problem of reproduction in plants and animals in great detail but could not offer a rational explanation for the origin of new generations. He thought in terms of "soul" which gives form to inanimate matter out of which living things then develop, rather than in mechanistic terms. To him an animal developing from an egg did so because "soul" was present. The egg, he believed, had a self-realization, called *entelechy,* to be an animal of a specific type. This type of abstract thinking had a powerful influence on future biology. It led directly to William Harvey's (1578–1657 A.D.) development of the concept of *epigenesis* whereby an organized adult develops from an unorganized egg or inanimate matter. Epigenesis, as such, had the fatal flaw of not being able to explain why offspring resemble their parents, for it essentially depended on a new creation for each new individual by spontaneous generation.

By the seventeenth century A.D., the existence of spontaneous generation as a general phenomenon began to be strongly questioned, and the way was left open for the development of a new theory, *preformation* or *evolution* (not to be confused with the generally accepted meaning of evolution). This theory originated in ancient Greece, too, but it was rejected by Aristotle and therefore was not found acceptable to the early biologists, such as Harvey, who were essentially Aristotelians in their thinking. Its rejuvenation was primarily the result of the speculations of J. Swammerdam (1637–1680) and C. Bonnet (1720–1793). As stated by Whitman (1075),* in a more recent commentary, preformation

. . . was the negation of all new creation. It was the dogma of special creation, according to which all real formation was completed at the beginning of the world. This was syngenesis versus epigenesis, original formation of all at one time in opposition to new formation all the time.

* Bibliographic references are compiled in the back of this book.

To the preformationist all was simple, provided one accepted the initial causes. The theory reached the peak of its development with Bonnet's encapsulation theory according to which the original individual of a species contained within it at the time of its creation the fully preformed individuals of all following successive generations, one within the other. This accounted for an orderly inheritance which insured that dogs come from dogs, cats from cats, and so on. It also explained development from an egg as a process of "unfolding" something already organized and preformed in all essential details in contradistinction to epigenesis which is organization from the unorganized.

C. F. Wolff (1733–1794) was probably the first observer to make it explicit that preformation as an explanation of development within the hen's egg had no basis in fact. The microscope was developed to a point where it became easy to see that the embryo did, in fact, develop epigenetically from materials which had a structure quite different from the structure in the completely developed organism. As a result of Wolff's work, and the observations of J. F. Blumenbach (1752–1840) and P. L. M. de Maupertuis (1698–1759), when the nineteenth century opened a large part of the scientific world had reversed its position and epigenesis was again a generally accepted theory. However, the question of origin of the egg was still undecided, and if spontaneous generation could not be used as an explanation, then there was no other suitable explanation. Furthermore, what was there to guide the egg's development? In a sense, little advance beyond Aristotle and Harvey had been made other than that all could now see with their own eyes the proof of the epigenetic development of the embryo.

Origin of Cell Biology. Thinking on this subject had reached an impasse which was to be partly broken by Schleiden and Schwann in 1839 when they made the first definitive statement of the cell theory. Their contribution gave new impetus to the investigation of the general problem of origin and development of living organisms, because they made it clear that the cell is the functional unit of life, and that the egg is a cell. R. Virchow and others followed soon after with the observation that all cells came from preexisting cells. To quote from Virchow's great treatise published in 1858,

Wherever a cell arises, there must have been a previous cell of like kind, so that the animal begets an animal, the plant begets a plant. In this way, although there are points in the theory where strict proof is wanting, . . . , the axiom is established that the whole spectrum of living organisms illustrates an eternal law of continuous development (1031).

This, according to Wilson (1081), is the first clear statement of the *law of genetic continuity* in which it is implied that there has been "an uninterrupted series of cell-divisions extending backward from existing plants and animals to that remote and unknown period when vital organization assumed its present form. Life is a continuous stream."

The realization of this concept, which seems obvious to us today, but which required centuries for its development, marked the beginning of a new, and the present great, epoch in biology. The question of origin which had plagued the epigenesists was answered by a theory with a definite preformationist implication stating that the embryo developed from the zygote formed by the union of sperm and egg, and that the zygote contained within itself *not the preformed embryo,* but the *capacity to produce the embryo.* In other words, the embryo is predetermined within the zygote. The origin of the sperm and egg was explained by cell divisions according to the principle that cells come from cells. Here we have preformation and epigenesis all over again. The determiners in the fertilized egg must have been formed in the previous generation, and, in order to produce something resembling the parents, they must have a preformed capacity to do so. The process of controlling development is, on the other hand, control of epigenesis.

To the embryologist it became clear that "the question is no longer whether *all* is preformation or *all* is postformation; it is rather: How far is postformation to be explained by preformation and how far as the result of external influences?" (1075) What had previously been thought to be antithetical theories now were in modified form welded together to form a very sound, new theory. It is to be noticed that the old preformationist idea of creation occurring only at the beginning, with no new creation following, had been cast out. In its place was substituted the idea of new creation every time a cell divided. This was new creation of a different type than spontaneous generation, however, for it meant the formation within the dividing mother cell of materials to be distributed to each of the new daughter cells. *Reduplication* took place instead of spontaneous generation. Reduplication of the essential parts of the mother cell was postulated, so that the daughter cells would be related to the mother cell physically, and not merely spiritually as Aristotle would have had it.

To August Weismann (1059) the presence of determiners in the cell was a necessary part of the concept of life through the germ line. Where were these determiners located? Were they to be found throughout the cell or only in certain parts? Or was the idea of discrete determiners only a help to the understanding of control of development, and no such things existed in the cell? Must a cell, in order to divide, reduplicate every particle of itself to be passed on to the next generation? An answer to these questions was given in part by the observation that the essential process in fertilization resulting in zygote formation appeared to be the union of the nucleus of the egg and sperm. This was concluded by O. Hertwig (437) working with animals and by Strassburger (977) with plants. It was based on the fact that the fusion nucleus formed from the gametic nuclei was the only part of the fertilized egg which seemed to have an equal share of material from each parent. The egg contributed considerable cytoplasm to the zygote, but the sperm little or none. On the other hand, offspring resemble both parents to a considerable degree, and this must therefore mean that the basis of inheritance resides in the nucleus. At about this same time Van Beneden (1028) observed the reduction of the chromosomes in maturing germ cells, a phenomenon which was speedily verified by others. This finding tended to draw still more attention to the nucleus and its chromosomes and their possible role in heredity, although the complete significance of the mechanism of meiosis was not then appreciated.

In 1884 Nägeli (730) postulated that the physical basis of inheritance resided in a cell stuff which he called idioplasm. The idioplasm was made up of *micellae* which were physiological units or determinants, and the cell processes were all supposed to be under the control of these determinants. This concept was, of course, purely an hypothesis with little or no basis in fact. But it had an important influence, because it emphasized the mechanistic point of view which then needed bolstering against vitalism. Almost simultaneously, Hertwig, Strassburger, Kölliker, and Weismann were led to hypothesize independently that the chromatin in the nucleus was the idioplasm postulated by Nägeli. The chromatin, as it could be seen, appeared as a diffuse network, but became manifest as discrete chromosomes which reduplicated just before the actual cell divisions (284). Roux pointed out as early as 1883 that by cell division half of the chromosomal material is distributed to each daughter cell in such a way that the daughter nuclei ends up with exactly equivalent portions of chromatin from the mother nucleus (856). His statement still stands today as the best description of the essential result of mitosis.

Ultimate protoplasmic units associated with the chromatin were postulated by various theorizers. These were given various names of which the pangens of De Vries and the biophores of Weismann received the most attention. De Vries' (218) pangens preexisted, according to him, in the germ cells and got into the body cells during division. Each kind of pangen controlled a unit character. Weismann called the pangens biophores, and assumed that they were grouped on the chromosomes, forming there the ultimate determinative particles which controlled the processes in the cytoplasm. He elaborated on this hypothesis at length and postulated, for example, that the biophores segregated during development so that the different tissue cells ended up with different particles with only the germ cells containing all the types of biophores necessary to direct the construction of a complete new organism.

As early as 1871 F. Miescher had isolated animal cell nuclei and extracted from them a substance he called *nuclein* (675). It became apparent after further purification that nuclein was really constituted of an acid, called nucleinic acid, and protein (15, 676). Nucleinic acid is now known as nucleic acid. Soon after the recognition of nuclein in the nucleus, it was postulated that the chromatin of the nucleus was really nuclein, and that the physical basis of inheritance resided in nuclein (1081).

The nineteenth century ended with the science of cytology having been well established as an important branch of biology and having made considerable contribution to the study of inheritance. There was no doubt in the minds of the prominent cytologists of the time, foremost among them E. B. Wilson, that the chromosomes carried a material which was duplicated when the chromosomes duplicated, and which determined the future functioning and development of the cell. What was needed was a good experimental approach to enable the biologist to either prove or disprove this decisively. There were really several problems recognized which may be classified under three categories: the nature of the determiners and their location; the way in which the determiners are transmitted from generation to generation through both the germ cells and the somatic cells; and the role of the determiners in controlling cell functions and development.

The first two of these problems have been essentially solved by the combined efforts of geneticists, cytologists, and chemists as described in the following paragraphs. The third problem has only been partly solved, and most of the rest of this book involves the discussion of it.

Origins of Modern Genetics. It is a remarkable fact that in the same year, 1900, in which the second and most influential edition of Wilson's monumental book, *The Cell in Development and Inheritance,* appeared, there also appeared three papers which were to set off the work which was to give partial answers to the questions posed by Wilson in his book. The three papers were by Correns (184), De Vries (219), and Tschermak (1021), the three well-known rediscoverers of "Mendelism." These men performed experiments similar to those performed by Mendel and reported by him in 1865 (665). They obtained the same results and arrived at the same general conclusions as Mendel, and, since at least one of them, De Vries, was widely recognized at the time through his theory of intracellular pangenesis, the results and interpretation became immediately and widely known. The contribution of Mendel and his successors was to show that starting with plant and animal individuals of the same species, but showing diverse opposite or alternative characteristics, such as tallness versus shortness, or colored integument versus colorless integument, and so on, one could, by making crosses between the opposites and subsequently their hybrid progeny, demonstrate the existence of *unit characters* which segregated during the formation of the germ cells and were reunited at the time of fertilization. There was implicit in the concept of the unit character the same connotation that was to be found in De Vries' pangens or Weismann's biophores. That is, it was conceived to be an ultimate determinant in the cell which determined a particular characteristic of the individual formed from that cell. The capacity of the unit character, or *factor,* as it was also frequently called, to determine the characteristic continued unaltered as the unit passed from one generation to the next. The genius of Mendel was his recognition of the fact that if one followed the inheritance of related "opposite" characters recognized as alternatives (or allelomorphs), one could demonstrate a rational basis for inheritance by experimental means. He was thus successful where others, such as Kölreuter and Gartner, had failed, primarily because they had not considered the inheritance of each character separately through several successive generations.

    The demonstration of the particulate nature of inheritance was the key to the puzzle that the essayists Weismann and De Vries had been vainly trying to solve with abstract thought. The logical next step, aside from the obvious continuation of experiments on inheritance using different characters and different organisms, was to give the unit character some physical basis in the cell. Immediately after the realization of the significance of Mendel's work, cytologists began to tie what was known of the unit characters to the then known facts about the cell. Chromosomal reduction, which we now call meiosis, was well known to the cytologists of the time, and the general interpretation was that the two divisions involved in the formation of functional gametes with half the number of chromosomes was a necessary prelude to fertilization when the full number was reconstituted. What was known about the process of segregation of the unit characters corresponded in all important respects with what was known about reduction of chromosomes. In 1902 Sutton (994) and Boveri (89) both advanced hypotheses which were almost identical in nature, and which united the Mendelian observations and the cytological observations into a coherent chromosomal theory of heredity. This theory was supported by conclusions drawn from observations on the sex chromosomes by Sutton (993), McClung (642), Stevens (970a), and Wilson (1082). These workers were quick to realize the significance of accessory sex chromosomes. If a particular sex was distinguished cytologically by the presence or absence of a particular chromosome, and it was assumed that sex was inherited, a fact which was quite evidently the case, then it appeared that sex was inherited through the chromosomes. Experimental evidence from a variety of sources gave reasonable proof that sex was inherited through the chromosomes, but the generalization that all or most inheritance was through the chromosomes was yet to be verified.

The union of cell biology and Mendelism shortly after the rediscovery of Mendelism must be considered as an event of the greatest importance in biology. The doctrines of the cell biologists became subject to direct experimental test, and the abstractions of the Mendelists became subject to direct observation in cells. From this union came modern genetics.

Despite the many signs pointing to the chromosomes as the bearers of the unit characters, there were serious obstacles. First was the fact that all the cells in an organism seemed to have the same complement of chromosomes. Another hindrance was that closely related species, morphologically quite similar, had different numbers of chromosomes. This obstacle has been eliminated with the understanding of polyploidy and chromosomal fusion and breakage, whereas the first objection has become a problem of gene action in differentiation which we will discuss later.

One obstacle to clear thinking during the first decade of development of heredity, or genetics, as it was beginning to be called, was that of terminology. The term *unit character,* or factor, was ambiguous because it was not always certain whether one was referring to cause or effect. Johannsen (493) relieved this difficulty considerably by inventing and applying the terms *genotype* and *phenotype*. The genotype referred to the postulated units carried by the gametes, and thus was constituted of the inherited determinants, or *genes,* which in turn determined the phenotype or final characteristic in the organism. The genotype was inferred from the phenotype and the phenotypic results obtained from crosses.

Johannsen coined the term gene, and defined it in this way:

The gene is nothing but a very applicable little word, easily combined with others, and hence may be useful as an expression for the "unit factors," "elements" or "allelomorphs" in the gametes, demonstrated by modern Mendelian researches (493).

Beginning in 1910 T. H. Morgan, later to be joined by A. H. Sturtevant, C. B. Bridges, H. J. Muller, and others, began to perform the experiments with *Drosophila* which were to locate definitely the genes on the chromosomes in the nucleus. The net result of the experiments using newly developed techniques was proof that the genes are carried on the chromosomes (106, 984), and that they are arranged linearly, like beads on a string, along the length of the chromosome (984).

By about 1930 a gene concept had been achieved which is presently frequently referred to as the "classical" gene. This gene was defined as a physical entity capable of mutating to a changed condition and maintaining the change through successive duplications. It was also conceived as a discrete segment of the chromosome which had a specific function, and within which no crossing over occurred. Thus the idea of a gene as a biological unit arose, a concept which still dominates genetic thought today.

Starting in about 1940, however, work began in the genetics of microorganisms which led to a necessity of a revision of the classical gene concept and a realization that things were not so simple as they had appeared earlier. Furthermore, the discovery of pseudoalleles in *Drosophila* by Oliver (754, 755) and Lewis (606) made it evident that in even the higher organisms the older concept was in need of revision. This matter is considered at length in Chapter 8, but here it can be stated that the idea that the gene is a unit within which recombination does not occur must be abandoned. Also it is necessary to discard the previous conception that the gene is a unit of mutation, because it is now known through studies with microorganisms that mutations may occur at many sites within a single gene.

Rise of Chemical Genetics. After the discovery of nucleic acid in the nucleus by Miescher, and the initial interest aroused by it, the matter was largely pushed to the background for almost 40 years by cell biologists and geneticists who turned to the problems of development, cell structure at the microscopic level, and the mechanism of inheritance. By 1940 the main outlines of the mechanism of heredity were clarified as much as they could be by the application of the Mendelian methods of analysis and cytogenetics, and many geneticists began to think more in terms of the functions of genes. Furthermore, the structure of the nucleic acids was now beginning to be understood, and it was recognized that two general types existed, DNA and RNA, with DNA being virtually confined to the nucleus and RNA present in both nucleus and cytoplasm. These discoveries on the nature of nucleic acids led to increasing speculation about the chemical nature of genes. Starting in about this period beginning with 1940, then, two main lines of thought and investigation arose and began to be vigorously expanded: inquiries into gene function as related to metabolic control, and inquiries into the chemical nature of the genetic material.

An important breakthrough occurred in 1941 when Beadle and Tatum (56) announced their findings with biochemical mutants of *Neurospora.* They demonstrated in this and succeeding reports that mutants of this mold could be isolated which were incapable of carrying out specific steps in metabolism. This led to the one gene-one enzyme hypothesis which is discussed extensively in Chapter 6. Actually, an earlier hypothesis of the same nature had been advanced by Garrod (313), but at a time when geneticists were not thinking so much in terms of function as of gross mechanisms of inheritance. The isolation of biochemical mutants in *Neurospora* was soon followed by the discovery of the same type of mutants in other microorganisms such as *Escherichia coli* (352), and it began to be realized that all mutants were basically biochemical mutants insofar as gene mutations caused alterations in metabolism by causing changes in enzymes, and other proteins.

The second type of inquiry, concerning the nature of the genetic material, obtained its impetus from several sources. First, the microspectrophotometric observations of Caspersson (136) made it clear that chromosomes contain high concentrations of nucleic acid, and the application of the Feulgen staining technique made it possible to determine that this nucleic acid is primarily deoxyribonucleic acid (DNA). The Feulgen stain, which is specific for DNA, was discovered in 1924 (277), and its discovery may be accounted as a major one in chemical cytology, for, in reality, it ushered in the new period of study of chromosome chemistry. Caspersson's technique, as well as specific staining techniques, was also used to show that a high concentration of protein occurred along with the DNA. Extraction of the material of isolated nuclei with various solvents then demonstrated anew that a considerable portion of the protein of the chromosomes was bound to DNA to form the nucleoprotein complex first described as nuclein by Miescher. The question was then asked, what constitutes the actual stuff of genes, protein, DNA, or deoxyribonucleoprotein? Initially, there was a distinct tendency to recognize protein as being the more important. The primary reason for this was the assumption current in the thirties and early forties that DNA was made up of tetranucleotides each containing adenylic, thymidylic, guanylic, and citidylic nucleotides. These tetranucleotides were thought to be repeated over and over, so that it could not be expected that DNA would have any specificity, that is, all DNA should be the same. Further chemical work in the early forties, however, made it evident that there could in fact be many different kinds of DNA, because the tetranucleotide structure was proved to be untenable (381).

Almost at the same time Avery, MacLeod, and McCarty (39) published a highly significant paper in which they demonstrated that the transforming factor in pneumococcus was probably DNA. These two new discoveries in this decade provided the major stimulus for the following further vigorous research into the idea that DNA itself was the genetic material. The idea received further support from the studies of the T series of bacteriophages infecting *Escherichia coli*. Work on these phages started by Delbrück (208) resulted by 1952 in the demonstration that the hereditary material of the phage was its DNA and not its protein (434). In 1950 it was established that the DNA content of vertebrate nuclei remained constant while the protein content did not. Further, the DNA content of sperm was shown to be half that in the diploid somatic cells (11).

These observations provided encouragement for a further concerted attack on the structure of DNA by physicists such as Wilkins (1079) using x-ray diffraction, and chemists such as Chargaff (158) who built on the earlier knowledge of nucleic acid structure provided primarily by Levene (595). This work provided the basic facts for the formulation of the structure of DNA by Watson and Crick in 1953 which is generally accepted today (1056).

When it was realized that the structure of DNA was such that it could be the genetic material, the attention of many geneticists, biophysicists, and biochemists was turned to trying to determine how this chemical substance functioned as such. It was at this point in the early fifties that it began to be realized that biochemical mutants could provide the material for answering this problem. For, if gene mutation affected enzymes, then they probably affected proteins in general. From this it was but a simple step to the hypothesis that the role of DNA is to determine the structure of proteins. This hypothesis received strong support when it was demonstrated by Ingram in 1956 that a biochemical mutation in man, which gives sickle-cell anemia, resulted in the production of an altered hemoglobin which was different from normal hemoglobin in one amino acid in one of its polypeptide chains (475).

The rest of the story of the gene is given in the succeeding chapters. But in retrospect it should be pointed out here that we are still dealing with the same problem considered by Aristotle over 2200 years ago. It was not until recent times, the beginning of the last century, that a real start was made toward a solution of the problem. This was the statement of the cell theory and the law of genetic continuity which followed from it. Out of the work in cell biology and Mendel's contribution came the early gene concept in the latter part of the last century and the beginning of the present century. Progress since that time has resulted not from the efforts of geneticists alone, but of cytologists, chemists, and physicists, among others, as well. At the present time we stand at about the climax of a long train of thought and observation which started with the ancient Greeks.

Some Elementary Considerations

Permeating the entire field of genetics are the concepts derived from studying the genetics of higher organisms. This derives from the fact that all the early work was done with cellular organisms with true nuclei and observable chromosomes. It is necessary, therefore, that the reader have some knowledge of certain basic phenomena which occur in the organisms with true nuclei.

Mitosis. One of the chief requirements for an understanding of genetics is an appreciation of the significance of the basic means of cell reproduction among cellular organisms. These processes, mitosis and meiosis, are discussed now to provide the uninitiated reader with a background sufficient to understand the fundamentals of genetics.

Mitosis is a type of cell division in which a cell divides into two daughter cells that possess the same number and same kind of chromosomes found in the mother cell. The process is diagrammed in Figure 1.1. The essential factors in the process are: (1) each chromosome

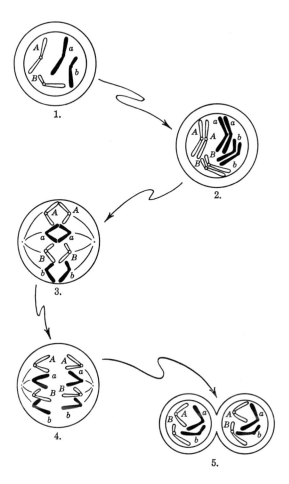

Figure 1.1. A diagrammatic representation of mitosis showing important stages in the division of a cell with four chromosomes. Stage 1: Early prophase. The chromosomes are visible in the nucleus. They have already duplicated, but the duplicates have not separated sufficiently to be distinguished. Stage 2: Late prophase, early metaphase. The chromosomes are now obviously duplicated and have separated. Stage 3: Early anaphase. The spindle fibers have attached and the chromosomes start moving to opposite poles. Stage 4: Late anaphase. Stage 5: Telophase and end of mitosis; cytoplasm dividing.

within the nucleus about to divide is duplicated—as far as can be determined by genetic means—exactly; (2) the duplicates separate and go to opposite poles of the dividing cell; and (3) the distribution of chromosomal material between the two new daughter cells is generally considered to be exactly equivalent because of the evident elegant mechanism operating to bring about an equal distribution. This is by no means true for the cytoplasm, which has a more homogeneous appearance than the nucleus and appears not to have any mechanism insuring its equal distribution. This point, concerning the distribution of the cytoplasmic part of the cell, is not of particular importance in the present discussion, but will receive more attention when its detailed consideration becomes important in Chapter 11 in connection with the discussion of cytoplasmic inheritance.

Meiosis. The higher organisms are, for the most part, diploid (*2N*), which means that they possess two similar haploid (*N*) sets of chromosomes in each of their cells. This fact has been established in part by cytological and in part by genetical observations. When the gametes of diploid organisms are formed, it can be demonstrated cytologically that the cells which give rise to them undergo a series of two cell divisions which result in a reduction of the chromosome number by one-half, thus producing gametes with a haploid set of chromosomes. The two successive cell divisions resulting in haploidy are together called meiosis. The more specialized terms, *spermatogenesis,* for meiosis which produces sperm cells, and *oögenesis,* for meiosis resulting in egg cells, are widely used when referring to the process in a particular sex of an animal. In seed plants the formation of haploid cells, which eventually give rise to sperm nuclei in the pollen grain, is called *microsporogenesis;* its counterpart process, producing female haploid cells, is termed *macrosporogenesis,* or *megasporogenesis.*

The diagram given in Figure 1.2 illustrates meiosis in an organism which has a diploid set of chromosomes numbering 4 (*2N* = 4). It will be noted that the chromosomes pair in the first stage of the first division (prophase I), thus, in this example, giving two sets of two chromosomes. The chromosomes which pair or *synapse* are described as homologues, because they can be shown to possess similar or identical sets of genes as well as being similar in appearance morphologically. It should be noted that the only way to demonstrate decisively the homology of two chromosomes is to note whether they pair in meiosis. The fact that they are similar in appearance and gene content is important—the latter point being particularly important to the geneticist —as will be discussed hereafter, but the cytological fact of homology

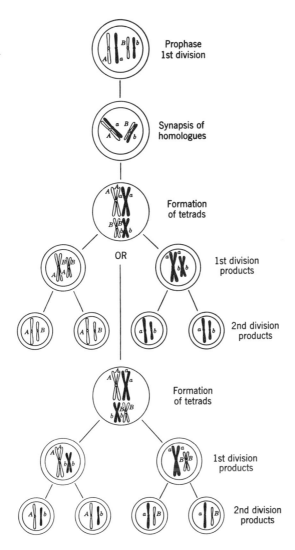

Figure 1.2. A diagrammatic representation of meiosis. Meiosis is an extremely complicated process, and this figure only indicates how the products are derived. The process of crossing over illustrated in Figure 1.4 and discussed on page 21 may be part of the process of meiosis, although not so indicated in this diagram.

between chromosomes in a diploid is based on the criterion of pairing rather than on genetic relationship.

At the time of pairing or shortly thereafter, the chromosomes can be seen to have doubled in number by each member of a pair becoming duplicated. The result is the formation of a pair of closely associated duplicates from each chromosome which are distinguished from the mother chromosome by the term sister *chromatids*. Thus each original pair of homologues becomes a tetrad, or a packet of four *chromatids* consisting of two sets of two identical (sister) chromatids. The members of each pair of sister chromatids remain attached to one another during this phase of meiosis because they have only one centromere (or spindle-fiber attachment) between them, for when the sister chromatids are formed, the duplication process does not extend to the centromere, at least not to an extent that becomes evident.

With the formation of tetrads the prophase of the first division draws to a close and the second phase, metaphase, commences with the aligning of the tetrads in a single plane perpendicular to the spindle fibers, and approximately half-way between the poles of the cell (see Figure 1.2). Metaphase ends with the homologous chromosomes separating and going to opposite poles. It will be noted from Figure 1.2 that this is equivalent to the homologues *segregating,* for each daughter cell then receives a representative of each homologous pair present in the original 2*N* germ cell.

In the second division of meiosis, the pairs of sister chromatids present in each of the two cells produced by the first division are broken up by the separation of the sister chromatids, which go to the opposite poles. The net effect of this is the formation of four cells each with a haploid set of chromosomes. It should be noted by referring to Figure 1.2 that each product of the meiosis receives a representative of each homologous pair. This is important. Gametes deficient in a chromosome type generally will not produce viable zygotes on fertilization. The second important point is that the segregation of homologues is *random*. Thus, if the homologues differ in gene content, from a diploid cell containing four chromosomes at least four different kinds of gametes result by meiosis. This can be readily appreciated by considering the combinations of *Aa Bb* by two's in which each pair contains one letter of each type, that is, *AB, Ab, aB, ab*.

When chromosomes synapse in meiosis it can be shown that they do so with like parts being located opposite each other. Hence, if chromosome A having genes *ABCDEFGH* arranged along it in that order were to synapse with its homologue, a, which has the gene order a b c d e f g $h,$ they would do so in the following way: $\dfrac{ABCDEFGH}{a\,b\,c\,d\,e\,f\,g\,h}.$ Like genes, or allelic genes, as will soon be made clear, are, in other words, located opposite one another. Failure of the apposition of like parts to occur results in absence of synapsis and the breakdown of the meiotic mechanism.

Mendelism. The phenomenon of Mendelism is recognized by the occurrence of certain ratios of phenotypes of offspring. It is the direct result of meiosis with its random segregation in the parents, as just described, and of the random reconstitution of diploid offspring by fertilization, as described hereafter.

When an individual breeds true for a certain characteristic (i.e., inbreeding to its siblings or family stocks produces more of the same phenotype in the offspring without exception), it is *homozygous* for that characteristic. The term *homozygous* refers to the genotype. It means that an organism so described not only possesses the genes for the characteristic but also possesses identical kinds of genes on both sets of homologues. As the result of meiosis a homozygous individual produces gametes which are identical with respect to the genes for which it is homozygous. A homozygote when bred to another with the same genotype will produce offspring identical to the parents and one another. This statement may be written in genetic shorthand: $AA \times AA$ (where A denotes a gene giving a particular characteristic) gives in the F_1 (first filial generation) AA offspring.

Two individuals with different phenotypes when bred together very often give offspring identical to one of the parents. If both parents are homozygous for their respective genes controlling these characteristics, the cross may be written as $AA \times aa$. By this means it is stated that one of the parents, $AA,$ is homozygous for $A,$ and at the time of meiosis one pair of its homologues could be marked thus: $\dfrac{A}{A}.$ The genotype of the second parent may be written $\dfrac{a}{a}.$ Now, if these genes are assumed to occupy equivalent positions on homologous chromosomes and hence are alleles, it is evident that the gametes produced by parent AA will be $A,$ those by parent aa will be $a,$ and the F_1 offspring will be $\dfrac{A}{a}$ or simply Aa. These offspring are described as *heterozygous* for the genes A and a, which is another way of saying that they are not pure breeding, since they will produce two types of gametes, A and $a,$ in equal numbers as a result of the segregation of the homologues in meiosis. If the F_1 offspring are inbred, therefore, three kinds of genotypes will appear in the F_2 generation, *AA, Aa,* and *aa.* As a result of random fertilization of equal numbers of *A* and *a* eggs by equal numbers of *A* and *a* sperms, these three genotypes should occur in the ratio of $1AA:2Aa:1aa.$

It was stated earlier that in this example the F_1 offspring are identical to one of the parents. Assuming that *Aa* heterozygous individuals are phenotypically identical to the parents designated genotypically as *AA,* then it is evident that the gene *A* masks the effect of its allele *a.* The more usual way to describe this effect is to call *A dominant to a* or, to put it another way, call *a recessive to A.* The F_2 genotypic ratio of $1AA:2Aa:1aa$ can be expressed phenotypically as a $3:1$ ratio, since *AA* and *Aa* individuals are phenotypically indistinguishable. The $3:1$ ratio is a *Mendelian ratio* generally described as the ratio expected from a cross between two individuals heterozygous for a pair of allelic genes both of which have something to do with the determination of the alternative phenotypes. In short, it is a ratio which tells the breeder that two alternative phenotypic characteristics are produced by allelic genes. If one allele is not completely dominant over the other, incomplete or no dominance will result, and the heterozygote will be phenotypically distinguishable from the homozygotes. This condition will be manifested by a $1:2:1$ phenotypic ratio identical to the expected genotypic ratio.

The $3:1$ or $1:2:1$ ratios are the basic Mendelian ratios. All other ratios are derived from them. Consider, for example, two animals of opposite sex which are heterozygous for two pairs of allelic genes, each pair located on a different chromosome, with the genotype *AaBb.* Figure 1.3 illustrates the types of gametes that would be expected from meiosis in each sex with details of how these gametes are derived. It will be noted that for each female diploid cell that undergoes meiosis only one of the four haploid products survives as a functional gamete. Oögenesis in animals is identical in principle to spermatogenesis with respect to the nuclear contents, but there is an unequal distribution of cytoplasm such that one nucleus retains all or nearly all of the cytoplasm to the end of the second division. This is the functional gamete; the others are polar bodies and disintegrate in time. Since, however, it is a random matter of chance which of the four nuclei resulting from the original diploid cell receives the cytoplasm, the same types of gametes should be expected in oögenesis as in spermatogenesis provided that the genotypes of the diploid germ cells are the same. The essential factor to be recognized in connection with meiosis involving

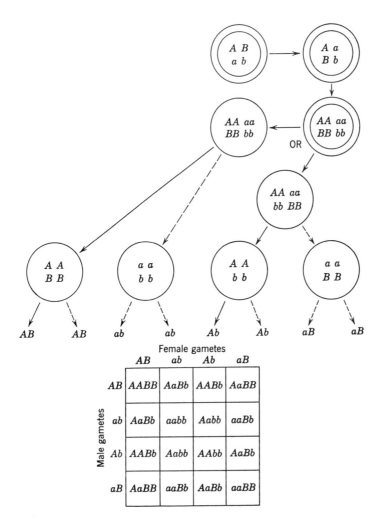

Figure 1.3. The results of the segregation of two pairs of homologous chromosomes. The chromosomes illustrated in Figure 1.2 are represented here by their letters. The dotted lines indicate the formation of polar bodies which occur during oögenesis. As indicated in the text, the principles involved in the formation of male and female gametes are identical. The Punnett square at the bottom of the figure shows the different possible genotypes obtained by crossing two individuals who are heterozygous for genes on two different chromosomes.

more than a single pair of homologues is that it is a matter of chance which nonhomologues accompany each other in the first division of meiosis. Thus, in Figure 1.3, in a cell with *AaBb* chromosomes, *A* can accompany *B* and hence *a* will go with *b,* or *A* and *b* going to one pole will result in *a* and *B* in the opposite pole. Hence, as already pointed out, four types of gametes are expected from a cell heterozygous for two pairs of genes on different chromosomes. Three pairs of allelic genes on as many different chromosomes will result in eight different gametes.

The results of fertilization with the gametes of the genotypes shown in Figure 1.3 are found by use of the Punnett square as illustrated. The genotypic ratio is, of course, complex because of the large number of different genotypes obtained. The type of phenotypic ratio obtained will depend on the type of dominance relationships between the alleles and also on possible interactions between the nonallelic genes. Such interactions need not concern us here but are discussed at length in Chapter 9. If complete dominance of *A* and *B* over their respective alleles and no gene interaction is assumed, a $9:3:3:1$ phenotypic ratio will result. This may be written as $9AB:3Ab:3aB:1ab$, since *AB* may be taken to be the phenotypic designation of *AaBb, AaBB, AABb, AABB,* and so forth for the others. Other types of ratios resulting from gene interactions are given in Table 35, Chapter 9.

The fact that more complex ratios can be derived from the $3:1$ ratio obtained when but a single pair of allelic genes are segregating can be appreciated by the application of simple probability. For example, in a cross of $Aa \times Aa$, the chances of an offspring having the gene *A* (that is, being either *Aa* or *AA*) are 3 out of 4, or $\frac{3}{4}$. There is a like probability that the offspring will have *B* from a cross $Bb \times Bb$. Since *B* and *b* segregate independently from *A* and *a,* the probability that any individual will have both *A* and *B* as a result of a cross between $AaBb \times AaBb$ will be $\frac{3}{4} \times \frac{3}{4} = \frac{9}{16}$. Similar rearranging gives the other components of the $\frac{9}{16}:\frac{3}{16}:\frac{3}{16}:\frac{1}{16}$ ratio by starting with the $3:1$ ratio.

It should be clear from the preceding discussion of Mendelism that if two organisms from pure-breeding homozygous stocks, but of different phenotype, are crossed and the F_1 inbred, the phenotypic results in the F_2 will determine the genotypic nature of the phenotypic differences. A definite $3:1$ or $1:2:1$ ratio obtained in the F_2 indicates that the difference is monogenic, that is, the result of the difference in expression between two allelic genes. If, on the other hand, a different ratio, such as $9:3:3:1$, and so on, is obtained, the conclusion must be that more than a single pair of allelic genes is involved.   Hence it can be seen that the definition of the gene rests on the kind of Mendelian ratio obtained, and, furthermore, that the number of gene pairs involved, if more than one, may be deduced from these ratios.

Linkage and Crossing Over.   Since each chromosome contains many genes arranged linearly, it is to be expected that a phenotypic difference may easily be the result of a difference in two nonallelic genes on the same chromosome.   Such a situation is described as linkage.   For example, assume that the genes M and N are located on the same chromosome and that m and n are their respective alleles.   If two individuals, $MMNN$ and $mmnn,$ are crossed, the F_1 genotype will be $MmNn,$ or $\dfrac{M\ N}{m\ n}$, and, according to the information given in the preceding paragraphs, the F_2 will consist of three types of progeny, $MMNN, MmNn,$ and $mmnn,$ in a $1:2:1$ ratio.   Since this is the ratio to be expected from the F_2 of a cross involving parents differing only in a pair of alleles, the conclusion must be that M and N are not recognizable as different genes.   However, the integrity of the chromosomes, although maintained in substance from generation to generation, is not absolute.   After the formation of chromatids, and prior to the separation of the elements of the tetrad, the phenomenon of *crossing over* occurs.   A diagrammatic representation of crossing over between two nonsister chromatids of a tetrad in the region between the loci of two genes M and N is given in Figure 1.4.   The net effect of the crossover

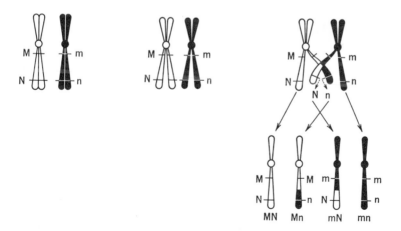

Figure 1.4.   Crossing over between the chromatids of a pair of homologous chromosomes.

in this region is the production of two additional types of gametes, *Mn* and *mN,* at the completion of meiosis.   The formation of these types proves, furthermore, that *M* and *N* and their respective alleles occupy different loci and therefore are nonallelic genes.

Crossing over is not a rare phenomenon but a general one, and to be expected whenever homologous chromosomes pair in meiosis and form chromatids.   The number of possible crossovers that can occur between the chromatids is limited by interference which certainly results in part from mechanical problems, but the points at which crossovers occur, and hence the gene loci they separate, are for the most part a random matter.   It is to be expected, therefore, that genes far apart on a chromosome will cross over more often than those close together.   This fact is the basis for the mapping of a chromosome by the location of its genes along its length.

Some Definitions and Symbols

The vocabulary of genetics probably presents more difficulties than the subject matter itself.   Therefore a number of important terms are defined below for the uninitiated reader, as well as a brief statement about symbols.

Polyploidy.   Occasionally a diploid cell doubles its chromosome number and becomes a tetraploid (4*N*) (Figure 1.5).   The gametes of a tetraploid will be diploid, and, if these fertilize haploid gametes, triploid (3*N*) individuals will result.   Higher forms of polyploidy, such as hexaploidy, octoploidy, and decaploidy, as well as their odd-numbered derivatives, are also known to occur.   As a general rule, polyploidy is uncommon among animals but common in cultivated plants and in microorganisms such as yeast, and possibly bacteria.   Its significance in connection with gene action is related to the fact that in polyploid cells it is possible to have a particular gene present more than twice. This is advantageous, as will be pointed out in Chapters 9 and 13, when it is found desirable to study the effects of genes in different doses.

Recombination.   When two different parental strains are crossed, the progeny may be identical to either one of the parents, or some may be different from either parent.   Those that are different are referred to as *recombinants,* and the process whereby they are produced called recombination.   Thus, if a strain designated as a^+b is crossed to one carrying ab^+, then four kinds of offspring are to be expected: $a^+b,$

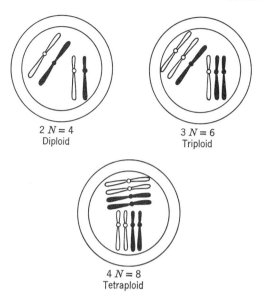

Figure 1.5. Different conditions of ploidy.

ab^+, ab, and a^+b^+. The first two have the parental genotype and the latter two are recombinants.

Two well-established mechanisms for the production of recombinants are independent segregation of genes present on different, nonhomologous chromosomes, as discussed earlier under Mendelism; and crossing over between linked genes, also discussed earlier. Other mechanisms probably exist for producing recombinants, particularly in the lower organisms, as discussed in Chapters 3 and 8. For this reason, the general term recombination will be used in this book except when a specific, identifiable mechanism is being discussed.

Genes and Alleles of Genes. The locus of a gene on a chromosome is its location as determined primarily by recombination data which can, when properly analyzed, give the approximate position of the gene relative to its neighbors on the same chromosome. At this same location, but not simultaneously, there may also exist allelic genes. Allelic genes may be thought of as different genes which occupy equivalent loci on homologous chromosomes, or, conversely, as different forms of the same gene. Both descriptions are in common usage, and both are used in the succeeding chapters.

Differences among the alleles at a given locus may be described by terms other than dominant and recessive. Terms such as standard allele, mutant allele, wild-type allele, and normal allele are commonly relied on to describe allelic relations. A gene described as "standard" is simply the allele taken by the experimenter as the one to which he compares all other alleles at the same locus. It may or may not be dominant. The "wild-type" gene is one which is found to give the "normal" phenotype in a wild or natural population. "Wild-type" and "normal" are used synonymously in this sense. Any deviation by mutation from the wild-type or normal condition results in a "mutant" allele or gene. In general the wild-type or normal allele is dominant to the mutant.

Symbols. The symbols used by the geneticist for the most part are the letters of the alphabet. In general, a capital letter designates a gene that is dominant; a lower-case letter, a recessive. If two genes are indicated by the same letter, they are generally alleles. Nonallelic genes which have similar or identical effects on the phenotype may be indicated by the same letter symbol, with the difference between them indicated by different subscripts. Thus w_1 and w_2 are nonallelic genes in maize (corn), both of which cause loss of chlorophyll. Their alleles (full chlorophyll) are W_1 and W_2, respectively. Capitals, lower-case letters, and subscripts are commonly used in this way by plant geneticists, as well as by mammalian geneticists studying inheritance in humans and other mammals.

A somewhat different symbolism is used by geneticists working with *Drosophila* and microorganisms such as fungi and bacteria. In these organisms the dominant, normal, or wild-type allele is given a lower-case letter symbol with superscript, $+$. Thus the gene for normal eye color in *Drosophila melanogaster* is given the symbol w^+. A mutant gene allelic to w^+ is given the symbol w (white). Other alleles at this locus are indicated by superscripts such as w^{co}, w^e, and so on. A mutant gene which is *dominant* to its wild-type allele is indicated by a capital letter, and the wild-type allele by the same capital, but with a $+$ superscript. When there is no doubt about which gene locus is being discussed, the wild-type allele is indicated simply by $+$. Thus the designation $w/+$ is equivalent to w/w^+.

References

Although an attempt has been made in this chapter, as well as in the succeeding ones, to provide the necessary background material for the reader's understanding of the rest of the book, it is probable that some will not find it sufficient. Among the many excellent general works on genetics that will be helpful are those by Srb and Owen (947), Sinnot, Dunn, and Dobzhansky (912), Sager and Ryan (873), and King (524). General and specific aspects of biochemistry are comprehensively treated by Fruton and Simmonds (304). References to review articles on specific aspects of the subjects treated will be found in the text.

2 Some Aspects of Cell Structure and Function

The continuity of life through cells is made possible only by the fact that cells have a definite organization which is duplicated at each cell division. With the rise of genetics in this century it was shown that the chromosomes maintain their integrity through successive cell generations and are responsible in a large part for the transmission of characteristics so that the specific organization of a cell comes about primarily through the transmission and functioning of the hereditary material of the chromosomes. Therefore any mechanistic interpretation of heredity must be based on the understanding of the chemical and physical nature of cell organization and function, including both the chromosomal and nonchromosomal material.

The cogent experimental evidence pertaining to structures and functions that are common to all living cells have come from two disciplines, which year by year explore more and more common ground. The first deals with what can be seen with the naked eye, with the light microscope, and with the electron microscope (95). The second begins at the other end of the scale of size, with structures and metabolism of small molecules and then with successively larger molecules and specific kinds of combinations of larger molecules. Already the line between these disciplines has begun to fragment, and an even more extensive fusion is to be hoped for and expected in the future. As a consequence, the division of the discussion in this chapter into structural morphology and composition and molecular morphology and function is one of convenience, and constant alertness for the common ground rather than the distinction should be exercised.

Cell Morphology and Function

Units visible in cells by microscopy include cell walls, cell membranes, membranes of the endoplasmic reticulum, ribosomes (Pallade granules) and microsomes, mitochondria, centrioles, the units which may include vacuoles, lysosomes, and pinocytotic inclusions, Golgi bodies, plastids, crystals and fibers, and the nuclei which include chromosomes, nucleoli, and spindles. This increasing number of visible and often physically separable cell components seems to provide more and more unit distinction and compartmentalization as a characteristic of cells, but in actual fact quite the reverse is true. In general, increased sophistication in methods of observation has shown less isolation of cell structures, fewer clear-cut distinctions between structures, and more potential for functional communication between structures. However, before considering these matters in detail, there is an important technical matter that should be kept in mind. With observations such as those made by phase contrast microscopy it is possible to observe living cells, and thus what can be seen is at least close to what exists. Other methods which require tissue fixation and staining are expected to give artifacts due to structure alteration, and this potential must always be evaluated carefully. In general, results with such different methods are compatible, and a comparison of an observation by phase contrast and one by electron microscopy is shown in Figure 2.1 (272). Delineation of several of the cell components, listed earlier, is shown in the figure.

Cell Wall and Membrane. Communication between cell contents and the external environment is a property of all cells whether they have only a plasma membrane, as in most animal cells, or also a cell wall, as in higher plants and microorganisms. Although the cell wall has structural importance and is laid down from within, the effective selective barrier around the cell is the plasma membrane. This barrier has remarkable specificity extending both to quantitative and to quali-

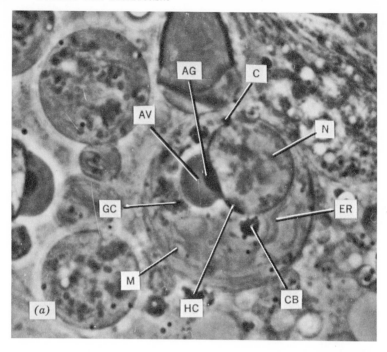

Figure 2.1.   A comparison of phase contrast and electron-microscope pictures of guinea-pig spermatids at similar stages.   (*a*) Phase contrast (living material ×2500).   Acrosomal granule, AG; mitochondria, M; Golgi complex, GC; chromatoid body, CB; centriole, C; head cap, HC; nucleus, N; Acrosomal vesicle, AV; endoplasmic reticulum, ER.   (*b*) Electron micrograph (section ×11,000).   Golgi complex, GC; head cap, HC; acrosome, Ac; nucleus, N; mitochondria, M; endoplasmic reticulum, ER.   From Fawcett and Ito (272).

tative control of materials moving both in and out of the cell.   It may be completely impermeable to certain low-molecular-weight compounds or ions and at the same time provide free passage to specific macromolecules, and thus it must be considered an important regulator of the cell activities.   The problem of an adequate description of a cell membrane has been considered extensively by Ponder (795), who has made it amply clear that it is presently possible to provide only superficial information about it.   Cells such as mammalian red cells can be plasmolyzed to yield "ghosts" which have a variable thickness when dry from the order of 50 to 100 Å.   Composition varies with the method of preparation, but in general it contains protein and lipid in a weight ratio of about 1.7 to 1.   Of the lipids, which include cephalins, lecithins,

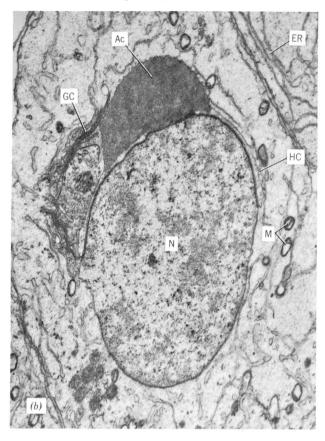

(b)

sphingomyelins, and cholesterol, some of the cephalin and all of the cholesterol is loosely bound.   On the other hand, a portion of the tightly bound protein is hemoglobin itself.   No completely satisfactory detailed arrangement of molecules of the membrane has been devised, and, indeed, since some components are loosely bound, it is reasonable to think that a functioning membrane may have even more loosely bound components that are lost in preparation.   This possibility is particularly significant in view of the fact that a great deal of membrane transport is energy requiring with specific combinations and enzyme actions.   Thus none of the prevailing hypotheses, the lipid layer, the sieve or ultrafilter, or the mosaic yet provides a satisfactory explanation for evaluating phenomena concerned with the genetic control of plasma membrane permeability.

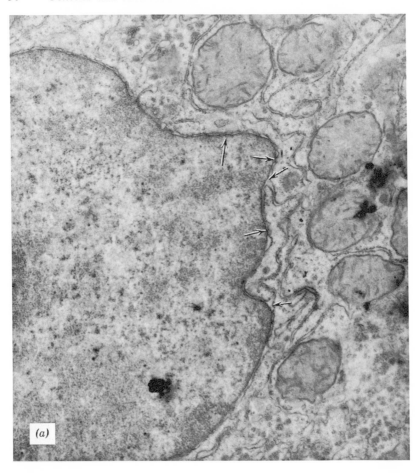

Figure 2.2.   Nuclear-cytoplasmic connections.   (*a*) Rat-liver cell (×16,600).
Arrows show nuclear pores in the nuclear envelopes and structural continuity
between the outer membrane of the nuclear envelope and the endoplasmic reticulum. From Porter (803). (*b*) Surface view from a newt oöcyte (×140,000).
The roughly circular areas of low density correspond to the pores in the nuclear
envelope (diameter 1000 to 1400°). From Gall (308).

Endoplasmic Reticulum.   By electron microscopy it has been firmly
established that most cells contain a very complex maze of membranous
material which assumes characteristic forms in different types of cells.
This is the endoplasmic reticulum, and its careful examination has made
clear important interconnections and actual inseparability in function
of certain observable cellular components.   This is perhaps most clearly

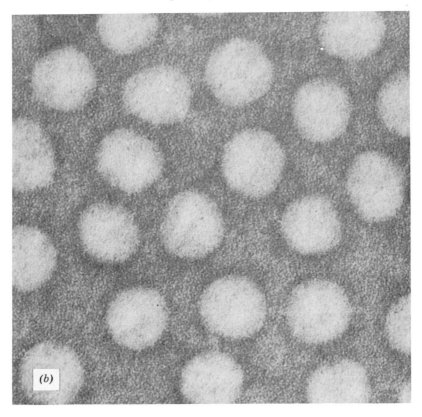

(b)

illustrated by the relation of the endoplasmic reticulum to the nucleus
(803). As shown in Figure 2.2, the nucleus is characteristically
bounded by a double membrane. The inner part of the nuclear envelope (Figure 2.2*b*) (308) contains perforations from 50 to several
hundred mμ in diameter while the outer membrane is continuous with
the endoplasmic reticulum (Figure 2.2*a*) (803). The latter then forms
tubules and other membrane-bounded inclusions taking a great variety
of forms which extend to the cell periphery where the reticulum may
form a double layer with the surface membrane as it does with the
nuclear envelope. Continuous connections with other cell structures
such as Golgi bodies, mitochondria, and plastids are not obvious, but
it is clear that all of these bodies are constructed basically from membranes (around 50 Å thick) similar to those of the endoplasmic reticulum, and a continuity in origin is possible.

In function the endoplasmic reticulum presents large surfaces on

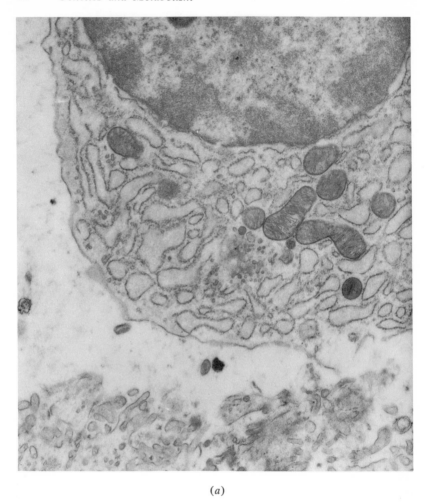

(*a*)

Figure 2.3. Forms of endoplasmic reticulum (ER). (*a*) Plasma cell from larvae of *Ambystoma* (×26,000). Granular form with abundant ribosomes attached to surfaces. (*b*) Muscle section from *Ambystoma* larvae (×42,000). Smooth form with tubules (SR = sarcoplasmic reticulum). From Porter (803).

which biosynthetic systems are ordered.   The granular or rough form is granular due to the presence, in sometimes very well-ordered patterns, of ribonucleoprotein spheres (ribosomes) which are clearly concerned with protein synthesis (Figure 2.3*a*) (803).   There is also evidence for a smooth form of reticulum carrying RNA on its surface. Here function is not known, but several other functions are indicated

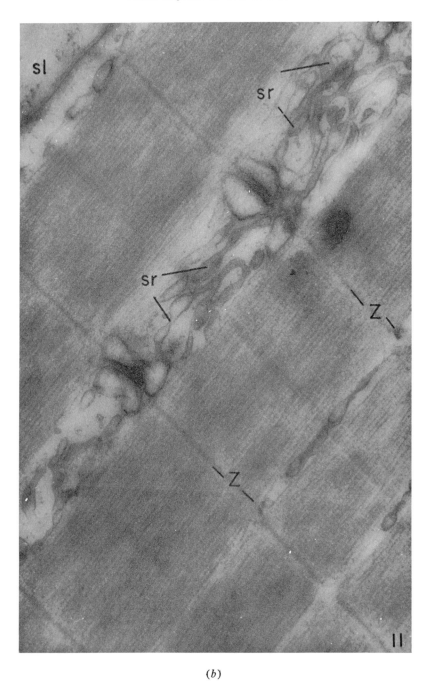

(*b*)

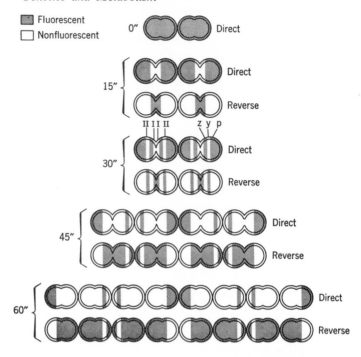

Figure 2.4. Diagrammatic representation of bacterial cell-wall growth (*S. pyogenes*) as followed by use of fluorescent antibodies. In the direct series new growth areas are not fluorescent (white areas). The time series shows growth at more than one band (I and II) per cell as cell chains are formed. Internal mechanisms are unknown. From Cole and Hahn (178).

for other areas of the smooth form as defined by studies of specialized cells. These include lipid synthesis (759), steroid-hormone synthesis (167), glycogen synthesis (1118), and cell-wall synthesis (804). One is led to the conclusion that the endoplasmic reticulum is, in itself, a highly differentiated cell structure with specific areas formed to carry enzymes or systems which carry out important physiological functions in localized regions. Although not yet associated with specific membrane function, an interesting example might be found in bacterial cell-wall formation (178) where new wall formation was demonstrated in two areas, that of finishing a wall in a completed division, and that of preparation for a new division (Figure 2.4) (178).

An interesting and important aspect to the picture concerning the endoplasmic reticulum is its behavior during cell division. At this time it is apparently fragmented as is the nuclear envelope (Figure 2.5) (803), and present evidence suggests regeneration from the nuclear area rather than reformation from fragments.

Microsomes and Ribosomes. With the increase in knowledge concerning the nature of the endoplasmic reticulum, the cell fraction previously known as microsomes has lost its meaning since it clearly arises from fragmentation of these membranes during disruption of cells. An extensive description and analysis of pancreatic microsomes has been presented by Palade and Siekevitz. As shown in Figure 2.6*a* and *b* (758), the components isolated by differential centrifugation after mechanical homogenization of pancreas cells resemble sections of the granular form of the endoplasmic reticulum. In this material there is relatively little smooth form, though patches continuous with granular membranes were observed in tissue sections. In any case, fragments of the smooth membrane would be lost in centrifuging as would be portions only slightly granulated with the heavy granules. Loss of granules by treatment of microsomes with RNAase is illustrated in Figure 2.6*c*. Thus microsomes represent variable-sized fragments of endoplasmic reticulum with lipid-protein membranes about 70 Å thick and carrying orderly arrays of ribonucleoprotein granules (the ribosomes) with a uniform diameter in the range of 100 to 150 Å.

Of the many studies on the nature of the ribonucleoprotein granules, a recent example is that of Tissieres et al. (1013) on ribosomes from *Escherichia coli*. Since the bacteria have little or no endoplasmic reticulum, the particles are free and easily isolated. Preparations were observed to contain particles with four different sedimentation constants in the ultracentrifuge: 100*s*, 70*s*, 50*s*, and 30*s*, corresponding to particle weights of about 5.9×10^6, 3.1×10^6, 1.8×10^6, and 1.0×10^6. The 100*s* material was shown to be aggregates of two ribosomes while 70*s* represents single particles. By control of Mg^{++} concentration, 70*s* particles could be dissociated into the unequal moieties 50*s* and 30*s*, and these could be reassociated again to 70*s*. Structurally the ribosome then appears to consist of a spherical particle (50*s*) combined with an asymmetrical cap (30*s*) like the cap on an acorn. Both components (30*s* and 50*s*) contain ribonucleic acid to protein in the approximate ratio of 60:40, and these seem to be the major significant components in terms of dry weight. Other investigations with ribosomes from microsomes of plant and animal tissues have yielded a qualitatively similar picture with, in general, 80*s* particles (ribosomes)

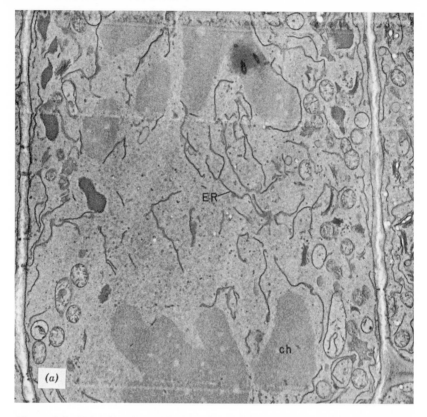

Figure 2.5. Behavior of endoplasmic reticulum (ER) in mitosis in onion-root tops. (*a*) Anaphase with chromosomes (Ch) at poles and ER in spindle area presumably by proliferation from sides and poles (×5300). (*b*) Two cells just after mitosis with cell plate (CP) and ER continuous with new nuclear envelope (at arrows) (×6000). From Porter (803).

yielding 60*s* and 40*s* particles by dissociation. Also, in general, the RNA from the larger of the ribosome subunits has been found to have a molecular weight of about 1.2×10^6 (23–25*s*), or about twice that of the RNA from the smaller units (15–18*s*).

Many further details are available (77), but a point of special interest and importance is that ribosomes from various sources yield very similar analyses with respect to nucleotide composition (932). The same is true for ribosome subfractions and, furthermore, protein components from ribosomes and fragments from different sources are strikingly similar (1022).

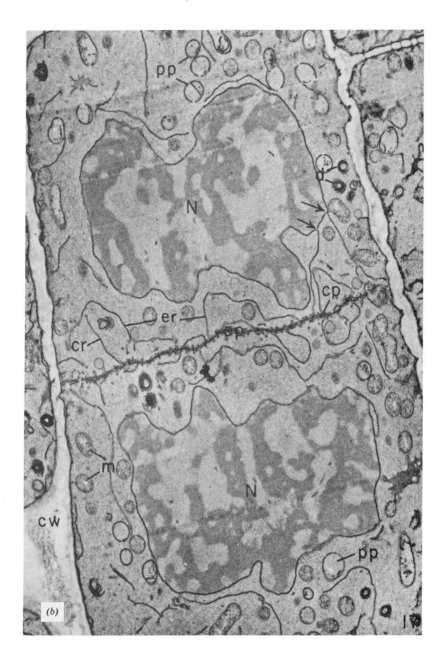

(b)

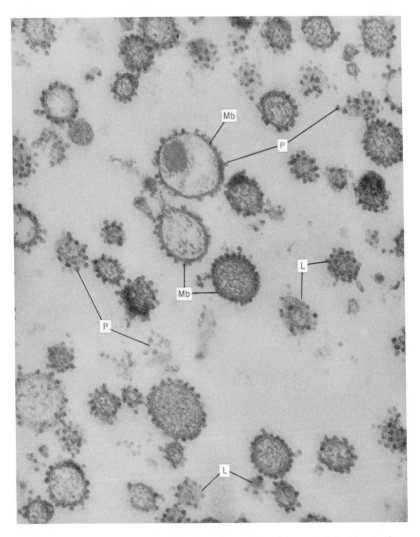

Figure 2.6. The preparation of ribosomes from microsomes. (*a*) Sectional micro-some pellet prepared by centrifugation from pancreas tissue ($\times$75,000). Membranes, Mb; ribosome particles, P; lateral sections, L. From Palade and Siekevitz (758).

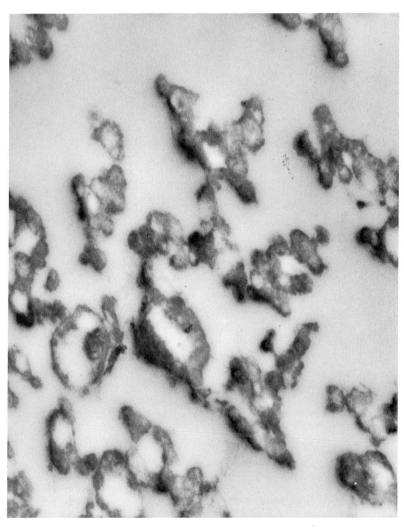

(*b*) Microsome material after removal of ribosomal particles with RNAase (×33,500). From Palade and Siekevitz (758).

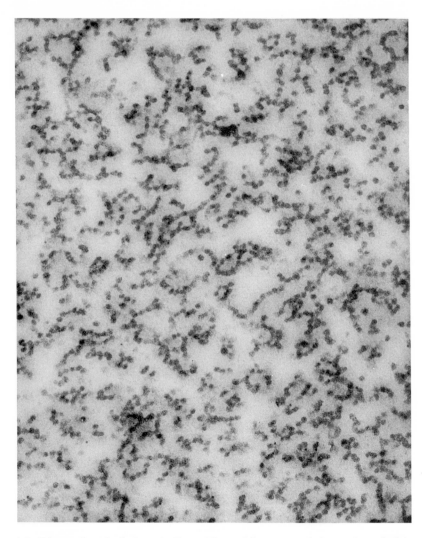

(*c*) Dispersed and chains of ribosomal particles prepared from deoxycholate treated microsomes ($\times 94,000$). From Palade and Siekevitz (758).

Mitochondria. A great variety of cells contain characteristic organelles, the mitochondria (or chondriosomes), which are large enough to observe in the light microscope, and hence something of their behavior in living material can be ascertained directly. Significant properties are that in many cells they move about rapidly and undergo very extensive changes in shape. In cell division they may be distributed in a random fashion, but in some cases they aggregate in chains or rings and appear to be distributed in a somewhat regular fashion associated with the more precise nuclear division cycle. (See general review by Novikoff, 746).

In the electron microscope mitochondria appear as bodies bounded by a smooth membrane and containing cristae (tubules) and folds or concentric layers of membranes. Some examples of various forms are shown in Figure 2.7*a*, *b*, *c*. Filamentous forms may vary in size from about 1 to 70μ in length and 0.2 to 2μ in width, and spherical forms from about 0.2 to 2μ in diameter. In general they contain about 20% to 30% lipid (dry weight) with a large part of the remainder as protein. The prominence of these structures is shown by the fact that in a tissue like liver where there are about 1000 mitochondria per cell, 35% of the total protein of the tissue is contained in the mitochondria. Numbers per cell vary as widely as 4 or 5 during spermiogenesis in snails to 5×10^5 in giant amoebae (746). Mitochondria, as such, are not observed in bacteria and some algae but there are perhaps equivalent simpler membranous organizations or microunits (715).

Of particular interest in the present discussion of structure and function in cells are the enzyme systems with clearly localized functions characteristic of mitochondria. The best known of these is the system involved in electron transport and oxidative phosphorylation (Figure 2.8 (579), and p. 229). This essential function for providing chemical bond energy from oxidation can be demonstrated in mitochondria in whole cells, in mitochondria isolated by fractional centrifugation, and most significantly, in fragmented mitochondria. These findings imply that this important series of enzymes is firmly fixed in a definite ordered structure which is essential to its normal function (355). Mitochondria can be fragmented by sonic vibrations, and, based on the three coupled functions, citric-acid-cycle oxidation, oxidative phosphorylation, and electron transport, function and structure have been related as follows (354). Citric-acid-cycle oxidation is lost with primary disruption of the outer mitochondrial membrane. Oxidative phosphorylation is retained as long as fragments retain double membrane structures, but it is lost at the single membrane level where only electron transport remains. These results imply loose binding and organization of components of the citric-acid cycle, stronger binding and organization in the components involved in oxidative phosphorylation, and extreme organization of electron-transport components. A graphic summary of this picture is given in Figure 2.8. Details here are not necessarily

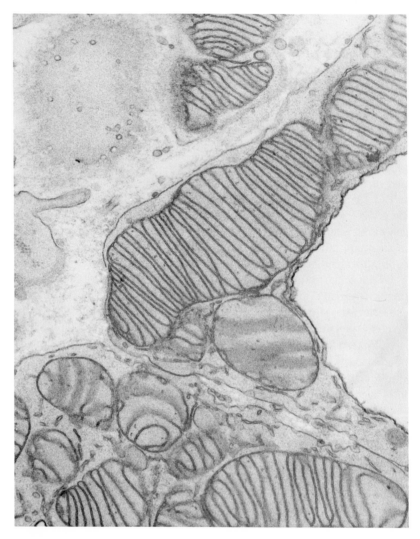

Figure 2.7. Several forms observed for mitochondria. (*a*) Mouse adrenal gland mitochondria with closely packed cristae (×26,000). From H. H. Mollenhauer, Electron Microscope Laboratory, University of Texas.

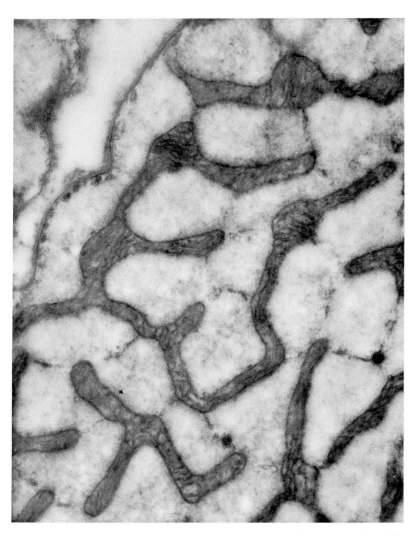

(*b*) Rat skeletal muscle ($\times$18,000). From Palade in Brachet and Mirsky (94).

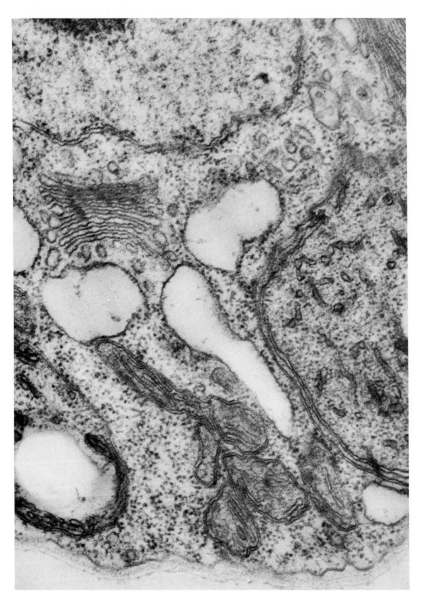

(*c*) *Chlamydomonas* with typical crista structures ($\times$45,000). From Sager and Palade in Brachet and Mirsky (94).

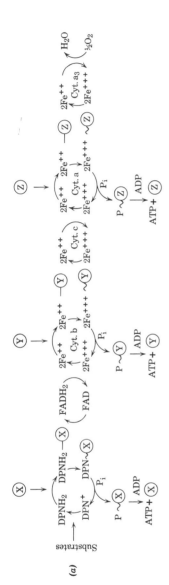

(a)

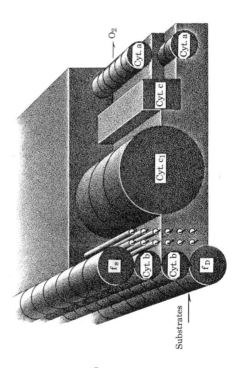

(b)

Figure 2.8. Schematic diagrams of energy-coupling systems in the respiratory chain and structural relationships. (a) Summary of reactions yielding ATP from oxidation through the respiratory chain. From Lehninger (579). (b) Suggested structural orientations of respiratory units of the more firmly fixed components of the electron-transport system (see a). Shaded areas represent lipid and other structural materials of the mitochondrial membrane. From Green and Lester (355).

complete or entirely correct, but the general picture is one which appears to present the shape of things to come in the problem of structure and functions. That is, lipo-protein membranes evidently form a structurally specified matrix which binds single or polymeric (in this case, of electron-transport components, perhaps 10,000 or more units) functional series of units in a highly organized fashion. All degrees of binding and organization of related functional units may be expected, and special localized membrane areas may be expected for different reaction systems. Such a general principle has already been suggested for membranes of the endoplasmic reticulum (e.g., rough and smooth forms), and it will also be evident as a strong possibility in the following discussions of other cellular components.

With respect to mitochondria, the oxidative system just outlined is only one of many present, though the others are likely related in terms of the building of the structure itself, providing substrates for oxidation, and in participation in processes concerned with change in shape. Other systems at the catalytic level include those concerned with active transport of molecules of various sizes, phospholipid synthesis, fatty-acid synthesis, amino-acid synthesis, porphyrin synthesis, energy-requiring proteolysis, and perhaps some protein synthesis (although the RNA content is probably less than 0.1%) and systems for oxidation for a great variety of normal metabolites.

There are also present, of course, a variety of coenzymes, vitamins K, E, and Q (which appear to be involved in electron transport), and ions such as Mg^{++} and Ca^{++}.

The question of the origin of mitochondria (746) has received a great deal of attention, but it remains open. The views of *de novo* origin, production from plasma membranes, endoplasmic reticulum, nuclear membrane and formation from preexisting mitochondria by division, have all been considered extensively but inconclusively. This matter and other important ones concerned with changes during development and cell division as well as dynamic aspects during formation will no doubt receive continued attention with improved technical means of observation.

Plastids. Structural elements of high complexity and prominence in the cells of plants are the plastids, which exist in various forms derived presumably by differentiation from a common type of protoplastid. A graphical picture of their origin (351) is given in Figure 2.9*a* and *b*. Here (Figure 2.9*a*) the amyloplasts are colored or colorless plastids filled with starch grains as in potato tubers, and the

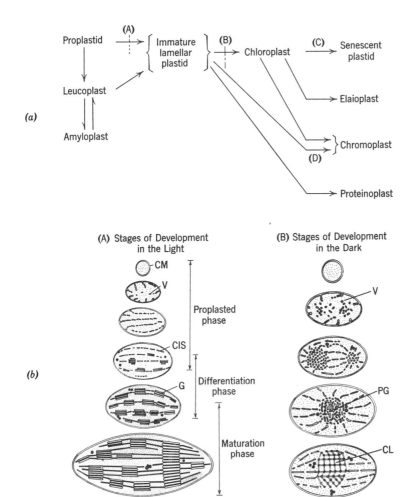

Figure 2.9. Plastid types and plastid development. (*a*) Plastid types (A) arrested in root cells, (B) arrested in leaf epidermal cells, (C) degeneration in autumn leaves, (D) chromoplasts of buttercup petals.

(*b*) Phases of development in light (A), in the dark (B). Double chloroplast membrane, CM; primary granum or prolamillar body, PG; vesicle, V; flattened cisterum, CIS; crystal lattice, CL; granum, G. From Granick (351).

elaioplasts are those which store oil droplets. Chromoplasts have special properties of pigment storage as in carrot roots, some algae and bacteria, and proteinoplasts are colorless but may contain clusters of needle-like protein crystals as in a variety of plant seeds. The chloroplast, the center of photosynthetic activity, has of course received the greatest attention, and a schematic representation of its development is shown in Figure 2.9b. The protoplastid stage represented by b is of similar size and structure as mitochondria, whereas the mature protoplast is far larger and more complex in structure. The latter is enclosed in a double membrane, each component of which is 30–50 Å thick. Inside are stoma lamellae (20–30 Å thick) and more dense grana lamellae (40–60 Å thick). The grana represent cylindrical piles of double membrane disks of 10 to 100, each of which is 0.3 to 1μ diameter. These structures are indicated in Figure 2.9b and shown further by electron-microscope photographs in Figure 2.10. These pictures, of course, provide only the smallest introduction and broadest generalizations. Chloroplasts exist in a great variety of forms and have also great variety in internal structures and kinds of inclusions, each of

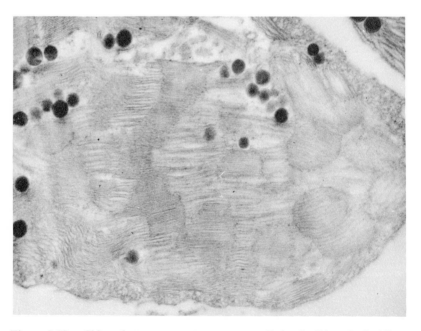

Figure 2.10. Chloroplast component structures. Isolated disks of *Aspidistra* (×27,000). From Steinmann and Sjostrand (959).

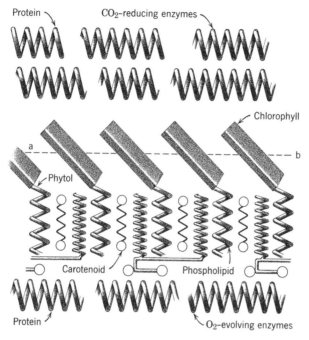

Figure 2.11. Model of a portion of disk membrane to show chlorophyll molecules in a fatty layer and the separation of oxidant from reductant. From Calvin (128).

which is of interest in connection with specific structure and function. As an example, a model representing some possible arrangements of functional components in a grana disk is shown in Figure 2.11.

Some further insight into the problem of structure-function relations in chloroplasts can be derived by consideration of some facts about composition and functional systems. In general, plastids contain, on a dry-weight basis, 35–55% protein (80% insoluble), 20–30% lipids, about 9% chlorophyll, about 5% carotenoids, 2–3% RNA, and a highly variable quantity of carbohydrate. With respect to the amounts and ratios of lipid and protein one again sees immediately the characteristics of a system built basically on membranes, as noted previously for mitochondria and endoplasmic reticulum. Within the chloroplasts are a number of whole synthetic systems to be accounted for in the fashion illustrated for the mitochondrial electron-transport system (p. 45). Of primary consideration are those concerned with the decomposition of water and with the fixation of CO_2 which occur in the grana disks. These two systems are outlined in Figure 2.12*a* and *b,* and each likely requires quite an elaborate fixed organization of enzymes perhaps oriented with membranes in an arrangement indicated in a general way as shown in Figure 2.11. Other systems to be accounted for include those which synthesize chlorophyll and heme, that which produces carotenoids, and those giving rise to starch and to lipids. Less well established are systems for protein and RNA synthesis. Although some of the enzymes involved in these many reactions may be loosely bound within the plastids, it is likely that many are fixed in definite arrays and loci to give an overall high degree of integration and order.

As already mentioned, there is good evidence that chloroplasts arise from the much smaller proplastids, and there is also evidence that in the early stages, when the proplastids are plastic, new ones arise by a process of pinching off pieces of those which exist. The evidence,

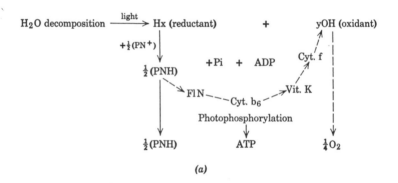

(a)

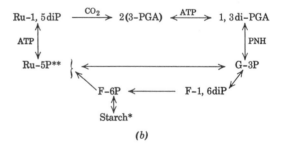

(b)

Figure 2.12. Starch formation from CO_2, H_2O, and light by chloroplasts. (*a*) Photochemical act and production of reduced pyridine nucleotides, ATP, and O_2. (*b*) CO_2 fixation using pyridine nucleotides and ATP to yield starch. From Granick (351).

then, is in favor of plastids coming from plastids and against any *de novo* formation. In certain mutants and in some organisms, by chemical treatment (e.g., streptomycin), plastids are lost and once lost they do not reappear. Furthermore, there are many well-studied examples in which inheritance of plastid characters (such as colorless or partly green) is independent of the nuclear genome and dependent on maternal transfer of preexisting plastids of one kind or another (see Chapter 11). The basis for such a genetic continuity in terms of composition, structure, and function is completely obscure.

Vacuoles, Lysosomes, and Pinocytosis. As noted previously, cell membranes serve as selective barriers to passage of low-molecular-weight substances, and, through special mechanisms of active transport, permeability can be controlled. On the face of it, there would seem to be little ground for confusion of these processes with a phenomenon like phagocytosis in an amoeba where pseudopodia may reach out and engulf a whole bacterium and transport it within the cell enclosed in a vacuole. Perhaps it is extreme to make such a comparison, but it is a fact that it is becoming increasingly difficult to know where to draw a line of distinction. In phagocytosis a vacuole is produced which has a single membrane derived from the plasma membrane, and it contains the visible capture victim. At a somewhat less dramatic level nonvisible material from the external environment can be transferred within the cell by pinocytosis. This occurs by invagination of the membrane and a pinching off again to form a smaller single-membrane-bounded vacuole. Micropinocytosis has also been described, and it remains an open question as to how far down into molecular dimensions such a mechanism really does extend. This question and that of the importance of the mechanism to the economy of cells in general are important ones for future studies. Obviously it is important to a phagocyte; that is how it eats.

At a more subtle level pinocytosis, or, as a more general term, cytosis (746), has been invoked as the origin of the cell structures known as lysosomes. These are vacuoles enclosed in a single membrane as shown in the photograph in Figure 2.13. By following the pinocytotic uptake of the enzyme peroxidase in rat kidney, Novikoff (745) obtained evidence for the sequence of transfer of the enzyme from the micropinocytotic vacuoles (p) to the larger luminal vacuoles (v) and thence to the lysosomes (L).

It is of interest to note that while the existence of vacuoles in tissues is hardly new, the recognition of lysosomes as cellular units of impor-

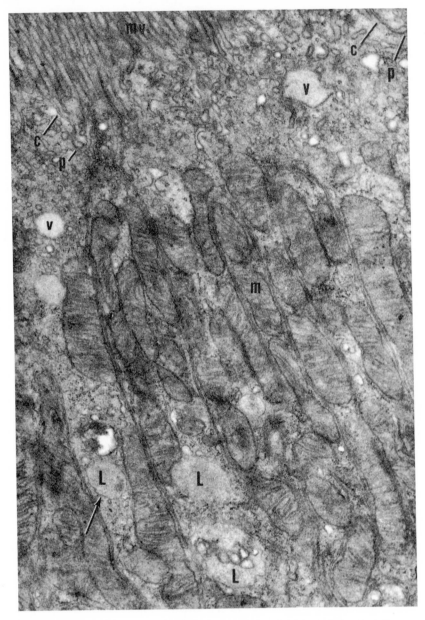

Figure 2.13. Section of rat kidney showing micropinocytotic vacuoles, p (×21,000). Microvilli, Mv; canalicular structures, c; mitochondria, m; lysosomes, L; luminal vacuoles, V. Vacuoles, p, appear to come from canalicular structures and fuse with the larger vacuoles, V. From Novikoff (746).

tance came as recently as 1951 from the work of de Duve and collaborators (79). By careful fractional centrifugation these investigators obtained, from liver, material slightly lighter than mitochondria, extremely rich in acid phosphatase but lacking in the cytochrome electron-transport system. By histochemical techniques for localization of acid phosphatase, the isolated particles were identified with cellular inclusions like those shown in Figure 2.13. Further investigations demonstrated the inclusion in these bodies of a variety of other lytic enzymes that have acidic pH optima. A summary of the picture is shown diagrammatically in Figure 2.14. Thus the lysosomes represent membrane-bounded compartments containing enzymes that would break down many cell components if they were all suddenly released into solution. The name "lysosomes" is derived from this biochemical property of being exceptionally rich in hydrolytic enzymes.

Functionally it is obvious that an initial action of the particles formed by pinocytosis could be digestion of large molecules taken in from the external milieu of the cell, but as lysosomes they frequently become associated with the Golgi apparatus and their continued function is not clear, nor is the origin of the enzymes they contain. There is also evidence that there exist different kinds of such particles having different enzyme complements such as those that contain uricase but not acid phosphatase. One possible role of the particles of this class is in intracellular self-regulation. That is, they may participate in turnover of intracellular components, and, as such, have a function in homeostasis and in normal resorption of tissue when it occurs.

The Golgi Apparatus. Since the initial description by Golgi in 1898 of a specialized reticular apparatus in animal cells, there has been much controversy and speculation concerning its structure and function. Although some of this doubt remains, it is now well established that the Golgi apparatus consists of a group of large vacuoles, a system of flattened sacs which may dilate to become vacuoles, and clusters of small vesicles coming from the edge of the sacs. Examples are shown in Figure 2.15a and b. Isolated preparations such as shown in the figures do not contain RNA and are high in content of phospholipid and acid phosphatase (193). Among the many hypotheses proposed for function of the Golgi apparatus, a promising one is concerned with storage and perhaps alterations of substances synthesized in the cell, and transported to the Golgi zone through the endoplasmic reticulum where they are deposited in granules. Secretory and zymogen granules are examples.

The Nucleus. Nuclei are usually prominent cell constituents with shapes varying from round to characteristic highly branched structures.

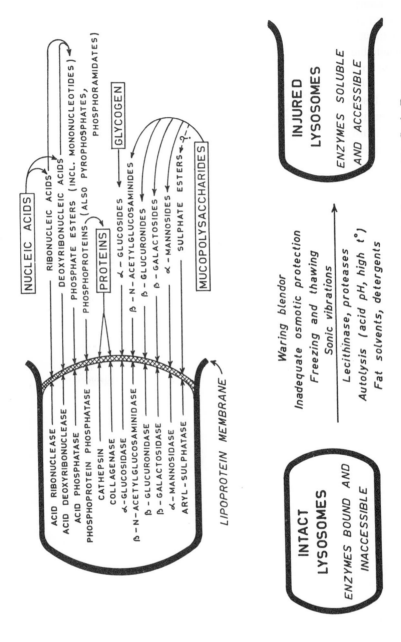

Figure 2.14. Schematic representation of the lysosome concept. Courtesy Dr. C. de Duve.

In many cells they occupy 10% to 15% of the cell volume, but in sperm nearly all of the material is nucleus. On the contrary, in eggs cytoplasm generally dominates, and the nuclear volume is relatively negligible. As noted previously, bacteria tend to lack membranes and thus bounded nuclei are not apparent. However, in the great variety of cells where a nucleus can be observed, the membrane is an important and fascinating structure. As discussed previously, page 27, the membrane is double with the outer layer continuous with the endoplasmic reticulum, and thus the space between the two membranes has connections open to the external part of the cell. There is, in fact, some evidence that materials can come from outside the cell into the nuclear space without going through the cytoplasm proper; see Mirsky and Osawa (680). Connections to the nucleus occur through pores where the inner and outer nuclear membranes are fused. As shown in Figure 2.16a and b, hollow cylindrical structures extending inside and outside the nucleus through the pores can be observed. Although these pores may provide for passage of large molecules, it has been demonstrated that such passage is restricted in some unknown fashion (273). One thing that is clear from these considerations of nuclear membranes is that the isolated nucleus, as it may be prepared for analysis or metabolic studies, or the exchanged nucleus as it may be used in transplant studies (p. 536), is never a really clear-cut entity entirely separated from structures or substances associated with its surfaces or those of nearby endoplasmic reticulum.

The major structures to be observed within nuclei are the chromosomes, nucleoli, and spindles. Here the relative morphological prominence of the structures is dependent on the time of observation, with the chromosomes and spindles being visible during cell division. In interphase nuclei, where metabolic activity is prominent, the chromosomes are diffuse and not visible in unfixed material, but structures do appear on fixation or appropriate changes in salt concentration (842). A notable exception is that of the lampbrush chromosomes of certain oöcytes (p. 65).

Nuclei of many kinds of cells can be isolated in aqueous or nonaqueous media (9) to provide a basis for informative analyses of composition and metabolic activities. A prominent constituent of all nuclei is deoxyribonucleic acid (DNA) which can be determined both by bulk analysis and cytologically. In a fowl-liver nucleus, for example, it comprises 10% to 15% of the dry weight and amounts to about 2.4×10^{-9} mg of DNA per nucleus. Erythrocyte nuclei from the same species contain essentially the same quantity of DNA while the

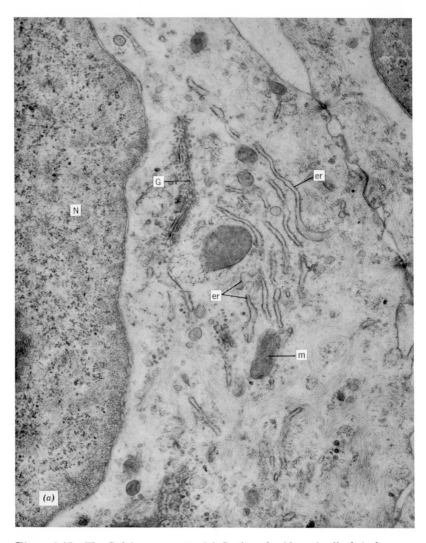

Figure 2.15. The Golgi component. (*a*) Section of epidermal cell of *Ambystoma punctatum* larvae (×14,200). Golgi complex, G; nucleus, N; mitochondria, m; endoplasmic reticulum, er. From Porter (803). (*b*) Isolated Golgi substance (×48,000). Flattened sacs (GM) delaminate into vacuoles (GV). Small vesicles (V) are also present. From Dalton (193).

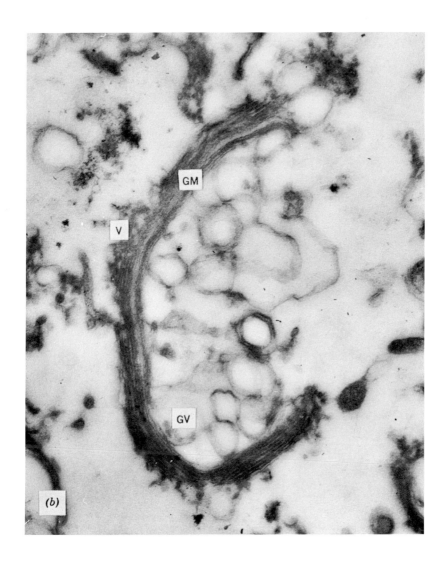

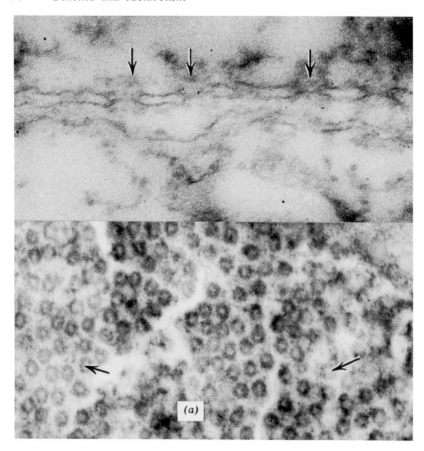

Figure 2.16. The nuclear membrane. (*a*) *Echinus* oöcyte ($\times$95,000). Arrows show annuli and cylindrical structures in transverse and tangential sections. From Afzelius (2). (*b*) Scheme of nuclear membrane ($\times$110,000). Dimensions are in angstrom units. From Afzelius (2).

haploid sperm contain half as much. This important relation, first observed by Boivin et al. (83) and Mirsky and Ris (681), is consistent in different species, even though the total amount of DNA per nucleus varies among species over a very wide range (about 0.04 to 160×10^{-9} mg per nucleus). That is, for a species having 0.04×10^{-9} mg of DNA per somatic cell nucleus, the amount in a sperm cell is 0.02×10^{-9} mg, and at the other extreme (160×10^{-9} mg per diploid cell), the amount per sperm cell is near one-half or 80×10^{-9} mg. Quantities of DNA in polyploid nuclei are multiples of the

N

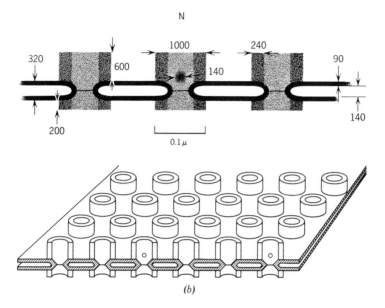

(b)

haploid amount. For example, in the salivary gland of *Drosophila* individual cells may contain as much as 1024 times the haploid quantity per nucleus.

In contrast to the constant values for DNA, other major classes of nuclear components may vary widely in amounts even in the same tissue. Ribonucleic acid (RNA) is found in a range of about 2% to 20% of the dry weight, and protein, including the basic proteins (histones or protamines), the soluble proteins (largely enzymes), and the insoluble residual protein (possibly of structural significance in chromosomes), may be present to the extent of about 70% to 85% in somatic cells. An even greater variation occurs in individual classes of all of these components. Several kinds of RNA and an indeterminate but very large number of molecular species of RNA exist. The basic proteins (protamines which occur in sperm and histones from somatic nuclei) are all complex mixtures of many molecular species and exist attached to DNA by salt-like linkages. Much of the soluble protein of nuclei is made up of enzymes. Their activities relative to activities in cytoplasm give some index to specialization of metabolic functions. Some representative data are summarized in Tables 1, 2, and 3. These data illustrate several important facts to be remembered in subsequent discussions. In several cases enzyme activities in nuclei are higher

Table 1. Intracellular Distribution of "Special" Enzymes *

Enzyme	Tissue	Units of Activity per mg in		
		Cytoplasm	Nuclei	100 N/C
Arginase	Calf liver	2.88	1.55	54
	Horse liver	1.21	0.6	57
	Fowl liver	0.04	0.025	63
	Fowl kidney	0.62	0.04	7
	Calf kidney	0.137	0.152	110
Catalase	Horse liver	41	29	71
	Calf liver	9.3	3.0	32
	Calf kidney	25	0	0
	Fowl kidney	33.5	0	0
Uricase	Horse liver	2.14	0.11	5
	Calf liver	0.41	0	0
	Calf kidney	2.9	0	0
Lipase	Horse pancreas	0.323	0.005	1.5
	Beef pancreas	0.655	0.055	8
Amylase	Horse pancreas	14.30	0.356	2.5
	Beef pancreas	36.05	2.51	7.1
Deoxyribo-nuclease I	Beef pancreas	330	2.5	0.76
Alkaline phosphatase	Calf intestinal mucosa	344	12	3.5
Adenosine deaminase	Calf intestinal mucosa	127	24	19

* From Mirsky and Osawa (680).

than in the cytoplasm on a weight basis (arginase, Table 1; esterase, adenosine deaminase, and nucleoside phosphorylase, Table 2; and several of the glycolysis enzymes, Table 3), indicating highly specialized metabolic activities for nuclei. Further, nuclei from different tissues of the same organism vary widely in activities for the same enzyme (Tables 1 and 2), but each has its own special pattern of enzymes. These data agree with other extensive evidence (p. 535) that nuclei, as well as cells and tissues, undergo a high degree of differentiation (Chapter 12). In addition to the examples of specific enzymes in

Table 2. Intracellular Distribution of Commonly Occurring Enzymes in Adult Tissues *

Enzyme	Calf							Horse
	Liver	Kidney	Kidney Cortex	Thymus	Heart	Intestinal Mucosa	Pancreas	Liver
Esterase								
C[a]	0.161	0.061	0.078	0.075	0.011	0.336	0.398	0.160
N[a]	0.122	0.006	0.018	0.026	0.014	0.029	0.128	0.128
100 N/C	76	10	23	35	127	9	32	80
β-Glucuronidase								
C	1.88	0.50	...	0.655	Trace	0.410	Trace	0.433
N	0.324	0.06	...	0.033	Trace	0.102	Trace	0.073
100 N/C	17	12	...	5	...	25	...	17
Adenosine deaminase								
C	52	Trace	Trace	80	7	127	0.81	...
N	108	Trace	Trace	53	42	24	1.32	...
100 N/C	208	...	...	66	600	19	163	...
Nucleoside phosphorylase								
C	52	15	17	19	18.5	19	28	15
N	52	19	17	7	82	4.5	21	41
100 N/C	100	120	100	37	440	24	75	274

[a] C: activity in cytoplasm; N: activity in nuclei.
* From Mirsky and Osawa (680).

Table 3.   Intracellular Distribution of Glyolytic Enzymes in Wheat Germ *

Enzyme	Units of Activity per mg in		
	Tissue	Nuclei	100 N/T
Aldolase	1.97	3.04	155
Glyceraldehyde phosphate dehydrogenase	1.16	1.14	99.4
Enolase	27.2	43.2	160
Pyruvate kinase	3.47	4.93	142

* From Mirsky and Osawa (680).

nuclei above, it is well established of course that both DNA and RNA are synthesized in nuclei, and a complete system for protein synthesis is also present with the highest activity in the soluble protein fraction (10).

The residual protein which remains after extraction of histones and soluble protein has been described by Mirsky and Ris (682) as material providing long threads serving as the backbone of chromosomes, but others (996) consider DNA to be the continuous component. Perhaps both are significant in this respect, but, in any case, protein of this type is present in nuclei to the extent of one-fifth to one-half that of the DNA.

Nuclei in general have prominent lipid components amounting to about 3% to 10% of the dry weight. Low-molecular-weight compounds such as amino acids, carbohydrate derivatives, and nucleotides are also present in significant quantities as would be expected from the indicated metabolic activities. Assays for the coenzyme diphosphopyridine nucleotide (DPN) in several tissues have shown higher nuclear levels than cytoplasmic levels (970), and there is evidence that all of the DPN-synthesizing enzyme is in the nucleus (444). These results are in accord with earlier observations that beef-heart nuclei contain from three to four times as much niacin as the whole tissue on a dry-weight basis, but the additional suggestive information that other vitamins such as pantothenic acid, riboflavin, thiamin, and folic acid are also in excess (749) in these nuclei has not been explored further. Perhaps several of the essential coenzymes are synthesized primarily in nuclei. Ions such as Ca^{++} and Mg^{++} are present in nuclei to the extent of about 1.3 and 0.1% respectively (dry weight), values which are about double those of whole tissue. Sodium ion is also a prominent constituent demonstrated to be important in the transport of amino acids and perhaps other substances into isolated nuclei (12).

Chromosomes. As carriers of genetic material, as objects with highly specialized and regular mechanisms for reproduction, and as carriers of molecules that need be represented only once in a cell, chromosomes are unique and of central interest. They are constructed primarily from DNA, RNA, and protein but may contain small amounts of a great many other substances, and it is important to note and remember in this connection that a biologically significant quantity of a substance may be a very small amount indeed. If one macromolecule is sufficient for gene expression, then it is possible that only a few specific smaller molecules may be sufficient to have a profound influence on gene reproduction and expression. Further, the situation demands a rigid control in structure in order that nonstatistical numbers of molecules can consistently react in the right way most of the time.

In living material the chromosomes are visible during nuclear divisions in different degrees of length and thickness, whereas in interphase they are usually diffuse and not visible in the light microscope. Pictures taken with phase contrast and a great many others obtained from fixing and staining followed by light or electron microscopy have yielded much important detail concerning chromosome structure, but a unitary description is not yet possible; see reviews: Ris (841); Gall (307); Beermann (68); Kaufman et al. (515), and Swift (996). Swift has outlined major hypotheses of chromosome structure as shown in Figure 2.17. Here, as in most current considerations of chromosome structure, the fundamental unit is assumed to be the DNA double helix (see p. 93) with histones, protamines, and perhaps other proteins closely associated by neutralization of the phosphate groups of the nucleic acid. Ribonucleic acid is also associated with the chromosome, sometimes in granules and usually in regions of the centromere (spindle-fiber attachment). The fundamental strand, then (as a DNA double helix), should be about 2 mμ in diameter (398), a dimension far below the limit of the light microscope; thus the visible chromosome is necessarily a highly condensed structure with perhaps coils built on coils to a higher order than shown at the more fundamental level in Figure 2.17. One acceptable general concept of chromosome structure is illustrated in Figure 2.18. Here individual strands are considered to carry double helices of DNA plus basic protein and RNA granules,

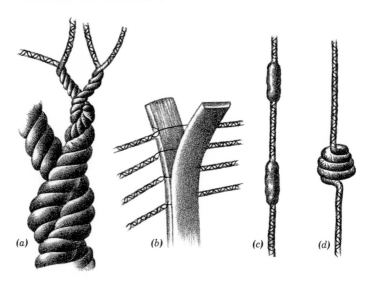

Figure 2.17. Hypothetical models of chromosome structure. (*a*) Multistrand or "rope" with eight helices of DNA-histone. (*b*) Protein backbone with lateral DNA-histone fibers. (*c*) Alternating DNA-histone fibers and proteins. (*d*) Differential coiling of single DNA-histone fibers. From Swift (996).

but the strands are thickened by coiling or by side branches (Figure 2.17) and coiling.   These diagrams must be taken only as reasonable representations of facts so far as they go, since the details on chromosome structures are far from complete.   Furthermore, structural changes during cell cycles and major problems concerned with mechanisms of breakage and fusion of chromatids and processes of separation of intertwined coils remain unsolved.

Except in bacteria, where DNA appears to be produced continuously, formation of new DNA and chromosomes is apparently confined to interphase.   Insight into this process has been gained by studies of incorporation of isotopically labeled substrates into DNA.   Taylor and collaborators (1001, 1002) made use of tritium-substituted thymidine (as a specific DNA precursor) and a combination of cytological and radioautographic techniques.   Due to the soft radiation of tritium it can be recorded locally on film by direct contact and observed in the light microscope superimposed on chromosome structures.   Some important results obtained by this method are summarized in Figures 2.19 and 2.20.   Photographs reproduced in *a* and *b*, Figure 2.19, show chromosomes of root cells of *Bellevalia romana* in first division anaphase

(1001): (*a*) after treatment with tritiated thymidine and in second division anaphase; (*b*) after continued incubation without label. A general conclusion is summarized in part *c* of the figure, from similar experiments with roots of *Vicia faba* (1003). Thus both daughter chromosomes (sister chromatids) from a chromosome duplicating in the presence of label were equally labeled showing doubleness of structure duplicated. After a second duplication in the absence of label,

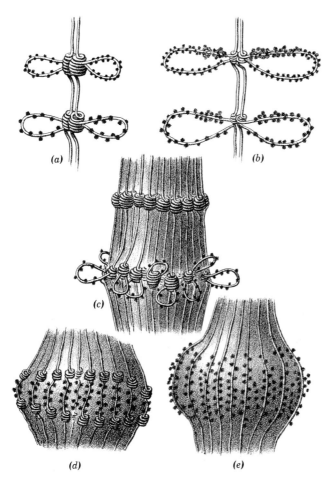

Figure 2.18. Lampbrush and polytene chromosome models. (*a*) and (*b*) Bivalents with coiled chromosomes and larger loops with more RNA-protein particles with increased chromosome activity. (*c–e*) Similar structures for polytene chromosomes with uncoiling in puff formation. From Swift (996).

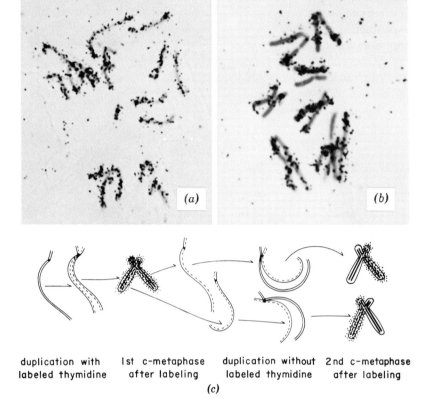

duplication with   1st c-metaphase   duplication without   2nd c-metaphase
labeled thymidine   after labeling   labeled thymidine   after labeling

(c)

Figure 2.19.   Distribution of tritiated thymidine to new chromosomes formed in
the absence of label.   (a) Complete label in first division anaphase in *Bellevalia*
(×990).   Focus at grain level.   From Taylor (1001).   (b) Second division
metaphase in *Bellevalia* showing unlabeled new chromosomes but with reciprocal
exchanges (×1420).   From Taylor (1001).   (c) Diagrammatic representation of
observed label patterns not including apparent reciprocal exchange of whole
chromatid segments as shown in some cases in (b).   From Taylor, Woods, and
Hughes (1003).

only one of each pair of chromatids carried label.   Thus, except for
sister chromatid exchanges, DNA is labeled throughout the extent of
a chromosome, and the units remain intact through replication of DNA
and nuclear divisions.   Detailed studies of the sister chromatid exchanges (Figure 2.19b) yielded the further important information that
while sister chromatid exchanges were frequent, exchanges between
one of the strands in each of two chromatids did not occur.   These

results lead to the conclusions that the chromosome is two-stranded with the strands not alike, and, further, all breaks are four-stranded and reunions or exchanges are limited by the unlike character of the two-stranded unit. This picture has been interpreted as being in accord with the Watson-Crick model of the double helix structure of DNA

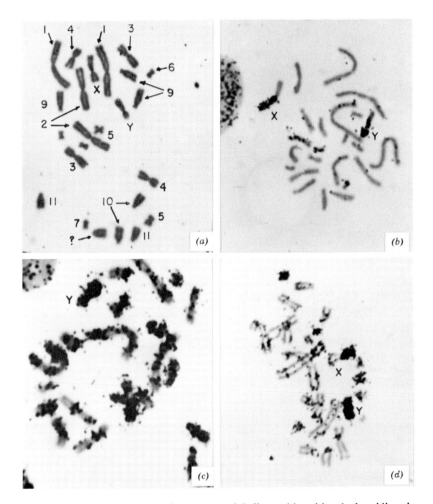

Figure 2.20. Asynchronous chromosome labeling with tritiated thymidine in the hamster. (*a*) The chromosome complement in the hamster (×1340). (*b*) Autoradiogram after short-term labeling (2 hours). (*c*) Autoradiogram after long-term labeling (4 hours). (*d*) Autoradiogram after short-term labeling carried out late in the period of DNA synthesis. The pattern is different from that shown in (*b*). From Taylor (1002).

Exp. no.   Generations

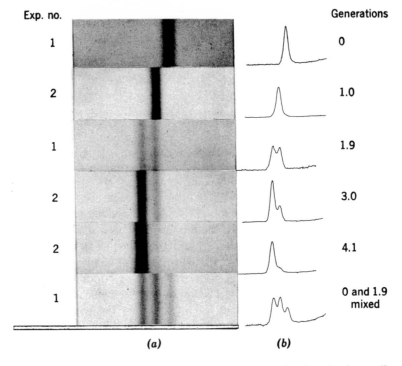

(a)   (b)

Figure 2.21. Ultraviolet photographs showing DNA banding in a density gradient in the ultracentrifuge. Preparations are from lysates of *E. coli* grown in the presence of N^{14} after initial labeling with N^{15}. (*a*) Direct photographs. (*b*) Densitometer tracings. From Meselson and Stahl (666).

(p. 93). A chromosome could thus be composed of a single double helix or a multiple thereof.

Another point of interest in thymidine labeling experiments is illustrated in Figure 2.20 (1002). As shown clearly, DNA replication is not necessarily synchronous or progressive down the length of a chromosome. In fact these and other observations have established that there is a definite pattern of order of synthesis in different regions on the chromosome during the period when replication occurs. Whether this extends down to the level of one gene at a time remains to be seen, but these and similar observations by other investigators emphasize the importance of a highly regulated time sequence of events, perhaps at the level of single molecules, in the process of replication of the genetic material.

The thymidine labeling experiments just cited indicate a linear continuity of DNA in a chromosome which is largely maintained in sequential chromosome duplications. These results agree with those obtained on a somewhat different basis by Meselson and Stahl (666). These investigators made use of the density difference of DNA containing N^{15} versus that containing N^{14} in studies of DNA replication in *E. coli.* The bacteria were cultured in the presence of an N^{15} nitrogen source so they contained only heavy N^{15} DNA, demonstrable by isolation and equilibrium centrifugation at high speed in a cesium-chloride density gradient. The bacteria were then washed and cultured in the presence of an N^{14} nitrogen source for several generation times. Some results are summarized in Figure 2.21. After one generation all of the DNA had a density intermediate between N^{15} and N^{14} nucleic acids, implying the existence of double strands, one containing N^{15} (old) and one containing N^{14} (new). At an average of 1.9 generation times, preparations contained approximately equal amounts of hybrid and N^{14} types, and subsequently more N^{14} was found but the hybrid was retained.

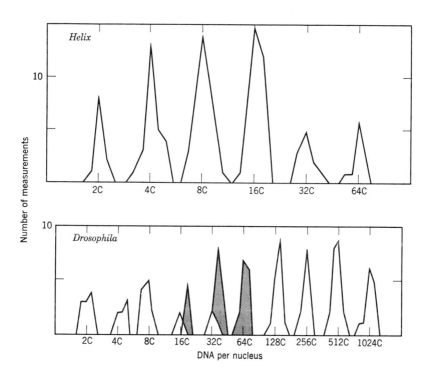

Figure 2.22. Relative amounts of DNA in individual nuclei showing polyploidy. 2C = diploid content. *Top: Helix pomatia* salivary gland. *Bottom: Drosophila melanogaster.* 2–32C, anlage and duct tissue; shaded, fat body; 128–1024C, salivary gland. From Swift (996).

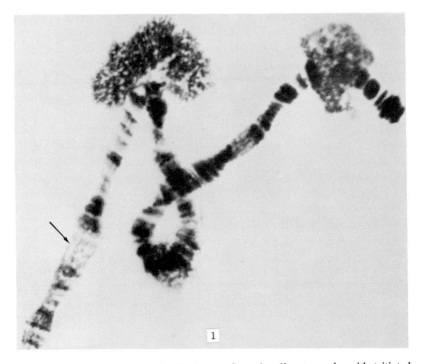

Figure 2.23. Puffs in *Sciara* by Feulgen stain and radioautographs with tritiated thymidine. 1: Chromosome 2 (×1060). Subterminal and medial DNA puffs. Arrow shows position of RNA puff. 2 and 3: Before and after DNA puff (×1450). 4a and b: Radioautographs of subterminal and medial puffs after incorporation of tritiated thymidine. From Swift (996).

Thus maintenance of linear integrity of DNA chains was clearly demonstrated, and again results are compatible with the DNA model of the complementary double helix (p. 93), as are similar data obtained by Sueoka (989) in studies of DNA from *Chlamydomonas* and mammalian cells. At this time, however, a direct interpretation of all of these data in terms of DNA structure must be considered with caution, since DNA is so easily altered in isolation by shear forces in mixing and grinding, and chromosome structure may be much more complicated than a single double-helix, basic unit. That the latter is indeed the case is indicated by studies of Cavalieri and collaborators (151) on the *E. coli* system.

The formation of new chromosomes is, of course, one aspect of function, and details are subject to variations in time sequences and specializations characteristic of different kinds of cells. Chromosome function is not separable from that of other parts of the nucleus and

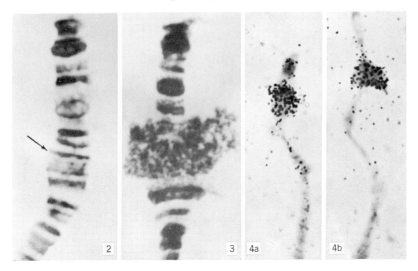

the cell, and for broader and more general aspects references should be made to more general reviews such as those by Brachet (94), Caspersson (137), and Mazia (664). Here two kinds of functions of chromosomes subject to direct observation are considered because of their special potential value in understanding chromosome functions. Certain tissues of some organisms contain cells which do not divide but only enlarge, but at the same time the chromosomes are duplicated regularly and they remain associated to form increasingly larger bundles with as many as 10,000 or more units in some cases. This process is called endomitosis and gives rise to polytene chromosomes as in the salivary glands of *Drosophila*. These contain 1024 times the haploid DNA content of *Drosophila* nuclei and have been extremely useful for cytological and cytochemical studies. Nuclear content of DNA in such cells is consistent with the origin of the giant chromosomes by successive doubling, as shown in Figure 2.22 (996); a diagrammatic picture of such bundles was shown in Figure 2.18. From studies of such chromosomes in *Rhynchosciara*, Breuer and Pavan (778) observed the occurrence of sudden expansions or puffs (Balbiani rings) in late larval and prepupal stages of development. These puffs occurred at specific cytological loci in the chromosomes, and they appeared and disappeared in reproducible time sequences. Beermann (69) has presented evidence that, in one case at least, puff formation is related to a single gene. The suggestion that puffs are a manifestation of chromosome or perhaps a single-gene function has gained support from many investigations—see review by Swift (996). It has been demonstrated

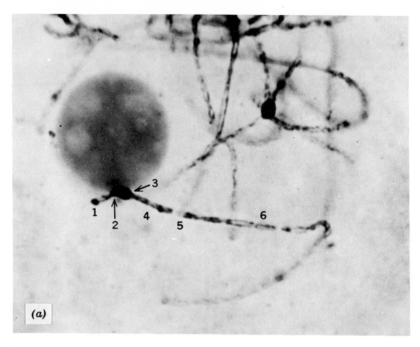

Figure 2.24. Nucleoli in light and electron microscopes. (*a*) Maize (×1400). Numbers 1–6 show satellite, connecting region, nucleolus organizer, the short arm, the spindle-fiber attachment region, and the long arm of the chromosome. From McClintock (636). (*b*) Nucleolus of a rat-liver cell (×37,250). From Bernhard and Byezkowska in Mirsky and Osowa (680).

that rapid nucleic acid and protein synthesis occurs in puff areas with both DNA and RNA puffs being formed in *Sciara* but only RNA puffs in *Drosophila*. Examples of the cytological appearance of puffs and DNA synthesis in one are shown in Figure 2.23. DNA synthesis in puffs is an exception to the constancy and doubling rule for DNA, and function of the products (which includes synthesis of an equivalent amount of histone) is not known. RNA synthesis in puffs is accompanied by synthesis of nonhistone protein and the production of nucleoprotein granules. As shown in Figure 2.18, Swift (996) has compared possible structures of polytene chromosomes and action in puffing with some observed structures and changes in the large "lampbrush" chromosomes in amphibian oöcytes—see Gall (307). Lampbrush chromosomes may be as much as one millimeter in length and are made up of a small number, perhaps only two, double strands. Chromomeres appear to be tight coils, and these carry lateral loops of thin fibers and

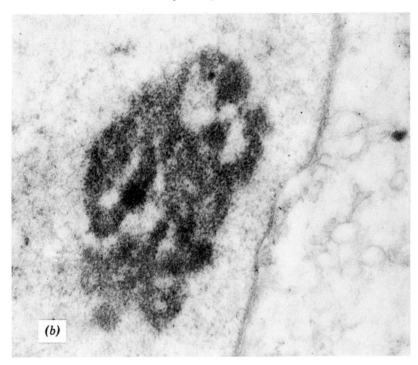

(b)

many RNA granules. In rapid growth the loops increase in size and chromomeres decrease. At the same time larger numbers of RNA granules appear with higher concentrations at one end of the loop as though their synthesis resulted as chromomere uncoiling occurred.

These two examples of correlation of chromosome structure and function may well set the pattern of things to come. That is, puffing in the polytene chromosome system possibly presents a much exaggerated model of the manner in which genes function in single chromosomes. If so, an important aspect of function is to produce new nucleic acid and protein at specific loci (genes) in a regulated and sequential order. Furthermore, it may be expected that the sequence of action is different in different tissues (65) and that the sequence can be altered through environmental changes (538). Perhaps this is a key to the nature of nuclear differentiation (see Chapter 12).

The Nucleolus. In general, each somatic nucleus contains a large, strongly staining body (nucleolus) attached to a chromosome at a specific region (nucleolus organizer). Light- and electron-microscope photographs are shown in Figure 2.24. The nucleolus is not bounded by a membrane, and it disappears during cell division to be formed anew at the end of the division. Evidence supports the contention that nucleolus formation involves some kind of organization of granular bodies coming from all areas of all chromosomes. But, once formed, the nucleolus shows high capacity for synthesis of new materials. As shown by McClintock (636), division of the nucleolar organizer by chromosomal translocation results in formation of two nucleoli, and loss of action of the organizer area results in formation of many nucleolus-like bodies at many loci in the chromosomes.

As to composition (680), nucleoli are dense with a dry weight in the order of 60–70%, and they contain largely RNA and protein with an excess of the latter. Amounts vary over a wide range with the stage in the cell cycle and among different tissues of the same organism. There are probably several kinds of RNA present and many kinds of protein including a wide selection of specific enzymes. These characteristics of composition along with knowledge of the presence of co-enzymes and other low-molecular-weight components designate the nucleoli as centers of biosynthetic activities. One of these activities of great importance is RNA synthesis. It is an established fact that RNA labeling in the nucleolus is faster than in any other part of the cell. Further, in rapid growth, high nucleolar activity correlates with high cytoplasmic activity, and indications are that nucleoprotein granules produced in nucleoli move into the cytoplasm where they become centers of specific protein synthesis (1002). Edström (249) demonstrated more similarity in composition between nucleolar and cytoplasmic RNA than between nucleolar and other nuclear RNA, observations that are in accord with electron-microscope observations such as illustrated in Figure 2.25. Also, emphasis on RNA synthesis is suggested by the fact that the nucleolus-organizer chromosomal region is highly heterochromatic. It thus seems clear that a major function of nucleoli (but by no means the only function) is the production of nucleoprotein. But it is only one of several specific sites of synthesis, since there is ample evidence for RNA synthesis in nearly all parts of cells.

Centrioles and Spindles. Toward the end of animal cell division the centrioles, cylindrical bodies (about 200×30 mμ) composed of tubular elements, appear in the cytoplasm and become the poles for chromosome movements of the next division (664). These are extranuclear, self-reproducing bodies which may contain RNA, but little else is known of their composition. The spindle fibers attach to them and to the chromosomes. The spindle fibers may be isolated with retention of

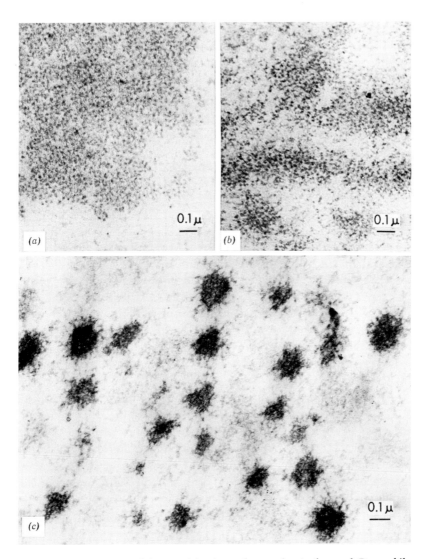

Figure 2.25. RNA-containing particles in nucleus and cytoplasm of *Drosophila*. (*a*) Nucleolus; (*b*) endoplasmic reticulum area; (*c*) RNA puff area with large granules. From Swift (996).

form and are found to be largely protein with a composition resembling that of muscle actin. RNA is present to about 6% and perhaps polysaccharide and lipid. ATPase and perhaps other enzymes may be real constituents of the highly organized system of fibers. In cell division spindle fibers contract and draw the chromosomes to the poles, but the whole process is highly complex, and evidence on physical and chemical mechanisms is fragmentary.

General Comments. The intent in the foregoing discussion has been to present some highlights of facts and thoughts derived from a great many studies which relate cell structure to function. It should be apparent that new techniques and tools are providing an ever-increasing magnification of the details of cell structures, and within the structures the jigsaw of individual molecules is becoming discernible. That great ordering in the arrangements of different kinds of molecules to give specific structures is present, and that the time of such ordering is fundamental to the characterization of function, is beyond question. Therefore continued development of understanding of structures as units of function is essential to all aspects of the problem of mechanisms of genetic control. As will be discussed in the ensuing section of this chapter, a major hypothesis in genetics considers that genes consist of linear segments of DNA and that they determine order directly only in a linear dimension. If this is true, how are these products put together to form the complex, three-dimensional structures that are so essential to cell functions? One way to do this is by orientation of gene products into monomolecular films to give membranes on which further specific structure can be built. On the other hand, preexisting three-dimensional structures may be essential for orientation of new molecular units into their own likenesses. If this is true, then formation of direct products of gene action would represent only accumulations of sophisticated junk, in the absence of preexisting cell structures. In any case, these are important matters for careful consideration when evaluating the facts and hypotheses presented in the section that follows where the problem is developed from the inside out—that is, from atoms to molecules to structures, rather than from cells to structures to molecules.

Molecular Morphology and Function

A great deal of work in biochemical genetics has been concerned with determinations of the nature of metabolic patterns (Chapter 7, for example) under genetic control. Thus the subject initially dealt with substances containing a few or relatively few atoms, and the rules of the game were primarily those of the organic chemist. That is, for proper identification, a new compound required isolation as a pure substance, elementary analysis, determinations of physical and chemical properties, degradations and preparation of derivatives, and finally confirmation of a deduced structure by total synthesis. Even with low-molecular-weight compounds this process is not always a simple, straightforward procedure, and there are many pitfalls that can lead to misinterpretation of structures if identifications are not carried to completion. A good discussion of such matters in relation to identification of new amino acids in tissues is given by Winitz (1090).

Now, in more recent years, much attention in biochemical genetics has turned to considerations of the higher-molecular-weight compounds of genetic importance such as the proteins and the nucleic acids. Here the problems of proper identification are multiplied enormously, so enormously, in fact, that in no case has it been completed and in the majority of cases even the first step, the isolation of a pure substance, is still open to question. This is especially true for the nucleic acids. In the discussion here, and in much of the literature, the abbreviations DNA and RNA are not singular but, on the contrary, they describe classes of compounds, often with thousands of different species of molecules represented. And yet we talk about the structure of DNA or RNA. Obviously, the rules of identification are being broken, and one reason is that this is the best we can do for the moment. The point here is not to belittle in any way the efforts and accomplishments in the field now known as molecular biology, but it is to emphasize the classical background and provide a basis for critical evaluations of deductions based on incomplete information. Even at the most conservative level, good educated guesses speed progress greatly by designating which experiments are the most critical and what information is essential to establish a given point. Thus molecular biology, the study of the morphology and function of macromolecules, needs no new set of rules but only a judicious use of the old ones.

In the present frame of things genetical, molecular biology presents the picture that DNA molecules are primary in terms of genetic material by being replicated linearly for a next generation, and by being mimicked linearly by RNA production for a function. In turn, the RNA provides for linear designation of amino acids to give proteins with appropriate biological functions in terms of structure and catalytic activities. A total proper evaluation of this picture is far from possible, but there are many important facts, deductions, and educated guesses that are important to consider. Highlights of some of these are presented in the following sections.

Proteins. At the present time methods for isolation and characterization of proteins are by far the most advanced among the various classes of macromolecules. In cells they normally exist to a great extent in combinations with a great variety of other substances such as nucleic acids, carbohydrates, and lipids, but once separated from fixed structures or loose combinations, preparations of quite homogeneous components are now quite practicable. For details in methodology of fractionation by precipitation, chromatography, electrophoresis, centrifugation, and so on, reference can be made to numerous books and reviews on the subject (96, 91, 179, 517). Criteria for purity are only moderately satisfactory, and one cannot really escape the rigorous analysis generally demanded for substances for lower molecular weights. For example, even in a small protein containing only 100 amino acid residues, substitution of leucine for an isoleucine residue at one position would probably not be detectable by any physical criteria for purity, and in the larger proteins containing several thousand residues the problem is even more difficult. In our favor is the remarkable specificity in biological synthesis which imposes strong limitations on the substitution of even closely related amino acids for each other. It remains to be established how near absolute this limitation obtains, but its acceptance is useful for the moment.

Proteins consist largely of combinations of some 20 amino acids linked together in chains by amide or peptide linkages (Figure 2.26), and their structures can be described in terms of the linear order of amino acids in a chain (primary structure), in terms of their conformation or three-dimensional arrangement in space (secondary structure and tertiary structure), and in terms of quaternary structure (specific combinations of more than one chain to form a specific unit). These fundamental characteristics are illustrated in Figure 2.26.

Methods have been developed to determine the amino-acid sequences in a number of the smaller proteins. The principles of these methods are simple, but their applications are tedious and time consuming. Hence progress in this field is slow. In general, proteins having single chains can be split into fragments of various sizes by nonspecific hydrolysis, as by acid or alkali, or by specific hydrolysis with enzymes. The proteolytic enzymes, trypsin and chymotrypsin, for example, yield fragments derived from splitting protein chains at the carboxyl groups

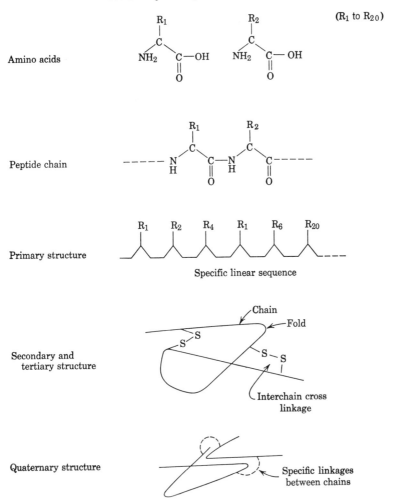

Figure 2.26. A diagram indicating the nature of basic concepts of protein structure.

of particular amino acids (for trypsin, lysine or arginine; for chymotrypsin, tyrosine, phenylalanine, tryptophan or methionine). The peptide fragments are separable by chromatography, electrophoresis, partition, and so on, and a variety of methods are available for the determination of amino acid sequences in each. Thus, by the determinations of sequences in small pieces derived by different methods of hydrolysis to give pieces with overlapping sections, it is possible to reconstruct a total sequence of a given protein chain. Obviously, technical difficulties increase rapidly with increasing chain lengths.

The first total sequence in a protein was that determined by Sanger (876) for the two chains of insulin. An example of a total sequence for an enzyme (ribonuclease) containing a single chain is shown in Figure 2.27. Such elegant results as these—from Stein, Moore, and collaborators (442)—are fundamental to an eventual understanding of structure and function of such molecules, and there is no substitute for information of this kind. However, it should be pointed out that for certain specific problems there are shortcuts that are very useful and informative. A case in point is that of the "fingerprint" technique developed most extensively by Ingram (478). Here it was possible to find a single amino-acid difference between two molecular species of hemoglobins (see p. 620) in a relatively simple way. The proteins are hydrolyzed to fragments enzymatically and subjected to separation on paper by chromatography in one direction and electrophoresis in the other. Treatment with ninhydrin shows locations of fragments. As indicated by the example in Figure 2.28 (475), all but one of the fragments in each of the two samples moved to the equivalent positions. Analyses of composition and sequence of the nonidentical fragments pinpointed a major difference in the two hemoglobins as a substitution of glutamic acid for valine at a given position in a portion of the total sequence. This is a method of many useful applications when it is used in a critical way. In this case a single difference was noted in the "fingerprint," but other substitutions having only a small influence on the behavior of any one of the peptides could also have been present but not detected. However, more complete analysis of apparently identical peptides has not revealed a second difference between these hemoglobins.

Another extremely important aspect of linear amino-acid sequence is that of forming helical, hydrogen-bonded structures, the secondary structure. Many fibrous proteins, such as keratin, have been shown to have unit structures of the form originally proposed by Pauling et al. (776), the α helix. Crystal structure studies of amino acids and small peptides yielded the atomic dimensions and angles shown in Figure 2.29a. The molecules lie in a flat plane around the $\diagdown\!\!\!\!\diagup \!\! C\!-\!N \diagup\!\!\!\!\diagdown$ bond, and successive amino-acid residues are rotated into a helix to give a cross-sectional and logitudinal picture as shown in Figure 2.29b and

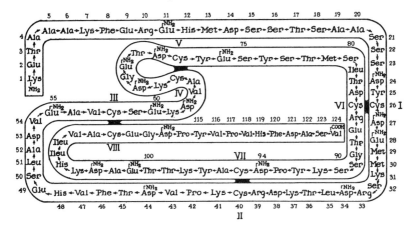

Figure 2.27.   A two-dimensional schematic diagram of the structure of bovine ribonuclease showing amino-acid sequences and interchain disulfide linkages. From Smyth, Stein, and Moore (920*b*).

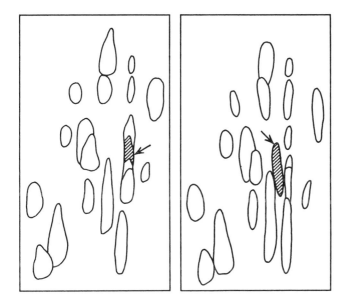

Figure 2.28.   "Fingerprints" of trypsin digests of hemoglobins A and S.   The significant differences are shown by arrows and cross hatching.   From Ingram (477).

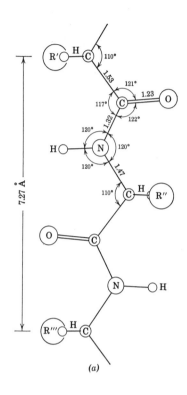

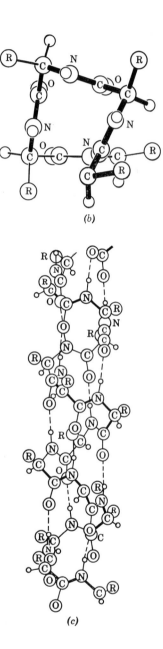

Figure 2.29. Hydrogen-bonded α-helix structure of a polypeptide. (*a*) Molecular dimensions in a polypeptide chain. (*b*) End view of 3.7 residue helix. (*c*) Side view of 3.7 residue helix. From Pauling, Corey, and Bronson (776).

c. This helix has 3.7 amino-acid residues per turn in a linear distance of 5.44 Å, and each amide group is hydrogen bonded to the third amide beyond it along the helix as $\overset{\diagdown}{\underset{\diagup}{\text{N}}}\text{—H---O}\overset{\diagup}{\underset{\diagdown}{=\text{C}}}$. This is not the only possible helical structure of this type, but it is clearly an important one among fibrous proteins and even in globular proteins (see Figure 2.27). Furthermore, it illustrates the important concept of the participation of hydrogen bonding (p. 253) in molecular structure to give special shape and dimensions.

It appears probable now that the α-helix structure is present to a considerable extent in most globular proteins and perhaps should be considered as a part of the secondary and tertiary structures.

An outstanding example of a study of total conformation is that of Kendrew and associates (522, 521, 520) on myoglobin, making use of high resolution x-ray diffraction patterns and the technique of isomorphous replacement. The latter requires that it be possible to replace bound water within the protein at a specific place by some component containing a heavy element, such as iodine or larger, without changing crystal structure. Several such derivatives (Ag, Hg, and I) can be achieved with myoglobin, which is a small protein containing about 150 amino-acid residues (mol. wt.—17,000) and a heme group. The structure derived from these studies is shown in Figure 2.30*a* and *b* (838). The schematic diagram shows the contribution of the α-helix structure as double lines, and the numbers represent the distribution of the amino-acid residues. The bend marked "O" in the diagram contains a proline residue which, since it lacks an amino hydrogen for helix bonding, would be expected to determine positions of bends in helices. Two other proline molecules occur in nonhelical sections. A further significant point from these experiments (838) is that resolution of the diffraction patterns was extended low enough to see the helical sections as hollow tubes with exactly the correct dimensions for the α-helix as predicted by Pauling et al. (776).

The above discussion provides a basis for a consideration of the relations of structure to function in proteins. In the fibrous types which lie in specific arrays to form large structures, periodicities in specific amino-acid residues allow for interfiber linkages to give reproducible geometric patterns (see Fig. 2.42, p. 109) providing for specific function. On the other hand, the globular proteins (which include many of the enzymes) have partial helical structures with folds determined by primary structure, as by proline, and tertiary structure, as

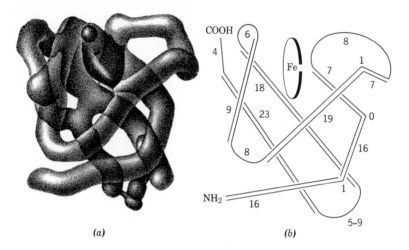

(a)   (b)

Figure 2.30. Tertiary structure of myoglobin. (*a*) Three-dimensional model. (*b*) Scheme representing folding of primary peptide chain as in (*a*). Double lines indicate regions of α-helix configuration and the numbers indicate the number of amino-acid residues per section. The plane of the heme is more nearly correct than in (*a*). From Rich and Green (838).

by —S—S— linkages between cysteine residues in different sections of the primary chain. Other kinds of cross linkages undoubtedly also occur, depending on relative positions of specific amino acids in the chains and orientations of functional groups in relation to helical structure. It is of interest to note that total conformation is thus determined by specific residues in certain positions. Hence it is possible for myoglobin from the whale and the seal to be almost identical in spite of marked differences in amino-acid composition (409). But in a very real sense primary order of amino-acid residues rules all of the structural potential of protein.

The structures derived for enzymes from x-ray studies of crystals, model building, and so on, are not necessarily the same as the structures existing in solution or in oriented combinations, although they certainly provide a sound basis for further considerations. For example, they will undoubtedly give a firm foundation for studies of functional mechanisms, the nature of catalytic active sites, and the nature of sites for combinations in oriented structures. Koshland (534) has presented interesting and useful concepts concerning active sites of enzymes based on the schematic diagram in Figure 2.31. The heavy black lines represent two portions of a primary chain folded to give specific shape and functional action. Here amino acids as R_1, R_2, R_6–R_9, and R_{163}–R_{165} are referred to as "contact" amino acids concerned with direct combination with a substrate. Others not in immediate contact, but which still may play a direct role (as 169), are labeled "auxiliary," while those concerned in maintaining a larger essential framework are "contributing" amino acids. This classification is arbitrary but useful in that it designates relative significance to different parts of a primary chain in terms of the three-dimensional orientation to give a small active site for function. A fourth category, "noncontributing" amino acids, comes from the knowledge that part of a structure can be digested away without loss of enzyme activity (e.g., for papain about two-thirds from one end, and for enolase about one-fifth from both ends). The case of ribonuclease is perhaps best known and most illustrative. Of the primary chain of 124 amino acids (see Figure 2.27) only three (122–124) can be removed without loss of activity. At the other end of the chain (N-terminal), splitting between amino acids 20 and 21 yields two pieces with no enzyme activity when separated (839). However, the fragments reassociate to give full activity when

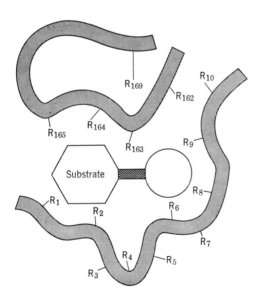

Figure 2.31. A schematic active site for an enzyme. Cross-hatched area indicates a bond to be broken in a reaction. The heavy line indicates portions of the primary polypeptide chains of the enzyme and the R groups side chains of various amino acids. From Koshland (534).

mixed together. This reassociation does not involve disulfide linkages, and indeed three of the four disulfide linkages of the original molecule can be broken without loss of activity. These observations demonstrate that the active site of ribonuclease contains portions from both ends of the primary chain, and further, the primary chain need not be intact or totally oriented in the native tertiary structure to yield an active site.

Studies of active sites have also yielded information on structural flexibility, a question that must be considered to be of great importance not only to this portion of a protein molecule but to the whole molecule in its associations in cells to form structures of higher orders. Reversible denaturation to give loss and recovery of enzyme activity has been demonstrated repeatedly, and Koshland has derived what he called an "induced fit model" of specificity. This assumes that the native conformation of an enzyme is not necessarily complementary to a substrate and that the substrate induces a change in conformation of the protein to yield a correct alignment for reaction to occur. Using a somewhat more rigid concept of an active site, Niemann and collaborators (424) have developed a three-dimensional picture of the active site of chymotrypsin which includes the important point that substrates can combine at the active site in more than one way. Some combinations do not lead to reaction, and thus a normal substrate can be an inhibitor and substrate at the same time with relative contributions dependent on the nature of the active site. Now, taking the potential for flexibility of all parts of a protein molecule into account, a possible environmental regulation of structure and function can be discerned. That is, an enzyme in combination in a cell structure may behave quite differently than in free solution, and further, a normal change of environment, as during cell division, could effectively turn off and on an enzyme activity on the foregoing basis alone.

Many other interesting and important aspects of protein structure and function are discussed in the literature cited, and some are considered further in subsequent sections of this book (i.e., pp. 261–279). At this point, in connection with cell structure and function, emphasis is to be placed on: (1) the great importance of the primary sequence of amino-acid residues in polypeptide chains; (2) the roles of hydrogen bonding, covalent inter- or intrachain (as in $-S-S-$) linkages, ionic inter- or intrachain linkages, and other less clearly defined repulsions and attractions; and (3) the potential dynamic aspects of structure and conformation as part of a cell's regulatory mechanisms. As will be evident in the discussion on nucleic-acid structure and biosynthesis that follows, genetic control of protein structure is only evident at the level of the primary sequence of amino acids in polypeptide chains. Secondary structure and functions (points 2 and 3 just mentioned) and higher orders of organization are possibly only consequences of this primary determination, but this remains to be established. In any case, though the primary sequence relation is fundamental, all the properties of proteins should be kept in mind during considerations of nucleic-acid structures and their known functions in protein synthesis.

Nucleic Acids. The nucleic acids, deoxyribonucleic acids (DNA) and ribonucleic acids (RNA), are, in general, much larger than proteins, and at present very little information exists on their exact primary structures. In fact, it is not possible to cite even a single case where a chemically pure (single-molecular species) nucleic acid has been prepared from a biological preparation which originally contained a mixture. As will become evident, there are excellent reasons why this is true, but it is very important to bear in mind that much of our knowledge of structure and function for nucleic acids has come from studies of complex mixtures. For example, DNA from a mammalian cell may have as many as 300,000 molecular species represented, and likely all have different primary sequences of the fundamental units, the nucleotides. Studies of protein structures in such mixtures could hardly be expected to yield more than average compositions with respect to the fundamental units from which the macromolecules are constructed and none of the unique details characteristic of individual molecules. The same situation obtains in nucleic-acid chemistry, and, though certain important unifying principles have been derived which simplify the problem to some extent, one must be prepared to make continued reevaluations of these principles until structure analyses and functional mechanisms of many different chemically pure molecular species are complete and unequivocal.

As in protein chemistry, a most important facet in studies of structures of nucleic acids has been the simplification of the purification problem by taking advantage of biological purification where it exists. With proteins this occurs in situations such as in keratin and silk production, and with nucleic acids it occurs especially in virus production (445). This point is well illustrated in Table 4. On the assumption of DNA molecular sizes of the orders of 10^4 to 10^5 basic units per molecule, the relative complexities of the cell and virus systems are clearly demonstrated (i.e., one molecule per virus and 60,000 to 600,000 per mammalian cell). Molecular weights for RNA are less than for

Table 4.   The DNA Content per Reproducing Unit from Various Sources *

Source and Unit	Deoxyribonucleotides per Unit
Mammalian haploid cell (cattle)	60,000 $\times 10^5$
Escherichia coli haploid nucleus	120 $\times 10^5$
Phage T2	4 $\times 10^5$
Phage λ	2 $\times 10^5$
Phage φX 174	0.05 $\times 10^5$

* Hogness (445).

DNA, but again viruses yield much more homogeneous material for study.   It should be noted that the long threads of nucleic-acid molecules break up easily when tissues are ground, and molecular weights are often difficult to determine.   They may go as high as 10^8 or more. Naturally, generalizations and details derived from studies of single systems must be evaluated with caution.

Deoxyribonucleic Acids.   The fundamental unit from which DNA's are built is the nucleotide containing an organic base, deoxyribose and phosphate.   Generalized structures and some of the potentials for isomerization are shown in Figure 2.32.   As indicated in the figure, the potential for possible isomeric structures in nucleotides is high, and many possible combinations of configurations in the sugar moiety exist. At present the structures indicated are those accepted, but the existence of other isomers in small amounts is not eliminated.   The finding, for example, of sugar substitution on the C_5 of pyrimidine in RNA's was observed comparatively recently, and total configuration determination on nucleotides from great varieties of DNA's has not been a favored subject for investigation in recent years.

Structural formulas for the heterocyclic bases that occur in DNA are shown in Figure 2.33.   In most DNA's the four bases, adenine, guanine, cytosine, and thymine predominate, but 5-methylcytosine is widespread.   A methylated adenine occurs in bacteria, while in the T-even bacteriophages cytosine is entirely replaced by hydroxymethyl cytosine and several different glucose- and disaccharide-substituted derivatives.

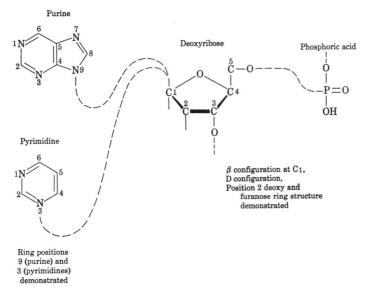

Figure 2.32. Components of deoxyribonucleotides showing kinds of isomers known to exist.

The occurrence of other naturally occurring bases is likely in view of the fact that cells will incorporate into DNA a number of unnatural bases such as the 5-halogen-substituted uracil. In addition, DNA preparations by simple salt extraction procedures are not always representative. For example, trypsin treatment is required in the extraction of DNA from human spermatozoa and alkali for bull spermatozoa.

The presently accepted general structure for DNA (p. 93) is the result of the establishment of some important basic principles by many investigators. One of primary importance was that of Chargaff and collaborators (158) derived from studies of base ratios of DNA preparations. It was found that in spite of large variations in the relative proportions of adenine (A) to guanine (G), the total purine (A + G) was always very nearly equal to the total pyrimidine (C + T). Furthermore, collecting analogs into the groups of the four major bases yielded the result that A = T and G = C. Although these near equalities might not be particularly significant in very complex mixtures from cells, they hold also for much simpler mixtures from viruses and for enzymatically synthesized DNA. Bendich et al. (70) has noted significant deviations in fractionated complex DNA mixtures and the ratios do not hold for DNA from the bacterial virus ϕX-174 (p. 94). Nevertheless, these original observations of base ratios were fundamental to the development of present concepts of DNA structures.

A second type of essential information derived from titration studies, such as those by Gulland and Jordon (380), Cox and Peacocke (187), and others (see Jordon, 496), was that DNA is characterized by a high content of hydrogen bonding. In solution DNA exhibits striking changes in viscosity and streaming birefringence below pH 4.5 and above pH 10. This has been fully confirmed by other physiochemical

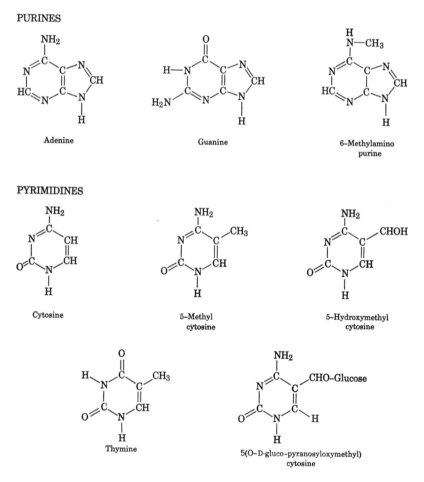

Figure 2.33.   Structural formulas for purines and pyrimidines that occur in DNA.

studies such as those concerned with heat denaturation and hypo-chromicity in absorption of ultraviolet light (see Figure 2.36, p. 95). Furthermore, titration data indicated that nucleotide chains of DNA were not branched.

A third area of investigation important to DNA structure analysis was that of x-ray analysis of DNA fibers that could be drawn from solutions of purified DNA. Astbury and Bell in 1938 (28) observed that such fibers showed prominent spacings at 3.34 Å along the axis and further repeating patterns at about 27 Å. On the basis of such observations these and other investigators (306, 774) proposed rea-sonable general structures to fit existing data. More extensive data was obtained and evaluated in essentially the now accepted form by Wilkins et al. (1079) and by Franklin and Gosling (293). In a fur-ther interpretation of the existing facts concerning properties of DNA, including H bonding and those from x-ray studies, Watson and Crick (1056) provided a generalized structure for macromolecules of this class which has received wide acceptance. The data on base-pair ratios were evaluated in terms of specific A—T and G—C pairing by hydrogen bonding, shown as modified by Pauling and Corey (775) in Figure 2.34. Such pairing then provided the basis for structure in terms of two complementary nucleotide chains wound about each other with sugars and phosphates on the outside and the flat H-bonded base pairs connecting the chains on the inside. This general picture is illustrated in Figure 2.35. Part *a* shows diagrams of two of several double-helix structures derived from the base-pair principle (the flat rungs on the spiral ladder) and x-ray diffraction patterns of DNA fibers drawn under different humidity conditions. Part *b* illustrates the second diagram as a molecular model with the bases packed on the inside and the sugar phosphate chains forming the longitudinal helical chains on the outside.

The participation of water molecules, cations, and protein molecules in influencing DNA structures is of great importance in evaluations of the postulated structures as biologically significant, since all are normal constituents in the biological environment of DNA. Four intercon-vertible structures for DNA have been described with dependence on water content (551). All are double helices, but nucleotide pairs vary from 9.3 to 11 per turn and the base-pair planes tilt at various angles to the helix axis (Figure 2.35). Highest humidity favors the nontilted structure. Nucleoprotamines, as in sperm heads, appear to have sim-ilar structure, with the protein wound about the DNA helix in the narrow groove with the basic amino-acid residues neutralizing the negatively charged phosphate groups. Nucleohistone structures (as in

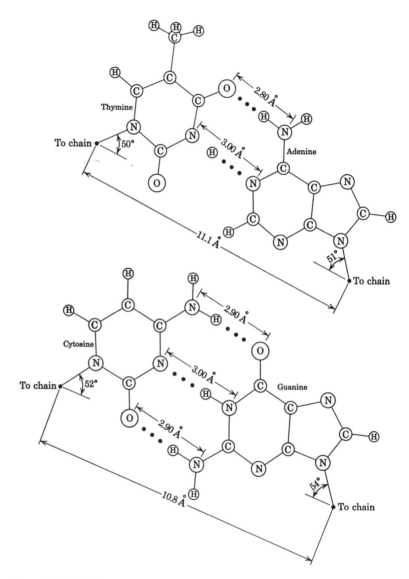

Figure 2.34.   Hydrogen-bonded complementary pairing of adenine and thymine, and guanine and cytosine.   From Pauling and Corey (775).

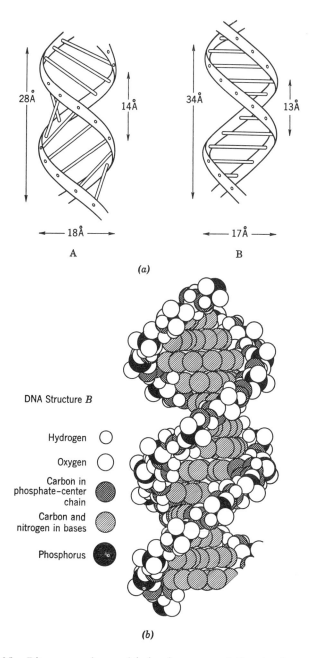

Figure 2.35. Diagrams and a model showing representative structures of DNA. (a) Diagrams showing two of the double-helix forms of sodium-deoxypentose nucleate obtained by drawing fibers of different water content. B is achieved at highest humidity. Cross rungs of the ladders indicate nucleotide pairs and the ribbons the sugar phosphate chains. From Jordon (496). (b) A molecular model for DNA corresponding to the B form in (a). From Feughelman et al. (276).

interphase nuclei) are more complex and not at all clear at present. However the information on the protamine combinations, taken from studies of intact sperm heads (276), provides the best direct evidence that the double-helix structure for DNA actually exists in a higher biological system.

It is perhaps just as important to note that all natural DNA does not exist in this form. As well established by Sinsheimer and collaborators (913), DNA in the bacteriophage ϕX-174 exists as a single strand. Furthermore, this type phage contains a single DNA molecule per particle and thus yields nearly a chemically pure product. Here the A/T and G/C unity ratios do not hold as might be expected. Although DNA of this type is considered atypical, appreciable amounts of this single-stranded form could be present in the very complex mixtures of nucleic acids derived from cells. Unequal purine to pyrimidine ratios have been reported for some fractions of DNA (70), and, in any case, the fact of existence of the homogeneous single-stranded DNA as in ϕX-174 demonstrates clearly that the double-helix structure is not a prerequisite to function.

Nevertheless, that most preparations have the forms as described above is supported by numerous and extensive physicochemical determinations. Calculations of density per unit length using molecular dimensions from electron-microscope pictures are consistent, and of special value have been studies of "denaturation" and "renaturation." As shown in Figure 2.36 (230), DNA in a given ionic environment gives an abrupt increase in the absorption of ultraviolet light in a critical temperature range and, at the same time, an abrupt decrease in optical rotation. Such changes can be induced by several means (pH, ionic-strength alterations, and organic solvents) and measured by a variety of other physical methods (150). Such changes are attributed to collapse of the rigid double-helix structure, primarily due to breakage of the interchain base-pair hydrogen bonding, though interplanar interactions may also be involved. The denaturation or "melting temperature" ranges are reduced in the presence of H-bond disrupting agents such as urea, and DNA with a high guanine and cytosine content (three H-bonds—Figure 2.34) has a higher melting temperature (989) than others.

Though many other interesting details are available (150), the important fact for the present discussion is that the denaturation is, at least to some extent, reversible. Though the picture is not entirely clear, it appears most likely that mild denaturation yields random

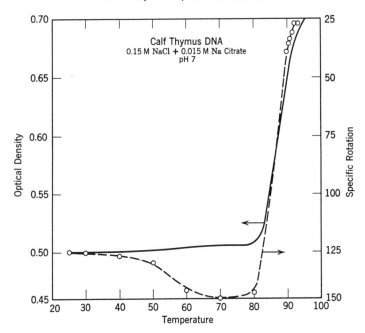

Figure 2.36. The variation in specific rotation (open circles) and optical density (solid line) with temperature of solutions of calf thymus DNA. From Doty et al. (230).

chains still coiled about one another with, not necessarily total, disruption of base pairing. In renaturation under conditions of lowered thermal agitation reorientation occurs, though a complete normal "rezippering" probably never obtains. Some abnormal pairing and single-chain looping probably occurs, and in actual fact complete pairing and maximum rigidity of structure may be only a transient limit in native DNA.

In any case, all of these considerations do give a consistent picture for the general form of DNA, that of two polynucleotide chains running in opposite directions, wound about each other in a double helix and held together by complementary AT and GC base pairing. No primary sequences have been determined as in proteins, but it is expected that the two chains of a double helix would not usually be alike in sequence. For example, the complement for a short chain such as ATCG → $PO_3^{\equiv}$ would be $PO_3^{\equiv}$ ← TAGC. This, too, is now well established through the method of nearest neighbor determination (498). This depends on incorporation of a nucleotide containing P^{32} in the 5′ position into DNA followed by isolation and hydrolysis with alkali or an enzyme which splits at the 5′ position to yield 3′ nucleotides. In this fashion the P^{32} appears in the nearest neighbor nucleotide to that originally incorporated. For the antiparallel double-strand structure, neighbor pairs and complementary neighbor pairs should occur with equal frequency, and this has been found to be the case in a number of natural and biosynthetic (see p. 97) DNA preparations.

As to the question of functions of DNA and its position in the hierarchy of organization of cells, the problem is considerably more complex. Some well-established facts concerned with a generalized static structure for DNA are outlined in the preceding paragraph, and certainly the principles of specificity through base pairing must be important in the functions of replication and control of metabolism assigned to DNA as genetic material, but, as emphasized by Cavalieri and Rosenberg (150), the structure alone is as yet insufficient to describe function. One view of function, as in replication, involves simultaneous helix uncoiling and synthesis of two new chains from one end with specificity determined entirely by base pairing. With nucleotides pointing in opposite directions this requires two kinds of coupling reactions for the two chains. This is possible, but there are two more formidable problems involved. First is the question of base-pairing specificity. The implication has been that A can only pair with T and G with C, but this is only a matter of degree of stability and fit into a rigid structure. As pointed out by Donohue and Trueblood (228), there are many possible H-bonded pairs, and likely an AG combination would even fit satisfactorily into the rigid helix structure. Thus it is probable that exact replication *in vivo* depends on orientation factors in addition to complementary base pairing. This may reside in protein, enzyme, or other structural factors. A second major difficulty concerning structure and function has to do with higher-order organization and the physical separation of very long strands of DNA wound about each other and further restricted by RNA and protein strands, as in the complex units which make up the chromosomes. Although a number of unique schemes have been invented, based on helix untwisting and controlled breakage and fusion, this problem remains unsolved (150). At the chromosome level, replication based on a nonrigid structure to give new DNA which is not wound about the old seems rather likely.

Although precise pairing and eventual strand separations present difficulties, it is firmly established that existing DNA structure can act as a template in biosynthesis. The function of replication in a simple system has been demonstrated by Kornberg and collaborators (530, 530*a*) by the demonstration that a highly purified enzyme from *E. coli* will catalyze the condensation of nucleotide triphosphates to give a net synthesis of DNA. The system requires, for maximum efficiency, the four most common deoxynucleotide triphosphates (nucleotides of A, T, G, and C), enzyme, and, most significantly, some preformed DNA as a primer. The DNA serves best if denatured. The native single-stranded form of ΦX-174 functions directly, and denatured forms from a great variety of sources are effective with about equal efficiencies. These results indicate that the synthesis occurs by the mechanism diagrammed in Figure 2.37 with the ordering of nucleotides in the production of new complementary chains directed by hydrogen bonding of bases to those in the primer chain. This essential point is illustrated by the data shown in Table 5, and is especially emphasized by the last

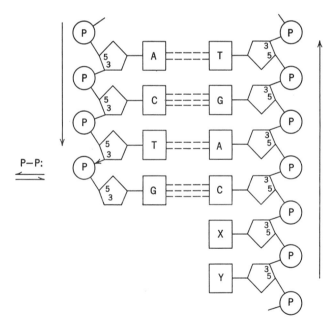

Figure 2.37. Diagram of a mechanism for enzymatic DNA replication. From Kornberg (530).

Table 5.   Chemical Composition of Enzymatically Synthesized DNA, Synthesized with Different Primers: A, Adenine; T, Thymine; G, Guanine; C, Cytosine *

DNA	A	T	G	C	$\dfrac{A+G}{T+C}$	$\dfrac{A+T}{G+C}$
Mycobacterium phlei						
Primer	0.65	0.66	1.35	1.34	1.01	0.49
Product	0.66	0.65	1.34	1.37	0.99	0.48
Escherichia coli						
Primer	1.00	0.97	0.98	1.05	0.98	0.97
Product	1.04	1.00	0.97	0.98	1.01	1.02
Calf thymus						
Primer	1.14	1.05	0.90	0.85	1.05	1.25
Product	1.12	1.08	0.85	0.85	1.02	1.29
Bacteriophage T2						
Primer	1.31	1.32	0.67	0.70	0.98	1.92
Product	1.33	1.29	0.69	0.70	1.02	1.90
A-T copolymer	1.99	1.93	<0.05	<0.05	1.03	>40

* From Kornberg (530).

item, the A-T copolymer.   This polymer is produced from A-T mixtures with a long lag, but after production it serves as an efficient primer in the production of more A-T even in the presence of all four nucleotide triphosphates (498).   Thus primer-enzyme interaction to provide base sequence control has been demonstrated clearly.   Nevertheless a paradox remains, since biologically functional DNA has not yet been produced by the system even though more than a twenty-fold net synthesis has been achieved and products always have compositions very close to that of primer DNA.   This is a prime problem for clarification before replication can be truly understood.

Besides the function of participation in its own synthesis, DNA also is concerned at least in the production of specific RNA, and some significant aspects of this problem are considered further on page 104 as part of the overall picture of the biosynthesis of order.

Ribonucleic Acids (RNA).   The fundamental unit in RNA, like that in DNA, is the nucleotide consisting of a heterocyclic purine or pyrimidine base, a sugar, and a phosphoric acid.   The four most common bases are adenine, guanine, cytosine, and uracil, but a variety of other bases are also present in small amounts. Structural formulae of the four common bases and a list of other known derivatives found in RNA are given in Figure 2.38. Although D-ribose is the most common sugar, nucleotides containing 2′-O-methylribose have been found (916), and an exception to the N_3 pyrimidine to sugar linkage is found in the compound pseudouridine which has the ribose linked at the C_5 position (915) on the uracil ring. In general, the ratios of A/U, G/C, purines/ pyrimidines, and 6-amino groups/6-keto groups tend toward unity, but in contrast to the situation with DNA, the ratios do not appear to be significant in relation to RNA structure. That base-pair hydrogen bonding does occur in RNA is beyond question, since it also shows hypochromicity and melting curves for optical rotation as does DNA, but melting occurs at a lower temperature, and a continuous double-helix structure is not shown by x-ray studies. Present evidence for isolated RNA favors partly helical-folded structures based on intermolecular interactions of single long chains (Figure 2.39a). On the other hand, the RNA of tobacco mosaic virus (TMV) is pictured as a regular, helical, single-chain coil embedded in protein as shown in Figure 2.39b. Since RNA isolated from TMV is not unlike other RNA preparations in structure, it is evident that the association with protein is highly significant. In a biological environment RNA is probably always associated with protein, and tertiary structure may be meaningful only in such combinations. That is, flexibility with potential alternative structures

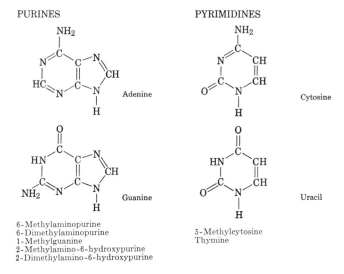

Figure 2.38. Purines and pyrimidines found in RNA.

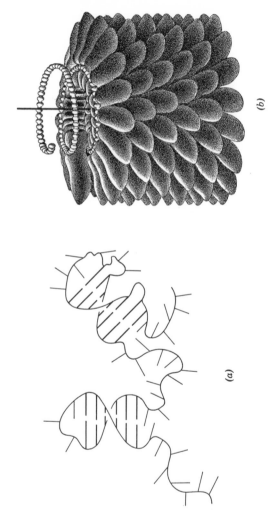

Figure 2.39. Models of RNA structure. (*a*) A diagram of a possible secondary structure of RNA showing internal base pairing to give some helical, double-stranded components. From Fresco and Straus (302). (*b*) A model for the structure of TMV showing a coil of RNA embedded in coat protein units. From Rich and Green (838).

and perhaps reciprocal influences with proteins and other macromolecules is a characteristic of RNA in contrast to the apparently more rigid conservatism characteristic of DNA.

With respect to detail of structure, little is known about the primary sequences of nucleotides in RNA, because technical problems associated with determining them are similar to those for DNA. The problem of first importance, that of separation of mixtures into chemically pure components, has not been solved here either. Biological separations are again useful and the categories of nuclear, ribosomal, soluble, and viral RNA form the principle basis for study. So far as known at present at least four types of RNA exist in nucleated cells: nuclear RNA; ribosomal RNA (rRNA); soluble, transfer, or adaptor RNA (sRNA); and messenger RNA (mRNA). Molecular weights are much less than those for DNA, with that for TMV being near 2×10^6. Ribosomal RNA may also be of this order, but there is evidence that it is made of subunits of about 1.2×10^5 (111). At present, the soluble RNA's (sRNA) which account for 10% to 20% of cellular RNA are the best known and provide the most favorable characteristics for analysis. Molecular weights are around 3×10^4 (80 to 100 nucleotides), specific biological tests for distinguishing components are available (specific amino-acid activation), and at least some methods for separation have been defined (673, 1128). These RNA's, at least 19 of which are concerned with activation of amino acids for protein synthesis, appear to have in common the sequences ApCpCp $\cdots$ Gp with the specificity of each corresponding to a like number of different activating enzymes residing in the primary sequence of nucleotides somewhere in the center portion of the molecule. Part of the specificity is apparently derived from the less common bases (Figure 2.41) most of which are found in the sRNA. Some statistical data concerned with this internal sequence have been obtained by Berg and collaborators (750), but no total primary sequence has been analyzed completely.

In function, RNA evidently plays a variety of roles. One of special interest is that of a carrier of genetic continuity as in the plant viruses (tobacco mosaic virus, for example) and in certain animal viruses (poliomyelitis) and bacteriophages (1, 602). In this role the first function of replication is of great interest in terms of mechanisms, but not much information is available. It has been pointed out that the rigidity of the DNA double helix is not necessarily essential for replication, and in view of the properties of RNA some other system involving nucleoprotein seems quite possible. If so, to draw a line between the replication function of nucleic acids will be very difficult since ribo-

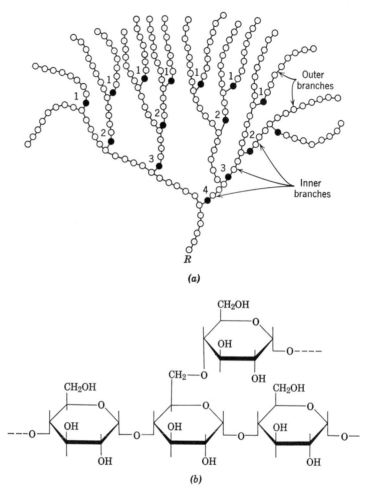

Figure 2.40. Model of a portion of a glycogen molecule. (*a*) Arrangements of glucose residues with branch points 1 to 4 having α-1$\rightarrow$6 linkage. R = reducing end group. From Harris (410). (*b*) Glucose residue chain and branch point.

nucleoproteins are not confined to chromosomes and nuclei and yet are capable of autonomy. It is even possible that the DNA system of replication and metabolic control represents an evolutionarily specialized mechanism derived from a broader, more mobile, and less stable genetic system based on RNA.

In functions other than replication, sRNA is concerned, along with specific proteins, in activation of amino acids for protein synthesis, while ribosomal RNA and nuclear RNA are concerned in protein synthesis by mechanisms not yet understood (see p. 105).

Other Macromolecules. Genetic characters are at least expressed in terms of macromolecular structures of classes of compounds other than nucleic acids and proteins. Included among these are a great variety of polysaccharides, polyuronides, chitins, hyaluronides, pectins, teichoic acids, muramides, and various mixed composition macromolecules such as the peptidoglycolipids. Little is known of such matters as the genetic control of specific branching of polymer chains, when it occurs, or of mechanisms of orientation of polymerization when two or more simple units enter into a polymer. An interesting case in point is that of glycogen, which has a molecular weight in the order of 1 to 4×10^6. The general form of the highly branched system is shown in Figure 2.40*a* (669), and the basic structural glucose units having 1–4 chain linkages and 1–6 branching linkages is shown in *b*. In studies of the enzymatic synthesis of glycogen, Leloir and coworkers demonstrated the requirement for a primer in the reaction (588). Glycogen itself was the most satisfactory, but some partially degraded glycogens and starches also had activity. Products had only 1–4 linkages, that is, the reaction apparently involved only terminal additions of glucose (from uridine diphosphoglucose) to chains of the primer. In this system there is no evidence for a copy mechanism concerned with the production of specific structure, but it does illustrate the requirement for a preexisting polymer in order to make more. This interesting property of a biological synthetic system was discussed previously (p. 97) in connection with the highly specific replication process for DNA where it is considered fundamental to transmission of genetic material. It is possible that the principle is also applicable with varying degrees of influence and specificity to the formation of other classes of macromolecules and through them to the formation of intracellular structures and to the formation of higher orders of cell organization.

The Biosynthesis of Order

So far in this chapter we have considered some highlights of structure and function of cells, their highly organized subunits, and a little about some of the macromolecules into which the subunits are divisible. This is all in anticipation of many subsequent discussions bearing on the questions of the nature of genetic units and how these can bring about the biosynthesis and regulation of such remarkable order. Obviously, we will not be able to answer all of these questions and obviously it will not be possible to present in detail all the evidence pertaining to them. Nevertheless it is appropriate at this point to anticipate somewhat further with a current general concept of how the biosynthesis of order comes about, and this can be continuously reevaluated as many pertinent details are developed in later chapters.

In short, it is considered that genes are DNA and each can be represented by a definite linear sequence of many nucleotides. Prior to cell division the sequences are copied exactly to maintain genetic continuity through new DNA. This is the first function of genes (replication), and in the second function the DNA sequences are copied exactly by another system which yields homologous sequences in molecules of RNA (messenger RNA). These unstable molecules are moved to various parts of the cell where, in collaboration with ribosomes, enzymes, and soluble RNA, they yield protein-synthesis systems. The linear sequence of amino acids in proteins synthesized is related directly to the linear sequences of nucleotides in the mRNA and hence back to the DNA. After formation of linear polypeptide chains, these can be folded and oriented in accordance with their linear order into three-dimensional structures, and the products thus formed can be further oriented into the complex structures that exist. This is a simplified picture of how genes function and control the biosynthesis of order. The important question is, how much of this story is true and what kinds of details must be introduced to account for life processes? Most of these matters will be considered later, but further background is desirable here.

The first part concerning the structure and replication of DNA has already been discussed (p. 93). Although biologically active DNA has not been produced using a purified enzyme and primer DNA, it is firmly established that the system can yield very near complementary copies of preexisting nucleotide sequences of DNA. Thus the system provides a sound model for the maintenance of genetic continuity. The next aspect of DNA function, that of providing primary sequence order to RNA, also has a firm foundation in experimental facts. Hurwitz (473a), Weiss (1060), Chamberlain and Berg (152), and others (374, 911, 129) have studied enzymatic (RNA polymerase) RNA synthesis from ribonucleotide triphosphates in the presence of native (double or single chain) DNA as a primer. As in replication, the product has a composition like that of the DNA used as primer, and it is presumed that the sequences are complementary, with uracil in the place of thymine for RNA. Though this is by no means the only enzyme system in which RNA can be synthesized (374), this one fits, as a reasonable model, for formation of messenger RNA. It evidently corresponds to the small fraction (3%) of cellular RNA which undergoes very rapid turnover (29).

If the specificity of DNA resides in its base sequence, it is to be expected that the mRNA formed in its presence will reflect that specificity in base sequence. The existence of RNA strands complementary to DNA strands has been demonstrated. Thus Spiegelman and his associates have made hybrid molecules with T2 phage DNA, and T2 RNA formed in its presence (939). These hybrids are double-stranded molecules containing one strand of DNA and one of RNA. That T2 phage DNA will not form hybrids with RNA not formed in its presence is taken as direct evidence that a complementarity exists in base sequence in the hybrids just as in the double-stranded DNA. The only essential difference is that adenine pairs with uracil in the hybrid instead of with thymine which does not usually occur in RNA. Hybrid DNA-RNA complexes have also been isolated from natural sources such as from *Neurospora crassa* (887).

This brings the picture up to the point of protein synthesis. Here, too, a number of important fundamental facts have been established even though little net synthesis has been achieved in isolated systems. The first two steps are clear-cut as summarized in Figure 2.41. In these reactions it is well established that specific activating enzymes and sRNA's are present for each amino acid, and in some cases there may be more than one sRNA. Moderate specificity is shown in step 1 and high specificity is shown at step 2 for the amino acid, but in subsequent reactions it appears that specificity resides entirely in the sRNA. This has been established in one case (157) by conversion of cysteine to alanine after coupling the former to sRNA. In subsequent steps alanine was incorporated instead of cysteine. The details of these subsequent steps are not so well established as steps 1 and 2, but it is presumed that the components shown in Figure 2.41 interact by orientation of complementary nucleotides of amino-acid-carrying sRNA's in a linear sequence on mRNA to bring the amino acids in adjacent positions for enzymatic transfer into a polypeptide chain. The messenger RNA thus may serve as a template corresponding to a portion of the sRNA which serves as an adaptor, and a nucleoside sequence corresponds to one amino acid. Further, a linear series of nucleotide sequences can thus determine a specific amino-acid sequence even though the actual physical orientation depends on base pairing.

Little is known of mechanism details and removal of formed peptide chains and regeneration of sRNA's. In view of the biological instability and rapid turnover of mRNA's, it is possible they are destroyed in the process. In any case the facts are that such isolated systems do

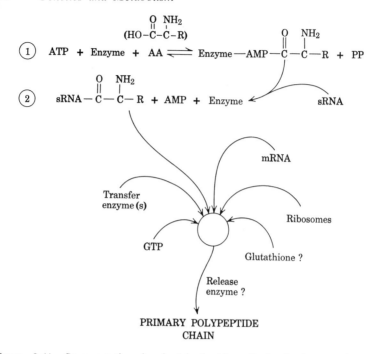

Figure 2.41. Some reactions involved in the biosynthesis of primary polypeptide chains.

yield amino-acid incorporation into polypeptides and the nature of the product is dependent on the nature of RNA supplied to orient the sRNA amino acids. Convincing and important evidence for this is summarized briefly in the following. It was observed initially by Nirenberg and Matthaei (740) that a system from bacteria containing an amino-acid mixture, sRNA, ribosomes, an energy source, and a synthetic polynucleotide containing only uracil (poly-U) was capable of polymerizing one and only one amino acid—phenylalanine. This indicated that the position of phenylalanine in a protein chain is determined by a sequence of uracil-containing nucleotides in RNA. With this beginning Nirenberg and Matthaei, Ochoa and collaborators (589), and others made extensive use of the nucleoside phosphorylase system first described by Ochoa et al. (749) to produce and test other homo- and mixed enzymatically synthesized polyribonucleotides. Poly-C yielded incorporation of proline only, poly-A was not active, and satisfactory preparations of poly-G were not obtained. Randomly mixed polymers of various kinds catalyzed incorporation of many other amino acids and a summary of some data is given in Table 6. It is to be

Table 6. Amino Acid Incorporation into Protein Stimulated by Randomly Mixed Polynucleotides*

Polynucleotide	UA	UC	UG	UAC	UGC	UGA
Base ratio	U = 0.87 A = 0.13	U = 0.39 C = 0.61	U = 0.76 G = 0.24	U = 0.834 A = 0.050 C = 0.116	U = 0.341 G = 0.152 C = 0.502	U = 0.675 G = 0.291 A = 0.034
Probability of triplet relative to phenylalanine (UUU) = 100%	UUU–100 UUA– 13 AAA– 0.3	UUU–100 UUC–157 UCC–244 CCC–382	UUU–100 UUG– 32 UGG– 10.6 GGG– 3.4	UUU–100 UUA– 6.0 UAA– 0.4 AAA– 0.02 UUC– 13.9 UCC– 1.9 CCC– 0.3 UAC– 0.8 AAC– 0.05 ACC– 0.12	UUU–100 UUG– 46.2 UGG– 21.0 GGG– 1.0 UUC–147 UCC–218 CCC–322 UGC– 68.1 GGC– 31.7 GCC–101	UUU–100 UUG– 43 UGG– 19 GGG– 8.1 UUA– 5.1 UAA– 0.26 AAA– 0.01 UGA– 2.2 GGA– 0.1 GAA– 0.01

Amino Acid	UA	UC	UG	UAC	UGC	UGA	Composition of coding units †
Phenylalanine	100	100	100	100	100	100	UUU · · ·
Arginine	0	0	1.1	0	*49.3*	2.9	UCG · · ·
Alanine	1.9	0	3.2	1.0	*40.4*	0.9	UCG · · ·
Serine	0.4	*160*	0	3.6	*170*	2.3	UCC · · · + UCG · · ·
Proline	0	*285*	0	0	*188*	0	UCC · · ·
Tyrosine	*13*	0	0	8		0	UUA · · ·
Isoleucine	*12*	1.0	1.0	*4.8*	1.0	*8.6*	UUA · · ·
Valine	0.6	0	*37*	0.4	*5.4*	*8.4*	UUG · · ·
Leucine	4.9	*79*	*36*	5.1	*29.8*	*75*	UUC · · · + UUG · · ·
Cysteine	4.9	0	*35*	0	*157*	*44*	UUG · · · or UGG
Tryptophan	1.1	0	*14*	0	*5.4*	*46*	UGG · · ·
Glycine	4.7	0	*12*	0.5	*1.6*	*23*	UGG · · ·
Methionine	0.6	0	0	0.6	*0.7*	*15*	UGA · · ·
Glutamic acid		0	0		*1.5*	*8*	UGA · · ·
Lysine	1.5			1.2	0.44	*6.2*	UAA · · · (?)

* From Matthaei, Jones, Martin, and Nirenberg (663).

† Sequence of nucleotides in a coding unit is not specified.

Note: The figures in the main part of the table represent the incorporation of any amino acid compared to phenylalanine incorporation expressed as percentages (mμ moles amino acid incorporated mμ moles phenylalanine incorporated × 100). Italic figures refer to the polynucleotide containing the nucleotides necessary to stimulate the incorporation of a given amino acid.

noted especially that each of the amino acids—arginine, alanine, and glutamic acid—requires a minimum combination of three different nucleotides for its designation.   Thus, on this basis and the assumption that each amino acid is designated by the same number of sequential nucleotides, the adequate number to account for all the amino acids (except hydroxyproline) is three.

Although this system is highly artificial, yielding neither extensive synthesis nor biologically functional polypeptides, it is conceptually of very fundamental significance.   It provides a means for designation of equivalence in specificity of a nucleotide sequence of about three to a specific amino acid (as its sRNA adaptor) and may well make the nucleotide sequences in genes directly understandable in terms of a specific and general code.   The present concept which is in accord with the hypothesis of Crick (189) assumes (1) that three successive nucleotides determine one amino acid (e.g., UUU = phenylalanine), (2) that the code is not overlapping (six nucleotides in two separate groups are required to determine two successive amino acids), and (3) that the code is partially degenerate (some amino acids are designated by more than one combination of three nucleotides).   However at the present time these points are, perhaps, not especially important since experimental methods are at hand to provide much factual detail (see discussions concerning amino-acid sequences in tobacco mosaic virus coat proteins, p. 276, and hemoglobin, p. 616), and very extensive information on code evaluations may be expected in the near future along with many revisions and extensions of present concepts.   Nevertheless, the principle is likely to survive, and we should arrive at a good understanding of how things go, from gene to RNA to polypeptide chain sequence, by means of it.

But now, if one can accept these beginnings (with enormous detail in mechanisms to be established), it is proper to give consideration to the adequacy of what has been accomplished.   Is the primary linear order of amino acids in polypeptides enough in itself to provide for organization at higher levels?   This question cannot yet be answered, but several examples are pertinent.   As noted previously, page 85 (409), a relatively simple enzyme, ribonuclease, can be split into a smaller protein and a peptide (20 residues) neither of which is active but activity is recovered on mixing.   Further, the four cystine bridges (Figure 2.27, p. 81) which bind the single chain in a tertiary structure can be broken with complete loss of activity, and reoxidation yields a product with 90% of the original activity (excluding a small amount of insoluble product apparently resulting from intermolecular linking).

By another criterion, that of breakage of hydrogen bonding by treatment with concentrated urea, the enzyme can be extensively unfolded, but it reforms in the presence of the substrate. From a consideration of each of these three treatments alone, it might be concluded that the enzyme can form spontaneously from one linear peptide or even from one long and one short one. However, combinations of treatments each of which disrupts a different kind of secondary or tertiary structure yields inactive material which does not reactivate. This indicates that single treatments do not cause denaturation to the linear-chain level, and thus there is little evidence for spontaneous folding of a linear chain to give the necessary three-dimensional structure for enzyme activity. On the other hand, there is no evidence for participation of native enzyme or other normal cell constituents in the process.

Another kind of example of an ordering process is that concerned with mineralization of tissues as in bone formation (1063). As shown diagrammatically in Figure 2.42, a solution of tropocollagen can be

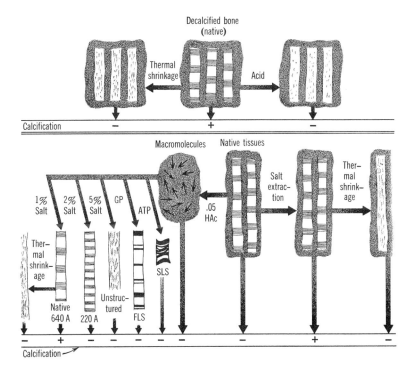

Figure 2.42. Diagram summarizing factors involved in orientation of collagen to permit calcification as in bone formation. From Weiss (1063) after Glimcher.

reoriented into several fibril patterns depending on solutes present in the solution. Furthermore, a systematic deposition of calcium phosphate, as in bone mineralization, occurs only when the collagen has assumed the native fibril pattern. Here a spontaneously formed structure provides the basis for formation of a structure of higher order, and one could think that the primary amino-acid sequence in collagen is sufficient to determine the entire subsequent structure. It should be noted that this would leave open the questions of the initiation and the termination of the processes, especially in view of the fact that collagen from noncalcifying tissues as tendon, skin, and swim-bladder will also serve in the calcification process. This places emphasis on the importance of "a sequence of events" for structure determination at all levels of order, a point considered extensively by Weiss (1063) in a broader discussion of this general problem. Thus there still remains the question of whether the genetic determination of linear sequences of amino acids in polypeptide chains is sufficient to give a living, dynamic system. If so, the answer must lie in regulatory mechanisms and self-limiting sequences of events, and our knowledge does not yet go that far.

3 Genetic Systems

The purpose of this chapter is to consider the life cycle of a variety of representative organisms which are of significance in the study of genetics. It is important to understand life cycles, for all life cycles are *genetic systems,* and these systems give genetic continuity to organisms and provide for variations to arise through recombination of existing genetic material.

Both continuity and increase in numbers can be adequately provided for by asexual reproduction which is superficially a fission of one into two. Basically, asexual reproduction occurs at the cellular level when one cell divides into two in the process of mitosis, or some similar process in which genetic material is equally divided. Recombination, on the other hand, can only occur when the genetic materials from two dissimilar strains of an organism are present in juxtaposition so that an exchange can take place, and a type of genetic material is produced in an offspring which is different in structure from the parental types. This ordinarily involves sex and sexual reproduction.

The word sex probably has more connotations than any other word in the language. The biologist ordinarily conceives of it as a description of the fact that in most organisms there are two different types of individuals, male and female, each of which provides genetic material to a fusion cell or zygote from whence the new generation arises. The cell fusion or fertilization process is sexual reproduction. Generally it involves the fusion of cells that are dissimilar in structure and physiological characteristics, but this is not essential. What is essential is that the genetic material present in the fusion cell or one of its mitotic descendants can recombine. This is the *sine qua non* of the sexual process, because sex has no other known meaning without it. Added to this in all diploid organisms is the further factor of the many different combinations that can be formed with maternal and paternal chromosomes by chance fusion of the many different types of gametes produced by meiotic recombination in the maternal and paternal parents.

Sexual reproduction is basically a process to insure recombinations. From this it follows that when recombinations occur, a sexual process must have occurred. The mechanism or form of the process may vary widely from one type of organism to another, as will be illustrated below, but its occurrence is made known by the appearance of recombinants.

Although the genetic systems of the greater majority of organisms have much in common with respect to the basic functions of reproduction and recombination, they vary greatly in detail—so much so that close examination of the details of life cycles is necessary to understand what is happening, its significance for that organism in particular, and for genetics in general.

Actually, the discussion of genetic systems in this chapter is primarily dedicated to those special microbiological systems which have yielded in recent years the most direct information concerning the molecular basis of heredity.

Viruses

Viruses are particles rather than cells which are incapable of reproducing except in living cells. Their structure is simple, consisting largely of protein and nucleic acid with the nucleic acid forming the core and the protein the cover. Some contain, in addition, lipid material. Figure 3.1 illustrates the structure of one of the better known viruses, the T2 bacteriophage.

The life cycles of only a relatively few viruses are known in anything approaching significant detail. Those infecting bacteria, the so-called *bacteriophages,* first discovered by Twort (1026) and d'Herelle (207),

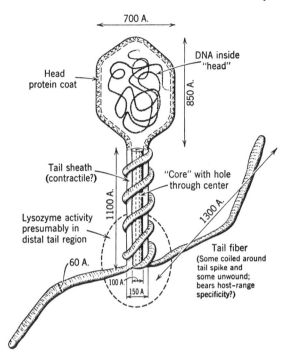

700 A.

DNA inside
"head"

Head
protein coat

850 A.

Tail sheath
(contractile?)

"Core" with hole
through center

1100 A.

1300 A.

Lysozyme activity
presumably in
distal tail region

60 A.

100 A.

150 A

Tail fiber
(Some coiled around
tail spike and
some unwound;
bears host–range
specificity?)

Figure 3.1. A schematic diagram showing the morphology of a bacteriophage particle. The dimensions are approximate. From Anfinsen (26).

are best known, and among these the T series of bacteriophages or phages infecting *Escherichia coli* have life cycles which have been studied in detail with a great deal of success. Seven phages in this series have been described (208), but we shall consider in detail only the phage T2, as an example. The others have similar or identical life cycles.

Life Cycle of T2. Phage exists outside of the bacterial cell in a non-reproducing state called the *resting* or *mature* phage. Upon coming into contact with a "sensitive" bacterium (i.e., a cell in which it can be reproduced), it adsorbs to the surface of the bacterial cell by its tail, and its nucleic acid plus a small amount of protein is injected into the host cell. The protein envelope remains outside as a "ghost." Within the bacterial cell the phage nucleic acid is reproduced utilizing many of the bacterial constituents. During this reproductive period the phage particles are apparently only nucleic acid and are called *vegetative* phage.

If the bacterial cell containing them is broken open prematurely, it is found that no infective particles are present. For this reason this is called the *eclipse* period. Phage protein is synthesized after the synthesis of phage DNA begins, and, during the period of maturation, the nucleic acid and protein are combined and mature phage are formed. The bacterial cell then lyses, that is, bursts, and the new phage particles are released. The entire process, the latent period, from adsorption to lysis may take about 20 minutes in T2, and up to several thousand particles may be released from a single cell. The particles so released are ready immediately to reinfect other sensitive bacterial cells which have not already been infected. The life cycle is diagrammed in Figure 3.2.

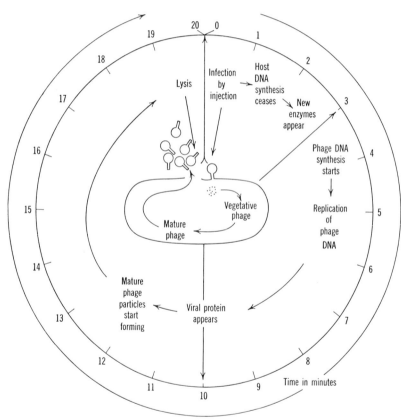

Figure 3.2. Diagram of the life cycle of a bacteriophage, T2. The numbers around the circle indicate minutes after infection of the bacterial host cell. Times given for indicated events are approximate and will vary with conditions.

Chemical Aspects of Phage Reproduction. The most fascinating facets of phage reproduction can only be appreciated by considering what is known about the chemical events that occur during the life cycle.

After the injection of the phage material there is an almost instantaneous reaction on the part of the bacterium. The synthesis of bacterial DNA comes to an abrupt halt, RNA synthesis is changed drastically, although not quantitatively, and the normal synthesis of bacterial protein is diverted into the production of other types of proteins (174). In essence, the metabolic machinery of the bacterium is converted to the production of more phages rather than of more bacterial material. All the steps in this transubstantiation are not clear, but it is clear that new phage particles are made out of bacterial substance in part, and exogenous material in part. The major part of the phage protein is synthesized from the nutrients surrounding the bacterial cell. Only about 6 to 12% comes from the bacterial cell itself (536). The bacterial DNA begins to degrade shortly after the entrance of the phage, and from this pool of fragments about 20 to 35% of the phage DNA is synthesized (536). The parent phage particle itself apparently contributes nothing to its offspring in the form of protein, but the phage-injected DNA can be accounted for in the progeny (430, 600).

The phage particles descended from the original single particle are therefore almost completely formed of new material from the surroundings and the stuff of the host itself. The parent phage contributes only a small amount of DNA which constitutes a fraction of the total DNA present in the progeny. This parental DNA is extremely important, however, for all evidence indicates that it is apparently the only genetic material provided by the mature phage to participate in the production of new phages. The basis for this conclusion rests on the observations made with phage particles which were doubly labeled with S^{35} and P^{32}. It can be shown that the P^{32} is in the phage DNA and the S^{35} in the protein as would be expected. When such phages are used to infect unlabeled *Escherichia coli* the labeled DNA enters and a considerable portion of it can be shown to survive in the progeny phage after lysis (432), but none of the S^{35}-labeled protein can be shown to survive in the progeny protein (434). One would expect any material which acted as a code to survive. Hence it may be concluded that DNA rather than protein is the genetic material. This is a reasonable conclusion, but obviously it is not proved by the data.

No interruption of total protein synthesis occurs after the infection of the host bacterial cell. On the other hand, the usual bacterial protein is not synthesized for the most part, and the bacterium cannot form adaptive enzymes (765). The host cell synthesizes two kinds of protein, viral protein and a nonviral, nonbacterial protein. About three to four minutes after the injection of the viral material into the bacterium, new enzymes begin to appear in the protein fraction, and rapidly reach high levels of activity (283, 531). Altogether at least 13 enzymes either increase greatly in activity after infection of the host cell or are synthesized *de novo* (174).

Some of these enzymes are required for formation of hydroxymethyl-cytidylic acid, its incorporation into the viral DNA, and for the conversion of deoxyuridylic acid into thymidylic acid. Viral DNA differs from host DNA by containing hydroxymethylcytosine instead of cytosine, and it is not synthesized until after these new enzyme activities begin to appear. It is then synthesized at a rate five times that which normally occurs for DNA in an intact bacterium. After the viral DNA is synthesized in the presence of polymerase it is glucosylated on the hydroxymethylcytosine bases in the presence of uridine phosphate glucose (UDPG).

Viral protein begins to form after viral DNA synthesis is initiated (436). This synthesis begins the period of maturation in which resting or mature phages are formed. About ten minutes after the initiation of viral-protein synthesis, the host cell lyses and the mature phage progeny are released. These observations are summarized in Figure 3.2.

While tremendous changes in DNA and protein occur within the infected bacterial cell during phage growth, there is no apparent great quantitative change in the RNA. Volkin and Astrachan (1040), however, showed clearly that there are qualitative changes in RNA, by demonstrating conclusively that a new RNA is formed which has base ratios similar to the DNA of the phage. This is the so-called phage-specific RNA which presumably acts as a messenger in carrying the code from the phage DNA to the bacterial ribosomes on which the phage protein is synthesized (371). The synthesis of the phage-specific RNA presumably precedes the synthesis of viral protein.

Genetics of Phage Reproduction. Strains of T2 and other T phages may be isolated which differ from one another in their ability to infect different strains of *E. coli,* or in the plaque type they form when growing on sensitive bacteria (see Figure 3.3). By infecting *E. coli* separately with recognizably different phage strains, it can easily be demonstrated that the progeny virus are identical to the parent virus in their characteristics. Hence phages have a genetic continuity.

If bacteria are infected simultaneously with two or more different

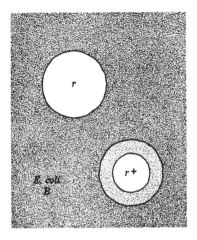

Figure 3.3. Plaques of T4 bacteriophage multiplying on a film of *E. coli,* strain B. Plaques of a mutant *r* and the wild-type *r+* are shown. A plaque is a clear area formed by the disintegration (lysis) of the bacterial host. Each particle of the phage seeded on the *E. coli* surface produces a plaque. From Sager and Ryan (873).

phages (multiple infection), the progeny phage particles include some which, on further testing, prove to have characteristics of two parents combined, or may be wild type if both parents were mutants. This is interpreted as being the result of recombination, since mutation has been eliminated effectively as the prime source of double mutants, or wild types, by executing the proper control experiments.

Two primary phenotypes that have been identified and found to be of value in phage genetics are *host range* and *plaque type*. Each of these phenotypes may be altered by mutations at many different loci. The many mutants which have been discovered have been effectively utilized to work out a linkage map for T2. The structural and functional organization of the loci involved are discussed in detail in Chapter 8.

The question of how recombination occurs in phage is not easily answered. One is tempted, of course, to assume that crossing over such as found in plants and animals occurs between phage "chromosomes" from different parents. The fact that recombinants of the expected complementary types are found in equal yields (435) might seem to argue for acceptance of a mechanism of standard crossing over by reciprocal exchange. However, complementary types in equal yields are obtained only when the results from a large number of similarly infected bacteria are averaged. Recombinants from a single bacterium form a random distribution as if they were produced independently from one another. The primary reason why data from phage crosses are difficult to analyze is that hundreds of offspring are obtained from the mating of a single pair of particles, and one does not know for certain precisely what occurs between the time the parent material is injected and the progeny burst out of the host. This situation is quite different from that in plants or animals where the meiotic process is fairly clear and can be followed.

If phage particles are assumed to be haploid, an assumption which seems to fit the observations best, and the parent chromosomes "mate" and exchange material between markers before chromosome duplication which then proceeds independently without further mating, one should expect only two types of recombinant progeny and no parental types. Such is not the case, however. Indeed, triparental recombinants are found when the bacterium is infected with three types of virus (433). This can only mean that mating (and exchange) occurs more than once during phage reproduction. The average number of rounds of mating is difficult or impossible to calculate. However, it is always certainly more than one. Actually, a phage cross is "an experiment in population genetics" (431) in which successive matings occur between vegetative phage particles as the population builds up within the host cell. If recombination occurs as the successive generations are formed, fewer generations or rounds of mating should give lower relative frequencies of recombinations in the progeny. This has been verified by lysing infected bacteria prematurely with cyanide and comparing the recombination frequencies with those found in progeny from crosses in which lysis is delayed. The recombination frequencies increase as lysis is delayed (226).

The data which have been accumulated on the reproduction of phage in *E. coli* indicate that the process of replication and the process of recombination are inseparably related, at least experimentally. If these two processes really are inseparable in fact, it is seen at once that a different situation exists here than is generally thought to exist in the higher organisms in which replication presumably precedes the process of recombination by crossing over. The mating event in phage in this light becomes a complex problem to analyze, and quite unlike the mating process in higher organisms in many respects.

Two hypotheses have been developed to explain the mating process: the pairwise-mating theory of Visconti and Delbrück (1035), and the group-mating theory of Hershey and Rotman (431, 435). In the pairwise-mating theory phage chromosomes are assumed to pair and recombinations occur by interaction between two chromosomes in a way analogous to the process visualized for higher organisms. The group-mating theory calls for interactions leading to recombination occurring between more than two chromosomes in close proximity. In addition, two alternative hypotheses have been advanced for the recombination process. One, called *copy choice,* visualizes a process in which a replicating chromosome copies from one or the other of two (or more) parent chromosomes (571). The parent chromosomes survive intact and the new chromosome may be a recombinant if it includes parts of the parent chromosomes that are different (see Figure 3.4). The alternative process is similar to crossing over in that it involves breakage, exchange, and reunion. The experimental evidence necessary to distinguish between these alternative hypotheses is indirect but suggestive. Luria and Dulbecco (628) have shown that relatively small doses of ultraviolet light can inactivate T2 particles and make them incapable of reproducing in the host cells which they infect singly. However, if the host cells are multiply infected with irradiated particles, many viable progeny are produced. This phenomenon is called *multiplicity reactivation.* It is presumed to be the result of the production of undamaged chromosomes by recombination during

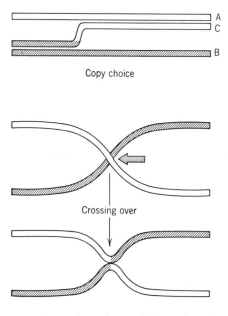

Figure 3.4. Comparison of copy choice and crossing over.

mating of phage chromosomes with irreversible damage at different loci (233, 227, 404). Since the radiation damage actually seems to increase the frequency of recombination between markers (262), it is assumed that "switches" are forced by radiation damage.

Thus, if copy choice, for example, is the mode of replication, one might assume that replication starts by copying at one end and proceeds along one of a pair or more of homologous chromosomes until a "lesion" produced by radiation is encountered. When this happens replication then switches to another homologue and proceeds until it in turn arrives at a lesion, and so on. In this way a perfect replica chromosome may be obtained from a pair or group of homologous elements each with lesions, but in different places along their lengths. The group-mating and the copy-choice mechanisms account for these observations best (431), although alternative schemes cannot be completely eliminated. The copy-choice mechanism of recombination is also independently supported by the fact that complementary phage recombinants from a single bacterium do not occur in equal numbers for either T2 (435), T1 (102) or lambda phage (504). As can be seen from Figure 3.4, copy choice does not produce complementary or reciprocal types as would be expected from crossing over.

Finally, a considerable amount of evidence indicates that mating, with its consequent result of recombination, is related to the replication of the genetic material of phages (960). The copy-choice hypothesis again fits this observation better than a crossover hypothesis in which interchanges are independent of replication. On the other hand, data also exist which seem to contradict these conclusions, at least so far as copy choice is concerned. It will be recognized from Figure 3.4 that a new chromosome arising by copy choice should be of entirely new material and not incorporate material from the old chromosomes, since it merely copies them. Experiments done with lambda phage (presumably not related to the T phage) infecting *E. coli* by Meselson and Weigle and others (667, 518) indicate that discrete amounts of parental DNA appear in recombinant progeny phage. This suggests that recombination occurs by breakage of parental chromosomes followed by the reconstruction of genetically complete chromosomes from the fragments. In addition, the data indicate that the phage DNA does not need to replicate in order to recombine. If these experiments are correctly interpreted, it means that the copy-choice mechanism is no longer tenable in its generally accepted form.

It should be clear from this discussion that the mechanism of phage replication and recombination is not yet completely elucidated.

When a bacterium is infected with two phage particles carrying a difference in a pair of alleles, say r^+ and r, about 2% of the progeny particles are found to have a + phenotype, but to actually carry both r^+ and r alleles (433). This can be demonstrated by infecting cells with single particles of this type and showing that both r^+ and r progeny are produced. The most obvious explanation of this phenomenon is that the particles are heterozygous and the r^+ and r alleles segregate after the heterozygous parent infects the bacterium. Considerable experimental work confirms the existence of a heterozygous condition (433, 599) and indicates further that the particles may approximate true diploids (244). The existence of a diploid or at least a quasi-diploid condition in phage makes it possible to test the effects of genes in combination and has led Benzer (73, 74, 75) and others to postulate some very interesting aspects of phage gene structure and function. These applications and the conclusions derived from them are considered in Chapter 8.

Other Viruses. The life cycles of the other T phages infecting *E. coli* are similar or identical to that described for T2. Certain of the T phages may differ from T2 in that their genetic material may mate with the bacterial host genome as well as with other phage material replicating in the same host (960). Other bacteria-infecting phages are described below in connection with their role in lysogeny and transduction.

The life cycles of the plant and animal viruses are by no means as well known as those for the bacterial viruses. In all, however, it would seem that the same fundamental aspects obtain. The viral DNA or RNA, depending on which is present, seems to be of overriding importance in determining phenotype, and in all cases infection is followed by an "eclipse" period within the host cell during which time no infective viral particles are demonstrable.

Bacteria

Prior to the year 1946 the bacteria were considered the prime example among organisms of a group that reproduced by an asexual process only. However, in this year Lederberg and Tatum (574) presented conclusive evidence that genetic recombination occurs in the K-12 strain of *E. coli*. The studies on linkage, employing about 15 different factors, indicated one linkage group in which the factors could be arranged linearly in a definite order (569).

Earlier in this same decade it was recognized that a phenomenon described originally by Griffeth in 1928 in pneumococcus, in which a nonvirulent, noncapsulated form of pneumococcus was transformed into a virulent, capsulated form in the presence of an extract of virulent, capsulated cells, was actually a transformation from one hereditary type to another. The transformation could in fact be brought about simply by putting cells of one genotype in the presence of highly purified DNA extracted from cells of another genotype (39, 634). This led Muller in 1947 to suggest that a fragment of a chromosome from a cell of one genotype is absorbed by a recipient cell of a different genotype which incorporates the fragment into its genetic material and becomes genetically transformed to the genotype of the donor. Subsequent work on transformation in pneumococcus and a number of other bacteria have verified this supposition and shown that in reality transformation is essentially a sexual process resulting in recombinations (260).

In 1952, Zinder and Lederberg (1136) announced the discovery of a process they called transduction, in which a phage particle carries a piece of bacterial genetic material from one bacterial cell to another. If the recipient receives a piece containing genetic material different from its own in a homologous region, it may substitute the new material and become changed in genotype. In essence, this phenomenon is similar to transformation except that in transduction a phage particle acts as the carrier for the genetic fragment.

In a period of less than ten years, then, bacteria were not only discovered to have the ability to recombine genetic material, but shown to possess, as a group, a number of different methods of achieving recombination. The discussion below considers these mechanisms in four distinct, but related, categories: (1) conjugation, (2) sex duction, (3) lysogeny and transduction, and (4) transformation.

Conjugation. The initial attempts to observe direct contact between cells, or the transfer of material between cells not in contact, were not successful in *E. coli* because only one recombinant cell out of a million was found when the original K-12 strain was used. It was not until Hayes (419, 420) and Lederberg and Cavalli (576) recognized the existence of two different types of *E. coli* cells—donor cells and recipient cells—that it became possible to resolve this question. It was shown that the donor cells (also variously described as male, F^+, or *Hfr*) and the recipient cells (female or F^-) make physical contact and that genetic material passes from the donor cell to the recipient (1098, 25). This process is called *conjugation,* although it must be recognized that apparently no *exchange* of nuclei occurs, as in the Ciliata, but a one-way transfer takes place as described hereafter.

Sex in *E. coli* has some of the characteristics of an infectious disease. Male cells (F^+) apparently carry a cytoplasmic factor, F, the fertility factor which they transmit readily to F^- cells by contact. Hence any culture containing a mixture of F^+ and F^- cells will become uniformly F^+. Cells carrying F may be caused to lose it, or be "cured" of it, and become F^-, female, cells by treatment with a variety of agents such as acridine dyes.

F^+ cells may become *Hfr* cells spontaneously at low frequency (circa 10^{-4} cells per division). The conversion is assumed to involve the attachment of the otherwise free-floating fertility factor, F, to the chromosome. *Hfr* (high-frequency recombination) cells conjugate readily with F^- cells and pass their genetic material with high frequency into the recipient cells where it undergoes recombination with the recipient's genetic material. Ordinarily, when *Hfr* cells are used in a cross about one recombinant in 10 to 20 parental cells is obtained. This is in contrast to the one in a million recombinant frequency obtained by Tatum and Lederberg in their early experiments with K-12. *Hfr* differ from F^+ cells in two observable ways: (1) they do not transmit the F factor to the F^- cell with which they conjugate, and (2) they produce many recombinants after conjugation.

Through a series of experiments by Jacob, Hayes, and Wollman (1098, 486), it was demonstrated by crosses between an *Hfr* and F^- strain that the *Hfr* "chromosome" is passed into the F^- cell over a period of time lasting up to about 30 to 40 minutes. Furthermore, the transfer always proceeds as though the *Hfr* chromosome is polarized, that is, the same end for any given *Hfr* strain always goes in first. This end, designated as O (origin) may be determined by the attachment of the otherwise free-floating F factor to the chromosome. To explain the data from experiments it has been postulated that the chromosome in F^+ cells is circular and that the attachment of the F factor causes it to break at the point of attachment. One free end becomes O, the other end has the F factor (486). See Figure 3.5.

The passage of the *Hfr* chromosome into the recipient cell can be interrupted by separating the conjugating cells by the shearing force of a Waring Blendor. When this is done at various periods after the onset of conjugation, it is found that F^- cells separated from their conjugating partners before eight minutes do not receive marker genes. After this, however, marker genes do appear, and in the same order as they are found to occur by standard recombination tests. Given in

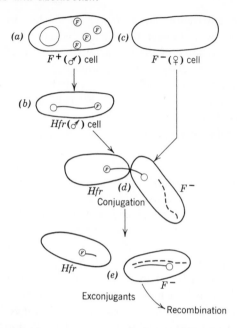

Figure 3.5. Conjugation in *Escherichia coli*. (*a*) A male cell showing a ring chromosome, and free episomes carrying the *F* (fertility) factor. (*b*) An *Hfr* cell produced by the attachment of an *F* factor to a chromosome. (*c*) A female cell carrying chromosomes not indicated in the diagram. (*d*) Conjugation and the polarized transfer of the *Hfr* chromosome into the female. (*e*) Exconjugants, each of which carries a piece of the broken *Hfr* chromosome. Recombination occurs in the female exconjugant.

Table 7 are some data reported by Jacob and Wollman (487) which show the correlation between the order of a certain group of markers on the *Hfr* chromosome as established by recombination tests after conjugation is completed, and the time at which the markers start penetrating into the F^- recipient as established by separating the conjugants prematurely. It will be noted that those markers assumed to be closest to O show the highest frequency of recombination and enter first.

The passage of the *Hfr* chromosome into the F^- cell generally appears to be incomplete because the chromosome presumably breaks before all of it is able to enter, or a portion is eliminated after entrance. Strains have been discovered, however, in which the *Hfr* transfers all, or essentially all, of the *Hfr* chromosome into the F^- cells (1000) showing that neither breakage nor possible post elimination are necessary aspects of the conjugation. The fate of exconjugants has been followed by Lederberg (572) and Anderson (23) who separated conjugants by micromanipulation and followed their descendants for several generations. The *Hfr* exconjugants are frequently viable—but they produce no progeny with recombinations and maintain the genotype of the parent *Hfr* strain. The viability of the *Hfr* exconjugant may be explained by the fact that *E. coli* contains not one nucleus but several identical nuclei. Each of these nuclei presumably contains one chromosome, and only one of these actually engages in conjugation. The lack of recombinants in the *Hfr* exconjugant descendants is, of course, easily explained on the basis that the transfer during conjugation is unilateral to the F^- cell only.

The F^- exconjugants produce many different types of individuals. The number of types depend, in part, on the degree of heterozygosity of the exconjugant zygote. Figure 3.6 (23) shows the pedigree of a family derived from a single pair of conjugants. Here the *Hfr* exconjugant divided regularly for at least four generations, the F^- exconjugant produced viable, nonrecombinant cells for several generations, then there began to appear nonviable cells and cells which gave rise to recombinant clones of several different types.

Results from many experiments such as the one illustrated in Figure 3.6 have shown that segregation of the characters in a merozygote does not occur in a regular fashion such as would be expected from true meiosis. Segregation may not begin until nine divisions after conjugation, and then only a few lines will show evidence of it while other lines

Table 7. **The Recombination Frequency for Various Characters in** *Hfr* **Compared to Time of Entrance of the Same Characters** *

			Genetic Factors							
	t	l	az^s	T_1^s	lac_1	T_6	gal	λ	21	424
Recombination frequency	...	...	90	70	40	35	25	15	10	3
Times at which factors begin entering F^- cells, in minutes	8	8.5	9	11	18	20	24	26	35	72

* Jacob and Wollman (487).

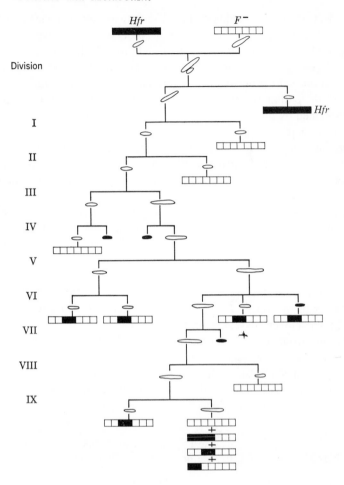

Figure 3.6. Segregation in *E. coli.* Pedigree of a family produced by conjugation of an *Hfr* and *F⁻*. Nonviable bacteria are indicated in black. The black and open bars indicate genomic differences; the black represents male genetic material. Note that recombinants did not appear until the sixth generation, and after the ninth division a cell produced the parental female genotype and three different types of recombinants. From Anderson (23).

may lead to clones with the *F⁻* parent genotype only. The *Hfr* fragment may exchange markers with the *F⁻* chromosomes over a number of generations and hence produce a number of different kinds of recombinants, some of which may be inviable. There is no clear evidence of complementarity among the recombinants descended from a single

F^- clone. The evidence indicates, in fact, that the *Hfr* chromosome may retain its identity over a number of generations while mating and engage in the formation of recombinant chromosomes with the aid of the F^- chromosomes in much the same fashion as successive rounds of mating apparently occur in the phage during its reproduction in a bacterium. While it is well established that a partial zygote is formed during conjugation in *E. coli,* the behavior following zygote formation does not correspond closely to any of the known phases of the life cycles of plants and animals.

As in the phage, the mechanism for the formation of recombinants in bacteria is not understood. There has been a general tendency to assume that the mechanism is copy choice, but no direct proof supports this assumption as yet.

In the foregoing discussion it has been assumed that *E. coli* is haploid and has one chromosome per nucleus. The evidence supporting this assumption is indirect. Lederberg (570) discovered early in his studies on *E. coli* that cells which appeared to be persistent diploids occurred rarely in certain crosses. These were recognized in the heterozygous form, and it was clearly demonstrated that the usual dominance, recessiveness relationships of alleles are expressed in them, and that they frequently produce stable haploid segregants during fission (1134). Ordinarily these heterozygous diploids are not complete diploids. They may in fact be either paternal- or maternal-deficient (739).

The evidence for the presence of one chromosome per "nucleus" is based solely on linkage data. There is but one known linkage group in which more than 40 characters have been mapped (487). As just stated, the attachment of the fertility factor of the F^+ strain causes the conversion of F^+ to *Hfr*. The point of attachment is apparently random, and a large number of *Hfr* strains should be expected depending on where the fertility factor attaches. These should act differently in crosses because their chromosomes have different origins. The data support this hypothesis (Table 8) which is a radical one, since it involves a circular chromosome (Figure 3.7) and an entity which can be either free or attached to a chromosome. However, when one recalls the action of the *dissociator*-controlling element in maize (see p. 182) as described by McClintock, one recognizes that events occur in higher organisms, too, that are not easily explained by orthodox genetic interpretations. Indeed, there may be a homology between the dissociator and other controlling elements found in maize and the fertility factor of *E. coli;* at the very least, an analogy exists.

Table 8. Linkage and Order of Genes in Several Different *Hfr* Types of *E. coli* *

Types of Hfr		Order of Transfer of Genetic Characters																	
Hfr H	O	*t*	*l*	*az*	T_1	*pro*	*lac*	*ad*	*gal*	*try*	*h*	*S-G*	*Sm*	*mal*	*xyl*	*mtl*	*isol*	*m*	B_1
1	O	*l*	*t*	B_1	*m*	*isol*	*mtl*	*xyl*	*mal*	*Sm*	*S-G*	*h*	*try*	*gal*	*ad*	*lac*	*pro*	T_1	*az*
2	O	*pro*	T_1	*az*	*l*	*t*	B_1	*m*	*isol*	*mtl*	*xyl*	*mal*	*Sm*	*S-G*	*h*	*try*	*gal*	*ad*	*lac*
3	O	*ad*	*lac*	*pro*	T_1	*az*	*l*	*t*	B_1	*m*	*isol*	*mtl*	*xyl*	*mal*	*Sm*	*S-G*	*h*	*try*	*gal*
4	O	B_1	*m*	*isol*	*mtl*	*xyl*	*mal*	*Sm*	*S-G*	*h*	*try*	*gal*	*ad*	*lac*	*pro*	T_1	*az*	*l*	*t*
5	O	*m*	B_1	*t*	*l*	*az*	T_1	*pro*	*lac*	*ad*	*gal*	*try*	*h*	*S-G*	*Sm*	*mal*	*xyl*	*mtl*	*isol*
6	O	*isol*	*m*	B_1	*t*	*l*	*az*	T_1	*pro*	*lac*	*ad*	*gal*	*try*	*h*	*S-G*	*Sm*	*mal*	*xyl*	*mtl*
7	O	T_1	*az*	*l*	*t*	B_1	*m*	*isol*	*mtl*	*xyl*	*mal*	*Sm*	*S-G*	*h*	*try*	*gal*	*ad*	*lac*	*pro*
AB 311	O	*h*	*try*	*gal*	*ad*	*lac*	*pro*	T_1	*az*	*l*	*t*	B_1	*m*	*isol*	*mtl*	*xyl*	*mal*	*Sm*	*S-G*
AB 312	O	*Sm*	*mal*	*xyl*	*mtl*	*isol*	*m*	B_1	*t*	*l*	*az*	T_1	*pro*	*lac*	*ad*	*gal*	*try*	*h*	*S-G*
AB 313	O	*mtl*	*xyl*	*mal*	*Sm*	*S-G*	*h*	*try*	*gal*	*ad*	*lac*	*pro*	T_1	*az*	*l*	*t*	B_1	*m*	*isol*

* Jacob and Wollman (487).

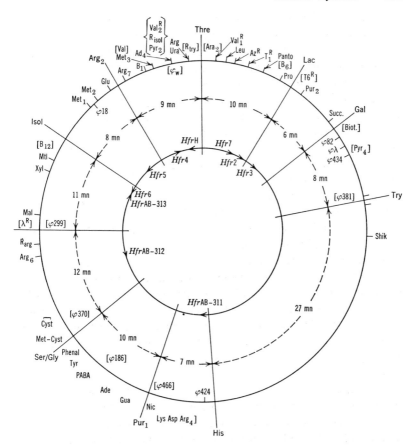

Figure 3.7. Schematic representation of the linkage group of *E. coli* K-12. The outer line represents the order of the characters (not their absolute distances). The dotted lines represent the time intervals of penetration between pairs of markers corresponding to the radial lines. The inner line represents the order of transfer of different *Hfr* types described in Table 8. Each arrow corresponds to the origin of the corresponding *Hfr* strain.

Symbols correspond to synthesis of threonine (thre), leucine (leu), pantothenate (panto), proline (pro), purines (pur), biotin (biot), pyrimidines (pyr), tryptophan (try), shikimic acid (shik), histidine (his), arginine (arg), lysine (lys), nicotinamide (nic), guanine (gua), adenine (ade), para-aminobenzoic acid (paba), tryosine (try), phenylalanine (phenal), glycine (gly), serine (ser), cysteine (cyst), methionine (met), vitamin B_{12} (B_{12}), isoleucine (isol), thiamine (B_1), valine (val); to fermentation of arabinose (ara), lactose (lac), galactose (gal), maltose (mal), xylose (xyl), mannitol (mtl); requirement for succinate (succ), aspartate (asp), glutamate (glu); resistance to valine (val^r), to sodium azide (az^r), to phages Tl (Tl^r), T6 (T6^r), λ (λ^r); repression for arginine (R_{arg}),

Sex Duction. The fertility factor, *F,* has been termed an *episome* by
Jacob and Wollman (486) to distinguish it as a separate type of cellular
entity, because it apparently exists free in the cytoplasm or attached
to a chromosome.   In addition to conferring the *F*⁺ condition on an
E. coli not previously carrying it, the *F* episome also introduces into
the cell receiving it characteristics from the donor cell from which it
was derived.   In other words, bacterial genetic material may be carried by episomes from one bacterial cell to another conferring new
characteristics on the receptor.   This is known as *sex duction.*   Presumably the male cells can produce many episomes, some or all of which
incorporate bits of host genetic material as they are formed.   When
F⁻ cells are infected with these they not only become *F*⁺, but may become *heterogenotes* for genetic material received from the donor cells.
For example, if *F*⁻ cells carrying the gene, *gal*⁻, for inability to utilize
galactose, are infected with an *F* episome carrying *gal*⁺, they will become capable of fermenting galactose but may continue to carry the
gal⁻ allele on their chromosome even after many cell generations (441).
These cells are galactose-positive only because they carry *gal*⁺ in their
episomes.   They are called heterogenotes because they are "heterozygous" for the fragment of the genetic material bearing the *gal* locus.
Some of the cells may cease being heterogenotes by incorporation of
the *gal*⁺ allele into their chromosomes by recombination in the place
of *gal*⁻.

Sex duction is obviously another method for transferring genetic
material from one bacterial cell to another.   Since the material transferred by the episomes can be incorporated into the host genome, it
is apparent that sex duction represents a sexual process in *E. coli* different from conjugation only insofar as a smaller amount of genetic
material is transmitted by episomes.

Lysogeny. Some bacteria are capable of producing, in the absence of
an external source of infecting phages, bacteriophages capable of infecting sensitive bacteria and lysing them.   Bacteria which act as sources
of infecting phage in this fashion, without themselves being lysed out
of existence, are called *lysogenic* bacteria.   The phages they carry are

---

isoleucine (R_{isol}), tryptophan (R_{try}), location of inducible prophages 82, 434,
381, 21, 424, 466, and of noninducible prophages 186, 370, 299, 18, and W.
Symbols in brackets indicate that the location of the marker with respect to neigh-
boring markers has not been exactly determined.

The figures given for time intervals between two markers correspond to the
average of several experiments of interrupted mating, using *Hfr* strains which
inject early the region involved.   From Jacob and Wollman (487).

generally described as *temperate* phage to distinguish them from the *virulent* phages which always cause lysis.

Temperate phage may exist in three different states: (1) an infectious state in which the phage is free of a bacterium and ready to infect (in this state it is comparable to the mature or resting stage of T2 described earlier, the particle consisting of a protein coat and a DNA core); (2) a vegetative state in which the phage reproduces at the expense of its host cell and causes its relatively rapid lysis (this is comparable to the vegetative state described for T2); and (3) the *prophage* state in which the genetic material of the phage is attached to the bacterial host genetic apparatus and is duplicated with it during the normal host cycle.

In contrast to a virulent phage which may only cause lysis, a temperate phage may follow one of two courses or responses. It may enter the vegetative state and lyse the host cell, or it may establish itself as a prophage and give rise to a stable clone of lysogenic bacteria. The first type of response is described as the *lytic cycle*. The second type is *lysogenization* or establishment of *lysogeny*. Whether a temperate phage enters the lytic cycle or the prophage condition in a sensitive cell depends on genetic and environmental factors.

Once in the prophage state, the temperate phage is not doomed to perpetual bondage to the host genetic material. Ordinarily, in lysogenic bacteria about one out of every thousand or ten thousand cells per generation contains a prophage which enters the vegetative state and the lytic cycle is started. In some cases the lytic state can be induced by the use of low doses of ultraviolet light or soft x-rays.

In Figure 3.8 the lysogenic and lytic cycles are generalized in a diagram. It should be clear from studying this diagram, and from the foregoing discussion of lysogeny, that the relationship between the host cell and phage is more than host-parasite in its broadest sense. The phage can actually become *part* of the bacterial genetic apparatus and be passed on from generation to generation as a prophage, which in this context may be thought of as a "gene" or group of genes. Looked at from a slightly different view, lysogeny may be thought of as an inherited ability to pass on a certain type of virus by infection. In addition, any bacterium carrying a specific prophage is immune to infection by infective particles of that prophage or closely related phages. This immunity is retained only as long as the prophage is present.

Many different kinds of phages exist which may be associated with a given bacterial host as prophage. They may be distinguished from one another by host-sensitivity tests. A host carrying prophage of one type is immune to infecting particles of that type, but not to other un-

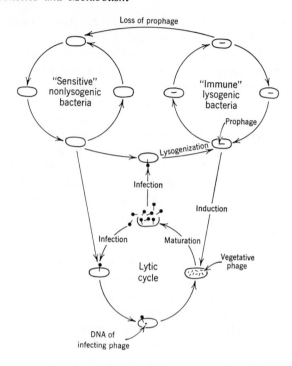

Figure 3.8. Life cycles for temperate phage showing lytic and lysogenic phases and the relation between them and nonlysogenic bacteria. Only temperate phage lysogenize. The others, like the T phage, have only the lytic cycle so far as is known.

related types. Additionally, they may be distinguished by the position of the prophage on the bacterial chromosome. Thus a number of prophages have been located on the chromosome of *E. coli* scattered out over its length (see Figure 3.7). Each one occupies a specific position. The existence and location of these can be readily determined by making the appropriate matings between lysogenic and non-lysogenic cells. Matings of this type are characterized by the peculiar phenomenon of zygotic induction. When an *Hfr* cell carrying a prophage, conjugates with a F^- cell not carrying that specific prophage, the prophage of *Hfr* becomes vegetative after infection and causes the destruction of the zygote. This is called *zygotic induction*. If, on the other hand, a nonlysogenic *Hfr* conjugates with a lysogenic F^-, the zygotes retain the prophage in that state and remain immune.

The close association of prophage with the bacterial chromosome naturally raises the question of how integral a part the prophage is of the host chromosome. When a bacterium becomes lysogenic its prophage does not seem to replace preexisting bacterial genetic material. Rather, the prophage seems to add on at some specific locus, thus constituting an addition rather than a substitution.

Transduction is probably understood most easily in the light of the phenomenon of lysogeny just described. The process may be simply described as the transfer of bacterial genetic material from one cell to another by a phage acting as a vector. The incorporation of the genetic material into the genome of the recipient cell completes the transduction.

Some bacteria such as *Salmonella typhimurium,* the rat-typhoid organism, have been found to possess transduction as the only known method of recombining their genetic material, while others such as *E. coli* carry out both conjugation and transduction.

Two readily discernible types of transduction have been described: (1) generalized, nonspecific, or common transduction; and (2) limited, specific, or restricted transduction.

In generalized transduction, the phage vector can apparently transfer *any* genetically determined character of a host, in which it develops, to a recipient, which it infects. In limited transduction, a specific phage known to occupy a specific locus in a bacterial host as prophage may, when induced to enter the vegetative phase, carry with it a *specific* piece of bacterial-host genetic material to a recipient which it infects.

Generalized transduction may be illustrated with *S. typhimurium* and certain of its infecting temperate phages such as P22. If a sensitive *Salmonella* strain carrying a particular character—ability to synthesize tryptophan (try^+), for example—is infected with the phage under conditions which promote the lytic cycle rather than lysogeny, a lysate will be obtained containing many phage particles. When these particles are used to infect sensitive cells of a strain unable to synthesize tryptophan (try^-), at the proper dilution, about 1 in 10,000 of the recipient cells will become changed in genotype from try^- to try^+. From this result it may be deduced that some of the phage particles from the try^+ host cells have, in the process of reproducing in these cells, incorporated genetic material containing the try^+ locus. This material is then injected into the recipient try^- bacteria, along with the viral genetic material, during the infective process, and the recipient cells incorporate the try^+ material in place of try^- by some mechanism of recombination. Their descendants are stable try^+ cells which may be transduced back to the try^- genotype with phage carrying try^-. All lines of evidence from many experiments support the interpretation that the phage vector particles carry only a small fragment of the bacterial genome, and that in the instance of P22 it may be a fragment from any part of the *Salmonella* genome. Therefore any genetic character of *Salmonella* should theoretically be transducible.

Two genetic characters may be simultaneously transduced provided they are close enough together on the bacterial chromosome to be incorporated in the same fragment in the phage particles. Distantly linked markers are only rarely incorporated into the same particle as two fragments, and simultaneous transduction of them is therefore much less frequent than that of closely linked markers.

The galactose-fermenting or nonfermenting characters (*gal*) in *E. coli* exhibit the limited or specific type of transduction via the lambda phage. The *gal⁻* mutants are, unlike the wild-type *gal⁺*, unable to use galactose as a carbon source. All the known *gal⁻* mutants are closely linked to one another and to the markers, λ (prophage lambda) and *succ* (succinate requirement). Any one or a group of the *gal⁻* markers may be transduced by the temperate phage λ, which is normally carried as a prophage by the K-12 strain of *E. coli* closely linked to the *gal⁻* loci. The ability of λ to transduce is limited to the λ-*gal* segment. Hence it is different from P22 discussed earlier for *Salmonella* and from P1, a general transducing phage of *E. coli* similar to P22.

Temperate phages are really quite complex in their behavior and characteristics. A particular one of them, after an injection of its DNA into a bacterial host, may (1) undergo reproduction rapidly and cause lysis, (2) persist in a state of slow reproduction in the host cytoplasm, or (3) associate with the bacterial nucleus and be replicated in synchrony with the replication of the bacterial DNA (629). Furthermore, they may incorporate, during vegetative growth and maturation leading to lysis, fragments of bacterial genetic material which may be either specific fragments as in limited transduction or any small fragment at random as in generalized transduction. These activities are summarized diagrammatically in Figure 3.9.

Since lysogenic phage may either attach to chromosomes as prophage or exist free in the cytoplasm as relatively autonomous units, they show some of the same characteristics as the *F* factors and hence may also be considered as episomes. The difference between sex duction and transduction seems to be rather minor, and it may be best to consider them as part of the same general phenomenon of transfer of genetic material in bacteria by semiautonomous units going under the name of episomes.

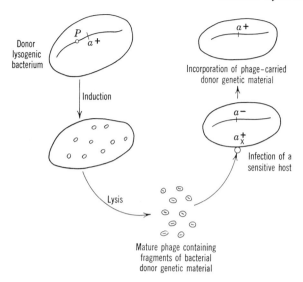

Figure 3.9. Transduction by a temperate phage.

Transformation. In many ways transformation resembles transduction, but without a viral vector. Naked DNA, instead of DNA encased in protein as in a virus, in some way enters the cells of bacteria related to the bacterium from which the DNA originated and becomes incorporated into the host genome. The exposure to the DNA needs be only a matter of a few minutes. By these means a bacterium of one specific genotype can be converted to the genotype of the strain donating the DNA (Figure 3.10).

Transformation has been carried out successfully with purified DNA from pneumococcus (*Diplococcus pneumoniae*) for such diverse characters as capsular type (39, 260, 38), mannitol-dehydrogenase activity, streptomycin and penicillin sensitivity (465, 468), and sulfonamide sensitivity (466, 467). Transformation has also been carried out successfully with *Hemophilus influenzae* (339, 580, 1129) for such characters as streptomycin resistance and capsular type. *Bacillus subtilis* has been found to be an especially valuable organism with which to study transformation, since its nutritional requirements are not as fastidious as pneumococcus and *Hemophilus*. It is possible to obtain stable nutritional mutants of *B. subtilis* with requirements for various amino acids and vitamins and then to transform these back to wild type by introducing DNA from wild-type *B. subtilis* (940). All in all, at

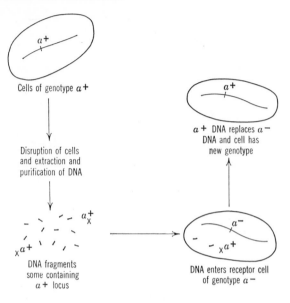

Figure 3.10.   Transformation in bacteria.

least 17 different species of bacteria have been shown to be capable of transformation, so it is obviously a phenomenon of general significance (826).

The DNA used as transforming principle has been purified to a high enough degree such that no protein or other compounds of high molecular weight such as RNA can be detected by the most sensitive tests. Solutions of transforming principle containing only 10^{-6} to 10^{-3} μg/ml of DNA are effective in transformation. Such solutions lose their transforming power when treated with deoxyribonuclease, but not when treated with various proteases or ribonuclease. Thus it is rather certain that the active transforming principle is DNA.

Molecular-weight determinations of the active transforming principle of *Hemophilus* indicate a large molecule with a weight of about 15×10^6 (339). Goodgal and Herriot (339) have calculated that about 120 molecules of this size may be taken up by a single *Hemophilus* cell. That the transforming DNA itself is indeed taken into the bacterial cell has been proved with the use of P^{32}-labeled DNA.

Transforming DNA is highly sensitive to radiations such as ultraviolet light (356), x-rays (590), and other ionizing radiations. The effect of the radiations is to cause an inactivation of the transforming ability of the DNA. Additionally, DNA loses transforming ability if it is heated to temperatures above 80°C. This is presumably because at high temperatures the DNA double strand unwinds and single strands are formed. If the heated DNA is slowly cooled to room temperature, it regains most of its transforming ability by formation of the double helices again. Rapid cooling, however, results in slight or no regaining of activity, probably because double-stranded helices do not reform properly (661, 229). The conclusion to be drawn from these observations is that only double-stranded DNA is active as transforming principle.

After the DNA with transforming activity enters the bacterial cell, it must in some way be incorporated into the bacterial genome. The two possible ways in which it could conceivably be incorporated are: (1) by actual physical exchange with the bacterial genome which would result in the ejection of a nonviable piece, and (2) by influencing the replication of the bacterial chromosome so that the infecting DNA is copied rather than that of the host cell. The first mechanism would be comparable to crossing over and the second to copy choice. The available experimental evidence indicates that the mechanism of recombination is by break and fusion rather than copy choice. The reason for this conclusion is that the integration of the transforming DNA into the receptor genome does not seem to require DNA synthesis. This would exclude copy choice as the mechanism if the experiments are being interpreted correctly (261, 826).

Fungi

Unlike the viruses and the bacteria, the fungi may be considered as eucaryotic organisms. That is, they all seem to possess "true" nuclei which contain discernible chromosomes. Furthermore, it is clear that a meiotic mechanism exists in fungi which is comparable or identical to meiosis in the higher plants and animals. Crossing over with the production of reciprocal products occurs regularly. In fact, in many fungi such as *Neurospora* all of the meiotic products from a single diploid cell may be isolated and characterized to demonstrate, better than can be done with most plants and animals, the classical interpretation of the meiotic process.

True diploids are produced in the life cycles of most fungi. Unlike the "diploids" of bacteria and perhaps phage, the homologous chromosome pairs are made up of complete chromosomes and not of pairs in which one or both partners may be incomplete. This is not to say, however, that incomplete diploidy does not exist in fungi.

The fungi as a group exhibit a great variety in life cycles. Raper (822) has broken the many different kinds down into seven basic types. These are diagrammed in Figure 3.11 from which it may be seen that three basic nuclear phases occur, the haploid, diploid, and the dicaryon. The dicaryon is found as a transient phase in certain of the Ascomycetes, but it may be a dominant phase in the Basidiomycetes. Essentially, it results from a delayed fusion of nuclei after a fusion of two haploid cells. After cytoplasmic fusion (cytogamy) the binucleate, dicaryotic cell divides mitotically in such a fashion as to produce two daughter cells each with the same dicaryotic constitution. The maintenance of the dicaryotic condition is carried to an extreme degree in the higher Basidiomycetes (mushrooms) where the haploid condition is extremely transitory and the diploid exists only as a zygote.

Only the life cycles of those few fungi which have been used extensively in genetic work are discussed hereafter.

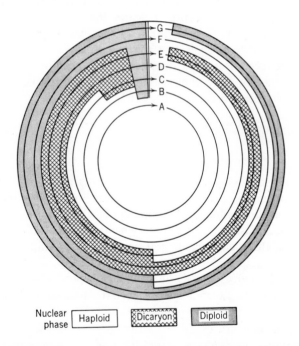

Figure 3.11. Schematic comparison of life cycles in fungi. Changes in nuclear phase are indicated by differences in shading, proceeding clockwise. The double vertical line at top represents meiosis. From Raper (822).

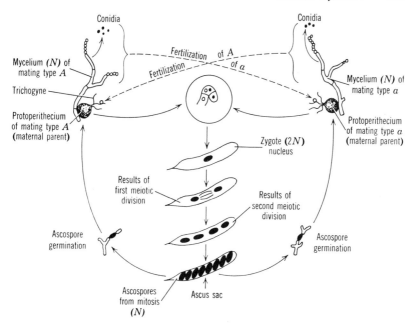

Figure 3.12. Life cycle of *Neurospora crassa*. The haploid mycelium reproduces asexually by conidia (macro- and micro-) which germinate to give more haploid mycelium, as well as by simple proliferation of existent mycelium. In this diagram the sexual cycle is illustrated. Two mating types are required for sexual reproduction. Fertilization takes place by the passage of nuclei of conidia or mycelium into the protoperithecia of the opposite mating type through the trichogynes. Fusion of the nuclei of opposite mating types takes place within the protoperithecia.

Neurospora. Three species of *Neurospora*—*crassa, sitophila,* and *tetrasperma*—have been employed in genetic investigations. The cycle of *N. crassa,* the most widely used species, will be described here. *N. sitophila* has an almost identical cycle, while *N. tetrasperma* differs from both by having four-spore ascus sacs and being homothallic.

Neurospora is an ascomycete fungus with a haploid vegetative phase. It is heterothallic with two mating types. The mycelium is septate, but the septa have holes through which the nuclei may migrate through the mycelium.

The life cycle of the fungus is diagrammed in Figure 3.12. The vegetative mycelium is normally haploid, and it can be propagated indefinitely through transfers of fragments of the mycelium or the asexual spores to fresh medium. The mycelium forms two kinds of asexual

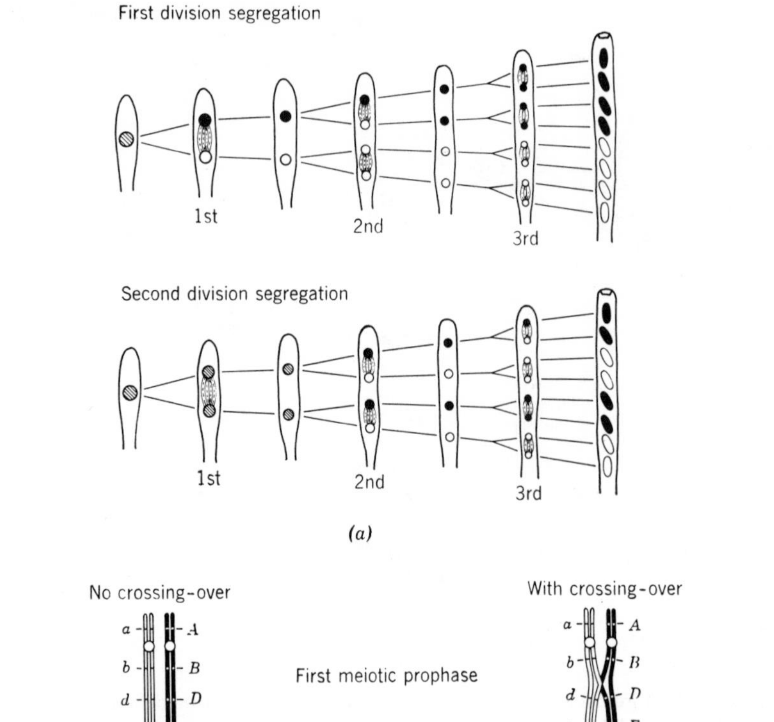

First division segregation

Second division segregation

(a)

No crossing-over · With crossing-over

First meiotic prophase

First meiotic anaphase

Second meiotic anaphase

Haploid products of meiosis

(b)

Figure 3.13. Meiosis and segregation in *Neurospora*. (*a*) Diagram to show how spores are aligned during the meiotic process in the ascus as a result of the orientation of the spindles. If a pair of characters segregate in the first meiotic division the distribution will be as shown above by black and white spores, if in the second meiotic division as shown below. (*b*) Diagram illustrating the effect

spores, the macroconidia, which contain four or five nuclei on the average, and the uninucleate microconidia. Of the two mating types which are designated *A* and *a,* either will produce fruiting bodies or protoperithecia on an appropriate medium (1071). After these have been produced, the sexual cycle is initiated by fertilization with conidia or mycelium of the mating type opposite to that of the protoperithecia. Even when a cross is made by mixing mold of opposite mating types, it is necessary for one or the other to produce protoperithecia before fertilization can take place.

After a nucleus has migrated through a trichogyne into the proto-perithecium, a dicaryon is formed which is reproduced mitotically to form a large number of dicaryotic cells each of which becomes a zygote nucleus upon the fusion of the pairs. Each fused nucleus then under-goes meiosis to produce four haploid nuclei. These undergo two mitotic divisions to give eight ascospores, each with two identical nuclei. Spore pairs are also identical in genetic constitution, since the last two divi-sions are mitotic and since the ascospores are contained in a thin-walled ascus sac which holds them in a definite order established by the di-rection of nuclear segregations during meiosis. The four primary prod-ucts of meiosis are also arranged in order, and for this reason one may distinguish a first from a second-division segregation (see Figure 3.13*a,* *b*). A perithecium may contain as many as 2 to 300 asci, each derived from identical fusion nuclei, and when the perithecia mature these asco-spores are ejected rather violently from the ascus sacs. The ascospores require heating at 60°C for 30 minutes to induce germination.

The entire sexual cycle requires from 10 to 15 days for normal strains. Reasonably homogeneous cultures can be obtained directly from ascospores, conidia, or bits of mycelium, but it should be noted that several million nuclear divisions take place during the development of a culture, and there is ample opportunity for the occurrence of spon-taneous mutations. Thus no two cultures, even those derived from identical ascospore pairs, are to be considered as being absolutely iden-tical, although they almost always appear to be by the relatively crude criteria ordinarily used for identification. Obviously, the relative iden-tity of two cultures will diverge in subsequent subcultures, since, even though the mutation rate of any single gene may be low, there are many

---

of crossing over on first and second division segregation of heterozygous genes. With no crossing over genes segregate in first division. With crossing over genes distal to the level of crossing over (*D* and *E*) segregate in the second division while those proximal (toward the centromere) still segregate in the first division. From Emerson (254).

genes that can mutate.   The spontaneous appearance of nutritional mutants is quite rare, and usually the mutations accumulated during subculturing only modify growth characteristics slightly.   Still, these spontaneous mutations must always be taken into account in descriptions of phenotypes in *Neurospora* and in other organisms.

Aspergillus.   The most commonly employed species of this ascomycete genus is *Aspergillus nidulans*.   Like *Neurospora,* its vegetative phase is normally haploid, but it is unlike *Neurospora* in a number of significant ways, some of which confer upon it certain advantages in genetic investigations.

The asexual life cycle is diagrammed in Figure 3.14.   Asexual reproduction is by conidia or mycelial fragmentation.   In contrast to *Neuro-*

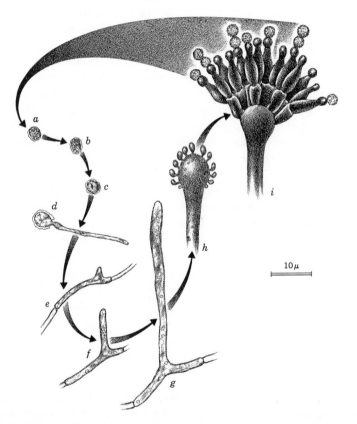

Figure 3.14.   Asexual cycle of *Aspergillus nidulans*.   Conidia (*a*) germinate and produce hyphae (*g*) which proliferate conidial heads (*h* and *i*).   The conidial heads produce more conidia to repeat the cycle.   From Emerson (254).

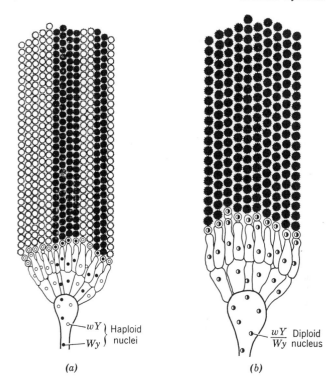

(a)   (b)

Figure 3.15.   Diagrammatic representations of conidial heads of *Aspergillus nidulans:* (*a*) conidiophore of a heterocaryotic strain, haploid nuclei *wY* and *Wy* both present in the coenocytic conidiophore, but occurring separately in the uninucleate primary and secondary sterigmata and in the rows of conidia developing from them (the conidia are white if *wY* and yellow if *Wy*); (*b*) conidiophore of a diploid strain in which all nuclei are heterozygous for *W/w* and *Y/y* and in which the conidia are green.   From Emerson (254).

spora, which is heterothallic and has two mating types, *Aspergillus* is homothallic, that is, it has no mating types.   Hence, conidia of a strain may fertilize its own protoperithecia.   The details of the formation of protoperithecia and the events involved in fertilization are not known. However, perithecia containing many ascus sacs are formed.   Eight ascospores are present in each ascus sac, but these are not arranged in order as they are in *Neurospora.*

The conidia of *A. nidulans* are uninucleate, and each row of conidia in a conidial head has its nucleus derived from the nucleus present at the base of the row of conidia (797).   See Figure 3.15.

Although *Aspergillus* is generally to be considered an organism with a haploid vegetative phase, it can diploidize. This was demonstrated for *A. nidulans* by Roper (852) and for *Aspergillus niger* by Pontecorvo (796). Diploidization occurs spontaneously. The ability to form diploids by *Aspergillus* confers upon it certain advantages not found in other fungi, such as *Neurospora.* First, it makes it possible to test the effects of genes in heterozygous combination. Second, it makes it possible to observe mitotic crossing over, a phenomenon which occurs regularly in diploid *Aspergillus* and has been used to map genes in *A. nidulans* (797, 800). The mitotic crossing over occurs at the four-strand stage, segregation is mitotic, and the expected reciprocal products are found (853).

Yeast. Yeast is an ascomycete fungus which has apparently degenerated to a unicellular condition in which the vegetative cell also acts as the sexual cell. The life cycle of *Saccharomyces cerevisiae,* a heterothallic species and the most widely used yeast species in genetics, is considered here.

The cytology of the yeast cell is incompletely understood, but the genetic evidence makes it quite clear that yeast has the standard mitotic and meiotic mechanisms in operation. At least the end results are comparable to those found in higher forms.

Yeast reproduces asexually by a process called budding which is apparently a mitotic division in which an equal division of chromosome material, but an unequal division of cytoplasm, occurs.

The sexual cycle in yeast was discovered by Winge (1089) and Kruis and Satava (540) who showed that in species such as *S. cerevisiae* there are two phases, haploid and diploid. The life cycle is diagrammed in Figure 3.16.

As in *Neurospora* and *Aspergillus,* the diploid zygote of yeast undergoes meiosis within a closed sac, the ascus sac, and hence all products of a single meiotic event are kept together. Like *Aspergillus,* the meiotic products are unordered. Only four rather than eight ascospores are produced within a single sac.

Polyploidy is apparently common in yeast. It arises as the result of a fusion of a haploid and a diploid or two diploid cells (see Figure 3.16). Although it has not been verified cytologically, it is evident from genetical analysis that degrees of ploidy up to tetraploidy exist (851, 617, 593).

Heterocaryosis. Many of the fungi show hyphal fusion or anastomosis when they are growing actively in culture (119). Fusion ordinarily results in the intermixture of the cytoplasm and nuclei of the two hyphae.

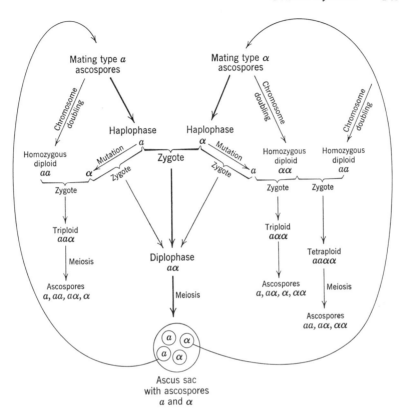

Figure 3.16. Life cycle of *Saccharomyces cerevisiae*. Ascospores germinate and may undergo division to produce more haploid cells (haplophase). These may also undergo chromosome doubling with the formation of diploid cells. The fusion of two haploid cells of opposite mating type (a and α) results in a zygote which may also undergo division to produce more diploid cells (diplophase). These diploid cells may under the proper conditions undergo meiosis producing four haploid ascospores. The diploid mother cell wall becomes the ascus sac. Triploid and tetraploid cells may also form by the mechanisms shown. Mutation of the mating type locus in *Saccharomyces* is fairly common. Emerson (254).

If the hyphae are different genetically, the fusion may result in the production of a strain which will have the cytoplasm and nuclei of both parent strains represented. Such a strain is called a *heterocaryon*. This is in distinction to a genetically pure, or homocaryotic strain, the homocaryon. Ordinarily only the nuclear constitution is considered of significance in heterocaryons, and ordinarily this may be adequate. However, it should always be borne in mind that a heterocaryon receives cytoplasmic as well as nuclear contributions from the parent homocaryotic strains which participate in its formation.

Heterocaryosis was first clearly described by Burgeff (121) in *Phycomyces* and later confirmed genetically by Dodge (225) with *N. tetrasperma* and by Beadle and Coonradt (54) with nutritional mutants of *N. crassa.* A heterocaryon is easily formed in *Neurospora* by allowing two different but compatible strains of the same mating type to grow on ordinary culture medium. If each strain has a different nutritional requirement due to gene mutation, the heterocaryon will have the wild-type allele of each represented and hence will grow on a minimal medium. Frequently, growth of heterocaryons is as good as that obtained from the homocaryotic wild-type strains.

Proof of the heterocaryotic nature of a strain is obtained by plating out conidia from it with the expectation that in a two-component system, for example, some of the conidia will be homocaryotic and distinguishable from the heterocaryotic conidia (810). Another method is to cut off the hyphae of a culture, growing on the surface of agar, near their tips with the objective of isolating tips with a single type of nucleus. For fungi such as *A. nidulans* which have uninucleate conidia, the matter of heterocaryon analysis is, of course, relatively simple. Indeed, if nuclei controlling different colors of conidia are present in a heterocaryon, this can be readily observed in the conidial heads after their maturation (see Figure 3.15).

The ratio of nuclei of diverse genotype within a heterocaryon depends in part on the input ratio, that is, the initial ratio of nuclei brought together from the heterocaryons, and in part on selection. It should not be expected to be 1:1 in a two component heterocaryon. Nuclear mixing during heterocaryon formation is ordinarily quite thorough (30), and after a heterocaryon has become stabilized in a given environment, it may be assumed that the ratio will be determined with certain limits by selection (491). It is important to recognize these facts because they emphasize the significant difference between heterocaryons and heterozygotes in diploids which, notwithstanding this, have an interesting resemblance to one another.

Early in the study of heterocaryons in *Neurospora,* it was proposed that the production of heterocaryons with wild phenotypes from mutant strains was proof of nonallelism of the mutant genes involved, whereas if a wild-type heterocaryon, or at least one approximating it, were not formed, it was proof that the genes were allelic (54). It became evident, however, after continued work on heterocaryons that this conclusion was in need of some revision. First, not all strains of *Neurospora* will form heterocaryons, because of the presence of certain incompatibility factors (311, 451). Hence the mere absence of the appearance of a wild-type heterocaryon when two mutants are tested together may mean only that incompatibility factors prevent the formation of any heterocaryon at all. Elimination of these factors by the appropriate crosses may result in the formation of a heterocaryon with a wild phenotype. Second, heterocaryons with wild or approximately wild phenotypes may be formed by mixing nuclei from strains which carry allelic mutations as established by previous criteria (694). The details and implications of this latter observation are discussed at length in Chapter 8 in connection with complementation.

The formation of heterocaryons among the fungi is widespread and may be universal among the Ascomycetes. It occurs among bacteria such as *E. coli,* and it certainly occurs in various species of the actinomycete genus *Streptomyces* (97, 897).

Heterocaryosis is undoubtedly of great significance in the organisms in which it is found. It is possible that those fungi which are capable of forming heterocaryons are frequently heterocaryotic in their natural habitats.

Heterocaryosis is a genetically significant phenomenon, since it enables the establishment of balanced genetic systems which have selective advantage over what may be possible in homocaryons. It may be thought of as a phenomenon midway in evolutionary development between syntrophism and heterozygosity in the stable diploid organisms. Syntrophism is a relationship between two different strains of an organism, resembling symbiosis, which enables both to survive in a particular environment whereas neither would survive alone. For example, a strain of *E. coli* requiring thiamin will grow quite well together with another strain requiring pantothenic acid in a medium containing neither vitamin, for each makes up the other's deficiency.

Diploidization, Haploidization, and Mitotic Recombination. The discovery of diploid nuclei in *Aspergillus* (852) has led to a series of interesting further discoveries in *Aspergillus* and to a new insight into the recombination potential among the lower organisms. Diploid nuclei are apparently formed as a result of fusion of haploid nuclei within the hyphae. If the strain within which the diploidization occurs is heterocaryotic, some heterozygous nuclei will be formed. These diploid nuclei may then undergo regular mitosis and produce more diploid nuclei of the same type. Diploid strains may be isolated, in fact, which produce 16 spore asci. But two other things may occur in addition: (1) the diploid nuclei may undergo mitotic recombination; and (2) they may haploidize (800, 503).

Mitotic recombination must be recognized as occurring in hetero-caryotic *A. nidulans,* because it leads to the formation of diploid nuclei which are homozygous. An example will make this clear.

A diploid strain of *A. nidulans* heterozygous for *y* (yellow), (y/y^+), is derived from a single conidium isolated from a heterocaryon between *y* and y^+. The phenotype of the diploid is green, since *y* is recessive. If now the diploid strain is propagated vegetatively, there will appear yellow conidia which must be of the genotype *y/y,* unless the diploid nuclei have haploidized. This possibility can be eliminated by showing that the conidia produced by the yellow strain are of the large diploid size, circa 4 μ in diameter, as compared to a diameter of circa 3 μ for haploids, and that 16-spore ascus sacs are produced by the yellow strain, instead of 8-spore sacs expected from haploids. The experiment may be extended by introducing another color gene, *w* (white, recessive), which is on a different chromosome than *y.*

A diploid synthesized from a heterocaryon containing w^+y and wy^+ nuclei will be heterozygous and green in phenotype but will produce some yellow and some white as well as green conidia. Cultures propagated from the yellow conidia will produce some yellow and some white.

It is quite well established that the explanation for the segregation

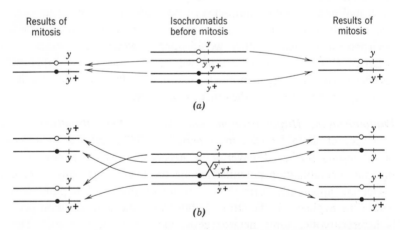

Figure 3.17. Homozygosity resulting from mitotic crossing over in *Aspergillus.* (*a*) Mitosis without crossing over. (*b*) Mitosis with crossing over resulting in the production of homozygosity for y^+ and *y* on the right side of the diagram.

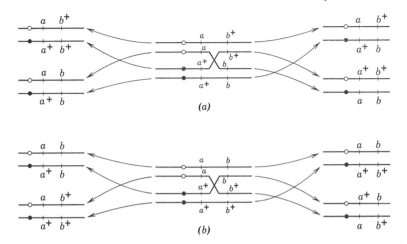

Figure 3.18. Results of mitotic crossing over between two markers *a* and *b*. (*a*) *a* and *b* on different chromosomes. (*b*) *a* and *b* on the same chromosome. Note that homozygosity does not obtain for *a* or *a*+ because crossover does not occur between this locus and the centromere.

of these characters in the heterozygous diploid *Aspergillus* is mitotic crossing over (800). Reference to Figure 3.17 will make this clear. If no crossovers occur between the marker genes and the centromeres, the mitotic products will be heterozygous, as generally expected from the results of mitosis in heterozygous diploids (Figure 3.17*a*). However, if crossovers occur as indicated in Figure 3.17*b,* depending on how the chromatids segregate, some of the resulting nuclei will be homozygous. The recessive genes can then be expressed, and from the heterozygote yy^+ww^+ (green) one expects to obtain *wwyy, wwy+*— (both white since *ww* is epistatic to y^+), and w^+—*yy* (yellow) conidia, as well as the parental genotype and genotypes that give the parental phenotype. All of the many different experiments carried out with *Aspergillus,* and reported by Pontecorvo and his coworkers using a variety of different markers, have yielded results consistent with this interpretation.

If two genes are linked and heterozygous in a diploid *Aspergillus* and a crossover occurs between them, one expects the results given in Figure 3.18; homozygosity results for each of them in the nuclei produced as indicated. Not only does homozygosity occur, of course, but recombinant chromosomes are formed, for in addition to the parental genotypes, ab^+ and $a^+b,$ there now also exist the recombinants, *ab* and $a^+b^+.$

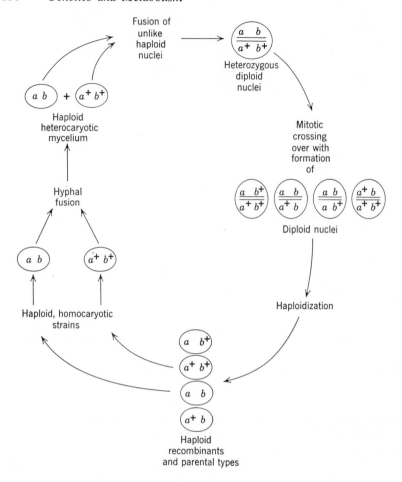

Figure 3.19. The parasexual cycle in *Aspergillus* according to Pontecorvo (799).

Diploid *Aspergillus* nuclei are relatively stable, but perhaps one out of a thousand of them become haploid, possibly by failure of regular mitotic separation of sister chromatids. Aneuploids so formed could gradually lose their disomic condition to become complete haploids (503). Whatever the mechanism, however, it is clear that recombinant haploid strains can be derived from diploid heterozygous and homocaryotic strains with no evidence of the involvement of a typical meiosis.

It will be recognized that a genetic system exists in *A. nidulans* that apparently allows for the formation of recombinants without meiosis.

To distinguish this type of system from systems in which meiosis occurs, it has been suggested that the term *parasexual* be used. The same kinds of recombinants can be formed in either system, parasexual or sexual, but the mechanism is somewhat different just as the mechanism is apparently different for bacteria or phage. *A. nidulans,* of course, has both parasexual and sexual cycles. Other fungi may have only a parasexual cycle, however. The parasexual cycle for *Aspergillus* is summarized in Figure 3.19.

The parasexual cycle has been found in other species of *Aspergillus, A. niger* (802) and *A. sojae* (480), and also in other fungi such as *Penicillium chrysogenum* (801) and *Fusarium oxysporum* (125). It is of considerable interest to note that *A. niger* and *P. chrysogenum* are known not to have a sexual cycle, at least under laboratory conditions. Hence they apparently depend on the parasexual cycle alone for the formation of recombinants.

Among the bacteria, a parasexual cycle seemingly similar to that in the ascomycetes has been found in a number of different species of the actinomycete, *Streptomyces* (99, 898, 875). Presumably, a transient diploid stage occurs, because recombinants are produced from various heterocaryotic combinations (454). The mechanism of recombination, therefore, may be by mitotic crossing over.

Protozoa

The Protozoa represent such a diverse group that it is quite impossible to make generalizations about life cycles within it. Genetic work has been done extensively only with certain genera of the Ciliata such as *Paramecium* and *Tetrahymena,* however, and the discussion will be limited to *Paramecium.*

The life cycles of all species of *Paramecium* seem to be grossly similar. Figure 3.20 illustrates the life cycle of *Paramecium aurelia,* a species which has probably been used more extensively than any other in genetic work, particularly by Sonneborn, his coworkers, and students.

Paramecium, like the Ciliata in general, possesses two types of nuclei, macronuclei and micronuclei. Each cell possesses a single macronucleus and, depending on the species, may have one, two, or four micronuclei. Ordinarily, all nuclei within a nonconjugating cell are diploid and of identical genotype. *P. aurelia,* the example organism in this discussion, possesses two micronuclei and what appears to be a multinucleate macronucleus.

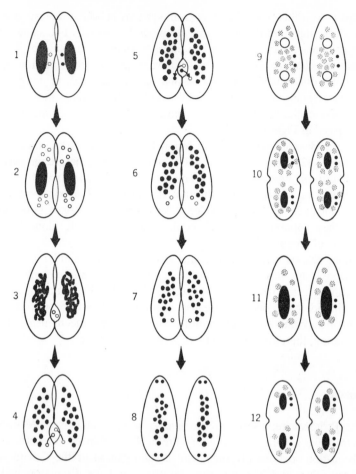

Figure 3.20.   Nuclear changes at conjugation in *Paramecium aurelia*.   (1) Two parental animals, each with one macronucleus and two diploid micronuclei.   (2) Formation of eight haploid nuclei from the micronuclei in each conjugant.   (3) Seven nuclei in each conjugant disappear; the remaining haploid nuclei pass into the paroral cones; macronucleus breaks up.   (4) The nuclei in the paroral cones divide mitotically, forming "male" and "female" gamete nuclei. The "female" gamete nuclei pass back into the interior of the parental animals.   (5–7) Fusion of "male" and "female" gamete nuclei from opposite mates.   (8) Each fusion nucleus divides twice, mitotically.   (9) Two products of each fusion nucleus differentiate into macronuclear anlagen (white circles); the other two into new micronuclei.   (10–12) Return to normal state of one macronucleus and two micronuclei per animal; fragments of old macronucleus gradually lost.   From Beale (61).

Three significant types of phenomena occur in *P. aurelia* which must be understood to understand the genetics of *Paramecium:* (1) binary fission, (2) conjugation, and (3) autogamy.

Binary fission constitutes asexual reproduction in *Paramecium* and is the only method for increasing the numbers of a population of *Paramecium*. In this process the animal simply divides into two. The macronucleus divides amitotically by simple fission. Each daughter receives a micronucleus. Since each of the original mother micronuclei are identical in genotype, the daughters, after forming two micronuclei by mitosis, are identical to one another in genotype. By binary fission one may obtain millions of animals starting with a single animal—all of identical genotype, except for differences that may arise through mutation. Animals derived from a single ancestor by binary fission constitute a *clone.*

Conjugation constitutes the primary sexual reproductive process in *Paramecium*. It occurs only between animals of different mating types. In Figure 3.20 an animal of mating type A is shown *in conjuguo* with one of type B. In this state they are attached along their oral regions. The details of the process are given in the legend to the figure. The net result of conjugation is that two animals, the exconjugants, are produced which are *heterozygous and identical in genotype.* However, they will not be identical in cytoplasms unless there is a complete mixing of the cytoplasms of the conjugating pairs across the bridge connecting the two. It can be shown that if the pairs remain *in conjuguo* for a long enough period some exchange does take place, but it is doubtful that that the exchange is ever complete. The exconjugants reconstitute new macronuclei, presumably of the same heterozygous genotype as their micronuclei, and they are then ready to multiply by binary fission.

After the achievement of heterozygosity by conjugation, *P. aurelia* ordinarily undergoes *autogamy.* This is an internal reorganization which results in homozygosity. The same process of macronuclear breakdown and micronuclear meiosis occurs in autogamy as in conjugation, but only a single animal is involved. The haploid micronuclei resulting from meiosis undergo a mitotic division and all but a pair of sister nuclei (derived from the same haploid nucleus by mitosis) degenerate. The two sister nuclei fuse to form a diploid which will obviously be homozygous. After this the steps are similar to those following conjugation. A new macronucleus is formed from the new micronuclei. After autogamy, animals which had earlier undergone conjugation are of one or another mating type and ready to undergo conjugation again.

Meiosis in *Paramecium* occurs in two different stages of the life cycle, conjugation and autogamy. Independent assortment and recombination of genes by crossing over can occur presumably at either meiotic event.

A number of other ciliate protozoa such as *Tetrahymena* and *Euplotes* have been employed in genetic investigation. The life cycles of these vary in details from *Paramecium,* but basically they appear to be somewhat similar.

Unicellular Algae

Many species of unicellular algae are adaptable to genetic work (605). Perhaps the one most utilized, and with the greatest success up to the present time, is *Chlamydomonas.* A number of species of this genus, such as *eugametos* (700, 701), *reinhardi* (596), and *moewusii*

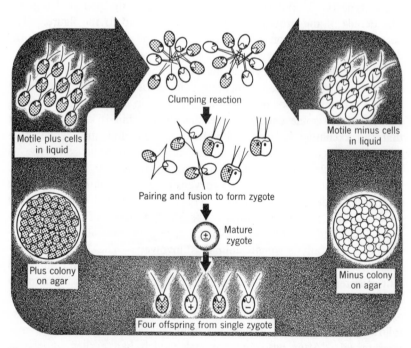

Figure 3.21. The life cycle of *Chlamydomonas reinhardi* showing the segregation of mating type, denoted by plus and minus signs, and of the marker y_1 (the dotted cells are y_1+, the undotted are y_1-). From Sager (872).

(604), have been used rather extensively. The life cycle of *C. reinhardi* is described here as representative.

Chlamydomonas is a motile biflagellate, uninucleate alga with a single chloroplastid. It carries out asexual reproduction by binary fission of the haploid vegetative cells (Figure 3.21).

The sexual cycle is extremely simple. When cells of different mating type are brought together, they clump and then pair off. The pairs fuse, the four flagella of the resultant zygote disappear, and the zygote matures for a period of about a week. Meiosis occurs within the zygote and results in the production of a tetrad of cells each of which grows a pair of flagellae and becomes a vegetative cell (Figure 3.21). Under the proper conditions any vegetative cell may apparently function as a "gamete."

Work with *C. reinhardi* has made it quite evident that this organism has a genetic system basically identical to that found in the higher organisms (597, 872, 240). A true meiosis with the production of reciprocal products from four-strand crossing over seems to occur.

Higher Plants and Animals

The genetic work done since 1900 with flowering plants such as maize and cotton, and animals such as *Drosophila* and the smaller mammals, forms the basis of our knowledge of present-day genetics. Their genetic systems are too well known to require a detailed discussion here, but a number of significant factors concerning them are considered for emphasis.

The active dominant phase of higher plants and animals is diploid. Meiosis occurs just prior to the formation of the gametophyte in flowering plants and results in the formation of the haploid gametophyte stages which produce the gametes (Figure 3.22). In this lies the chief difference in the life cycles of flowering plants and animals. In animals the haploid phase is transitory, existing only as sperm and egg.

Recombinations in plants and animals have for many years been assumed to occur in the prophase of meiosis by crossing over between linked genes and during anaphase for unlinked genes by independent assortment. Recently, evidence has been found in maize (651) that crossing over or at least some type of recombination mechanism may occur earlier in interphase at the time of chromosome duplication.

Plants reproduce asexually as well as sexually. Asexual reproduction occurs by runners, rhizomes, man-made cuttings, and so on. Some

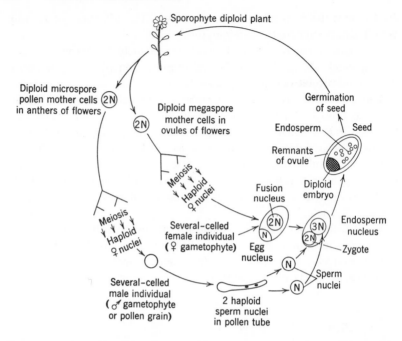

Figure 3.22. The life cycle of a flowering plant. The flowering plant is called a sporophyte because it reproduces asexually by means of spores which, although they never leave the mother plant, give rise to large numbers of haploid nuclei by meiosis. The diploid microspore pollen mother cells are produced in the male parts of the flower, specifically the anthers, and give rise, as shown in the diagram, to the pollen grains. Each pollen grain upon maturing contains three haploid nuclei, two of which are involved in the fertilization process. The diploid megaspore mother cells are produced in the ovule of the ovary of the flower and give rise to specialized structures within the ovule known as the embryo sac or the female gametophyte. The important parts of this structure are: (1) a haploid egg nucleus at one end and (2) the fusion nucleus which may be diploid or of higher ploidy toward the center of the structure. This nucleus results from fusion of two or more haploid nuclei within the embryo sac. In corn this fusion nucleus is always diploid.

Pollination consists in the pollen grain being deposited on the stigma of the female part of the flower and the growth of a tube from the pollen grain through the tissue of the stigma and the style, a structure which connects the stigma with the ovary, into the ovary where it comes into contact with the embryo sac. Thereupon two of the haploid nuclei within the pollen tube, designated as the sperm nuclei, enter the embryo sac and one of them fuses with the egg nucleus to form a zygote; the other fuses with the fusion nucleus to form the endosperm nucleus. The zygote nucleus by a series of mitoses gives rise to the embryo. The endosperm nucleus gives rise to many nuclei which together with the surrounding cytoplasm constitute the endosperm tissue. This tissue has a purely

plants have developed an asexual method of reproduction by apomixis which involves the formation of seeds without the egg being fertilized and nuclear fusion taking place. Animals, on the other hand, reproduce primarily by sexual reproduction, except for some of the lower forms such as Coelenterates. Among the insects and some of the primitive phyla, parthenogenetic development from unfertilized eggs is common.

General Observations and Conclusions

The discussion of genetic systems in this chapter has been limited primarily to those systems in phage and microorganisms which have yielded the most direct information concerning the molecular basis of heredity. This is a matter of expediency and should not be taken to mean that microorganisms are relatively more important than plants and animals in genetic research. From a more general biological and biochemical viewpoint the well-known genetic systems in such organisms as *Drosophila,* corn, mice, and man are at least as important as the microbial and fungal systems. It should be recognized that the genetic systems in microorganisms would probably not have been worked out had it not been for the prior existence of a large body of genetic information gained from the study of organisms such as *Drosophila* and corn. On the other hand, it should also be appreciated that the study of the genetics of microorganisms at the biochemical and molecular level has taught us a great deal about fundamental aspects of genetics that we would have found extremely difficult or impossible to discover first in the higher organisms.

The Genetic Material. A study of the transformation process in bacteria has brought to the fore the important fact that in those bacteria in which transformation is demonstrable, DNA is the genetic material. By extension we can assume, until it is proved otherwise, that this is true for bacteria in general. Analysis of phage replication has also made it apparent that DNA is the only significant genetic material in phage, for although some protein is injected into the host bacterial cell along with the DNA, there is no evidence that this protein acts as genetic material by incorporating a chemical specificity for the forma-

---

nutritive function. The embryo, the endosperm, and the remnants of the ovule within which the embryo sac develops, together constitute the seed. Upon germination of the seed, the embryo develops into a new sporophyte diploid plant. The remains of the ovary form the fruit, which is not shown in this figure.

tion of more protein, or even DNA. But all the evidence definitely indicates that only DNA carries such a specificity from which more phage DNA and protein are synthesized. Analysis of the viruses which do not contain DNA, but do have RNA, makes it evident the RNA is the genetic material in these entities, as will be made clear in Chapter 6.

No doubt now exists that the nucleic acids, DNA and RNA, have in their structural arrangement of linear arrays of nucleotides the capacity to function as the material determining genetic continuity, and to direct the metabolism of the cells within which they occur, by virtue of their control of specific protein synthesis as discussed in detail in Chapter 6.

The question of whether this is their universal function is now generally answered in the affirmative for these reasons. First, they are universally present in organisms, and particles such as phage that maintain a genetic continuity. Second, the role of DNA in the genetic continuity of bacteria, phage, and other viruses and the role of RNA in the RNA viruses can now be taken to be proved sufficiently well so that the burden of proof now rests with those who would hold a contrary view. Third, the assignment of the role of maintaining genetic continuity in the higher organisms which carry out meiosis is most logically made to DNA rather than protein or other macromolecules.

The basis for making this last statement is as follows. The chromosomes of all meiotic organisms that have been analyzed always contain DNA in addition to RNA and protein, but the DNA content per nucleus remains relatively constant in a single species, unless there is an increase in ploidy, whereas the RNA and protein contents vary over wide ranges depending on the phase of the cell cycle, the physiological state, and the kind of tissue. Second, an increase in ploidy generally causes a corresponding multiple increase in the amount of DNA. Thus the haploid sperm cells of an organism will have one-half the DNA found in the diploid cells, and triploid and tetraploid cells of the same organism will have $3\times$ and $4\times$ the DNA content of the haploid cells, respectively (680). Third, the turnover of the components of DNA, as measured primarily by following the exchange with isotopically labeled bases and phosphate, is extremely low as compared with RNA and protein in most types of cells (680). Finally, it may be stated that, with a few exceptions found primarily in egg cells, DNA is confined to the nucleus, whereas other macromolecular constituents, such as RNA and protein, are widely distributed in both nucleus and cytoplasm. Reports that DNA also exists in the chloroplastids of green plants (351) and possibly also in mitochondria have never been well supported.

In fact, it now seems clear that mitochondria do not contain DNA
(746). That some chloroplastids do contain DNA remains a possibility.

While none of these aspects of DNA constitute direct evidence
that it is the genetic material in the higher organisms, none of the facts
stated are in contradiction with DNA playing this role. Obviously,
the other two principal candidates, RNA and protein, cannot be entirely eliminated, but there are no compelling reasons for assigning such
a role to them at present. Of course, it should be recognized that even
if DNA is the genetic material by virtue of its acting as the physical
link between generations, it is not able to function properly unless
protein and RNA are present.

Protocaryotes and Eucaryotes. One of the major impressions that one
receives on making a survey of life cycles in the organic world is the
considerable gulf that seems to exist between the protocaryotes, or
premeiotic organisms, and the eucaryotes, or meiotic organisms.
Roughly, the phage, bacteria, and possibly the blue-green algae are
protocaryotic, and all other organisms are eucaryotic by virtue of their
possession of a defined nucleus with morphologically demonstrable
chromosomes, and, most important, a meiotic process. The existence
of these differences (which are difficult to evaluate at present because
so little is known about the precise structure of the genetic elements
at both levels, and practically nothing is known about the mechanics
of recombination and segregation at the protocaryote level) is real
enough to cause a certain amount of caution to be exercised in making
generalizations about all organisms based on what is learned in either
one of these two major groups.

Despite these strictures, however, it should be apparent that studies
in comparative genetics over a broad spectrum of the organic world
has yielded considerable information, and that much more is to be
expected as these studies are continued and extended.

4   Mutation and Its Effects

Mutation Defined

Mutation is an event occurring in cells which results in heritable change. The change may result in a new phenotype, or it may merely be manifested as some visible structural alteration in the chromosomes or other inherited particles within the cell. The term *mutation* describes any inherited change which is not due to segregation or the normal recombination of unchanged genetic materials such as occurs in sexually reproducing organisms. It should not be applied to noninherited environmental modifications, which are discussed in Chapter 10.

The capacity to mutate is as important a part of the properties of genetic material as its stability. Mutation is assumed to occur in all organisms and to be the basic source of all heritable natural variation. Therefore we shall consider here the process and its results in some detail, placing particular emphasis on those aspects of the subject which are generally believed to bear on the nature of genes and chromosomes.

A rigorous proof of chromosomal mutation is possible only in organisms with a sexual phase. However inherited changes which may be chromosomal occur in asexual forms as well as in the somatic cells of multicellular organisms. Somatic mutations may sometimes be demonstrated to be chromosomal by carrying the mutation over into the germ line, but in the bacteria and fungi in which only asexual reproduction has been observed it may only be inferred that the mutations are indeed of chromosomal origin from the results with organisms having sex cycles, or parasexuality.

The fact that mutations do occur in unicellular asexual species and the soma of higher forms is good evidence that the mutation process is one which involves a basic type of change not equivalent to segregation brought about by independent assortment and crossing over as part of the sexual process. Proof of the occurrence of mutations in sexual organisms may be had from experiments in which isogenic strains (all individuals completely homozygous for the same genes) are isolated from any possible contamination and closely observed for a number of generations to detect the occurrence of an inheritable change. East (239), who performed such an experiment with tobacco (*Nicotiana rustica*), found that after a period of time his homozygous population gradually reverted to the normal state of variability found in natural, inbreeding populations. By starting with a haploid strain of tomato which was doubled to produce the diploid, Lindstrom (618) produced a completely homozygous strain which he carried under close observation for ten years. In contrast to the results obtained by East, Lindstrom found a high degree of stability in the tomato diploid lines; nonetheless, spontaneous changes did arise which could only have been the result of changes in the genetic material of the chromosomes, since there was no heterozygosity to start with, and the changes were inherited according to a Mendelian pattern.

The Spectrum of Changes in Phenotype. When genetic material mutates, the effect on the phenotype may be anywhere from unobservable to an immediate lethal effect which causes the death of the cell within which the mutation occurs. Between these two extremes lie an almost infinite number of phenotypic conditions, and an almost equal number of terms to describe them. Here we shall restrict ourselves to definitions of only the more important and most generally used terms and the phenotypic conditions to which they refer, emphasizing convenience rather than fundamental differences.

Four general types of phenotypic effects—lethal, morphological, physiological, and chemical—are described most frequently by geneticists. The spectrum of effects may perhaps be best discussed by considering each of these in turn.

(1) *Lethal mutations* result in the death of the organism prior to the reproductive stage (388, 391). This may occur at any time from in the egg before cleavage to a time after the adult structure has been

formed.   Lethal mutations may be quite specific in their action producing death as a result of the cessation, or sometimes regression, of
development at specific stages.   Hence, *early-* or *late-*acting lethals may
be further defined on the basis of their action in biological time (see
Chapter 12).   Lethal mutations may be dominant or recessive.   Dominant lethal mutants are obviously impossible to maintain in stock, but
their occurrence is easily scored in many organisms and used as an
indication of mutation rate.   Recessive lethal mutants may be easily
maintained in artificial or natural populations, although the heterozygous
lethal may actually not be as well adapted as the homozygous normal,
that is, the lethal mutant gene may not be completely recessive to the
normal or wild type.

(2) *Morphological mutations* may affect the phenotype in many
ways ranging from changes in shape, growth habit, color, color pattern,
and size.   The term is used with reference to the higher plants, animals,
and microorganisms in which latter group it ordinarily applies to type
of colony formation.

(3) *Physiological mutations* are considered in terms of changes in
functions resulting in alterations of growth rate, ability to withstand
certain environmental conditions, and other types of reactions to the
environment which may range from temperature changes to chemical
stimuli such as antibiotics or salt concentration.

(4) *Chemical mutations* cause changes in the organism which are
readily traced to some chemical change.   Chemical changes may be as
simple as failure to produce a particular type of pigment or vitamin or
amino acid.   If the organism is unable to produce a specific substance,
such as an amino acid, by means of its own metabolism because of the
mutation, it is called a biochemical or nutritional mutant.   Such mutants require the addition of the essential compound to their growth
medium.   Frequently they are described as *auxotrophs* to distinguish
them from the wild type or *prototroph*.

As an example of nutritional mutants, mutants of *Neurospora crassa,*
an ascomycete fungus the life cycle of which has been described in
Chapter 3, may be considered.   The wild-type *Neurospora* will grow
on a medium consisting of inorganic salts, inorganic nitrogen in the
form of nitrate or ammonium, a carbon source such as sucrose, and
a single vitamin, biotin.   Nutritional mutants, or auxotrophs, will not
grow on this medium unless supplemented by one of the amino acids,
vitamins, purines, pyrimidines, or other organic compounds which may
be required.   Supplementation with the required compound gives a
response curve.   Some more or less typical growth-response curves for

an amino acid and a vitamin-requiring mutant are shown in Figure 4.1. Many mutants grow at nearly a wild-type rate and to about the same total dry weight as the average wild-type strain in the presence of the required metabolite, but a great many do not give normal growth even in the presence of an excess of a required metabolite. Many mutants have been isolated that grow only about 10% as well as the wild type.

As discussed in more detail in Chapter 7, the acquisition of a nutritional requirement is accompanied by a variety of physiological changes. A very common one is reduced viability. Many of the *Neurospora* mutants will not produce ascospores in self-crosses, and there is a great variation in the time of maturity of perithecia.

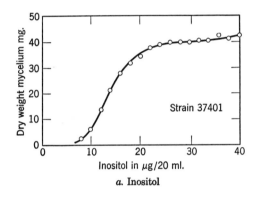

a. Inositol

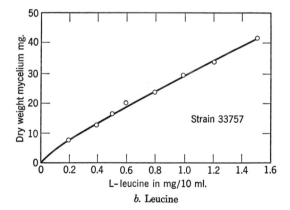

b. Leucine

Figure 4.1.   Growth curves for two nutritional mutants of Neurospora.   (*a*) Inositol; after Beadle (52).   (*b*) Leucine; after Regnery (828).

Neurospora is by no means the only organism in which an intensive study of nutritional mutants has been made. Among the other fungi, *Ophiostoma* (300, 301), *Aspergillus* (797), *Ustilago* (779), and *Penicillium* (84) have all yielded nutritional mutants of types identical or related to those found in *Neurospora*. The Lindegrens (616) and others (794, 416) have induced and utilized nutritional mutants in yeast, and many workers (202, 203, 569, 940) have found *E. coli* and other bacteria to be particularly useful organisms from which to obtain nutritional mutants for use in the study of metabolic patterns (see Chapter 7).

It should be recognized that this classification of types of phenotypic changes brought about by mutation is highly artificial and used only as a matter of convenience. Basically, all mutations cause a chemical change in the organism of some type which may become manifest in one or more of thousands of ways in the phenotype. Furthermore, more than one aspect of the phenotype may be affected by a single mutation. For example, mutation of the gene *P* to the allelic p^s in the mouse results in: (1) a decrease in pigment in the hairs; (2) slightly uncoordinated behavior; (3) difficulty in chewing; (4) sterility in males; (5) premature senility; and several other conditions (450). Plural phenotypic effects such as these are ordinarily referred to as *pleiotropy*. Pleiotropy is not uncommon and, in fact, is to be expected.

The vast majority of mutations which are observed to occur in both plants and animals are generally considered to have detrimental effects on the phenotype. This is to say that under natural conditions in the organism's natural environment mutant deviations from wild type are disadvantageous and selected against. Of course, under certain conditions which are "unnatural"—cultivation by man for his own purposes, for example—a mutation may be advantageous, whereas it would not be so "in nature."

Types of Mutations. Two general classes of mutations are recognized by geneticists. One class is associated with visible chromosomal changes and the other with phenotypic change unaccompanied by visible change in the nuclear material. Mutations in the first class are referred to as *chromosomal mutations* or aberrations, and they may be divided into a number of subgroups, depending on the type of structural change. Some of the different types are illustrated in Figure 4.2. Not all these mutations result in detectable phenotypic changes in the organism in which they occur, but it is highly probable that even those which do not are of importance in providing the genetic variability which is potentially capable of manifesting itself as phenotypic variability under

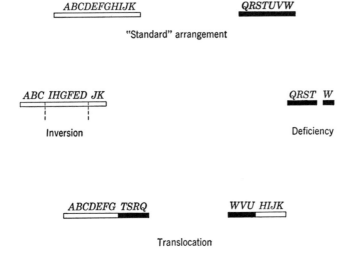

Figure 4.2. Types of chromosomal aberrations.

the proper conditions. Mutations of the second class, occurring as they do in the absence of any visible alterations in chromosome structure, are considered to be the result of changes at the submicroscopic, chemical level and are described, therefore, as *gene mutations* or *point mutations*. By definition, when a gene mutates it is transformed from one allele to another at that locus.

Some chromosomal mutations are considered to produce phenotypic change as a result of rearrangement of the spatial relationship of genes to one another on the chromosomes. They are therefore arbitrarily described as *extragenic* changes, with the implication that there is not necessarily any modification of the chemical constitution of the chromosome which would in itself result in a new phenotype. They are for this reason distinguished from gene mutations, which are assumed to be *intragenic* in nature.

In *Drosophila,* translocations and inversions frequently result in somatic instability in the expression of certain genes which have been moved to new regions as a result of the rearrangements. This instability is generally expressed as a variegation; that is, some of the cells show mutant characteristics while others around them appear normal. Since it is a variegation resulting from the change in position of genes relative to one another on the chromosomes, the phenomenon has been called *position effect* of the variegated type (V-type) by Lewis (608).

Figure 4.3 illustrates an inversion in the X-chromosome and a translocation between the X and fourth chromosome which results in variegated-type position effects. In the inversion ($N^{264\text{-}52}$) the wild-type genes rst^+, fa^+, dm^+, ec^+, bl^+, and peb^+ have been moved to the right end of the X-chromosome and situated next to the heterochromatin at that end. Five of them produce a variegated phenotype, as a result, while peb^+, the farthest from the heterochromatin, appears unaffected in its expression. Owing to the X-4 translocation, $w^{258\text{-}18}$, the genes w^+ and rst^+ are brought into close proximity to heterochromatin of the left arm of 4, and both of them exhibit an instability in their expression.

In all these examples of variegation it is difficult to determine whether

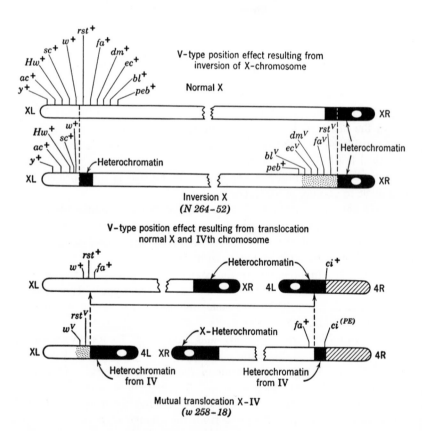

Figure 4.3. Types of chromosomal aberrations that result in V-type position effects. After Lewis (608).

the genes mutate in certain somatic cells to produce the mutant phenotype or their expression is modified by other circumstances. There is a close relationship between the production of variegation and the location of the dislocated gene near the heterochromatin, but it cannot be asserted that location next to the heterochromatin is necessary to produce variegation (368), nor that all genes moved to an abnormal position next to the heterochromatin respond by showing an unstable expression.

Since the rearrangements resulting in position effect are nearly always produced by radiations such as x-rays, there exists the possibility that the relocation of the gene alone is not responsible for the mutant effect, but that the gene's structure has been changed independently of the rearrangement. Proof that this is not so, in some position-effect rearrangements at least, has been obtained by moving known + alleles into a translocation by crossing over with the production of a V-type position effect (235a, 499). Moreover, the position effect may be caused to revert to normal by further rearrangements. Thus the roughest phenotype, rst^3, associated with an inversion may be caused to revert to normal rst^+ by reinverting the inverted segment back to the normal position (255, 375). In general, the position effect disappears if genes showing position effect are moved by further rearrangement to a euchromatic region (401).

V-type position effect has also been described for the evening primrose, *Oenothera blandina,* by Catcheside (140, 141) and in the mouse by Russell (863). It is probable that the phenomenon is more widespread than indicated at present, and therefore possibly an important source of variation.

In addition to variegated phenotypes, rearrangements involving inversion and translocation may also produce lethal effects and cause sterility in *Drosophila melanogaster* (722). The rearrangements involving the fourth chromosome (see Figure 4.3) frequently have an effect on the expression of the ci^+ gene, which is involved in normal wing venation. When a break is produced in the heterochromatin near ci^+ and a translocation results, the ci^+ gene responds by giving a mutant phenotype in heterozygotes with the mutant allele *ci.* Under normal conditions ci^+ is dominant to *ci* (see Chapter 8).

Cytologically detectable deficiencies in chromosomes invariably produce "visible" mutations or lethality. Some cause a dominant mutant phenotype when heterozygous and act as lethals when homozygous. In a very few cases undoubted deficiencies may be viable homozygous and produce a recessive mutant phenotype (see Chapter 8). Duplications, on the other hand, of single genes or groups of genes may result in lethality or in mutant phenotypes.

To summarize: (1) a mutation accompanied by phenotypic effects may be intragenic (a point or gene mutation); (2) it may be due to obvious structural changes, such as loss or duplication of genetic material; or (3) it may be due to rearrangement of the position of genes resulting in position effect, which may be an *extragenic* change. Actually, no methods are available for always making satisfactorily clear distinction among these types. In organisms such as *Drosophila* and maize, in which certain cells contain chromosomes large enough for detailed microscopic analysis of possible structural changes, uncertainty may still exist as to whether one is dealing with an intra- or extragenic mutation or deficiency mutation, for many minute rearrangements or losses may lie well beyond the limits of observation with even these relatively large chromosomes. We can therefore only assume that a mutation may be an intragenic change when there are no visible chromosomal aberrations and when the change is inherited as a single gene. This important point was most ably stated by Stadler (952):

Mutations in which the altered phenotype is produced by a gene mutation (that is, by the production of a new gene form) cannot be distinguished from these extragenic mutations by any positive criteria. All observed gene mutations therefore are merely presumptive; we can only say that a new allele seems to have arisen, because we cannot detect any of the various extragenic phenomena that might have produced the mutant effect observed.

The significance of this difficulty in establishing the structural basis of mutations becomes evident when it is recognized that our knowledge of genetics and our recognition of genes is based to a very large extent on the study of mutations. A gene is recognized because it has an allelic form which gives a different phenotype. The action of the gene in determining the phenotype is ordinarily deduced from a comparison of contrasting phenotypes.

The Detection of Mutation

Mutant strains are the working material of the geneticist. Without them he can do practically nothing. Many of the mutants found in the higher plants and animals have been found by chance and not by any organized effort. Methods exist, however, for obtaining mutants of specific types, particularly in microorganisms, in large numbers by certain types of screening methods. It is particularly important to understand certain of these methods, because many of the conclusions drawn by the geneticist regarding gene mutation and action are influenced by the operations he uses in obtaining his raw material, mutant genes.

When the objective is to obtain mutants for further experimental exploitation, selective techniques are generally employed for the screening of the desired type of mutant. For example, if one desires to obtain nutritional or biochemical mutants of *Escherichia coli,* one uses the penicillin technique developed originally by Davis (201) and Lederberg and Zinder (575). This method depends on the fact that penicillin-sensitive bacterial cells are not sensitive to penicillin unless they are in active growth. Mutations are induced in the cells by some mutagen, the treated cells are allowed to proliferate for several generations to allow the mutations to be expressed, and then the cells are treated with penicillin on minimal medium. All cells with no more than the minimal nutritional requirements start growth on this medium and as a result are killed by the penicillin. The nutritional mutants, however, cannot grow, and as a result they survive. Removal of the penicillin, either by washing or the action of penicillinase, enables one to test the screened mutants free from "wild types" and nonnutritional mutants. Further testing of the screened nutritional mutants to determine their specific requirements may be done by the replica-plating technique employing a velvet stamp as described in Figure 4.4. If a

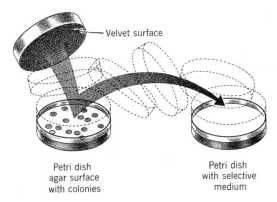

Velvet surface

Petri dish
agar surface
with colonies

Petri dish
with selective
medium

Figure 4.4. The printing method for testing colonies of bacteria or yeast on selective media.

specific type of mutant is sought, for example one requiring pyrimidines, many of the other types of nutritional mutants may be eliminated by adding all known growth factors with the exception of pyrimidines to the penicillin medium. By such means as these hundreds of mutants having very specific requirements may be obtained with little expenditure of effort.

A number of techniques have been developed for the isolation of nutritional mutants of *Neurospora,* as well as other fungi, and it is now a simple matter to obtain a great variety in quite a short time. The original method devised by Beadle and Tatum (58) and the procedure of Lein et al. (585) both make use of ascospore isolations. By these procedures conidia from one mating type are treated with a mutagenic agent in a dosage sufficient to produce more than about 90% killing. These conidia are then used to fertilize protoperithecia of the opposite mating type. After mature perithecia are obtained, one spore is removed from each perithecium and transferred to a test tube containing a complete medium (i.e., the minimal one which will support growth of a wild type plus a complex mixture of metabolites such as are provided by yeast extract, hydrolyzed casein, and nucleic acids). Cultures thus obtained are tested for growth on the minimal medium, and if they do not grow in the absence of supplement they are retained as nutritional mutants.

By the Lein et al. (585) procedure, crosses are made on petri dishes, using treated conidia, and after spores begin to be ejected from the mature protoperithecia (2–5000 per plate) the plates are inverted for a short interval over plates of agar minimal. The several thousand spores thus collected on the minimal plate are then heat-activated and allowed to germinate and grow for about 20 hours. At this time mutants can be distinguished from wild types as shown in Figure 4.5. Those assumed to be mutants are then transferred to a complete medium and tested back on minimal as already described. Specific nutritional requirements of the mutants are then ascertained by systematic tests on minimal medium plus individual pure metabolites such as amino acids or vitamins.

It should be noted that most of the mutants obtained in experiments like these are ordinarily discarded on the basis that they do not fulfill the arbitrarily chosen criteria for desirable nutritional mutants. Some grow too much on the minimal medium; some revert after being subcultured; and many are slow in growth and not sufficiently stimulated on a complete medium. This selection is shown clearly by data obtained during isolations of nutritional mutants by both of the foregoing

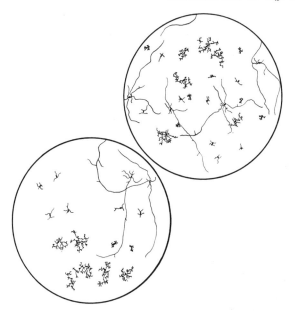

Figure 4.5. Germinating ascospores of *Neurospora crassa*. Camera lucida draw-ings showing the segregation of two genes in a cross of *Neurospora crassa* mutant 70007 (colonial) and 38502 (pyrimidine). Spores were isolated on minimal agar medium in petri dishes, heated at 60° for 30 minutes for activation, and incubated at 25° for 18 hours. This nutritional mutant (pyrimidine) grows sufficiently on minimal medium to permit identification of *pyr-co* double mutants.

The figure at the lower left shows segregations in two asci with spore pairs isolated in order. From top to bottom the two asci show, respectively, genotypes *pyr, pyr, co, co,* and *wild, pyr, co-pyr, co.* The figure on the right shows results from plating spores from the same cross at random. Genetic analyses can be made directly, using either procedure or by transferring geminated spores to appro-priately supplemented culture tubes for further testing. Much larger numbers can be observed conveniently using the random method.

methods. For example, using the original method (58), 7049 asco-spores germinated out of 8795 isolated in one series of experiments. Of these, 54 gave rise to clear-cut nutritional mutants. A much greater number, 489, were distinguishable from wild type because of their slower growth on complete medium. Of these, 159 did not grow sufficiently to permit testing. The remaining 330 were discarded be-cause they did not give sufficiently clear responses to added nutrients.

These same principles undoubtedly apply in procedures for isolating nutritional mutants using direct platings of asexual spores of *Neurospora* as well as in similar methods used in other organisms. With *Neurospora,* nutritional mutants have been obtained directly by plating treated macro- or microconidia on minimal medium containing sorbose to induce colonial growth (999). After wild-type conidia have grown, nutrients are added to the plate, and new colonies that appear are isolated. In another adaptation of this method, the wild types are permitted to germinate in liquid minimal medium (1106). They are then filtered off, the remaining spores are plated out, and the mutants are isolated and identified as described previously. These methods permit testing a very large number of individual conidia, and they are especially useful in selecting strains that have desired nutritional requirements. To select for a desired nutritional mutant it is only necessary to plate the filtered spores on minimal medium plus the growth factor for which mutants are desired. All colonies which grow should be of the desired type. A modification of this technique by Lester and Gross (591) allows a more efficient disposal of wild-type spores which escape through the filtration procedures, because they germinate late. The procedure, known as "inositolless death," depends on the rather paradoxical finding that double nutritional mutants may be maintained alive, although without active proliferation, in minimal medium longer than single inositolless mutants, which tend to die rather rapidly in minimal medium. This is particularly true of mutants with a requirement for inositol. Hence, when conidia of an inositolless strain are irradiated or treated with chemical mutagens, double (i.e., inositol plus some other newly induced nutritional mutant) mutants survive more readily than the inositolless conidia which are otherwise wild type nutritionally.

Techniques which screen for a desired mutant initially in the presence of the required compound are in one important respect superior to the older methods which utilized a preliminary screening in the presence of a mixture of known growth factors. It has been observed that a number of types of nutritional mutants are inhibited by constituents of "complete" media, and it is frequently necessary to select for them directly by using minimal medium containing one metabolite as the only adjunct.

When studying mutation rates or mutation as a process without any desire to obtain specific types of mutants for further biochemical or physiological work, a somewhat different approach than those just described is taken, although the basic techniques are the same. To study mutation rate, "back" or "reverse" mutation rather than "forward" mutation is frequently tested. This method is particularly useful with nutritional mutants of microorganisms such as bacteria and fungi. Cells of a particular auxotroph are treated with a mutagen and then plated on a minimal medium. If back mutation from auxotrophy to prototrophy (wild phenotype) has occurred, those cells in which genes have back mutated at the specific locus will form colonies. This method has been used with great success in *Neurospora* by Kölmark and Westergaard (1067) for studying mutation rates in the presence of various chemicals. The major danger inherent in the method is that prototrophs may be obtained from auxotrophs, not only by reverse mutation but by mutation of another gene which suppresses the mutant gene (see Chapter 9). The suppressor versus the reverse mutation origin of prototrophs can be resolved only by making the appropriate crosses as described in Chapter 9. Reverse mutation from inability to ability to reproduce in a given host has been studied in the T-series of bacteriophage (295, 296).

Reverse mutation rate can also be studied in higher organisms despite their diploid condition, because the "wild type" is generally dominant to the mutant. Hence any wild-type *Drosophila* appearing among the progeny of two recessive mutant parents can be assumed to have arisen by

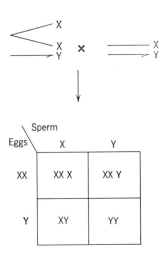

Figure 4.6. Results obtained with attached X (X͡X) female *Drosophila melanogaster.* The males produced will always carry the X-chromosome of the paternal parent rather than the maternal parent.

mutation either at the mutant locus or at a suppressor locus. The latter possibility is less likely, however, since suppressors are generally recessive. A more accurate method in *Drosophila* employs attached-X females. Male progeny from a cross between attached-X females and normal males receive their X-chromosomes from their fathers as shown in Figure 4.6. Any mutation occurring on the X-chromosome in the male parent which gets into a sperm cell may be detected in the male offspring whether it is recessive or dominant, since the male is haploid for most X-chromosome genes. This method has been used extensively by Green (360, 364) to demonstrate back mutation at a number of different sex-linked loci in *Drosophila melanogaster.*

By far the most powerful tool for the study of mutation rate in any organism is the method for detection of sex-linked recessive lethals in *D. melanogaster* developed by H. J. Muller. Two variants of this method are in use, the ClB test and the Muller-5. The tests are similar in principle and only the Muller-5 is described here, because it is the most useful. Muller-5 refers to a specially constructed X-chromosome which carries two inversions and the markers, apricot (w^a) and Bar (*B*). Figure 4.7 describes the test. By means of it, the X-chromosome recessive lethal mutation rate in the male germ cells can be determined with a high degree of accuracy (936). Historically, the method of determining X-chromosome lethals in *Drosophila* is of great importance, because the development of the technique by Muller was the first major step toward his demonstration of the mutagenic effects of x-rays. With-

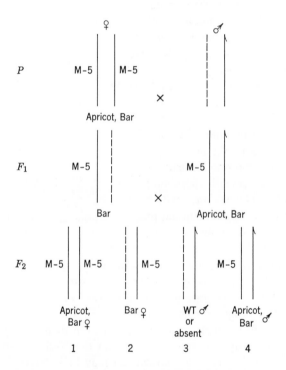

Figure 4.7. The Muller-5 technique. The Muller-5 females carry apricot (w^a), a recessive, and Bar, a dominant, as well as inversions which prevent crossing over with non-Muller-5 chromosomes. If a new lethal arises on a male test X-chromosome all class (3) males will be absent in the F_2. If a new visible arises all class (3) will show it. Class (2) females will carry any recessive lethal from the original male parent so that it is not lost and may be tested further.

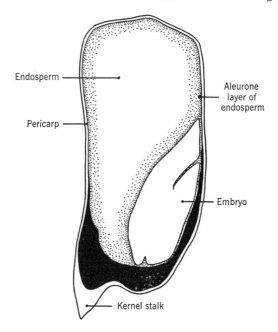

Figure 4.8.   A diagram of a corn kernel.   After Fuller, *The Plant World,* Henry Holt and Co., New York.

out a *quantitative* method such as this, such a demonstration would have been quite difficult and unconvincing.

Maize is another organism in which it is easy to make observations on the occurrence of mutations.   This is particularly true of those genes which control the coloration of physical appearance of the endosperm or pericarp of the kernels (Figure 4.8).   Mutations of either dominant or recessive genes can be recognized immediately, if the endosperm is affected and the cross producing the kernel is properly constituted. Since each kernel is the result of a fertilization by a single pollen grain, the mutation rate in the pollen can be measured simply by scanning the kernels on the cobs.   By this means millions of pollen grains may be tested for mutation of certain genes with much less labor than would be involved even with sperm in *Drosophila.*

Spontaneous Mutation Rate

The study of mutation as a process is ordinarily restricted to (1) considering the types of visible structural rearrangements of the chromosomes, (2) observing changes in phenotype, (3) determining the rate at which mutations occur under various conditions, and (4) studying the process at the level of chromosome chemistry. The more important of these is the study of mutation rate and the factors which influence it, particularly those which increase it over the normal or spontaneous rate. Before considering such factors, however, the spontaneous rate as such will be considered first.

The "normal" mutation rate is that obtained under ordinary conditions of observation without the use of abnormal external agents or conditions, such as extreme temperatures, high intensities of radiations, and unnatural chemical conditions. It is generally described as being a spontaneous rate, although there is no necessary implication that spontaneous mutations occur completely independent of influences outside of the genetic material which undergoes mutation. The spontaneous rate is what one might assume to be an approximation to the mutation rate in any natural population of organisms. It is usually stated to be low, but this needs to be qualified as to whether one is speaking of the rate of mutation of a specific gene or of the total mutation rate of all the genes in the genome. According to Muller (721), the average mutation rate per gene is about 1 in 100,000 to 1 in 1,000,-000 in any given cell cycle (generation) of *Drosophila melanogaster*. Thus, for each gene, mutation is a rare event. On the other hand, the overall rate in this organism is probably about 1 mutation in 10 gametes to 1 in 30, if all genes and all possible mutations are considered.

Spontaneous mutation rates have been determined for a number of specific genes in several organisms, and, although the data are not extensive, they clearly show that not all genes in the same species mutate at the same rate (see Table 9). Furthermore, the overall average rate is different for different species and may vary within different populations of the same species. There seems to be a rather great range of mutabilities, ranging from those genes which are not observed to mutate at all to those which mutate at excessively high rates.

It is to be expected that the number of mutations accumulated by a cell will be proportional to the age of the cell. This expectation has been verified by experiments with plant pollen and seeds and with *Drosophila* sperm. Plant seeds stored for long periods show a large increase in the number of accumulated mutations when compared to unaged seeds (198). Stubbe (982), for example, detected an increase from 1.5% to 14% in recessive mutations in snapdragon plants (*Antirrhinum majus*) derived from seeds aged from five to ten years. The aging of *Drosophila* sperm in the male or in the sperm-storage recepta-

Table 9. Spontaneous Mutation Rates at Specific Loci

Organism	Mutation	Number Gametes Tested	Frequency per 10,000	Reference
Maize	$Wx \rightarrow wx$	1,503,744	0.000	948
	$Pr \rightarrow pr$	647,102	0.11	948
	$Sh \rightarrow sh$	2,469,285	0.012	948
	$Su \rightarrow su$	1,678,731	0.024	948
	$I \rightarrow i$	265,391	1.06	948
Columbia	$R^r \rightarrow r^r$	20,984	6.2	950
Cornell	$R^r \rightarrow r^r$	43,416	18.2	950
D. melanogaster	$+ \rightarrow$ yellow	51,380	1.16	327
(male germ	$+ \rightarrow$ brown	102,759	0.29	327
line)	$+ \rightarrow$ ebony	102,759	0.20	327
	$+ \rightarrow$ eyeless	102,759	0.58	327
	$+ \rightarrow$ yellow	70,000	2.7	483
	$+ \rightarrow$ white	70,000	0.29	483
	$+ \rightarrow$ lozenge	70,000	0.29	483
Man	Retinoblastoma		0.18 to 0.06	649
	Congenital aniridia		0.04	903
	Achondroplasia		0.14	903

cles of the female results in an increase in the number of lethal mutations of the X-chromosome, as shown in Table 10.

Although these data clearly indicate that mutations accumulate in
time, they do not necessarily mean that the time rate of occurrence of
mutations is constant. Olenov (753) has shown that *Drosophila* raised
on a deficient food medium which increased their developmental period
from the normal 10 days, obtained on complete medium, to 30 days
gave sperm with approximately the same number of X-chromosomes
with lethal mutations as those with the shorter period of development.
The simplest explanation for these facts seems to be that the occurrence
of mutations is dependent on physiological factors, and that mutations
are not simply accidents occurring independently of internal environmental factors. Muller (720) and Lamy (545) have reported experimental results in agreement with this hypothesis. They found that
the rate of sex-linked mutations in *Drosophila* is different in different

Table 10.   Spontaneous Rate of Occurrence of Sex-Linked Lethals in Young and Aged Sperm of *Drosophila melanogaster*

Sperm Not Aged			Sperm Aged 15 to 20 Days			
Number Tested	Number Lethals	Per Cent Lethals	Number Tested	Number Lethals	Per Cent Lethals	Refer- ence
9,751 *	10	0.102	8,637	21	0.243	1009
13,481 *	14	0.104	18,659	49	0.263	819
3,545 †	5	0.141	3,471	11	0.317	514

* Stored in male.
† Stored in female.

stages of the germinal cycle.   For example, the sperm accumulated during the preimaginal life of a male show a two to three times greater mutation frequency than those produced six to nine days later.   In agreement with the results shown in Table 10, Muller did find that the mutation frequency rises considerably in sperm stored in the adult male, but this increase is not sufficient to account for the high frequency obtained in the very young males.   The X-chromosomes from females when tested for lethals show an almost constant frequency of lethals whatever the age of the parent, which would indicate that mutations occur only in the preimaginal life of the females.   Muller (720) concluded, therefore, that most mutations, particularly in the female, arise at some definite stage in development, such as early cleavage.

It is clear that at least in *Drosophila* the time rate of mutation is not constant in all cells of the individual during its life span, and that it probably varies not only from one stage of development to the next but even within the different stages.   It is also evident that the term mutation rate, as used by the geneticist, is best considered as being the number of mutations per generation, unless the experimental conditions are such that it can be shown that the rate for a specified period during the generation time can be measured.

It has been assumed by many that cell division is required for mutation (979).   This assumption is based on the hypothesis that mutation results from errors in gene duplication.   This may be true for some mutations and some genes, but it does not seem to be a general explanation for mutation, or its time of occurrence.   If it were true, then the mutation rate should be proportional to the rate of cell division.   This is not true for *E. coli* maintained in a chemostat (744, 541). Novick and Szilard (744) showed quite clearly that the number of mutations to phage resistance per generation in this organism is not constant, but increases proportionately with the absolute time. Thus the longer the generation time, the more mutations that accumulate per generation. As for *Drosophila,* it has been shown with a fair degree of certainty that mutation can occur in the already existing gene without the intervention of cell division (725). None of these comments should be taken as evidence that mutations do *not* occur at the time of gene duplication, however. It is possible, for example, that mutation may occur in viruses only at the time they are duplicating in the host, but even this point has not been proved.

Genetic Factors Influencing the Spontaneous Mutation Rate

Laboratory stocks of *D. melanogaster* collected from different sources have long been known to show different rates of mutation (717, 724, 598). Table 11 lists a number of stocks which have been analyzed for lethal mutations, particularly sex-linked recessives. It will be noted that the range for sex-linked recessives runs from as low as 0.07% for the Oregon R strain to as high as 1.09% for the Florida stock. The increase in rate is probably general for all chromosomes, for a high rate in the Florida stock is found for both the X and second chromosomes, as compared to the somewhat lower rate in the Lausanne stock. Several of the highly mutating strains have been analyzed in detail. Demerec (211) carried out genetic studies with the Florida stock which revealed that a recessive factor on the second chromosome was responsible for the high rate of 1.09% in this strain. Elimination of the factor from the stock resulted in a lethal rate of 0.074%, almost a fifteen-fold difference in the strain carrying the factor homozygous. Similar mutator genes have been described in *D. melanogaster* by Neel (736) and Ives (483), and by Mampell (652) in *Drosophila persimilis*. Mutator genes have even been discovered in the bacterium, *Escherichia coli* (1019) and *Salmonella typhimurium* (698) in which latter a gene has been located by transductional analysis. It increases the spontaneous mutability of most of the loci tested in *Salmonella,* although to different degrees.

In addition to mutator genes, other genetic factors may be involved in increasing the mutation rate in *D. melanogaster*. The presence of multiple-inversion chromosomes causes an increase in the spontaneous

Table 11. Spontaneous Mutability in Different Stocks of *D. melanogaster* Raised at 22°–25°C

Sex-Linked and Second Chromosome Recessive Lethals

Stock	X-Chromosome			Second Chromosome			Reference
	Number Tested	Number Lethals	Per Cent Lethals	Number Tested	Number Lethals	Per Cent Lethals	
Florida inbred	2,108	23	1.09	...	...	...	211
Wooster	1,266	8	0.63	...	...	...	211
Oregon R	3,049	2	0.07	...	...	...	211
Florida No. 10	916	10	1.09	516	9	1.74	791
Lausanne	955	2	0.21	436	3	0.69	791
Leningrad	8,614	14	0.16	...	...	...	1138
Sukhami	2,309	24	1.04	...	...	...	1138

lethal-mutation rate in normal chromosomes coupled with the inversion chromosomes in heterozygotes (1007). This phenomenon has been verified for the second and third chromosomes in the presence of inversions on these chromosomes. When second- or third-inverted chromosomes are present, they affect only their normal homologues and not the X-chromosome.

Perhaps the most interesting examples of genetic control of the mutation rate have come from the investigations made on maize. These studies not only have confirmed the earlier finding from the work on *Drosophila* that the rate of mutation of some genes is under the control of other genes, but have also provided further data supplemented by cytological observations which may eventually lead to an understanding of the phenomenon of spontaneous mutation.

Historically, the work of Rhoades (833, 834) was the first important investigation in this area. Rhoades discovered that the a_1 gene of the A_1 series of alleles in maize, although ordinarily very stable, can be caused to mutate at a high rate to other alleles in the series. This gene is located on the third chromosome, and when a_1 is present homozygous, no anthocyan pigment is formed in the aleurone of the endosperm or in the plant (see Figure 4.8). However, in the presence of dominant *Dt,* a gene on the ninth chromosome, it mutates to the other alleles in the A_1 series which are dominant to a_1 and allow the production of anthocyan pigment. Mutation of a_1 occurs both in the germ cells and in the somatic tissues. Somatic mutations show up as a variegation of small spots of anthocyan in the aleurone and as narrow stripes of pigment in the plant parts. No other allele in the series is affected by the *Dt* gene in this fashion. The mutation rate of a_1 to the higher members, such as A_1 (which produces the greatest amount of pigment), is most easily measured by observing the appearance of colored spots in the aleurone of genotype *aaa Dt Dt Dt.* During development of the endosperm tissue, the a_1 genes mutate to A_1, starting off centers of growth of colored tissue which become visible as spots when they become large enough. Each spot is assumed to arise from a single mutation.

The fact that the spots are usually of about the same size shows that the mutation takes place at a definite period of development, and the relatively small size of the spots proves that this period is quite near the end of the development of the aleurone. The same is true in general for the other tissues in the plants, including the sporogenous tissue in the anthers which leads to the production of male gametes. There are interesting dosage effects. As the dosage of the *Dt* allele is increased from *Dt dt dt* to *Dt Dt Dt* in the aleurone, the number of mutations observed per seed increases as follows:

dt dt Dt gives 7.2 mutations per seed
dt Dt Dt gives 22.2 mutations per seed
Dt Dt Dt gives 121.9 mutations per seed

The increase in the number of sensitive a_1 alleles in the presence of a constant dosage of *Dt* causes, as would be expected, a linear increase in the appearance of mutations.

That the *Dt* effect is not restricted to the a_1 allele of Rhoades has been demonstrated by Nuffer (747) who has shown that another allele, $a^m - 1$ derived from A^r:*Cache* mutates to a variety of different dominant and recessive alleles in the presence of *Dt,* but is stable in its absence. It is different from Rhoades' a_1 because it mutates 67 times more frequently in the presence of *Dt dt dt.*

The significance of factors such as *Dt* which seem to control action via mutation of other genes, but yet seem to have no direct phenotypic effect themselves, has been considerably extended by the work of McClintock on what she has called *controlling elements* in maize (638, 639, 640). Controlling elements are elements associated with the chromosomes but which do not remain at one position in the chromosome complement. They may move about from one part to another of the same chromosome or from one chromosome to another. Their existence is recognized by their action as moderators of gene action. When a controlling element is inserted at or near the locus of a particular gene, it may act as a modifier of the gene's action or a mutator. When the controlling element, *Ac* (Activator) for example, is inserted at or near the bronze locus (*Bz*) on chromosome 9, the phenotypic effect is that of the recessive *bz* at this locus, but the genotype is considered to be *BzAc.* (In the presence of *Bz* a purple anthocyanin pigment develops, both in the plant parts and in the aleurone layer of the endosperm. The recessive, *bz,* gives a bronze rather than purple color in these areas.) It can be shown that mutations occur at this locus when *Ac* is displaced from the *Bz* locus, *BzAc* → *Bz* + *Ac,* to another region of the genome, or is lost from the cell. In this event, the phenotype changes from bronze to purple. If this occurs during the development of aleurone of the genotype *BzAc/bz/bz,* variegation, that is, purple spots on a bronze background, will result. Each spot will have developed from a cell in which *Ac* was transposed from its site at the *Bz* locus. The *Bz* thus freed of *Ac* is stable in continuing to give a purple phenotype. The "mutation" may be transferred into the germ line and carried through successive generations as *Bz. Ac* is not confined to acting on *Bz* alone, but it has a similar activity on a number of other genes in maize. In fact, it appears that in the presence of *Ac* genes not only immediately adjacent to it, but those located some distance to either side of it, may be influenced by its proximity.

In addition to its direct action on genes, *Ac* also controls other controlling elements which are grouped under the term Dissociation (*Ds*). The *Ds* elements are of several types, distinguishable by their ability or inability to cause chromosome breaks and their direct effects on genes. In the absence of *Ac*, *Ds* cannot be detected, because it has no effect. In the presence of *Ac*, however, an *Ac-Ds* system is constituted in which *Ds* acts very much like *Ac* alone, as described earlier for the bronze locus. *Ac* controls the action of *Ds*, with respect to both time and location. For example, the higher the dose of *Ac*, the later in time of development does the alteration caused by *Ds* occur. Like *Ac*, *Ds* can be transposed from one site to another. *Ac* does not need to be situated adjacent to *Ds* to control it, but *Ds* must be at or adjacent to a locus to influence it. The *Ac-Ds* system acts similarly to the *Dt* system with respect to the A_1 locus. The recessive allele, a_1^{m-3}, gives, like the standard a_1 just described, a kernel and plant without anthocyanin pigment. In the absence of *Ac* it is completely stable. In the presence of *Ac* it mutates regularly to full A_1-type expression (i.e., anthocyanin in kernel and plant), caused by the transposition of *Ds* from the A_1 locus to other parts of the genome.

The similarity between the a_1-*Dt* system of Rhoades and the a_1^{m-3} *Ac-Ds* system is evident. McClintock has been prompted to suggest that *Dt* is comparable to *Ac* and that a *Ds*-type element is associated with a_1. *Dt* is not identical to *Ac*, however, because it will not replace *Ac* in the a_1^{m-3} *Ac-Ds* system.

Other controlling elements have been discovered in addition to *Ac*, *Ds*, and *Dt*, such as the *Suppressor-Mutator* (*Spm*) by McClintock (639), *Modulator* (*Mp*) by Brink (115, 113), and *Enhancer* (*En*) and *Inhibitor* (*I*) by Peterson (782). *Modulator*, when associated with the pericarp gene, P^{rr}, to form the complex $P^{rr}Mp$, produces a variegated pericarp, red spots on a light background. When P^{rr} is present alone, it produces a self-red pericarp and cob. Evidently, *Mp* is similar, if not identical, to *Ac*. It is assumed that *Mp* suppresses the action of P^{rr} in the $P^{rr}Mp$ complex, and that the variegation is produced when transposition of *Mp* to other sites occurs. Transpositions to many different sites have been demonstrated (1030). Like *Ac*, *Mp* has a dosage effect. $P^{rr}Mp + tr\text{-}Mp$ (transposed *Mp*) in the same genome gives a much lighter variegation than $P^{rr}Mp$ alone.

The complete and general significance of controlling elements cannot be evaluated at present. That they are of significance in maize there can be no doubt, and they may have a counterpart in the regulators and operators proposed for bacteria as discussed in Chapter 9. Little doubt exists that the further study of controlling elements will lead to a better understanding of the mutation process, as well as of gene function and control.

Another phenomenon in maize, apparently unrelated to the activity of controlling elements, is of great interest in connection with the mutation process as affected by the genotype. This is *paramutation,* found to occur by Brink et al. (114, 116) when the R allele, R^r, is present in heterozygotes with the stippled allele, R^{st}, or the marbled allele, R^{mb}. The R^r bearing gametes from R^r/R^{mb} or R^r/R^{st} plants always carry an R^r allele which has an altered expression. Ordinarily, R^r determines self-colored aleurone when present in two or three doses and dark-mottled aleurone when present once. However, R^r extracted from the indicated heterozygotes produces only a very weak pigmentation in the kernels. This condition is stable and is inherited. Hence, R^r is in some fashion altered in the presence of R^{st} and R^{mb}.

Extrinsic Factors Influencing the Mutation Rate

Besides the considerable influence of the genotype on the mutation rate, there are three major environmental factors: the temperature, certain radiations, and certain chemicals. The latter two factors, in particular, have been extensively applied not only in the study of mutation, but also to provide a source of new mutations for genetic work.

Temperature. The effect of temperature on the mutation rate has been investigated primarily in *D. melanogaster.* As shown in Figure 4.9, increasing the temperature at which the fly develops increases the mutation rate, particularly at temperatures above 15°C. If the generation is used as a biological time unit, the temperature coefficient, $t°Q_{10}$ (the ratio of the rate constant at one temperature to the rate constant at a temperature 10°C lower), is calculated from these data to lie between 2 and 3. This is what would be expected for most biological processes and chemical reactions. It is to be recognized, however, that this is an average coefficient for a class of mutations, lethals on the X and second chromosome of *D. melanogaster.* What the coefficients may be for specific loci is not known, but it is quite possible that there may be a considerable degree of variance from the indicated mean of 2 to 3.

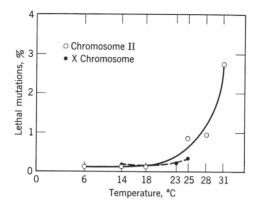

Figure 4.9. The effect of temperature on the lethal mutation rate in *Drosophila melanogaster.* From Plough (791).

Temperatures outside the normal range for the organism are not feasible in experiments in which the complete course of development is at a constant temperature, as in the examples just described, but extreme temperatures may be applied for short periods as temperature "shocks." The lethal and visible mutation rate is approximately doubled by treating three-day-old larvae of *D. melanogaster* at temperatures of 36° to 38°C for 12 to 24 hours (117, 118). Low temperature shocks of $-6°C$ for 25 to 40 minutes have been reported to triple the lethal rate of mutation for the X and second chromosomes (81). Sperm can be heat treated conveniently in the sperm receptacles of *Drosophila* females after insemination by subjecting the females to a moderately elevated temperature or by high-temperature shocks. In both cases the mutation rate in the sperm is significantly increased (126).

There is no evidence that the mutations produced by the temperature shocks are any different from those occurring under prolonged conditions of moderate temperature, and, like the prolonged temperature treatment, heat shocks seem to have no effect on the translocation frequency (791).

While the general mutation rate can be significantly increased by the elevation of temperature in *Drosophila,* it remains an open question whether this is true also in actively growing bacteria. In both *Serratia marcescens* (510) and *E. coli* (744, 1092), the $t°Q_{10}$ has been found to be about 2. When these findings were reinvestigated by Ryan and Kiritani (868), using the mutation rate from histidine requirement to histidine independence $(h^- \rightarrow h^+)$ in *E. coli* as an indicator, they found again that $t°Q_{10}$ is about 2, but that the rate of mutation per *generation* is the *same* at *all* temperatures from 15° to 37°C. On the other hand, if the cells of *E. coli* or the spores of *Bacillus subtilis* are first dried and then heated in this inactive state to 135° to 155°C in a vacuum, about 1 to 10% auxotrophic mutants are obtained (1130). The evidence seems clear that these mutations arose because of the heat treatment and not because of a mistake during replication.

Genes with extremely high mutation rates, the so-called unstable or mutable alleles, seem to respond differently to temperature from the wild-type genes mutating to lethals in *D. melanogaster*. For example, it has been found that the a_1 allele in maize, when in the presence of the specific mutator gene *Dt,* mutates to recognizably different alleles at a rate four to five times higher at 15.5°C than at 27°C (833). The effect of temperature on the mutation rate of a_1 in homozygous *dt* plants is not known. A similar inverse relation between mutation rate and temperature has been reported for the unstable flaking gene in *Portulaca* (264) and an eversporting eye-color condition in *D. melanogaster* (347). On the other hand, Demerec (210) was not able to discover any measurable effect of a 10° change in temperature on the mutation rate of several unstable genes in *Drosophila virilis*.

It is clear that no generalizations can be made at the present time based on the known data concerning the effect of temperature on mutation rate. Attempts have been made to reconcile the differences in response shown by the "stable" genes in *Drosophila* to temperature by the application of quantum theory, which dictates a smaller increase in rate of change of unstable molecules with increase in temperature than with stable molecules (884). The theory, however, is not consistent with all the known data. The observed stability of a gene is a function, not only of the gene's thermodynamic state but also of its immediate environment in the cell and the mechanism by which it replicates, and it will be necessary for hypotheses attempting to explain changes in mutation rate with temperature to take these factors into consideration.

Radiation. x-Rays, α-rays, β-rays, and ultraviolet light are proven mutagenic agents capable of changing the genes and chromosomes. Of these, ultraviolet light is the only nonionizing radiation. Ionizing radiations are assumed to cause their primary biological effect by producing ionization within the tissue and to have secondary effects resulting from thermal agitation or excitation of the tissue molecules

(303). It is difficult to assess the relative effectiveness of ionization and excitation (271), although some workers have assumed for theoretical purposes that changes due to excitation without ionization are inconsequential (565). In either case, whether the transformations are caused by ionization or excitation, the molecules hit by particles or quanta of energy may be expected to undergo chemical change. The probability of the change occurring increases with the amount of energy transferred from the radiant energy particles or quanta. Ultraviolet light, not being an ionizing radiation, produces its effect only by excitation of the molecules of compounds which absorb it. Compounds which do not absorb ultraviolet light are, of course, unaffected by it *directly,* for there is no transfer of energy and thus no cause for molecular agitation which might result in a chemical and hence genetic change. Compounds which do not absorb ultraviolet light may, however, be affected by it indirectly, as discussed in succeeding sections.

All these radiations are presumed to exert an effect on genes and chromosomes by providing energy for chemical changes resulting in mutations. The nature of the chemical changes is unknown, but many of them are stable, as shown by the fact that mutant phenotypes produced by irradiation maintain a stability from generation to generation characteristic of naturally occurring mutations. In addition to point mutations, chromosomal breaks are also produced, resulting in inversions, translocations, and deletions, as well as loss of whole chromosomes and other types of abnormal conditions. The extent of change induced in the mutation rate depends on the type of radiation, the dose, and various environmental factors discussed hereafter.

The dosage of ionizing radiation applied to a tissue is measured in terms of roentgen (r) units for x-rays and γ-rays. The roentgen unit represents the number of ionizations produced per unit volume of matter irradiated. For water and tissue irradiated by a dose of 1r, approximately 1.8 ionizations are produced per cubic micron, the exact value depending on the type of radiation and the composition of the tissue. α-Rays, β-rays, and neutrons are radiations emitted as atomic particles. Measuring the dosages of these radiations is more difficult than for x-rays and γ-rays, but they may be converted to r units for comparative purposes, since all produce ionizations.

A direct proportionality exists between effectiveness of ionizing radiations in inducing "point" mutations and dosage of the applied radiation. This relationship is illustrated in Figure 4.10 for sex-linked recessive lethals in *D. melanogaster.* The proportionality has been found to hold for visible mutations as well as for lethals. The number

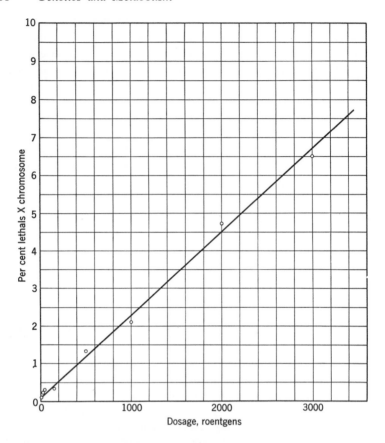

Figure 4.10.   The relationship between dosage of x-ray and the percentage of X-chromosome lethals induced in *Drosophila melanogaster*.   Data from Spencer and Stern (936).

of mutations produced by ultraviolet light is also proportional to dose (ergs/cm²), but only at low dosages; at sufficiently high dosages the mutation rate fails to increase with increasing dosages, or may even fall off.

The frequency of chromosomal breaks produced by ionizing radiations is proportional to the dose (49, 132, 535, 878, 1004), but the frequency of chromosomal rearrangements resulting from broken ends rejoining in new combinations is not related to the dose of x-rays in a simple linear fashion.   This is attributable to their origin from two or more breaks, each break being produced by a different ionizing

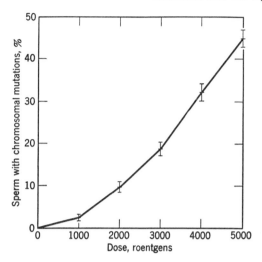

Figure 4.11. The relationship between the dosage of x-ray and the percentage rearrangements induced in *Drosophila melanogaster*. After Bauer (48).

particle. The dose-frequency relationship of rearrangements is a complex function which does not correspond to a straight-line function; see Figure 4.11 (241, 148, 43, 48).

The mutation rate induced by ionizing radiations for the most part is independent of the time rate of application of the dose (generally called the intensity) of the radiation (769, 1010). Figure 4.12 gives some data which show that this independence between intensity and

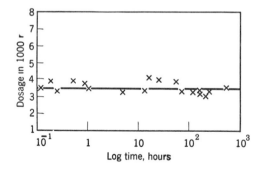

Figure 4.12. The effect of the intensity of application of x-ray on the mutation rate of *Drosophila melanogaster*. From the data of Patterson (769) and Timofeeff-Ressovsky and Zimmer (1010).

mutation rate holds over a very wide range of intensities, provided that the total dose is kept constant. There are certain exceptions to this generalization, such as, for example, the production of translocations in *Tradescantia*. Here, as might be expected, spreading the dose causes a reduction in the number of aberrations due to two breaks (879), for the use of low intensities limits the number of breaks available for forming new unions at any one time. In addition, the production of translocations by x-ray of *D. virilis* sperm is greatly enhanced by the application of very high intensities. The same dose of x-rays delivered at 2000r per minute produces 60% more translocations than when delivered at the rate of 2000r per 20 minutes (386). It is difficult to explain the intensity effect on the basis of the number of broken ends of chromosomes available at any one time, for there is considerable evidence to indicate that chromosomes broken in the *Drosophila* sperm do not rejoin until the time of fertilization. In addition, the types of translocations which result indicate definitely that it is not merely a matter of availability of broken ends (386). If this is so, then the explanation of at least part of the intensity effect must be sought elsewhere. The answer is probably to be found in the fact that radiation causes indirect effects in the cell by producing free radicals, and so forth, out of nongenic material, which can then act as mutagens by reacting with the genic material. This indirect effect of radiation and chemical mutagens is discussed in the succeeding sections. In the present context it is evident that the concentration of mutagenic material produced by high intensity would be expected to be much greater than that produced by low intensity if the mutagens are labile.

Although all ionizing radiations cause the production of mutations and breaks in chromosomes, they do so with different efficiencies. Table 12 gives some quantitative estimates of the relative efficiencies of several types of ionizing radiations in producing lethals in *Drosophila* and chromosome (isochromatid) breaks in *Tradescantia*. These data have been collected from several sources, and hence a certain amount of divergence is to be expected as a result of differences in experimental technique. However, the differences shown in the table are so great as to indicate definitely that the ion intensities of ionizing radiations are of some significance in determining their effectiveness. The ion density of a radiation is a measure of the distribution of ionizations produced along the paths of its ionizing particles. Radiations with high ion density, such as α-rays, neutrons, and soft x-rays, have their ionizations spaced close together, whereas the ionizations are more widely spaced

Table 12. The Relative Efficiencies of Various Ionizing Radiations in Producing Lethals in *Drosophila* and Chromosome Breaks in *Tradescantia*

Lethals in *Drosophila*

Radiation	% Lethals per 1000r	Relative Efficiency	Reference
x-Rays from betatron			
(23 mev)	1.7	0.59	827
(20 mev)	1.4	0.48	555
β-Rays, γ-rays, and hard x-rays			
(12.4 kv to 2.2 mev)			
(1 Å and lower)	2.9	1.0	1135
Soft x-rays (2.2 Å)	2.54	0.96 *	1135
Neutrons (Li + D)	1.9	0.66	1011
α-Rays (radon)	0.84	0.29	1052, 565

Isochromatid † Breaks in *Tradescantia*

Radiation	No. Breaks per 100 Cells/r	Relative Efficiency	Reference
x-Rays			
(0.15 Å)	0.27	1.0	1005
(1.5 Å)	0.26	1.0	145
(4.1 Å)	0.44	1.6	145
Neutrons (Li + D)	0.99	3.7	145
α-Rays (radon)	2.10	7.8	1005
Thermal neutrons	3.02	11.0	180

* Compared to 2.65% lethals at 0.94 Å found by these authors.

† Sister chromatids which have presumably been broken simultaneously at the same location along their length during early prophase, or earlier in interphase after duplication.

in the "hard" radiations, such as short-wave x-rays, γ-rays, and β-rays (see Figure 4.13).

The stage of a cell at the time of radiation may have a great influence on the susceptibility of the cell to mutation. In the male germ line of both *Drosophila* and the mouse, ionizing radiations are least effective on spermatogonia, most effective on spermatocytes and spermatids, and intermediate in effect on mature sperm (303, 7, 867). There has been no satisfactory explanation offered for all of these differences, but

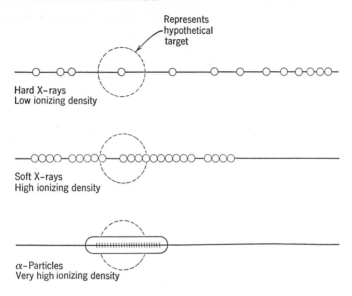

Figure 4.13. Diagram of the ionization paths of three different kinds of ionizing radiations. After Gray (353).

Schmid (881) has advanced an hypothesis to explain the differential effect on sperm and spermatids. This is based on the fact that in sperm the chromosomes are packed together, whereas in the spermatids they are not, but are distributed in the nuclear sap. If the ionizing radiation is assumed to produce mutagenic products as well as cause direct mutation by direct hits on the genetic material, as discussed hereafter, then it might be expected that the spermatid chromosomes would be more susceptible, since they are surrounded by considerably more material capable of producing mutagenic substances than the sperm chromosomes.

The effectiveness of ultraviolet light in inducing mutations is closely related to the wavelength employed. In general, the greatest mutagenic activity is found in the range between 2500 and 2800 Å, but the specific wavelength with maximum effectiveness is different for different organisms, as shown in Figure 4.14. The relative efficiency curve (action spectrum) for maize compares favorably with the absorption spectrum curves of nucleic acids, which also have a peak in the region of 2600 Å. This correspondence prompted many to suggest that it is positive evidence that nucleic acid is an important functional constituent of the gene. Nucleic acid is, as would be expected, rapidly decomposed by disruption of the purine and pyrimidine structure when irradiated by ultraviolet light in the region of 2600 Å. Furthermore, it is well known that the bactericidal activity of ultraviolet light is greatest at 2600 Å. As for the T1 and T2 phage, the action spectrum for inactivation also shows a definite peak at 2600 Å (1133). These facts taken together would indicate some validity for the hypothesis that ultraviolet mutagenic and lethal activity is primarily the result of the modification of nucleic acid within the cell. However, not all organisms, as demonstrated in Figure 4.14, show the same response to ultraviolet light as measured by mutation rate; nor has the possibility of other substances which absorb ultraviolet light in the cell being implicated been properly considered in the interpretation of these data.

The Target Hypothesis. In order to explain the tremendous mass of data which accumulated with great rapidity after the discovery by Muller (716) in 1927 of the genetic effects of ionizing radiations, a hypothesis was propounded which has been variously described as the direct-effect, hit, target, or Treffer hypothesis. The hypothesis begins with the assumption that simple excitation of the atoms of the irradi-

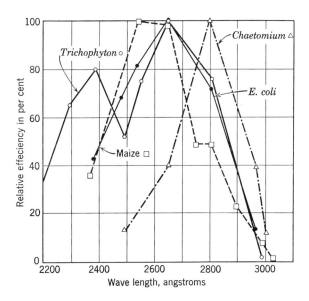

Figure 4.14. The relative efficiency of various wavelengths of ultraviolet light in inducing mutations in four different organisms: Trichophyton (449), Chaetomium (635), Maize (953), and *Escherichia coli* (1132*a*).

ated material is of little consequence in inducing chemical change, but that the observed effects are primarily induced by ionization. This is an assumption for which there is no direct supporting evidence from gene-mutation studies, and it has been subjected to considerable criticism; see Fano (271) and Gray (353). Next, the assumption is made that genes and chromosomes, on being "hit" or traversed by an ionizing particle, are caused to mutate with a probability close to 1. Though acknowledging that ionizing radiations cause molecules in the cell other than the chromosome molecules to undergo transformation, the hypothesis in its purest form does not concede that the altered extragenic material will in turn affect the chromosome constituents to any significant degree (565). To allow for the possibility that an ionization very near the chromosome may have an indirect effect on the chromosome, it is generally assumed that there exists a sensitive volume which includes the chromosomes. An ionization must occur within the sensitive volume to produce a point mutation or a chromosome break. Therefore the target is the sensitive volume, and hits on the target may be detected as mutations.

The target hypothesis is mainly supported by the following evidence: (1) the number of induced "point" mutations or chromosome breaks observed appears to be directly proportional to the dose as determined by the number of ionizations measured; (2) the yield of "point" mutations and breaks is independent (with the exceptions noted) of the intensity or time rate of application of the dose; and (3) the efficiency of an ionizing radiation in inducing mutation and breakage seems to be related to its ion density. These observations taken together definitely indicate, according to the proponents of this hypothesis, that the effect of an ionizing radiation is an all-or-none phenomenon, and a single ionization or a cluster produced by a single primary ionizing particle acts as the effective mutagenic unit. A linear relationship between effect and dose, and independence of the effect and time rate of application of the dose, would be expected only if each mutation or break were caused by a single primary event or "hit." If two or more independent hits were required, as, for example, in producing large rearrangements, the linear relationship would not hold, nor would the effectiveness of the dose be independent of the intensity of application.

That the "hits" may be single ionizations or clusters is indicated by the data given in Table 12. With reference to isochromatid breaks in *Tradescantia,* it will be noted that these are produced most efficiently by α-rays. Neutrons, very soft x-rays, and hard x-rays follow in effectiveness in decreasing order of efficiency. This corresponds to the order of decreasing ionization density. It has been concluded that the direct relationship between decreasing efficiency and decreasing ionization density indicates that the breaks in *Tradescantia* chromosomes are produced by a number of ionizations in a group (566) and not by single ionizations. The ionizations produced by hard x-rays are so spread out along most of the track that only the tail ends are effective, whereas most or all of the track range of α-particles and protons produced by neutrons would be effective (see Figure 4.13).

The efficiency of x-rays, neutrons, and α-particles in inducing lethals in *Drosophila* decreases with increasing ion density. Lea (565) and Catcheside (142) have interpreted this as meaning that recessive lethals in *Drosophila* are produced by single, or at most a very few, ionizations. The bombardment of a small object, such as the gene is assumed to be, with ionizing particles from α-rays producing ionizations spaced so closely together along the track that many occur almost simultaneously within the target or sensitive volume, would result in some of the ionizations being wasted, if only one is necessary to produce the effect (see Figure 4.13). On the other hand, with radiations of lower density, the efficiency would be increased, since the ionizations are more widely distributed, and each will have a higher probability of making a "hit" in a different location and hence produce a different mutation. It will be noted that the extremely high-energy x-rays produced by the betatron do not give results which conform to this interpretation, despite the fact that they have the lowest ion density among the sources listed in Table 12.

Despite the fair degree of correspondence between the target hypothesis and some of the observed facts, it is clear that it must be used with caution in the interpretation of radiation data in genetics, for, while it has been useful as a working hypothesis, results from investigations to be discussed in the succeeding paragraphs make evident the fact that it may well be incorporated into a more general theory in the future.

Chemical Mutagens. The possibility of producing mutations by chemical treatment had been extensively explored by numerous workers before and after the discovery of the mutagenic action of radiation. The first reported positive evidence was presented by Thom and Steinberg (1006) who found that nitrous acid caused mutations in *Aspergillus*. This finding was largely ignored until the work of Auerbach and Robson (34, 35), begun in 1940, established that chemical treatment could be as effective as radiations in inducing mutations. The original results were obtained by treating *D. melanogaster* with allyl isothiocyanate (mustard oil) or mustard gas, and then testing for sex-linked lethals. With mustard gas it was found possible to increase the percentage of sex-linked lethals from about 0.2% in the controls to as high as 24% in the treated flies (35). Subsequent work by these and other workers on *Drosophila,* certain of the higher plants, fungi, bacteria, and bacteriophage soon made it apparent that many other chemicals besides the mustard vesicants and nitrous acid are mutagenic. A list of the chemical compounds which have been found to induce mutations and chromosome breaks would probably include several hundred entries and serve small purpose except to show that there is little relationship, chemically, among mutagens. Simple inorganic salts such as $MnCl_2$ and complex organic compounds such as caffeine have both been shown to be effective mutagens. For simplicity, chemical mutagens will be considered here under three categories: (1) purines and purine and pyrimidine analogues; (2) alkylating agents such as the nitrogen and sulfur mustards, epoxides, and alkyl sulfates and sulfonates; and (3) all other effective chemical mutagens.

In general, chemicals which are mutagenic also cause breaks in chromosomes. This so-called radiomimetic property of most mutagens enables one to use the rather quick cytological test rather than a slower breeding test with many of the chemicals to test their effectiveness in the higher organisms.

A compound may be mutagenic for one organism and have no effect on another. Thus hydrogen peroxide and diazomethane cause mutations in *Neurospora* but produce no evident effect on plant chromosomes. On the other hand, urethane, phenols, and others produce mutations in *Drosophila* and chromosome aberrations in plants but no reversion mutations in *Neurospora.* The reasons for these differential effects remain to be elucidated. The explanation may well be that some chemical mutagens are active only during a particular stage of the development of the germ cells or their ancestors. For example, formaldehyde seems to act mutagenically only at an early stage of the developing sperm of *Drosophila* which Auerbach (31) calls the "sensitive" stage. Furthermore, only the male gametes are affected. The treatment of female larvae or adults with formaldehyde has no effect whatever on the frequency of lethal mutations in the eggs (33). This is to be contrasted to the action of urethane, which is effective on the mature sperm of *Drosophila* but has no apparent effect on the spermatogonia.

Since usually only one stage or part of the life cycle is treated in experiments testing chemical mutagens, it is quite possible that the stage of an organism sensitive to a particular chemical may not have been the one treated, thus giving negative results which might have been positive had the treatment been given at the proper stage. One must also consider the possibility that the ability of these compounds to penetrate into the cells is different from one organism to another, or that the compounds are broken down by active enzymes before they are able to produce mutations. Negative results should not therefore be taken as final evidence that the substance tested is not mutagenic without thorough testing under different environmental conditions and at different stages of the life cycle.

Since the recognition of the role of the deoxyribonucleic acids as hereditary material, considerable emphasis has been placed on the study of mutation by the use of compounds which can be expected to alter DNA specifically. Among the first compounds tried were the purines and pyrimidines of natural occurrence and their analogues.

Purines and Pyrimidines. Natural purines such as adenine and caffeine are mutagenic, at least in some organisms, but in general natural pyrimidines are not. A partial list of mutagens in this category is given in Table 13. Analogues of pyrimidines, such as 5-bromouracil, have been used with considerable success. This compound and its nucleoside, 5-bromodeoxyuridine, are of great interest because they are analogues of thymine and deoxythymidine, respectively, and have been shown to replace thymine in the DNA of bacteria (1131, 901), bacteriophage (621), and mammalian cells (221, 527). The effect of 5-bromodeoxyuridine incorporation into mammalian cells in tissue culture is to cause visible chromatid breaks and translocations (472). Surprisingly enough, the breaks induced in the cells of the Chinese hamster by this compound were found not to occur at random but occurred most frequently at the natural constriction regions of the chromosomes (centromeres and secondary constrictions) and near their ends. These regions are also the ones at which the highest concentration of heterochromatin is supposed to be found.

Whether 5-bromodeoxyuridine also causes point mutations in mammalian cells is not known, but it does induce mutations in phage T4, along with 2-aminopurine, 2,6-diaminopurine, and 5-bromouracil (295). Freese (296, 297) has found that the mutations produced in the *r*II region of T4 in the presence of 5-bromouracil and its deoxynucleoside are generally located at the same sites, that is, they give the same spectrum. Spontaneous mutations and mutations induced by the aminopurines tend to cluster at somewhat different sites. A degree of rela-

Table 13. **Purines and Purine and Pyrimidine Analogues with Mutagenic Activity**

Adenine

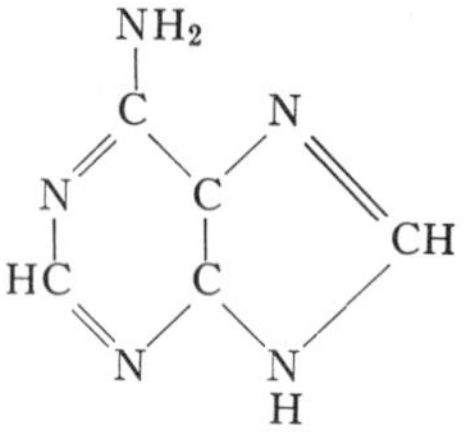

$$
\begin{array}{c}
NH_2 \\
| \\
C \\
\end{array}
$$

Adenine structure:
NH₂ — C; N; C; CH; HC; C; N; N—H

2-Aminopurine

H—C; N; C; CH; C; C; H₂N; N; N—H

Caffeine

O; CH₃; C; N; CH₃—N; C; CH; C; C; O; N; N; CH₃

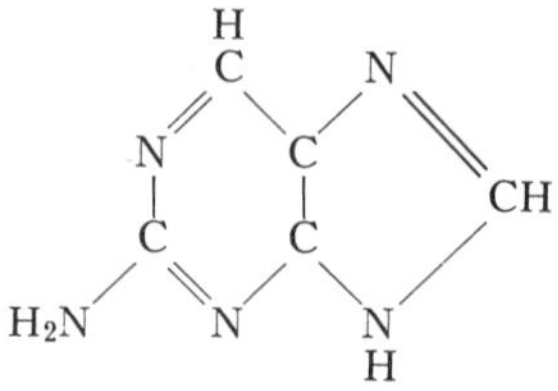

Theobromide

O; CH₃; C; N; HN; C; CH; C; C; O; N; N; CH₃

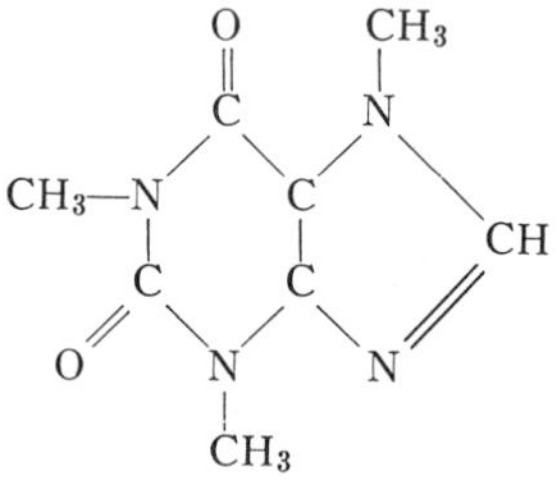

2,6-Diaminopurine

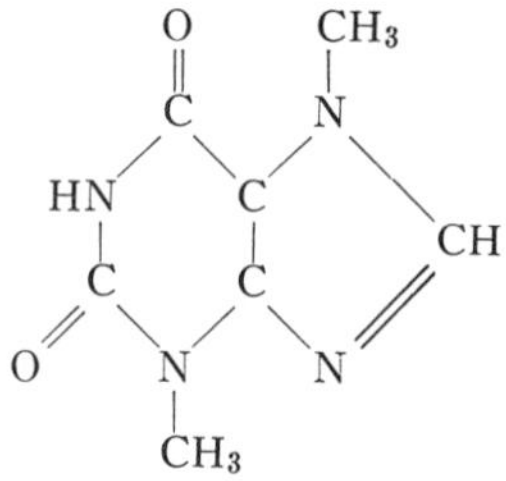

NH₂; C; N; N; C; CH; C; C; H₂N; N; N

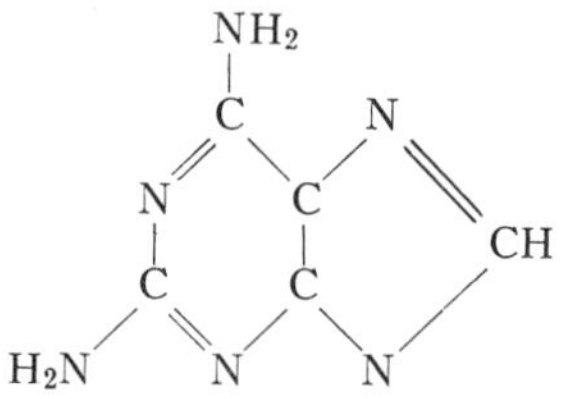

5-Bromouracil

OH; C; N; C—Br; C; C; HO; N

198

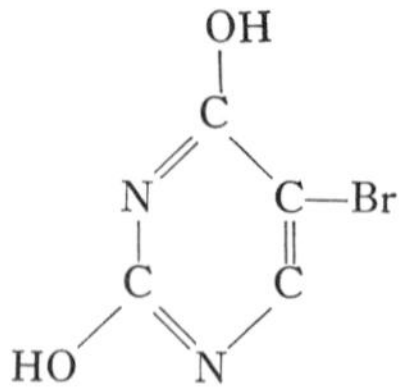

tionship among the mutants produced in the presence of bromouracil and 2-aminopurine is indicated, however, because these are caused to revert readily back to *r*II⁺ in the presence of 5-bromouracil. This is not true of the spontaneous *r*II mutants or those induced by an unrelated mutagenic compound, proflavine (3,6-diaminoacridine). Mutations which have a spontaneous origin or are induced by proflavine do *revert* spontaneously—at low frequency, however—indicating that they are not necessarily deletions. The fact that this difference in inducible reversion exists has led Freese to postulate that the mutations induced by these analogues are basically different from those induced by proflavine. The data given in Table 14 taken from Freese (298) strongly support this conclusion.

If a 5-bromouracil molecule replaces a thymine in a DNA chain, a BU-A nucleotide pair will be formed with the bases joined by hydrogen bonding (BU = 5-bromouracil). This would mean that when the two chains of the double helix separate and duplicate (assuming the correctness of the Watson-Crick model and mechanism of duplication), one of the single strands will carry a BU base. It is conceivable that this chain might not function to produce a continuing strand, but on the other hand, the BU in it may form a hydrogen bond with another adenine, BU-A, or with a guanine to give BU-G. If thymine and cytosine (or hydroxymethylcytosine for phage) are made available, the BU-A combination would be expected to revert to the original T-A condition and the BU-G combination might become C-G. This would cause a change in base sequence at this point from T-A to C-G, and presumably be manifested as a mutation. It will be noted that this results in a purine replacing a purine and a pyrimidine a pyrimidine. Such a hypothetical change has been called a *transition* by Freese. It should also be noted that the bromouracil may theoretically replace a cytosine to give a BU-G double nucleotide, which by the same hypothetical process might become BU-A, and then T-A, resulting in another type of transition: C-G to T-A.

Transitions induced by analogues, whether purine or pyrimidine derivatives, might be expected to be induced to revert by these same analogues by the processes just described. However, if the mutagen induces a *transversion,* that is, the replacement of a pyrimidine by a purine or vice versa, then one should not expect such mutations to be induced to revert by means of the purine or pyrimidine analogues. On these grounds, Freese postulated that proflavine induces transversions rather than transitions, and hence these are not revertible by mutagens that induce only transitions. These arguments have only heuristic

Table 14. Reversion Induction of Spontaneously Reverting Mutants in the *r*II Mutants of Bacteriophage T4 *

*r*II Mutants Induced by	Number Mutants Tested	Per Cent of Mutants Found		Approximate Spontaneous Background of Nonreversion Inducible Mutants, in %
		Inducible to Revert by 2-Aminopurine and/or Bromouracil	Noninducible to Revert by Either 2-Aminopurine or Bromouracil	
2-Aminopurine	98	98	2	2
5-Bromouracil	64	95	5	2
Hydroxylamine	36	94	6	4
Nitrous acid	47	87	13	15
pH5, 45°C	115	77	23	15
Ethyl ethane sulfonate	47	70	30	15
Proflavine	55	2	98	⋯
Spontaneous	110	14	86	⋯

* Freese (298).

value, as indicated by the data given in the next paragraph, but they indicate the type of thinking current in this field.

A direct test of the ability of 5-bromouracil to act as an analogue of thymine and to pair by hydrogen bonding, with adenine primarily and guanine secondarily, in a double-stranded DNA helix has been made by Trautner and associates (1018). These workers have utilized synthetic DNA polymers containing, in one case, alternating adenine and thymine deoxyribonucleotides and, in the other, alternating adenine and 5-bromouracil. The first type may be referred to as dAT, and the second as dA$\overline{\text{BU}}$, since in the double-stranded condition adenine pairs with thymine in the one and with bromouracil in the other. The two polymers were used as primers (see p. 97) in an *in vitro* system which synthesizes DNA in the presence of triphosphodeoxynucleotides and the Kornberg enzyme, DNA polymerase.

Theoretically, when dAT is used as a primer, only adenine and thymine deoxynucleotides should be incorporated in the presence of guanine deoxynucleotide. This proved to be true to the extent that no detectable guanine was incorporated in the experiments. On the other hand, guanine was incorporated in the presence of dA$\overline{\text{BU}}$ primer

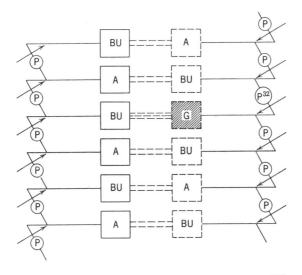

Figure 4.15. Scheme for incorporation of dGTP[32] into a dA$\overline{\text{BU}}$ polymer as suggested by the mutagenesis hypothesis. Bases in newly synthesized dA$\overline{\text{BU}}$ are represented by dotted lines. The arrows indicate the linkages cleaved by micrococcal DNase and calf spleen phosphodiesterase. The P[32] label introduced with dGTP[32] is thus transferred to the adjacent BU. From Trautner et al. (1018).

at a frequency of one per 2000 to 25,000 adenine and thymine nucleotides polymerized. This is as expected according to theory, but the nearest-neighbor frequency determinations (see p. 96) were not in agreement with the hypothesis that guanine pairs with bromouracil only. If only bromouracil paired with guanine, then, with the polymer of the type employed here as primer, guanine should be exclusively next to bromouracil in the chain (Figure 4.15). This was not found to be true. Guanine was found to be associated with bromouracil, guanine and adenine as nearest neighbors in the chain at frequencies of 41, 42, and 17%, respectively. Hence, the hydrogen bonding of guanine to bromouracil only, in the complementary chain, is not supported by the evidence.

Alkylating Agents. The alkylating agents are another group of mutagens whose effects are being studied extensively in phage, microorganisms, and the higher forms. These compounds must act quite differently than those which presumably cause mutation by being incorporated into DNA. The alkylating agents must be considered to be acting either directly on the DNA to change its bases by chemical reaction or to be reacting with other compounds in the cell which are thereby converted to mutagens. It is possible that ethyl ethane sulfonate, for example, causes the ethylation of guanine and adenine in DNA with the result that the ethyl derivatives are then eliminated and either replaced by different bases, or not replaced to produce deletions. This compound shows considerable mutagenic specificity in phage, since it causes a reversion of those mutants induced by 2-aminopurine, but not those induced by 5-bromouracil (50). On the other hand, about 70% of the *r*II mutants in phage T4 induced by ethyl ethane sulfonate are induced to revert by 2-aminopurine and 5-bromouracil (Table 14).

Kolmark and Westergaard (1067, 1069) have tested various alkylating agents on *Neurospora crassa*. As a quantitative test for determination of effectiveness, they measured the reversion of a specific adenine, inositol double mutant. This enabled them to observe the reversion rate of two genes simultaneously and to make comparisons of the relative effectiveness of the different mutagens tested, as shown in Table 15.

It is clear that the different mutagens have rather specific mutagenic effects. Compare, for example, the effects of bromethyl methane sulfonate in inducing reversions. It can hardly be said to be effective on the *inos*⁻ gene at all, but it is maximally effective in inducing reversions of *ad*⁻ to *ad*⁺. Ethyl methane sulfonate, on the other hand, is about

Table 15.   The Effects of Various Alkylating Agents in Producing Back Mutations in *Neurospora crassa* °

Mutagen	Mutations per 10^6 Macroconidia		Proportion $ad^+/inos^+$
	$inos^+$	ad^+	
Spontaneous rates	.02	.2	10
Ethyl methane sulphonate	11.3	17.4	1.5
Diethylsulphate	4.3	16.8	4
Dimethylsulphate	3.4	84.0	19
Chloroethyl methane sulphonate	.3	51.0	190
Diepoxybutane	.2	89.0	445
Bromoethyl methane sulphonate	.04	152.0	3800
Ultraviolet light	7.1	3.5	.5
x-Rays	.2	3.2	16

* Westergaard (1069).

equally effective on each.   It is important to observe that the back mutation patterns of ad^- and $inos^-$, as presented in Table 15, are specific only for *ad-3* (38701) and *inos* (37401).   Other *ad-3* or *inos* alleles at these same loci would give different patterns (1067).   The mutagenic effects are, therefore, *allele* specific rather than *locus* or *gene* specific.   In other words, the *type* of mutation produced in the first place probably determines how it will respond to mutagenic agents with respect to reversion in the second place, a conclusion in agreement with Freese's back-mutation results in phage which have already been described.   Implicit in this point of view is the assumption that any gene may mutate in many ways to give many different types of alleles. If this is so, then one should expect just as many different mechanisms for reversion.   The allele specificity of mutagens, while not proving this, is certainly in agreement with it.

The study of the effect of alkylating agents on the mutation rate in *D. melanogaster* has brought to light a number of interesting phenomena which would be difficult, if not impossible, to uncover using micro-organisms as experimental material.   Treatment of *D. melanogaster* with mustard gas results in the same types of mutations produced by x-rays, but in somewhat different ratios (34, 35, 31).   Thus Auerbach and Robson (34, 35) found that "equivalent" doses of x-rays and mustard gas, as determined by the number of lethals, gave 30% and 6.7% lethal translocations, respectively, and about twice as many large deletions are produced by x-rays as by mustards (35). Another significant difference between the effects of x-rays and mustard gas is that mustard gas seems to have delayed effects in *Drosophila* on the production of mutations. It may produce a mutation in a cell a generation or more removed from the treated cell (31), resulting in mosaics. x-Rays do not cause a similar delayed effect detectable by the techniques employed.

The Fahmys (265–270) have done extensive analyses on *D. melanogaster* treated with a number of different types of nitrogen mustards and have concluded that the proportion of visible mutations to lethals obtained depends on the specific mutagen used. In general, they found that chemical mutagens produce a somewhat higher proportion of visible mutations than ionizing radiations, and they have proposed that two classes of genes exist which differ in their response to mutagens, the *alpha* and *beta* loci. This conclusion has been questioned, however, by those who maintain that the data are not sufficient to establish two classes of loci based on mutagen specificity (36).

Esters of methane sulfonic acid do not appear to affect the spermatogonia of *Drosophila,* but spermatocytes, spermatids, and mature sperm respond with the production of mutations. The mature sperm were most susceptible to the mutagenic action (268). In direct contrast to the methane sulfonic acids, a cysteine derivative, S-2-chloroethylcysteine $[Cl(CH_2)_3SCH_2CH(NH_2)COOH]$, demonstrates a marked selective mutagenicity for the spermatogonia.

The data obtained so far from phage, *Neurospora,* and *Drosophila* relative to the effects of alkylating agents are clear in pointing to differential effects of the different agents. But the data do not by any means indicate that it will be possible to cause specific genes to mutate and leave all others unmutated. Directed mutation may be possible in the future, but the results at hand indicate only that some types of genetic structure may be more susceptible to mutation in the presence of certain inducers than others. If this response is allele specific rather than locus specific, which seems to be the case, then the possibility for specific directed mutation would seem to be slight. One should expect that in a complex genome many alleles will always be present simultaneously with the particular type of structure which will respond to a given chemical mutagen. In simpler genomes such as found in phage, the process may be possible to carry out, but even here it has not been established to occur even though each chemical mutagen seems to produce mutations most often at specific points called "hot spots." There are always a number of these hot spots, however, produced by a given mutagen within a given locus.

One of the chemical mutagens which is neither an analogue of a base nor an alkylating agent deserves special mention because of its widespread usage at the present time. Nitrous acid has been shown to produce mutations in the tobacco mosaic virus (726) and in *E. coli* (513). Nitrous acid reacts readily with amino groups converting them to hydroxyl groups. Therefore, besides reacting with amino acids, bound and unbound within the cell, it should be expected to react with other amino-containing compounds such as the nucleic-acid bases: adenine, guanine, and cytosine. The reaction of nitrous acid with these bases is expected to produce hypoxanthine, xanthine, and uracil in the cell. These could presumably act as analogues and cause mutation in the same fashion as 5-bromouracil, provided that the analogues are formed *in situ* in the nucleic acids, for these compounds have not been shown to be mutagenic in the free state. That mutations can be induced in the tobacco mosaic virus by the treatment of the RNA alone has been demonstrated by Mundry (726), so the possibility that DNA can also be altered in the intact cell should certainly be entertained. Formaldehyde, like nitrous acid, is also expected to react with certain of the bases constituting the nucleic acids. However, it is doubtful that it causes mutation by acting directly with the bases in the intact DNA. To induce mutations in *Drosophila* with formaldehyde, it is necessary to add formaldehyde to the larval food. (Formaldehyde itself has no effect on the mutation rate in the adult (33).) Alderson (6) has made the interesting discovery that formaldehyde added to larval food which does not contain adenosine or adenylic acid is not mutagenic. Addition of either of these compounds in the presence of formaldehyde recreates the mutagenic environment. These results are interpreted to mean that formaldehyde reacts with these compounds to produce a metabolite which is introduced into the DNA of the germ cells. Formaldehyde, in other words, may form a substance analogous to 5-bromouracil in function.

As a result of the discovery of chemical mutagens, a new point of view has evolved with respect to mutation. The genetic material had, during the 1930's, been looked upon, almost mystically, as being intrinsically exceptionally stable and subject to chemical modification only by the use of extremely large amounts of energy carried directly to the gene and chromosomes by penetrating radiations. It is now necessary to recognize that it is no more stable than any chemical compound. It is apparent that the previously assumed extreme stability of the genetic material is a result of the protection afforded by the nongenic material surrounding it. Indeed, it should not have been surprising to find that the genetic material reacts chemically and is changed thereby, for the fact that mutation is under genetic control is in itself strong evidence, if not proof, that extragenic chemicals in the cell can bring about mutation. Otherwise there would be no rational explanation for the phenomenon of one gene having an effect on the mutation of another.

This view is strengthened particularly by the observation that naturally occurring compounds such as allyl isothiocyanate, hydrogen peroxide, formaldehyde, and various purines are effective mutagens. Indirect evidence that hydrogen peroxide produced during the course of aerobic respiration may be a factor in determining part of the spontaneous mutation rate is provided by the results of experiments with catalase and inhibitors of catalase (490, 1116). If catalase is added together with hydrogen peroxide in the treatment of *Neurospora* conidia or bacteria, the mutagenic activity of the peroxide is stopped, presumably because the H_2O_2 is destroyed by catalase. If catalase poisons such as KCN or sodium azide are added, the mutation rate increases even in the absence of added peroxide, presumably because the hydrogen peroxide synthesized by the organism accumulates instead of being decomposed by the cell's own catalase. The addition of catalase poisons together with hydrogen peroxide, as would be expected, increases the mutagenic activity many-fold in *Neurospora*.

It is evident that the living cell possibly provides for its own mutations by means of certain of its metabolic products reacting with its genic material. In addition, there may also be found here an explanation for the observations just described that mutations may occur more frequently at one stage of the life cycle than at others, or that there is a relation between the physiological state and mutation rate. As differentiation proceeds, one would expect metabolic conditions to change within the cells and tissues, and hence mutagenic chemicals may be produced in effective concentrations at certain stages.

Environmental Effects on Action of Mutagens. The action of mutagenic radiations and chemicals in inducing mutations is not independent of environmental conditions at the time of treatment. Some of the factors which affect the mutagenicity of chemicals have already been noted. The effectiveness of both ionizing radiations and ultraviolet light in producing "point" mutations and chromosome aberrations is considerably altered by the physiological state of the cell being irradiated. If the effects of extrinsic and intrinsic factors on the sensitivity of organisms to x-rays alone is considered, it is found that such physical factors as temperature, centrifugation, infrared and ultraviolet light before, after, or during radiation with the x-rays may have a drastic effect, either enhancing or reducing the mutagenicity or the formation of chromosome breaks (303). The number of chemicals which have been demonstrated to influence the mutagenicity of x-rays is probably now over one hundred (933).

There may be a number of reasons why these factors cause a change in the observed rate. (1) They may be effective in increasing or decreasing the rate independently of x-rays. (2) They may modify the cell's metabolism in such a way that the cell recovers more easily from "extragenic" radiation effects, and hence more mutations are observed. (3) They may predispose the genes to the action of x-rays by getting them into a labile state. (4) They may create conditions within the cytoplasm and other extragenic materials which allow for the production of compounds which, when irradiated, form mutagenic chemicals. (5) The gene may be placed in a labile, intermediate "semistable" state by the action of the mutagen, from which state it may revert to the stable original state or proceed to a stable mutant state (646). The influence of the cell environment might be expected to be considerable in determining in which direction the semistable gene may go, and hence it may have a considerable effect on the mutation rate (646).

The first possibility has been eliminated as an explanation for most of these effects by the use of the proper controls, even though some of these factors have an independent effect on the mutation rate. The last four hypotheses are all, to one degree or another, possible explanations for the observed phenomenon; at least, no one of them has been completely eliminated. But, in any case, it is now firmly established by such data as these that the effect of x-rays, and perhaps other ionizing radiations, may be far from being a purely physical phenomenon involving only direct ionization or excitation of the genic material with no influence of the extragenic environment. The direct effect of radiation quanta or particles is still probable, but to this direct effect there must be added the indirect effects.

The indirect effect of radiation received direct proof from experiments performed by Stone, Wyss, and colleagues. These workers were able to demonstrate that mutations in the bacterium *Staphylococcus aureus* are induced by placing the cells into a nutrient medium previously irradiated with ultraviolet light (974, 975). Mutations resulting in resistance to streptomycin and penicillin as well as inability to ferment mannitol were found to occur at a rate of 10 to 500 times higher than the spontaneous rate. Irradiation of the medium instead of the cells also produces mutations in *Neurospora*, although not at such a high rate as detected in *S. aureus* (1051). It appears very likely that the mutagenic agent produced by the irradiation of the nutrient medium is hydrogen peroxide or perhaps some organic peroxides (1051, 975). Peroxides are mutagenic. Although similar experiments have not as yet been reported with x-ray or other ionizing radiations, it is highly probable that similar direct evidence of indirect effects will be obtained, since these, as well as ultraviolet light, are known to produce hydrogen peroxide, the free radicals, [OH], [H], [HO$_2$], and probably other peroxides and free radicals in aqueous solutions (41, 44, 1061).

The demonstration of an indirect effect for radiations in producing mutations brings their mutagenic effect into the realm of chemical mutagenesis. Although it is hardly likely that radiations act only through the medium of chemical mutagens in producing chromosomal and genic changes, it does seem probable that a considerable part of their activity is expressed in this indirect fashion.

On the Mechanism of Mutation

Despite the tremendous amount of work which has been done, and continues to be done, in the study of mutation, little is known about the process of mutagenesis which leads to mutation. What can be said with some conviction is that it is a relatively complex process which is highly sensitive to factors in the environment, both within and without the cell. An attempt to convey some idea of the degree of this complexity, particularly with respect to the breakage of chromosomes which precedes the formation of extragenic mutations, is given in Figure 4.16. With respect to intragenic mutations even less is known. The most obvious explanation for these mutations is, of course, that they are the result of a relatively restricted change in base sequence in a DNA chain. Such changes as (1) the deletion of one or a few nucleotide pairs, (2) substitution with other bases in one or more nucleotide pairs, (3) minute inversions involving a few nucleotide pairs, and (4) rotation of bases (725) may be visualized as producing sequence changes, as diagrammed in Figure 4.17. Any one or all of these may be the cause of intragenic mutations, but the mechanisms whereby they come about are not clear.

That changes brought about in nucleic acid can result in alterations in protein, which in turn result in mutant phenotypes, has not been definitely established for DNA, but it has been established for RNA. The experiments relating to this were performed with the tobacco mosaic virus (TMV) and are described in Chapter 6.

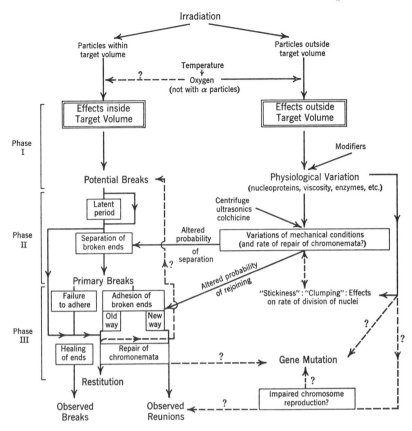

Figure 4.16. The paths of action of ionizing radiations and the factors that influence chromosome breakage. From Thoday (1005).

A number of workers have made attempts to investigate the mechanism underlying the mutation process in bacteria. Among the more successful of these endeavors is the work of Witkin (1093, 1094) and Haas and Doudney (231, 385). These workers have demonstrated that the mutation rate induced by ultraviolet light in *E. coli* can be drastically altered by pre- and postirradiation treatment of the cells with certain growth factors. Cells incubated with adenine, guanine, uracil, and cytosine, or their ribosides, for an hour prior to irradiation yield a much higher number of mutants following irradiation than cells previously incubated on an unsupplemented minimal medium (384). Similarly, if cells are incubated with an amino-acid mixture immediately following irradiation, a much larger number of mutants is obtained than

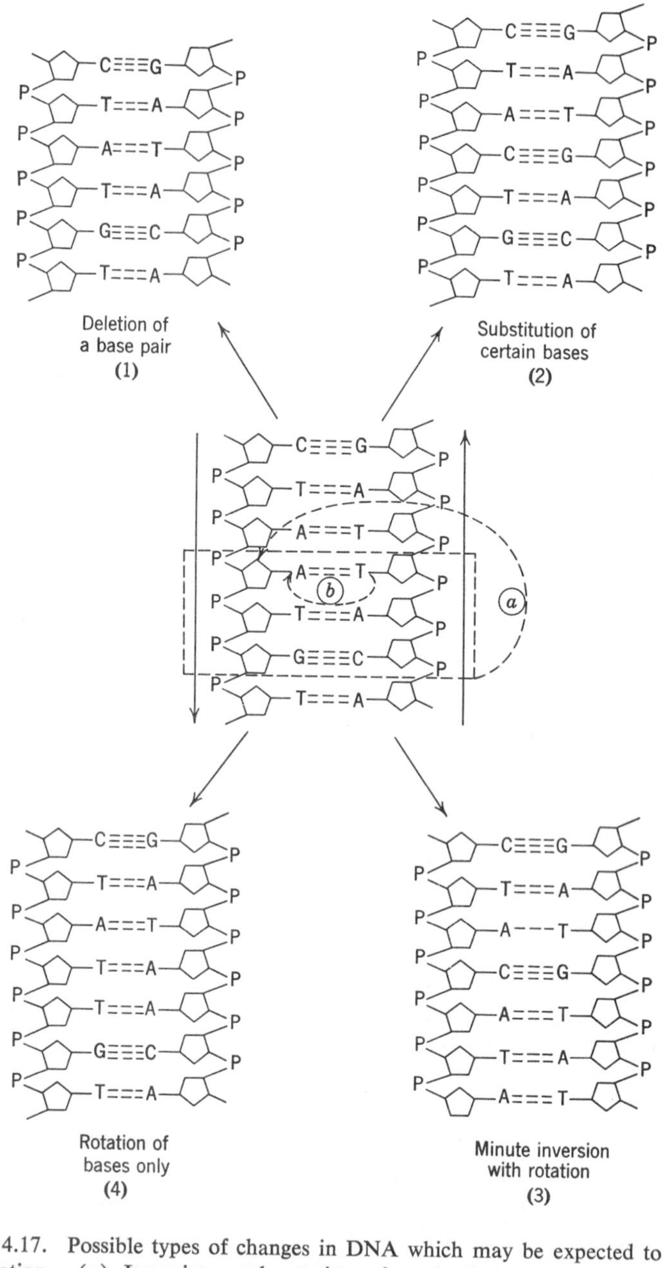

Figure 4.17. Possible types of changes in DNA which may be expected to result in mutation. (*a*) Inversion, and rotation of nucleotides to preserve polarity. This requires the breaking of four bonds simultaneously. (*b*) Rotation of the bases only. This requires breaking only two bonds.

if the cells are incubated in a nitrogen-deficient medium as shown by Figure 4.18. The addition of chloramphenicol, a protein-synthesis inhibitor, along with the amino acids causes a definite decline in mutation frequency similar to the decline obtained in nitrogen-deficient medium. Dinitrophenol, a compound which inhibits oxidative phosphorylation, also causes a reduction in the number of mutations obtained in the presence of amino acids as shown in Figure 4.18. As a result of experimental results such as these, Witkin (1093) has suggested that immediate protein synthesis is necessary following irradiation with ultraviolet light in order for mutant expression. If such synthesis does not occur, or is depressed, many mutations are not "fixed," and mutants that would otherwise come into being are lost. In other words, there is a definite time interval between the absorption of radiant energy

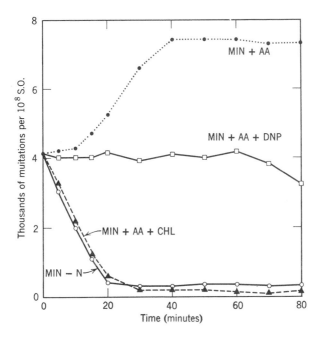

Figure 4.18. Effect of postirradiation incubation on the reversion of tryptophan-requiring *E. coli* to tryptophan independence. Tryptophan-requiring cells were incubated in the different types of media shown for the indicated periods of time. Following this the number of revertants was determined. MIN = minimal medium. AA = casein hydrolysate + tryptophan. CHL = chloramphenicol. DNP = dinitrophenol. M − N = deletion of ammonium sulfate. From Haas, Doudney, and Kada (385).

and the final stabilization of a genetic change initially caused by that energy, and during this interval certain chemical processes must occur in order for this stabilization to be reached.

Not all mutations induced by ultraviolet light are capable of modification by postirradiation treatments. Actually, there seem to be two general types of mutations produced in *E. coli* by both ultraviolet light (1094) and x-rays (385): one in which mutation induction is unstable and sensitive to postirradiation treatment; and one which is relatively insensitive. Therefore, the foregoing discussion of mutation stabilization concerns only the first type of mutation and throws no light on the induction process in the stable ones. Of course, it is quite possible that the distinction between the two types may be that the unstable mutations are the result of the indirect effect of the radiation as discussed in previous sections, whereas the stable ones are the result of "direct hits" on the genetic material by the quanta of energy. No evidence exists, however, to support this hypothesis.

Just what role protein synthesis plays in the fixation of unstable mutations is not known, but it is not an independent one, because the process is also correlated with the synthesis of RNA as shown in Figure 4.19. As can be seen from this figure, the "mutation fixation" process is closely correlated with RNA synthesis in the irradiated cells. Mutation fixation is defined as the state arrived at by an irradiated cell in which mutation is fixed because it no longer responds to factors such as chloramphenicol or nitrogen deficiency which may prevent the unstable mutant state becoming stable. What is of further interest in these data is that the mutation fixation process occurs before measurable DNA synthesis occurs. Ultraviolet irradiation characteristically causes a brief cessation of DNA synthesis, which in the example experiments cited here ends at about 40 minutes after irradiation. This is 20 minutes after mutation fixation occurs. However, at the time of onset of DNA synthesis, the phenotypic expression of the mutations begins to occur as shown in Figure 4.19. Thus it is apparent from these experiments that the processes leading to mutation induction and fixation occur before new DNA is formed, but new DNA is required for their expression.

These findings pose a rather interesting question relative to the functions of RNA and DNA with respect to the unstable type of mutation. The indications from these data are that "mutant" RNA is produced first by irradiation, and this mutant condition is then passed secondarily to DNA.

The discussion in this and the preceding sections presents ample

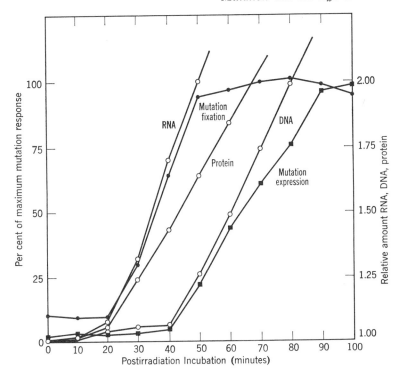

Figure 4.19. Relation of mutation fixation and mutation expression to synthesis of RNA, DNA, and protein in ultraviolet irradiated tryptophan mutants of *E. coli*. See text. From Haas, Doudney, and Kada (385).

evidence that the mechanism of mutation is far from being understood. It may be assumed that the final effect is a change in DNA, but how that change is brought about remains a mystery.

Mutation and the Nature of the Genetic Material

What has been learned about the genetic material as a result of the study of mutation? The answer to this is relatively simple, if one takes the superficial view, because not much has been learned directly. But from the broader point of view it must be recognized that the existence of the genetic material would have been practically impossible to establish in the first place without the use of mutant forms. Once the existence of such a material was demonstrated, however, further analysis of it by the use of mutation studies has proved to be disappointing in certain respects, probably mostly because too much was expected before the complexity of the mutation problem was recognized.

The original, somewhat naive, idea was that by studying the chemical group or groups that produce mutations we would learn something about the nature of the other partner in the reaction, the genetic material. This, I'm afraid, has proved an illusion. We now know that there is not one magic group which confers mutagenic ability on a compound. Instead, vastly different compounds may have mutagenic abilities, and closely related ones may differ in this respect. On the other hand, . . . our knowledge of the chemical nature of the genetic material has advanced spectacularly, but the advance did not come from mutation research. The situation is now almost exactly the reverse of what we expected. Instead of inferring the chemical nature of the gene from the nature of the substances which make it mutate, we tend to interpret the action of mutagenic compounds on the basis of what we know about the nature of genetic material.

This quotation from Auerbach (32), a pioneer in the field of chemical mutagenesis, serves to emphasize that the present research is primarily aimed at interpreting the effects of mutagens on DNA, the genetic material. By so doing, it may be possible to learn more about the genetic material. Certain rather significant steps have been taken in this direction. In the first place, it is evident from mutation studies that what we call a gene may mutate at a number of different points in its structure, a matter which will be discussed in more detail later. Furthermore, mutation studies have shown that mutagens, whether physical or chemical, have different spectra of effects. Thus a particular mutagen may cause more mutations at certain specific points along a linear array than at others. This demonstrates that the genetic material is by no means homogeneous along its length.

A dream of many investigators has been to develop techniques whereby specific mutations could be produced at will. Although it is apparent now that each mutagen has a degree of specificity as already described, it is also apparent that it may not be possible to produce one type of mutation and no others with a particular mutagen. The reason for this appears to be that although the genetic material is heterogeneous along its length, it consists of repeating elements, and it is difficult to see how mutagens can be produced that will affect only certain of similar or identical elements and not others.

5 Kinetics and Dynamics of Metabolism

Through continually increasing sophistication in methods of analysis of cells and their molecular components, it has been possible to obtain an ever more penetrating view of structure and composition of living material. At the same time, it has been quite apparent that this structure and composition does not describe life. It is necessary, in addition, to have knowledge of mechanisms of reactions and the rates of interconversions of cellular substances. Actually, it is this dynamic aspect of biochemistry that has the more fundamental significance. In the enormous complexity of the total pattern of metabolism in a cell, even a small change in the rate of one reaction can result in a large change in composition and extensive readjustments of other rates to maintain a steady state, or an orderly change of a balanced state, leading to cell division. As recognized long ago (333, 1108), gene mutation usually results in the change in rate of one or more biochemical reactions, and it is now firmly established that the range of such alterations can extend from the extreme of total loss through the normal to extensive enhancement of rate. Thus it is clear that a careful consideration of the various factors that can influence reaction rates is essential to an understanding of gene function and gene interaction.

The Nature of the Problem

A great deal of experimental and theoretical work has been done on chemical-reaction kinetics, and at first glance it would appear that biological systems present complexity in understanding only in that they contain a very large number of different kinds of interrelated reactions to be analyzed. This, however, is not the case, as has already been anticipated by the discussions in Chapter 2 where some of the principles of organization and order were introduced. Chemical kinetics is based on systems which contain statistically significant numbers of molecules interacting predictably through random collisions. Biological systems do not fit entirely into this kind of analytical picture. At the level of the gene itself, the essential specific molecular structure (DNA, or RNA in some situations) need be represented in a cell only once, and the reactions of replication need yield only one more molecule per cell division to maintain the system. Here the kinetics involved would be those of a microsystem were it not for the remarkable organization which provides for directed reactions a limited space, as in a pipe or an assembly line.

An introduction to principles involved has been given earlier (p. 96) in connection with the enzymatic replication of DNA. In this case a reaction product (an existing DNA chain) evidently reacts directly with substrates through hydrogen bonding to organize the polymerization reaction that is to occur. Here there may be statistical numbers of nucleotides in the "pipe" to provide adequate probability for correct base-pair combinations, but incorrect pairs can also hydrogen bond in various ways, and it will be surprising if there are not other organizational restrictions in addition to base pairing (such as enzyme specificity, substrate orientation, etc.) which are present *in vivo* as an aid in avoiding mistakes. A potential mechanism here, for which there is no evidence whatever, is separation of complementary strands of DNA at replication, and coupling of nucleotide triphosphates to adaptors so that base pairing to one chain and adaptor pairing to the complementary chain could serve as a double restriction on synthesis. In any case, many details of the replication mechanism *in vivo* remain to be established and we do not know how to handle the kinetics of such processes.

Proceeding from replication itself and production of first gene products (a process in nearly the same category as replication), to actions of gene products in producing enzymes, and so on, and to the actions of enzymes in innumerable metabolic reactions, we approach more closely situations which deal in known ways with predictable reaction kinetics. However, it will become evident that unknown factors are apparent here, too. For example, in a discussion of the nature of active sites of enzymes, Koshland (533) emphasized the potential for the existence of enzymes in different conformational states giving flexibility for substrate combinations and appropriate proximity and orientation for reaction. Flexibility of this kind provides for more control of re-action rates than random collisions based solely on enzyme and sub-strate concentrations. Thus it is likely that structural organization exerts an influence on reaction rates at all levels in biological systems.

Reaction Rates

It is not within the scope of this book to present an overall account concerning enzyme action, and very extensive details can be found elsewhere (4, 894, 422, 5). Here the purpose is to select information that seems to have special significance in relation to genetic potential, interaction, and regulation. Thus the interest in isolated enzymes is not directed toward their intrinsic properties but toward the potential importance of these properties in the complex interactions which occur in living, changing metabolic systems.

General Properties of Enzymes. Several hundreds of enzymes are known in terms of specific activities and well over a hundred have been obtained as essentially chemically pure substances. So far, all that are well characterized are primarily protein in nature, and though many are conjugated with other classes of substances, it is evident that specific amino-acid sequences (primary structure), folding and weak bonding as in helices (secondary structure), and chain cross linking by —S—S— bonds, H—bonds, ionic attractions, and accumulations of hydrophobic groups (tertiary structure) are all highly significant in the determination of catalytic potentials. Further, all enzymes are not formed from single primary chains as is ribonuclease (see p. 81). Glutamic acid dehydrogenase, for example, has a quaternary structure in which the enzyme is a tetramer derived from four like molecules. This case is especially interesting in that it has been established that the monomer has alanine dehydrogenase activity (1016). Another important basis for difference is illustrated by tryptophan synthetase, which is made up of two different primary chains (p. 270). These last two examples, and numerous others that can be cited, are of special interest in the present context since they represent a beginning of understanding of organization and may be expected to give an importance to portions of protein molecules not necessarily involved directly in formation of active catalytic sites. Perhaps detailed extension of the involvement of total structure in formation of "assembly line" units such as in the cytochrome system (p. 45) is not so far off.

Proteins, as catalysts, are considerably more complex than solid-state particles such as metals or metal oxides because of their highly specific, and at the same time less rigid, structures. As discussed by Lumry (626), thermal vibrational motion in a long α helix would be relatively simple, but in globular proteins (such as enzymes) with helical and random sections, vibrational behavior must be extremely complex and of sufficient magnitude to break and remake continually some of the weak secondary bonds. This kind of flexibility, and other kinds concerned with evidences for conformational changes accompanying combinations with substrates (see discussions by Lumry, Westheimer, Koshland, and Linderstrom-Lang and Schellman (626, 1072, 533, 619), provide a basis for a picture of the catalytic protein as a dynamic unit in itself.

From such considerations, Linderstrom-Lang and Schellman listed four ways in which multifunctional catalysis can arise in an enzyme: (1) the catalytic groups are connected to a secondary structure which is relatively rigid; (2) the catalytic groups are part of a system of random chains, and the groups arrange themselves into a catalytic site with a certain probability which is partially determined by the binding of substrate; (3) the catalytic groups are brought into the proper position by a transient secondary folding of portions of the molecule also determined by binding with substrate; and (4) the first three possibilities occur in combination. Here the first possibility has long served as a fundamental postulate, and it embodies the concepts of the lock-and-key and the template. It now seems reasonable to accept the fourth possibility just listed—the combination—as a general concept. To do so invalidates little and broadens the horizons. Our analyses of genetic influences on enzyme activities thereby may gain some new and perhaps more realistic parameters.

It is evident even from these brief considerations of the properties of enzymes as catalysts that much remains to be discovered concerning specific mechanisms of action and control of action. Furthermore, it is just as apparent that the commonly used criteria for evaluating the effects of environmental factors which influence the enzymatic control of reaction rates are, in fact, complex and composite in nature. Some of these important variables are:

1. Concentration of enzyme.
2. Concentration of substrate or substrates.
3. Concentration of inhibitors (including reaction products).
4. Concentration of various ions.
5. Temperature.
6. Oxidation-reduction potential of the environment.

These are all significant variables, and even though each is complex in nature with respect to the possible number of different origins, from the practical standpoint, all are useful for comparative purposes and as aids in discovering exact details of precise mechanisms. It is not to be expected that data from quantitative measurements of enzyme activities measured *in vitro* can be applied directly to *in vivo* systems, but the principles involved are certainly applicable. For example, temperature and pH optima have been used extensively for characterization of enzymes, and yet it is quite obvious that these are influenced extensively by the nature of buffers used, ionic strength, and a multitude of other environmental factors. Even under one set of conditions, measured reaction rates represent a summation of characteristics of substrate and enzyme. Niemann and collaborators (424) have made extensive studies of kinetics of hydrolysis of synthetic substrates, using chymotrypsin, which indicate that a given substrate can combine with the enzyme at the active site in a number of different ways, some of which lead to hydrolysis and some of which lead only to inhibition by reduction of effective enzyme concentration. Relative contributions of different kinds of combinations are expected to be different with different substrates, and an observed rate is no measure of intrinsic capacity for reaction in the most favorable mode of combination. This principle should apply whether an active site is considered to have rigid geometry or flexibility influenced by combination with substrate.

In spite of all these complications, experimental measurements of reaction rates in isolated enzyme systems are very useful in many ways, and, for present purposes, they delineate many possible parameters for consideration in evaluations of genetic effects on metabolism. Some discussion of enzyme kinetics is therefore pertinent. Very extensive descriptions can be found in recent literature (91) but a large part of this is concerned with limiting cases where substrates are always in excess. As will be pointed out, this is not necessarily true *in vivo,* and, for our purposes here, principles based on a more general situation, with E (enzyme) and S (substrate) in any ratio, are outlined briefly.

Estimation of Reaction Rates. Consider a simple isolated system containing a pure enzyme and a pure substrate dissolved in water or buffer. For a hydrolytic reaction the following equation can be written:

$$ROR' + H_2O \xrightarrow[E]{} ROH + R'OH$$

(Substrate, S_f)   (Enzyme, E_f)   (Products, P)

Since the total concentration of H_2O does not change significantly during the reaction, this component can be neglected. It will be assumed that the reaction proceeds by formation of a complex between the substrate and the enzyme, and that the complex can decompose to reform the reactants, or to form enzyme plus products, in accordance with the following equation:

(1) $$S_f + E_f \underset{k_2}{\overset{k_1}{\rightleftharpoons}} ES \xrightarrow{k_3} E_f + P$$

The subscript f refers to the free (uncombined) substances, and k_1, k_2, and k_3 are reaction-rate constants. By application of the mass law and several reasonable assumptions, Michaelis and Menten (670) derived an equation which relates the concentration of the substrate (S) and the reaction velocity, v, for reaction 1:

(2) $$v = \frac{V_m(S)}{K_m + (S)}$$

Here the apparent dissociation constant $K_m = k_2/k_1 = (E_f)(S_f)/(ES)$, and it is numerically equal to (S) at half maximum velocity. The maximum velocity, V_m, is a theoretical value approached by v at large substrate concentrations. Under these conditions, an enzyme molecule is surrounded by substrate molecules, and with every dissociation of the complex ES a new substrate molecule is immediately ready to enter into combination. Although an enzyme in such a system is often referred to as saturated, it is obvious that complete saturation cannot be reached. The question of the formation of ES is a complex one, since there is good evidence that E and S can combine in more than one way. This is probably especially significant in reactions where S is a large, complex molecule. Each combination possible contributes to the reaction velocity in its own characteristic way; that is, the reaction may not go at all with some combinations, and these in effect reduce the enzyme concentration, and thus the velocity, of the overall reaction (290).

Graphical representations of equation 2 are shown in Figure 5.1. The curves of graph *a* are perhaps most familiar, but those of graph *b*

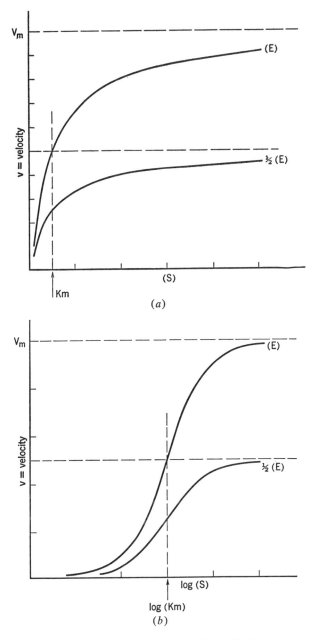

Figure 5.1. Graphical representations of the Michaelis-Menten equation. (a) Plot of reaction velocity versus substrate concentration (S) showing $K_s = (S)$ at $v = V_m/2$. (b) Plot of reaction velocity versus log (S) showing that the velocity approaches limits at both ends of the curves.

221

are more instructive since they show that v approaches zero with decreasing substrate concentrations just as it approaches V_m with increasing substrate concentrations.

Equation 2 agrees remarkably well with many experimental results obtained with isolated enzymes and at relatively high substrate concentrations. However, as pointed out by Straus and Goldstein (978, 337), equation 2 does not contain the term (E_f), the enzyme concentration, and in addition the equation should contain the term k_3 as part of the dissociation constant. Introduction of these terms into the equation provides a more generalized description of an enzyme-catalyzed process. That such a general relation is obtained is particularly important to the present discussion, since in an *in vivo* system the ratio $(S)/(E)$ may have any value within limits approaching $+$ or $-$ infinity. For these reasons the formulations of Straus and Goldstein are preferable to equation 2, and some of the details of their derivations are now presented.

The reaction shown by equation 1 is considered to be a part of a whole chain of reactions operating in a steady state. That is, (ES) remains constant, $d(ES)/dt = O = k_2(ES) + k_3(ES) - k_1(S_f)(E_f)$, and S is supplied and converted to P at a definite velocity, v. Such a steady state perhaps approximates the conditions of *in vivo* enzyme reactions, at least over short periods of time in constant physical, chemical, and genic environments. Referring to equation 1 and neglecting the possibility of a reverse reaction from P back to S, we find the constant

$$(3) \qquad K_s = \frac{(E_f)(S_f)}{(ES)} = \frac{k_2 + k_3}{k_1}$$

and

$$(4) \qquad (E_{\text{total}}) = (E) = (E_f) + (ES)$$

while

$$(5) \qquad (S_{\text{total}}) = (S) = (S_f) + (ES)$$

As noted previously,

$$v = k_3(ES)$$

and

$$V_{\max} = k_3(E_{\text{total}})$$

since $(E_{\text{total}}) = (ES)$ at maximum velocity.

Now, for convenience, by definition the fractional activity

(6) $$a = \frac{v}{V_m} = \frac{(ES)}{(E)} \quad \text{and thus} \quad (ES) = a(E)$$

Substitution of the equivalents of (E_f), (S_f), and (ES) from equations 4, 5, and 6, respectively, into equation 3 yields the following equality:

(7) $$(S) = K_s \frac{a}{1 - a} + a(E)$$

In order to simplify equation 7 further, for practical manipulations, Straus and Goldstein introduced the idea of specific concentrations which, like specific gravities, are dimensionless. The specific concentration of E is $E_s' = (E)/(K_s)$, and that of S is $S' = (S)/K_s$. Substitution in (7) gives the dimensionless general equation

(8) $$S' = \frac{a}{1 - a} + aE_s'$$

Equations 7 or 8 can, of course, be expanded by substitution of equalities already shown and written:

(9) $$(S) = K_s \frac{v/V_m}{(1 - [v/V_m])} + \frac{v}{V_m} (E)$$

If (E) is assumed to be negligible relative to (S), then the second term can be dropped, and rearrangement gives the familiar Michaelis and Menten equation 2. It is at once evident from equation 9 that a reaction velocity v is dependent simultaneously on the concentration of the substrate, the concentration of the enzyme, and the properties of the enzyme described by the constants K_s and V_m. These two constants are evidently related to the "affinity" of the enzyme for the substrate and the rapidity with which the enzyme is able to convert the substrate to the products. The equivalent equations 7, 8, and 9 are represented graphically in Figure 5.2a and b. The limiting curve at the left in Figure 5.2a, where (E) is much smaller than (S), is essentially the same as that obtained from the Michaelis and Menten equation and shown in Figure 5.1. In this region the fractional activity is independent of the enzyme concentration. The series of curves to the right show an increasing dependence of the fractional activity on (E) until $a = (S)/(E)$ and the curves become nearly identical in shape and are equally spaced with equal changes in log (E) or (S). The

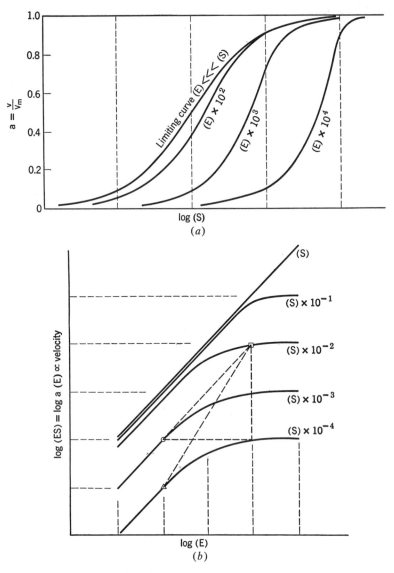

Figure 5.2.   Graphical representations of the Straus-Goldstein equation.   (a)
Plots of velocity versus log (S) at different relative concentrations of E.   The
limiting curve at the left where (E) is much smaller than (S) corresponds to the
curves in Figure 5.1.   (b) Plots of log velocity versus log (E) at different concentrations of S.   The symbols, $\square$ = initial state of a system, $\odot$ = expected state
by dilution of the system 100-fold, and $\triangle$ = the actual state of the system after
dilution, illustrate the fact that the magnitude of the effect of an imposed change
is dependent on the prechange state of the system.   In the illustration, if the initial
state had been at (S) instead of (S) $\times$ 10^{-2}, then the reaction velocity change
would have been proportional to the dilution.

curves of Figure 5.2*b* illustrate clearly the relations existing between (E), (S), and the rates at which products are formed. It should be observed that a tenfold change in (E) or in (S) can result in a change in the reaction velocity of any amount from zero to tenfold, depending on the state of the operating system prior to the change. A related property of such an enzyme system, the dilution effect, is also illustrated in Figure 5.2*b*. As shown in the figure, if an enzyme system operating under conditions shown by the square is diluted a hundredfold, the reaction velocity will decrease, not a hundredfold as might be expected, but nearly a thousandfold as shown in the figure. On the other hand, with an initial high (S) instead of $(S) \times 10^{-2}$, the velocity change would be proportional to the dilution.

The foregoing discussion has been concerned with an analysis of the simplest kind of an enzyme-catalyzed system, but it serves as an adequate illustration of principles that are certainly quantitatively applicable to somewhat more complex systems and probably qualitatively applicable to *in vivo* biochemical reactions. *Particular note should be made of the fact that the magnitude of a biochemical reaction depends not only on the magnitude of a change in concentration of a component that enters into the reaction but also on the existing operational state of the system.* The dilution effect (p. 224) serves as a good example of this important point. That is, if the cytoplasm of a living cell can become diluted (or concentrated) with respect to water, it is expected that some reaction rates will be altered in proportion to the dilution while others will not. Which result is obtained depends on the relative concentrations and specific properties of the components that enter into each reaction. For these reasons a change as simple as a dilution requires extensive revisions in an existing balance of reactions and concentrations of cellular components. These conclusions are necessarily qualitative in nature, since they are based on observations of simple isolated systems. Still, they should be considered along with the discussion of more complex systems on the pages that follow.

Enzyme Inhibitors and Inhibition. The inhibition of enzyme-catalyzed reactions is a very common and well-known phenomenon that has been used extensively in describing the action of enzymes as well as in investigations of enzymatic reactions in complex systems and whole tissues. Inorganic ions, such as F^-, CN^-, N_3^-, $S^=$, $AsO_4^=$, and heavy metal cations (Cu^{++}, Ag^+, and Hg^{++}) owe at least one part of their toxicities to inhibition of enzyme systems. Similarly, a great number of organic substances (e.g., iodoacetate, malonate, various alkaloids, etc.) act as inhibitors for both *in vitro* and *in vivo* enzyme systems. This field of endeavor became expanded enormously following the suggestion of Fildes (1101) and Woods (1100) that sulfonamides exert their inhibitory actions because of their chemical structural relations to that of the B vitamin, *p*-aminobenzoic acid. That is, the drug competes with the vitamin for combination with an enzyme, the action of which is essential for growth of the cells that are inhibited by the drug. As a consequence of these observations, a very large number of analogs of normal metabolites have been prepared and tested as inhibitors and chemotherapeutic agents. There is no doubt of the validity of the principle involved, that is, that structurally related substances do act as antagonists for specific actions of enzymes. On the other hand, it has not been possible to predict accurately the structural changes that can be made in a normal metabolite in order to produce a good antagonist.

These substances just mentioned, for the most part, do not occur or function as inhibitors in living systems, but it is most important to note that a very large number of inhibitors do occur naturally. Actually, it is probable that nearly every substance produced in an organism can act as an inhibitor for the same, or another, organism under some set of conditions, though attention is usually focused on the more highly active inhibiting compounds such as the antibiotics, the regulatory hormones, and antibodies or other specific protective agents.

Some good examples of inhibitors of the last type are to be found in the work on the isolation from pancreas tissue of a protein that interferes with the action of trypsin (743) and in the early work on the intestinal parasitic roundworm, *Ascaris* (554). It has been found that extracts of the worm inhibit the action of the pepsin and trypsin that are found in the intestinal tract of the host. The extracts do not, however, inhibit the action of the proteolytic enzyme, papain, and the living worms can be digested with this enzyme, which is of plant rather than of animal origin. The naturally occurring inhibitors such as the specific ones mentioned are undoubtedly extremely important in the processes of metabolic control, and some of them will receive attention in the subsequent discussions. However, now it is more appropriate to consider the phenomenon of multiple substrate inhibitions.

It is well known that many enzymes catalyze a type of reaction, and thus they act on a variety of naturally occurring substances that frequently occur together in the same tissue. Amylases, phosphatases, proteolytic enzymes, transaminases, hexokinases, and many other enzymes can be cited as examples. When two or more substrates exist together in the presence of an enzyme that will act on both, each substrate can act as an inhibitor for the reaction involving the other. A good illustration of these facts comes from the work of Foster and Niemann (289), who studied the kinetics of the simultaneous hydrolysis of acetyl-L-tryptophanamide and acetyl-L-tyrosinamide in the presence of chymotrypsin. The reactions involved are:

Acetyl-L-tryptophanamide $\longrightarrow$ Acetyl-L-tryptophan $+ NH_3$

Chymotrypsin

Acetyl-L-tyrosinamide $\longrightarrow$ Acetyl-L-tyrosine $+ NH_3$

Here the two substrates appear to combine with the enzyme at the same site, and there is a quantitative dependence of each reaction velocity on the relative concentration of the two substrates and their relative affinities for the enzyme. (Some aspects of inhibition kinetics are considered in the next section.)

It is not known to what extent multiple substrate inhibitions contribute to the normal balance in metabolism, but there is reason to believe that such phenomena do occur *in vivo* and that they account for the regulation of the velocities of many reactions.

Inhibition Kinetics. It is appropriate at this point to examine some of the quantitative aspects of the action of inhibitors on reaction velocities of relatively simple enzyme-catalyzed reactions. Reviews by Wilson (1087), McElroy (644), Hearon et al. (422) provide many details that will not be represented here. Inhibitors have been classified into three general categories of simple isolated systems.

1. Competitive: The inhibitor competes with the substrate for a specific active site on the enzyme.

$$E + S \rightarrow ES$$
$$E + I \rightarrow EI$$

2. Noncompetitive: The inhibitor and the substrate combine with the enzyme independently at different positions on the enzyme surface.

$$S + E + I \rightarrow S\text{---}E\text{---}I$$

3. Uncompetitive: The inhibitor combines with the enzyme substrate complex but not with either of these components alone.

$$ES + I \rightarrow E\big\langle{}^I_S$$

These somewhat arbitrary classifications do not delineate all of the possibilities, but they present useful concepts. Again, referring to Goldstein's general formulation and to competitive inhibition only, it is necessary to consider simultaneously the concentrations of the substrate (S), the enzyme (E), and the inhibitor (I). Thus, as in the derivation of equation 8,

$$K_s = \frac{(E_f)(S_f)}{(ES)}$$

and, in addition,

$$K_I = \frac{(E_f)(I_f)}{(EI)}$$

By combining these two equations through the common term (E_f) and by appropriate substitutions, as described previously, the following general equation can be obtained

$$(10) \quad I' = \left[(S' - aE_s') \left(\frac{1-a}{a} \right) - 1 \right]$$

$$+ \left[(1-a) \left(1 + \frac{1}{S' - aE_s'} \right) \right] E_I'$$

Here $I' = (I)/K_1$; $S' = (S)/K_8$; $a = v/V_m$; $E_8' = (E)/K_8$; and $E_I' = (E)/K_I$, and the equation describes the quantitative relation between E, S, K_8, K_I, v, and V_m. In experimental practice the equation can be simplified by setting up conditions such that some of the terms become negligible. For example, in a system with an excess of inhibitors but not of substrate the equation becomes

$$I' = \left[(S' - aE_s') \left(\frac{1-a}{a} \right) \right] - 1$$

At another extreme, when the enzyme is present in great excess over both the substrate and the inhibitor, equation 10 can be reduced to

$$I' = (1 - a)E_I'$$

Under these circumstances there is more than enough enzyme to combine with all of the substrate and inhibitor, and competitive inhibition does not exist.

The general equation 10, which describes competitive inhibition over a broad range of values for the variables it contains, is analogous to the general equation 8, where (I) and K_I are not included. The additional complexities resulting from the introduction of this variable and constant do not alter the principle already set forth, that is, that the magnitude of change of a reaction velocity is dependent not only on the magnitude of change of one of the variables (S), (I), or (E) but also upon the specific preexisting relations between these variables and the constants K_s and K_I. The dilution effect and other quantitative aspects of the problem have been discussed by Goldstein, and they will not be elaborated here since many *in vivo* systems are too complicated to be analyzed quantitatively by use of equation 10. For example, some reactions involve the interaction of a number of additional components, as indicated:

$$\left.\begin{array}{l} S_{1,2}\cdots \\[4pt] I_{1,2}\cdots \\[4pt] \text{Coenzyme} \\[4pt] \text{Metal ion} \end{array}\right\} + E \rightleftharpoons Me\!-\!\overset{\overset{I}{|}}{\underset{\underset{Co}{|}}{E}}\!-\!S \rightleftharpoons E + P_x + Co + Me + I$$

In this example each of the reactants will have some characteristic affinity for the enzyme as in the simplified case. That some reactions are probably this complex does not necessarily introduce any new principles or ideas. It is clear, however, that the variables and combinations of variables that can influence the reaction velocity will be enormously increased. That is, the reaction velocity can become altered in many more different ways than in the simple system.

Enzyme Complexes

From the foregoing discussion it is obvious that even the simplest enzyme system is subject to influence and control by a great variety of factors, and when the enzyme is part of a biological system the complications are compounded. As discussed in Chapter 2, such systems are the rule rather than the exception. Green and Hatifi (354) have described this situation in terms of "biological machines," and their partial list of energy "transducing machines" is reproduced in Table 16. Of these, the mitochondrial system involving citric-acid-cycle oxidation, electron transport, and oxidative phosphorylation is virtually universal in biological systems, and some details are especially informative with respect to the nature of the enzyme-complex problem. A generalized

Table 16.   Biological Transductions *

Energy Transduction	Biological Transducing System
Sonic to electrical energy	Ear
Radiant to electrical energy	Eye
Mechanical to electrical energy	Skin
Chemical to electrical energy	Nerve
Radiant to chemical energy	Chloroplast
Chemical to radiant energy	Luminescing organs (firefly)
Chemical to osmotic energy	Kidney, cell membrane
Chemical to mechanical energy	Muscle
Chemical to sonic energy	Vocal cords
Chemical to electrical energy	Electric organs in electric fish
Oxidative to utilizable chemical energy	Mitochondrion

* From Green and Hatifi (354).

overall pattern of the reactions involved here is shown in Figure 5.3. At the experimental level, intact mitochondria isolated from tissues will oxidize components of the citric-acid cycle through the cytochrome electron-transport system and generate ATP. Fragmented mitochondria (sonic disintegration) which still retain double membrane structures perform electron transport and oxidative phosphorylation, while more drastic disintegration to give single-membrane fragments yields material having only electron-transport function. The overall picture, then, is one of relative order and confinement of active components. The units of electron transport are the most firmly fixed in position (but to different degrees among themselves), evidently by specific orderly embedment in phospholipid-protein membranes. At the next level, oxidative phosphorylation, at least one essential component appears to be relatively loosely bound between membranes while citric-acid-cycle organization is disrupted the most easily. Loose binding by some components may be essential to the dynamics of such systems, but structurally directed order involving portions of enzyme molecules not directly concerned with functional active sites is probably important even with the most mobile components. This concept of total structure significance may be applicable to all macromolecules whether they are apparently confined in cell structures or not. If so, the number of potential variables concerned with genetic regulation and control becomes expanded considerably. Experimental observations on structures of hemoglobin (p. 614) and tryptophan synthetase (p. 270) should be considered with the foregoing concept in mind.

The kinetics of reactions in complex-enzymes systems poses problems of great interest, but factual information is limited. Artificial systems of sequential reactions have been dealt with to a considerable

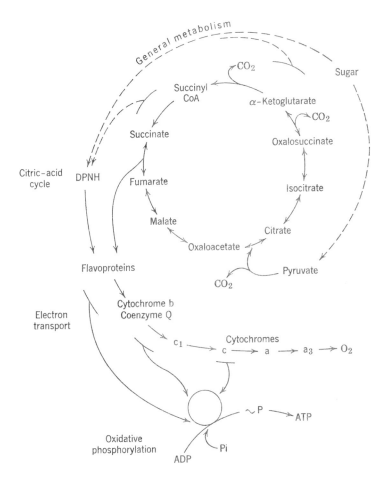

Figure 5.3. A diagram showing the general pattern which relates citric-acid cycle oxidation to the electron-transport system and oxidative phosphorylation.

Table 17. Metabolic States of Mitochondria and the Associated Oxidation-Reduction Levels of the Respiratory Enzymes *

			Characteristics					Steady-State Percentage Reduction of Components		
State	(O_2)	ADP Level	Substrate Level	Respira-tion Rate	Rate-Limiting Substance	a	c	b	Flavo-protein †	DPNH
1	0	Low	Low	Slow	ADP	0	7	17	21	90
2	0	High	0	Slow	Substrate	0	0	0	0	0
3	0	High	High	Fast	Respiratory chain	4	6	16	20	53
4	0	Low	High	Slow	ADP	0	14	35	40	99
5	0	High	High	0	Oxygen	100	100	100	150	100

* From Chance (153).

† These values are based on the amount of flavoprotein that is reduced on addition of antimycin A to the mitochondria in state 2 which is $\frac{2}{3}$ the state 5 value.

extent, and the principles derived are useful but not necessarily applicable to normal systems, where relative enzyme and substrate concentrations are not usually known; nor is it usually possible to evaluate the other variables of organization. A more direct approach to this problem of kinetics in complex systems is that achieved by Chance and collaborators (154, 155). Following the technical development of methods for very rapid measurements of absorption spectra in turbid suspensions, extensive studies were made of kinetics in the electron-transport system of isolated mitochondria and whole cells of various kinds. The methods permit estimation of total quantities and the states of oxidation—reduction of cytochromes *a, b,* and *c,* flavoprotein, and DPN. Examples of information gained are given in Tables 17 and 18. Several states of the system in mitochondria are shown in Table 17 where state 3 represents that of active rapid respiration and phosphorylation. Complete oxidation is shown by state 2 (no substrate) and complete reduction by state 5 (no oxygen). Relative velocity constants for the separate reactions with and without phosphorylation are shown in Table 18. These are very significantly different, and a mechanism of control of respiration by the oxidative phosphorylation system is indicated. Further limitations in whole cells as opposed to isolated mitochondria were demonstrated by the turnover ratio of about $1:3$ in the electron-transport system. Additional definition of interrelations of complex systems in cells has been made by application of computer analysis of the complex data derived from metabolism of substances quite far removed from the electron-transport system. Results indicate, among other things, an important control of metabolism of mitochondrial membrane permeability and an intracellular separation of ATP in at least two functional compartments. By use of various substrates and alterations of various other environmental factors such

Table 18. Relative Values of Reaction-Velocity Constants for a Phosphorylating and a Nonphosphorylating Respiratory Chain as Calculated from Steady-State Data on the Basis of the Equivalent Homogeneous System °

Phosphorylating chain (state 3)
$$O_2 \rightarrow a_3 \xrightarrow{4} a \xrightarrow{1.0} c \xrightarrow{0.7} b \xrightarrow{0.3} fp \xrightarrow{0.2} DPNH \rightarrow substrate$$

Nonphosphorylating chain
$$O_2 \rightarrow a_3 \xrightarrow{2} a \xrightarrow{1.0} c \xrightarrow{1.3} c_1 \xrightarrow{0.9} fp \rightarrow DPNH$$

* From Chance (153).
Note: The velocity constant for the reaction between cytochromes a and c is taken to be unity.

as specific inhibitors, it can be expected that a much more elaborate definition of organizational control of metabolism can be obtained, and thus evaluations of genetic and developmental influences on organization should become more approachable on a biochemical basis.

In the foregoing discussion of organized enzyme complexes, only part of the fundamental systems of mitochondria have been used as specific examples, and only some aspects of interrelations and control of these are described. From the discussion of cell structure in Chapter 2, it is evident that what has been said here is only a small part of the whole picture. Organizational details are less well known, for example, in protein-synthesis systems associated with microsomes or in nucleic-acid-synthesis systems in nuclei, but in these, too, order in structure must be an all-important factor in the regulation of reaction rates.

The Dynamic State

An important principle for consideration in evaluating gene-metabolism relations is that anticipated by Borsook and Keighley (87) and further characterized, in an early use of isotopic tracers (N^{15}), by Schoenheimer (882). At the beginning of a series of lectures prepared for presentation in 1940, Schoenheimer stated: "The general title, *The Dynamic State of Body Constituents,* designates an attempt to consider . . . some results of modern biochemistry which suggest that all constituents of living matter, whether functional or structural, are in a steady state of rapid flux." This concept along with introduction of the use of isotope tracers instigated a revolution in biochemical thought in that it became clear that chemical constituents taken in from the environment of an organism could no longer be considered only as fuel. Rather, it quickly became accepted that molecular constituents of food undergo rapid interchange with like constituents of cells and the particular molecules that become oxidized to provide energy can come directly from food, but they also come by exchange from the apparently unchanging structures of the organism. It was expected that exchange might be relatively unimportant during rapid growth.

In the present context, the first aspect of this generalization of the dynamic state that is of particular interest is its applicability to the macromolecules, the proteins and nucleic acids. Is it necessary to consider genetic mechanisms of control of turnover of these components as well as mechanisms of initial synthesis? With respect to proteins, an affirmative answer is clear. Eagle and collaborators (238), from studies of human cell cultures, have established an incorporation rate of amino acids into protein of about 1% per hour. This level is essentially the same in resting and growing cells. In resting cells, in the absence of net synthesis, the rate was constant for 72 hours, at which time more than half the total protein amino-acid residues had exchanged with those in the medium. General, but not necessarily uniform, renewal of different kinds of protein is indicated. In growing cells with a protein-synthesis increase of about 4% per hour, it is evident that turnover (1% per hour) is an important factor in the overall processes concerned. These results with human cells, in some respects, are in sharp contrast to observations made with bacteria. At one point evidence seemed uncontestable that turnover of enzyme protein does not occur in *E. coli* (447), and this is evidently true for rapidly growing cells in the log phase. In these experiments rapidly growing cells were isotope-labeled and transferred to a medium containing unlabeled amino acids and an inducer for the enzyme β-galactosidase. After synthesis, the enzyme was found to be *not*-labeled in contrast to the other proteins present in the bacteria. However, it was subsequently established that resting cells of *E. coli* do exhibit renewal (653) to the extent of about 5% per hour, a rate much higher than in the mammalian cells. At the same time, this is only about 1% of the very rapid rate of protein synthesis that occurs in log-phase bacterial cultures.

The foregoing examples are taken from many which yielded similar results, and turnover or renewal of protein on a continuous, dynamic basis is clearly a factor to be taken seriously. In relation to genetic control, there are at least three important questions to be answered. (1) Is the degradation of proteins in turnover the reverse of synthesis? (2) Is total hydrolysis to amino acids a characteristic of the turnover process? (3) Do different kinds of proteins differ greatly in the degree to which they undergo turnover? There is no answer to the first question, a little evidence indicating total degradation in answer to the second, and a good deal of evidence, in answer to the third, that major differences in turnover rates among different proteins do exist (653).

The turnover situation with respect to RNA has similarities to that of proteins. That is, in rapidly growing bacterial cells RNA is stable, but in resting cells high rates of replacement have been demonstrated. Turnover in RNA of mammalian cell cultures has also been shown, and different rates for different kinds of RNA are indicated. For example, the so-called messenger RNA, a prime essential in genetic expression, is evidently quite unstable and continuous renewal is necessary (77,

Table 19. Incorporation of Adenine into Nucleic Acids *

Experiment	Desoxyribonucleic Acids			Ribonucleic Acids		
	Total Purines †	Adenine †	Guanine †	Total Purines †	Adenine †	Guanine †
1. Mixed viscera, adenine fed 10 days	...	0.55	0.38	...	15.9	9.1
2. Nongrowing liver, adenine fed 5 days	0.17	0.29	...	...	21.2	8.7
3. Regenerating liver, adenine fed 5 days	8.7	16.3	3.2	11.8	22.7	8.1
4. Regenerating liver, 26 days, no adenine after 5 days	6.5	...	...	1.8	2.7	1.7

* From Furst, Roll, and Brown (305).
† Calculated on the basis of 100%-labeled adenine fed.

703), whereas ribosomal RNA is relatively stable. DNA, on the other hand, has been generally found to be quite stable in most situations. Overall comparisons of synthesis and turnover of the nucleic acids in regenerating rat liver made several years ago by Furst et al. (305) is illustrative and still quite valid. As shown in Table 19, N^{15} adenine when fed to rats is rapidly incorporated into liver RNA whether the liver is nongrowing or regenerating, and the label is lost from the liver when feeding is stopped. In contrast, extensive labeling of DNA occurs only with regenerating liver, and most of the label is retained for a long period after termination of the supply. A great many more recent results with various kinds of biological material have yielded similar results. A noteworthy exception is that of *E. coli* at the time of infection with phage. In this case, bacterial DNA is degraded and utilized for phage production, and, at the same time, there is a rapid production of both stable and unstable RNA (77).

Metabolic Pools

An expression of a genetic change in terms of chemical composition (e.g., a pigment change) can be an extremely complex phenomenon in terms of metabolic reactions that give rise to an observed phenotype. Relative reaction rates in simple and complex enzyme systems and the dynamic state are all subject to limitations by compartmentalization, both within cells and among tissues of higher organisms. Thus there exist metabolic pools of various kinds and dimensions and the results of a mutation in one may be altogether different from the results in another. A good illustrative example of a general case for a whole organism, the rat, is shown in Table 20. The amino-acid leucine was synthesized using deuterium and N^{15} as H and N labels to give the substance with the following composition:

$$
\begin{array}{c}
CH_3 \qquad\qquad N^{15}H_2 \\
\diagdown \qquad\qquad\quad | \\
\quad C-C-C-COOH \\
\diagup \;|\;\;\;|\;\;\;| \\
CH_3 \;\; D \;\; D_2 \; D \\
\quad\;\; \gamma \;\; \beta \;\; \alpha
\end{array}
$$

After feeding the labeled leucine to rats the animals were sacrificed, the proteins were extracted from various tissues, and pure leucine was isolated from each preparation. The product was then degraded step-

Table 20. Metabolic Changes in Deuterium-N^{15} Ratios of
α-, β-, and γ-Deuterium Atoms of L-Leucine *

| Ratio D:N^{15} | Leucine | | | Isolated/Fed | |
	Fed	From Carcass	From Internal Organs	Carcass	Internal Organs
D:N^{15}	1.46	0.38	1.53	0.26	1.05
D:N^{15}	3.72	8.05	5.53	2.17	1.49
D:N^{15}	1.15	2.45	1.83	2.13	1.59

* Sprinson and Rittenberg (943).

wise in order to determine the relative interchange of N^{15} to D and H in the γ, β, and α positions. The data in the table demonstrate, first, general incorporation in accordance with principles of dynamic equilibrium. Further, they demonstrate not only interchange of the whole molecule with leucine already in protein, but, in addition, interchange of the nitrogen atom within the molecule relative to the more stable β and γ C-D bonds. D in the α position is interchanged even more than the nitrogen atom, and the extent to which all these changes take place is not the same in different parts of the animal. At this level there is no definition of the nature of specific metabolic pools involved, but there is a clear demonstration of exchanges of atoms and whole molecules—results which demand the use of the pool concept. Differences here could lie in different balances of reactions in different groups of like cells, but even within single cells there is ample organization and structure to provide for extensive physical separation of pools by membranes, as well as others limited more by local concentrations and relative diffusion and reaction rates. The pool concept, then, is only a recognition of the existence of localized conditions for reaction in which a metabolite is altered and is regenerated. The change may be through some directly reversible system or through circuitous reaction series, but in either case there can be genetically controlled complications in just returning to the starting point of the reaction excursion.

The idea of separate metabolic pools within single cells has been invoked many times to account for various kinds of experimental data, as for example in ATP synthesis and metabolism in ascites tumor cells.

An extensive discussion of the problem with respect to amino-acid metabolism can be found in reports of a recent symposium on the subject (448).

Biochemical Reactions

The discussion so far in this chapter has been based on the contention that the units of heredity primarily control the relative rates of reactions. It has been pointed out that there are many possible ways in which rates can be affected even in relatively simple systems and that even more variables must be considered when dealing with enzyme complexes. It is not known to what extent a reaction rate must be altered in order to produce an obvious phenotypic change in an organism. However, it is reasonable that a very small change in rate can result in an extreme morphological or biochemical alteration. For example, if substance A is an observed structural unit in a cell and it exists because it is produced at the same rate it is degraded, then a small reduction in the rate of its formation will result in the disappearance of substance A if the rate of degradation remains unchanged and if no other biochemical compensations exist.

In the discussion of reaction rates little attention has been given to the kinds of biochemical reactions that occur or to the mechanisms by which they take place. It is not within the scope of this book to discuss these problems in great detail, but some aspects of them have a direct bearing on the subject of reaction velocities and the genetic control of these velocities. Some of these points are discussed in the remainder of this chapter.

Energetics of Biochemical Reactions. When the chemical reactions of a cell reach a thermodynamic equilibrium, the cell is dead, and it is well decomposed. Cells exist in life at the expense of the energy supplied by the environment and by virtue of their capacities to extract and store, for a future time, energy and materials from the environment. The cell wastes rather than conserves energy, and its apparent defiance of thermodynamic laws by its extraordinary degree of organization is an artifact arising from the dynamic state of its existence.

As already noted, many investigators in the field prefer to consider a living system as existing in a steady state with respect to its biochemical processes. This, of course, can be true literally only over a short period of time since cell divisions, adaptation to environmental changes, and aging certainly introduce elements of unsteadiness in cellular processes. In actual fact, it must be considered that the rates of reactions will not be precisely the same over a finite interval of time.

In the photosynthetic organisms, the energy required for maintenance of the system is derived from light energy, whereas in other organisms it is obtained from the energy of the chemical bonds of the substances taken up from the environment.

In both kinds of cells the reactions by which carbohydrate is formed and broken down are of fundamental importance since they provide the pathways by which much of the light or chemical bond energy can be conserved and redistributed among a great variety of new chemical bonds. A general pattern of carbohydrate metabolism is presented in Figure 5.4. This scheme is by no means complete, and further details can be found in recent literature (1017). The pattern as given is adequate to provide a general reference for the discussions that follow. Particular note should be made of the fact that several of these reactions involve the formation of phosphate esters or anhydrides. As will be discussed later, such substances are extremely important in the transfer

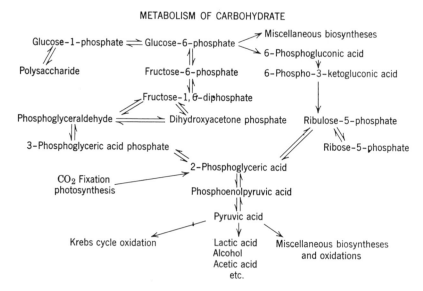

Figure 5.4. A general scheme showing some pathways of metabolism of carbohydrate.

of energy in metabolic processes. On the other hand, it should be kept in mind that the organic derivatives of phosphoric acid are not the only substances that are used in energy transfer.

Coupled Reactions. The biochemical reactions of living cells occur in consecutive series with a very large number of divergences and convergences of the metabolic pathways. Although some reversible reactions may reach equilibrium, in general there is a more or less steady flow of materials from those taken in from the environment to those that are excreted back into the environment. Nevertheless, the potential equilibrium state of each reaction in any series is important in determining the course of the reaction pathway as well as the composition of the tissue itself. Some basic principles concerned with reaction equilibria should therefore be examined (494). Consider the following hypothetical reaction:

(1)
$$(A) + (X) \underset{k_2}{\overset{k_1}{\rightleftharpoons}} (B) + (Y)$$

Here k_1 and k_2 are specific rate constants for the reversible reaction which occurs among the four substances A, X, B, and Y. Although the rates may be affected by a catalyst such as an enzyme, the rates in both directions are affected equally, and the reaction can be described in terms of its equilibrium constant:

$$K_{eq} = \frac{k_1}{k_2} = \frac{(B)(Y)}{(A)(X)}$$

Assume that $K_{eq\ 1} = 0.01$, and that in dilute solution actual concentrations equal active concentrations. In this example, then, at equilibrium: $(A)(X) = 100\ (B)(Y)$, and a high concentration of A or X is required to force the production of a much smaller amount of B or Y. This necessity for a high concentration of one or the other or both reactants can be avoided by addition of a second reaction to the system as shown below:

(2)
$$(B) \underset{k_4}{\overset{k_3}{\rightleftharpoons}} (C) + (D)$$

Here,

$$K_{eq\ 2} = \frac{k_3}{k_4} = \frac{(C)(D)}{(B)}$$

and it is assumed, for the present purpose, to have a value $K_{eq\ 2} = 1000$.

These reactions, 1 and 2, can now take place in series through the common substance (B). Since the equilibrium in 2 is very much in favor of formation of the products C and D, the concentration of B in the system can be reduced to a very low value and reaction 1 can proceed without the necessity of having a high concentration of A or X. The overall reaction will then be:

(3) $$(A) + (X) \rightleftharpoons (Y) + (C) + (D)$$

where $K_{eq\ 3} = (K_{eq\ 1})(K_{eq\ 2}) = 10$

In quantitative terms and in units of energy it can be shown that the standard free-energy change during a reaction $F° = RT \ln K_{eq}$. This value represents the required change in free energy of the system for a conversion of the reactants in a standard state (in a molal solution) to the products in a similar standard state. For the foregoing reactions, at $25°C$:

(1) $F° = +2700$ cal

(2) $F° = -4110$ cal

(3) $F° = -1370$ cal

By convention, $F°$ is negative for a spontaneous (exergonic) reaction and positive for an energy-requiring (endergonic) reaction. Obviously, any single reversible reaction can go in either direction, depending on the concentrations of the reactants and products and on the equilibrium constant. No reaction can proceed in a direction away from equilibrium, but an endergonic reaction can be "driven" by coupling with an exergonic reaction through a common reactant and through consecutive steps, as shown in the example just given. It should be noted that the reaction velocities in the single or consecutive reactions are not dependent on the equilibrium constants but are functions of the properties of the individual catalysts of the system. That is, the catalysts determine the actual course and velocity of a reaction while the overall free-energy change indicates the potentiality for the occurrence of a reaction. As discussed later (p. 246), the mechanism of an enzyme-catalyzed reaction can be very complex, and coupling increases this complexity.

Coupling is an essential feature of biochemical processes. It makes possible the flow of materials through long series of reactions without the necessity for the accumulation of excessive amounts of intermediates just prior to a reaction in which the equilibrium is unfavorable for the direction of flow of substances.   In any given series of biochemical reactions there may be only one that requires coupling or there may be many, either immediately following one another or placed at intervals along the chain.   A consideration of these facts about the energetics of chemical reactions yields some important conclusions pertaining to the effects of genetic changes on metabolism.   First, with regard to any reaction that normally operates in the vicinity of equilibrium, it is obvious that a genetic change which alters the concentration of one of the reacting substances can change the direction of spontaneous flow of materials.   The result might be the accumulation of an intermediate, a reversal of the direction of flow of materials, and perhaps a loss in adequate capacity for synthesis of some essential substance.   A second point of significance arising from a consideration of coupled reactions is that a genetic change which results in the loss of the exergonic step of a coupled system results also in the loss of one or more dependent endergonic steps.   If the series of dependent endergonic steps is long, a number of different mutations which block the series in different ways and at different points can be expected to yield essentially the same phenotype.

In living cells there exists a wide variety of substances which, through their oxidation or reduction, hydrolysis, or other chemical changes, can yield energy for coupling.   These include hydrogen carriers, peptide bonds, various kinds of esters, hemiacetals, and many other kinds of compounds.   The esters, anhydrides, and amides of phosphoric acid are particularly important in the transfer of energy in biological systems. Hydrolysis of these substances serves as the driving reaction for coupling with a great many endergonic systems.   Some of the biologically important derivatives of phosphoric acid are shown in Figure 5.5.   These are derived from the energy of oxidation and serve as major intermediates in the biosynthesis of order of many kinds.   Genetic changes which directly alter capacity for their production would be expected to have profound influences of many kinds.

Reaction Mechanisms.   In the earlier discussion of enzyme kinetics the assumption was made that a substrate and enzyme combine to form a complex which either dissociates to the starting materials or to products plus enzyme.   Such a process can be slow or fast with turnover numbers (moles of substrate converted per mole of enzyme per minute) ranging from less than 10^2 to more than 10^6, depending on the nature of the reaction system but not on the equilibrium constant.   That such

Group 1. High-energy anhydrides and amides 7 to 14,000 cal per mole. The symbol $\sim$ indicates a high-energy bond (toward hydrolysis).

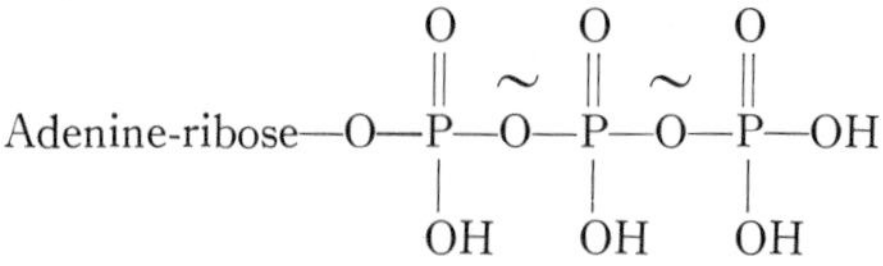

$$\text{Adenine-ribose—O—}\underset{\underset{OH}{|}}{\overset{\overset{O}{\|}}{P}}\text{—O}\sim\underset{\underset{OH}{|}}{\overset{\overset{O}{\|}}{P}}\text{—O}\sim\underset{\underset{OH}{|}}{\overset{\overset{O}{\|}}{P}}\text{—OH} \qquad \textit{Adenosine triphosphate}$$

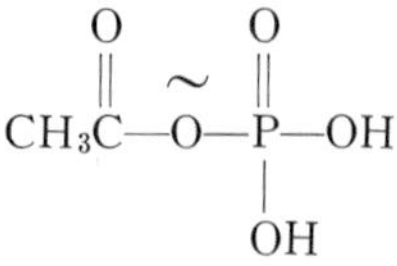

$$\text{CH}_3\overset{\overset{O}{\|}}{C}\text{—O}\sim\underset{\underset{OH}{|}}{\overset{\overset{O}{\|}}{P}}\text{—OH} \qquad \textit{Acetyl phosphate}$$

$$\underset{\underset{COOH}{|}}{\text{CH}_2}=\text{C—O}\sim\underset{\underset{OH}{|}}{\overset{\overset{O}{\|}}{P}}\text{—OH} \qquad \textit{Phosphoenol pyruvic acid}$$

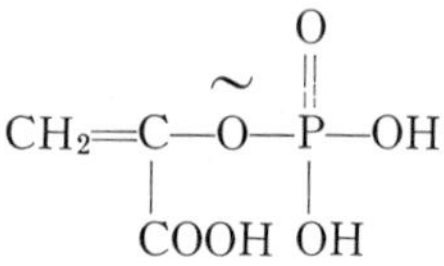

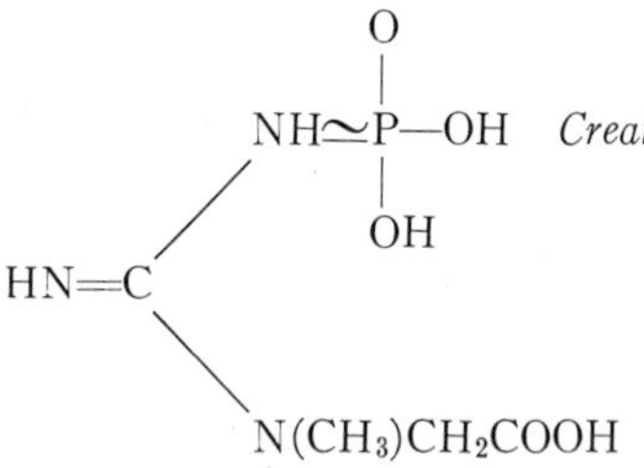

Creatine phosphate

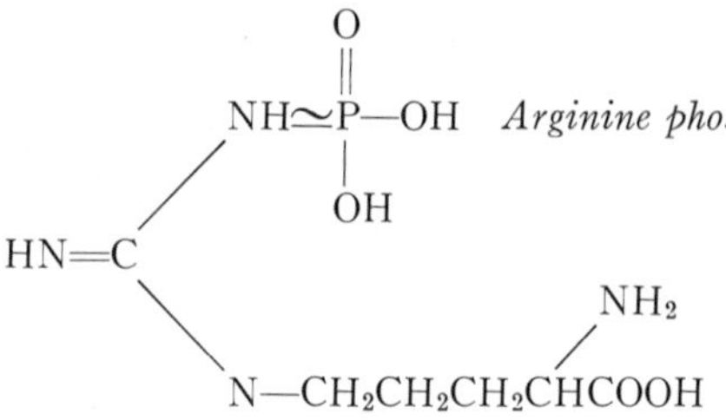

$$\text{HN}=\text{C} \begin{cases} \text{NH}\simeq\underset{\underset{OH}{|}}{\overset{\overset{O}{|}}{P}}\text{—OH} \\ \\ \text{N(CH}_3)\text{CH}_2\text{COOH} \end{cases}$$

Arginine phosphate

$$\text{HN}=\text{C} \begin{cases} \text{NH}\simeq\underset{\underset{OH}{|}}{\overset{\overset{O}{\|}}{P}}\text{—OH} \\ \\ \text{N—CH}_2\text{CH}_2\text{CH}_2\overset{\overset{NH_2}{|}}{C}\text{HCOOH} \end{cases}$$

Group   2. Low-energy esters. 2000–3000 cal per mole.

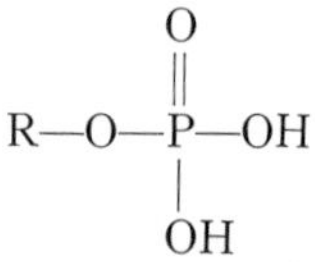

$$\text{R—O—}\underset{\underset{OH}{|}}{\overset{\overset{O}{\|}}{P}}\text{—OH}$$

R represents a variety of substances including sugars, the hydroxyamino acids, and phospholipid constituents such as ethanolamine and choline. The sugar phosphates include glucose-6-phosphate, fructose-1,6-diphosphate, the ribose and desoxyribose esters in nucleic acids and coenzymes, and the phosphate esters of the three carbon intermediates in glycolysis.

Figure 5.5. Some biochemically important phosphate esters, anhydrides, and amides.

processes are not this simple is clear from the facts that there is still relatively little information on the exact nature of enzyme-binding sites and that it is not yet possible to decide whether such sites are geometrically fixed or flexible in relation to substrates and reaction mechanisms. The importance of these matters has been emphasized by a variety of studies on model systems (1072). It has been shown, for example, that pyridoxal plus certain metal ions will catalyze most or all of the reactions in which pyridoxal phosphate participates as a cofactor with specific enzymes. As in many other similar model systems, specificity is lacking but perhaps certain parts of the overall reactions are the same.

Another model system that has special interest is that of Swain and Brown (995). As shown hereafter, tetramethyl glucose undergoes

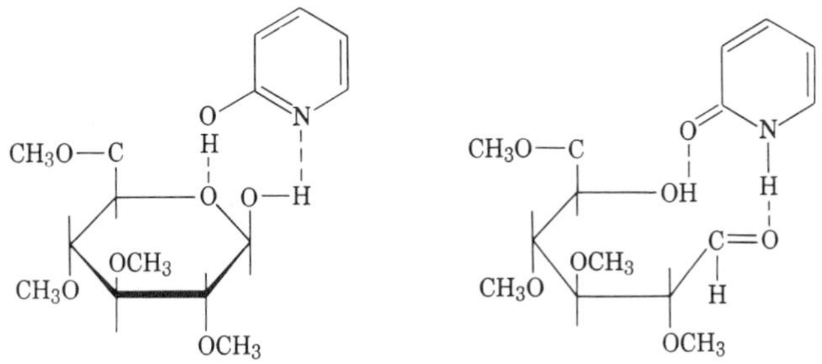

mutarotation in benzene by ring opening as the first step. The reaction is catalyzed by acid and base, and, in benzene, pyridine is an effective base and cresol an effective acid. *o*-Hydroxy-pyridine (α-pyridone) is a less effective proton donor and acceptor than cresol and pyridine respectively, but it is, nevertheless, a far more effective mutarotation catalyst than are the two compounds combined. This example, then, demonstrates the advantage of a fixed structure with appropriate dimensions for simultaneous donation and acceptance of protons. Perhaps some enzyme-substrate combinations and reactions are of this nature and this simple. Whether or not this, a directed sequential mode or other possible mechanism, obtains, it is certain that "multipoint" attachment is a characteristic of enzyme-catalyzed reactions. Substrates do combine with enzymes in specific ways, and in the process there is a reduction of activation energy. The activation can involve a single energy barrier, but on the other hand, more than one is probable in most cases. An illustrative generalized example presented by Lumry (626) is given briefly here.

Consider a reaction:

$$A + C \underset{\rightleftharpoons}{\overset{E}{}} D + G$$

and as an arbitrary sequential mechanism:

$E + A \rightleftharpoons E{-}A$ (first ES bond)

$E{-}A \rightleftharpoons E{=}A$ (second ES bond)

$E{=}A + C \rightleftharpoons C{-}E{=}A$ (second substrate)

$C{-}E{=}A \rightleftharpoons D{-}E{-}G$ (reaction to bound products)

$D{-}E{-}G \rightleftharpoons E{-}G + D$ (first product)

$E{-}G \rightleftharpoons E + G$ (second product)

Free-energy profiles that might accompany such a reaction sequence are shown in Figure 5.6. Obviously, either a fewer or greater number of steps could have been introduced, but this is sufficient to illustrate the probable multistep nature of what one might consider a single enzyme-catalyzed reaction. It should be clear, also, that the possible ways in which a genetic change can influence such a reaction are many indeed. Included are those which may cause significant amino-acid substitutions in the enzyme itself and those which may alter immediate enzyme environment to influence one or more energy barriers. This

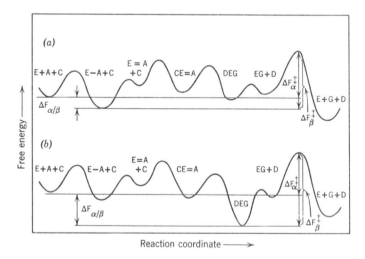

Figure 5.6. Possible theoretical free-energy diagrams which demonstrate unusual interpretations of the empirical rate parameters. From Lumry (626).

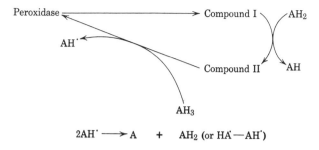

$$2AH\cdot \longrightarrow A \quad + \quad AH_2 \text{ (or } HA\cdot-AH\cdot)$$

Figure 5.7. Mechanism of peroxidase-catalyzed reaction. From Morrison (713).

multistep aspect within a reaction is no doubt more directed in sequence than is a series of different reactions, but it is perhaps not so different from a series of reactions in an organized system. At the experimental level, where there is only limited information on the nature of binding sites and the numbers of bonds formed in enzyme action, detailed evaluations, as in the examples just given, are not usually possible at present. Investigations by Chance and others (713), making use of spectroscopic differences among *ES* complexes, have provided an example relating to the action of peroxidase as shown in Figure 5.7. Compounds I and II represent different *ES* complexes, and in the presence of excess peroxide, two other complexes which are noncatalytic have been observed. Even though noncatalytic, these should participate as components in the overall kinetic picture of peroxidase action.

General Comments

The information presented in this chapter was selected for the purpose of illustrating some fundamental principles of metabolic processes that are important in understanding phenotypic effects of gene mutation. Thus, in the most general sense, the important concept to be emphasized is that systems affected by genes are those characterized by dynamic action. They consist not only of long-reaction series leading to production of chemical energy and synthesis of essential substances, but also of innumerable cross reactions and cycles within and between such series. On this basis, mutation of one gene can have many and drastic observable results far removed from initial action. With enough knowledge of overall patterns and kinetics, it can be expected that such changes can become describable in detail and entirely predictable. At present this is not possible in any example, but some factors that must be taken into account are:

1. Reaction kinetics at all levels from gene replication to the most remote metabolic process.

2. Quantity, quality, and organization of specific macromolecules.

3. Detailed mechanisms of action and interaction of molecules of all sizes.

6 Mutation and the Agents of Metabolic Control

The fundamental concept that the hereditary units (the genes) function through control of the existence and specificity of the biological catalysts (the enzymes) came into being slowly. Garrod brought forth the idea in about 1908 (313), and it received only modest attention from others for some 30 subsequent years. Wright (1109), Goldschmidt (336), and Grüneberg (376) made use of the concept in studies in physiological and developmental genetics, but probably the greatest impact of its application was made in connection with genetic studies using microorganisms. Here the initial observations by Beadle and Tatum (56, 57) using *Neurospora*, Lederberg and Tatum (574) using *E. coli,* and Delbrück and Luria (209) using bacteriophage provided the impetus, and much of modern developments on the chemical basis of heredity has developed from these beginnings.

In 1945 Beadle (53) summarized an accumulation of genetic and related biochemical information "as a framework on which to arrange conveniently the varied observations and inferences bearing on the nature of genes and their action" and set forth a general hypothesis called the one gene-one enzyme hypothesis relating inheritance to biochemical processes. In essence this suggested: (1) that a gene has an autocatalytic function in its own reproduction; and (2) that a gene has a single heterocatalytic function, and, in relation to enzymes, it determines enzyme specificity directly. This means that each gene is unique and each enzyme or other molecule with a biologically specific function must obtain its specificity from a corresponding gene. It is presumed that mutation can give rise to loss or change in autocatalytic and heterocatalytic functions.

At the time of preparation of the first edition of this book in 1955, the "one gene-one enzyme" hypothesis was accepted as useful in the context in which it was originally presented, but it was not taken to be either proved or disproved. The reasons for this indecision were listed as a series of questions repeated here.

1. Exactly what is a single gene in physical and chemical terms? Experimentally, it is a unit defined by crossing over, but is it justifiable to equate this to a unit of mutation or a unit of function?

2. Do we know exactly what is meant by the term specificity? Can two macromolecules that are not identical have the same biochemical specificity?

3. Are the enzymes and other specifically functional substances that we have been able to isolate produced only under the direct influence of genes? Are these substances as we know them in the isolated state identical in structure and action to the same materials *in vivo?*

4. Is the expression of a mutant phenotype as a deficiency or change in the properties of a substance, such as an enzyme, an adequate reason to consider the substance produced in the normal state to be a direct product of gene action?

5. Is it a necessary postulate that all genes function by the same mechanism?

As already evident from preceding discussions, some partial answers to these questions have been obtained during the intervening years. At the same time, although the continued usefulness of the concept has been demonstrated repeatedly, complete acceptance remains subject to more detailed definition of units involved and mechanisms of their actions. That this is the situation will be clear from a consideration of a present concept of gene-enzyme relations. In the total nakedness of oversimplification (see Chapter 2), this is illustrated diagrammatically in Figure 6.1. As shown, overall there is indeed good evidence that a simple relation exists between genes and enzymes. At the same time there are important details that require clarification, and some of these are indicated by the numbers 1–8 at the left of the diagram. Some questions still to be raised with reference to each component and step indicated are as follows.

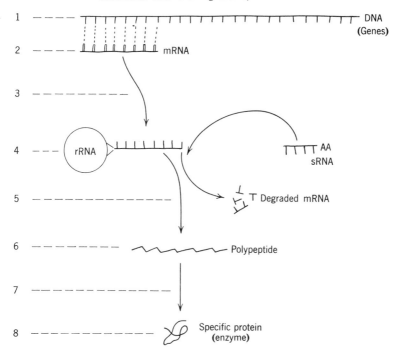

Figure 6.1. A schematic outline of probable components and steps concerned with the relation of DNA nucleotide sequence to polypeptide amino-acid sequence. RNA's: m = messenger; r = ribosomal; and s = soluble or transfer.

1. If it is accepted that DNA alone represents the material of genes, what are the sizes of the pieces that are genes; do all segments of all sizes have specific functions; do identical base sequences always yield the same result regardless of tissue or organism? What is a mutation and how is it related to mechanisms of replication and messenger RNA production?

2. If it is accepted that messenger RNA (mRNA) is a primary product of the heterocatalytic activity of a gene, how large and how small are the significant products and what controls are exerted to designate the beginning and the end of a given molecule?

3. Is all messenger RNA directed toward protein synthesis or does some of it enter into other processes (specifically or in addition to) such as control of gene action by combination with DNA of other genes or by undergoing replication itself at points remote from the DNA? What changes and interactions of mRNA can occur during its transport from the site of production to the site of action?

4. What is the role of the ribosome in combination with mRNA and sRNA? Is a considerable part of the nucleotide sequence of mRNA designated for essential specific combinations with rRNA (ribosomal RNA), and are the sequences designated for combinations with sRNA of the size and composition corresponding to the DNA?

5. What is the role of mRNA instability in the synthesis of polypeptide chains?

6. Is a designated primary amino-acid sequence sufficient alone to determine final conformation of protein?

7. Are primary chains altered in any way as by transpeptidation or specific hydrolysis in the process of becoming recognizable specific products?

8. Is this product itself the recognizable enzyme or is alteration or combination with other products required? In the whole process of formation where are mistakes in sequence possible, and how many and what kinds can be tolerated?

From these considerations and others which take into account the normal milieu in which all of this and more must occur, it is clear that the gene-enzyme relation is still difficult to define and it still must be thought of in terms of complex processes as well as clearly defined chemical structures at each end.

Some partial answers to some of the foregoing questions have been obtained and others will be considered further in this chapter. Most of the problems concerned with base sequences in nucleic acids depend on the development of methods of obtaining pure components or interpretation from model systems (Chapter 2), whereas such methods are now available at the protein end of the picture. Still, in relation to earlier stages of the overall processes, and to the question of whether all immediate gene products are concerned with protein synthesis, observations of McClintock (639) on corn, and Jacob and Monod (484) on bacteria are highly significant (see Chapter 9). The latter investigators have presented a detailed interpretation of results which includes a distinction between two classes of genes, structural and controlling. Controlling genes are postulated to produce RNA which functions at the gene level by combination with DNA and inhibition of production of a specific mRNA (see p. 102). Such interaction at the gene level is reasonable, but it is difficult to rule out the possibility that this kind of regulation might represent a dual function in which mRNA or a product or derivative might involve protein production and gene inhibition as well. A point of special interest to note in the Jacob-Monod scheme is the direct participation of specific low-molecular-weight compounds (inducers) in the regulation of activity at the gene level. There has been little previous evidence for augmentation of macromolecular specificity in this way at the nucleic-acid levels.

The preceding oversimplified outline of relations between genes and enzymes is now sufficiently well founded to provide a general directional picture of progress. Information is available on the chemical nature of mutation (p. 195), and the agents of metabolic control as considered may include all the gene products down to the functional catalytic units, the enzymes. Though little more can be said at present concerning the RNA intermediates and mode of function, there are many significant facts available concerning the overall processes in terms of the nature of products, and a selection of examples will be considered in detail in the remainder of this chapter.

Macromolecules

A general discussion of structures and synthesis of several kinds of macromolecules was presented in Chapters 2 and 5, but certain extensions are pertinent to the present problems. As has been noted, in nucleic-acid chemistry hydrogen bonding represents an especially important phenomena since it provides a basis for nucleotide pairing and thus replication as well as synthesis of RNA. It is, however, an important point that H-bonding is not confined to macromolecules, but it occurs between hydrogen and all of the more electronegative elements. For example, H_2O exists in a highly associated state at ordinary temperatures. Hydrogen thus forms quite stable bridges between a great variety of simple and complex molecules (773, 784), and when the potential exists, competition between them exists to provide the most stable combinations as dictated by chemical structures. The bonds are essentially ionic in character and have bond energies in the vicinity of 5 kcal/mol for —O—H---O= and ⟩N—H---O= as compared to 110 for O—H and 83 for N—H. Some of the simpler H-bonded structures are shown in Figure 6.2, and these may be compared with the more complex systems shown previously (Figure 2.34).

In the present context, that of evaluating potential mechanisms of metabolic control in the stages involving functions of nucleic acids, knowledge is limited to positive action. That is, A-T and G-C pairings are important in replication and A-U, G-C, and T-A pairings are important in RNA production, but little is known of H-bonding interactions that give limitation of action or inhibition. Obviously, the potential here is very great, especially with RNA which is not protected by a complementary partner or by isolation in the chromosomal complex as is DNA. Furthermore, short complementary-sequence pairing, single-phase pairing, and even noncomplementary-base pairing by single H-bonds between RNA molecules may be sufficient to affect functions. The potentials here for interactions among DNA, mRNA, rRNA, and sRNA are so multitudinous that a prediction of nonrandom sequence restriction seems reasonable. That is, for example, all mRNA molecules may have long complementary sequences within each molecule so that much internal pairing occurs. This in combination with association with protein should be sufficient to provide enough limitation of intramolecular interactions to permit positive function to occur. True assessments of these factors of potential interactions are important matters for the future.

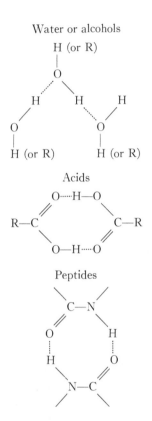

Figure 6.2. Some illustrations of hydrogen bonding. *R* = an organic radical.

Hydrogen bonding is also an important factor in protein conformation, interaction, and function, along with other linkages concerned with tertiary and quaternary structure (see p. 79). In this case it is possible to be more specific and to see what kinds of changes can and do occur within the limits of compatibility with function.

Variations in Specific Macromolecules. It has been known for a long time that enzymes obtained from different species are not necessarily alike, even though they may catalyze the same reaction. They may differ in size, amino-acid composition, and in numerous related functional properties. Some early observations bearing on this question are summarized in Table 21. These differences, of course, do not describe details of structure, only criteria for distinction, and a more sophisticated comparison is illustrated in Figure 6.3.

These data show amino-acid sequences in the vicinity of the heme

Table 21.   Some Species Differences in Crystalline Enzymes

Enzyme	Origin	Activity Difference
Glutamic-acid dehydrogenase	Yeast or *E. coli*	Requires TPN as hydrogen acceptor
	Higher plant	Requires DPN as hydrogen acceptor
	Animal	Requires TPN or DPN as hydrogen acceptor
Alcohol dehydrogenase	Yeast	Completely inhibited by 0.001 M iodoacetate
	Animal	Not inhibited by 0.01 M iodoacetate
Glucose dehydrogenase	*E. coli*	Inhibited by toluene
	Animal liver	Not inhibited by toluene
Pepsin	Bovine Swine	Crystalline products are identical immunologically and solubilities are the same. However, solubilities are independent in mixtures (i.e., one of the proteins will dissolve in a saturated solution of the other).
Hexokinase	Yeast Sheep brain Rabbit muscle Rat muscle, liver	Yeast and brain enzymes act on glucose or fructose as substrates. Others act on glucose or fructose and do not show multiple-substrate inhibition.

group of cytochromes c from several species. Similarities are shown in spacing of cysteines which are attached to porphyrin and in the locations of histidine and threonine, but a variety of variations are also shown. Many other variations are also likely in regions not shown (over 90% of the total number of residues). These data show clearly that a considerable variation in primary amino-acid sequence is compatible with normal enzyme activity. On the other hand, some essential similarities may be universal, and even different enzymes may have significant similarities around certain points. For example, there is some evidence that eight enzymes (chymotrypsin, trypsin, thrombin, elastase, ali-esterase, pseudocholine-esterase, phosphoglucomutase and phosphoglyceromutase) all have at the active site Asp·Ser·Gly or Glu·Ser·Ala (534). Such data obtained so far suggest a common origin in evolution of related enzymes with retention of some essentials of primary structure and variation in nonessential parts or parts concerned more with species variation than with enzyme activity as such.

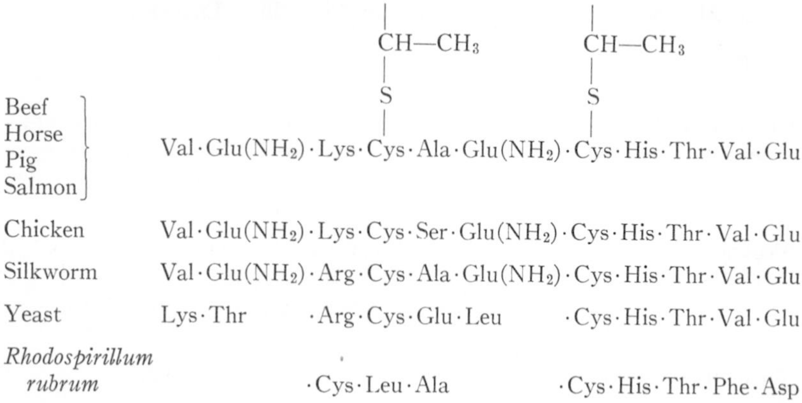

$$
\begin{array}{cc}
\overset{\displaystyle |}{CH-CH_3} & \overset{\displaystyle |}{CH-CH_3} \\
| & | \\
S & S \\
| & |
\end{array}
$$

Beef ⎫
Horse ⎪
Pig ⎬   Val·Glu(NH₂)·Lys·Cys·Ala·Glu(NH₂)·Cys·His·Thr·Val·Glu
Salmon ⎭

Chicken   Val·Glu(NH₂)·Lys·Cys·Ser·Glu(NH₂)·Cys·His·Thr·Val·Glu

Silkworm   Val·Glu(NH₂)·Arg·Cys·Ala·Glu(NH₂)·Cys·His·Thr·Val·Glu

Yeast   Lys·Thr   ·Arg·Cys·Glu·Leu   ·Cys·His·Thr·Val·Glu

Rhodospirillum
   rubrum   ·Cys·Leu·Ala   ·Cys·His·Thr·Phe·Asp

Figure 6.3. Amino-acid sequences near the heme group in cytochromes c from different species. From Koshland (534).

A second kind of normal variation that is perhaps of great significance is that of the existence of different molecular species in the same tissue with the same or very similar enzymatic activity. An example is that of chymotrypsin. Since the original isolation of this enzyme in crystalline form by Kunitz and Northrop (542), numerous investigators have demonstrated that the enzyme is separable into at least six components, all of which have similar or greatly overlapping specificities. These six components (B, α, β, γ, δ, and π chymotrypsins) are distinguishable by one or more criteria which include solubility, electrophoretic mobility, composition, and enzyme specificity. That at least some of these enzymes are metabolic products of chymotrypsinogen has been demonstrated (see p. 261). This kind of variation may be a very important and a very common one since in most enzyme isolations minor components are very often encountered and discarded in efforts to obtain a major component in a pure form.

A different type of variation in enzymes as they occur naturally is shown by frequent observations that in one tissue there may be several active forms with the same specificity. Markert and coworkers (658, 659) have distinguished these from "families" of enzymes, as in the chymotrypsin case, and referred to them as "isozymes." These are considered to be different molecular forms having the same enzymatic specificity, and justification for the distinction is based on observations that in the case of lactic-acid dehydrogenase the active enzyme can be made up of combinations of two different subunits in units of four.

Cahn et al. (127) extended these observations to studies of lactic-acid dehydrogenase in different tissues of a wide variety of species. As an example, it has been established that two distinct lactic-acid dehydrogenases (LDH) occur in the chicken. In the adult, one of these is found principally in breast muscle (CM) and the other in heart muscle (CH). These two enzymes differ markedly in electrophoretic behavior, and in the rabbit they produce antibodies which do not cross-react. Each enzyme is dissociable into four subunits, and each contains four molecules of the cofactor DPN. However, the subunits, each with one molecule of cofactor, are not enzymatically functional. Studies were conducted to determine the types of LDH present in chicken embryos at different stages of development resulting in the observation that either or both CH and CM types and three others can be present. In six-day embryos nearly all enzyme has the CH character, but by eight days after hatching nearly all of the LDH of breast muscle is CM type. At intermediate stages mixtures containing CH, CM, and three other components which crossreact immunologically with antibodies to CH and CM can be found.

The five components are separable electrophoretically on starch gel with CH and CM at the extremes in migration. From this and additional varieties of studies it has been concluded that the CH enzyme is made of four identical subunits HHHH, CM from four different identical subunits MMMM, and the three intermediate types from combinations of the two to give HHHM, HHMM, and HMMM combinations. Cahn et al. (127) prefer to call these forms hybrid enzymes rather than isozymes. With this interpretation of four subunits from two types of protein there should be only five components, and this has been observed in several species as shown in Figure 6.4. As shown, in this particular system, the two limiting types in the chicken do not separate enough to show the hybrid classes, but the five components show clearly for mouse, rat, beef, and human. A number of other species variation characteristics have been observed, including that of only one enzyme type in the flatfish.

It now seems rather probable that a number of enzymes have characteristics related to that illustrated by lactic-acid dehydrogenase (279). That is, they are made up of subunits which are more likely to be the most direct polypeptide products of gene action than the functional enzyme. Presumably, in this case CM and CH types are derived under the influences of two different genetic loci which yield different subunits which are still sufficiently alike to form stable mixed combinations. Another point of interest is that of species differences. Although the electrophoretic characteristics shown in Figure 6.4 are not absolute,

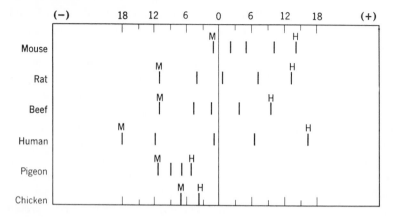

Figure 6.4. Starch grain electrophoresis of lactic dehydrogenases of different animals. M, most negative band or muscle LDH; H, most positive or heart LDH. From Cahn, Kaplan, Levine, and Zwilling (127).

it seems clear that individual subunits in the different species must be quite different in composition. Thus, many different primary chains can yield products with the same kind of enzyme activity.

Enzyme Derivatives. In the preceding section examples have been cited to show that a great deal of natural variation in primary structure of enzymes can be tolerated without loss of activity. This conclusion has also received strong support from studies of activities of chemical derivatives of enzymes. An extensive review of this subject has been made by Frankel-Conrat (292), and a generalized summary of observations of the effects of a number of reactions on activities of several enzymes is given in Table 22. Clearly, a great many changes can be made without loss or only partial loss of activity. The use of such derivatives is a most valuable adjunct to methods of study of structure.

Macromolecules as Substrates

Many of the general aspects of synthesis of macromolecules have been considered previously, but relatively little attention was given to significant changes that normally occur after initial formation. When this does happen, it obviously provides an additional parameter for metabolic control. It seems likely that this will become a very significant problem in the future, though known examples are limited in number at present.

Table 22. **Enzyme Modifications** *

Enzyme	Type of Reaction	Groups Involved	Effect of Activity
Amylase (Pancreas)	Acetylation	Amino	Inactive
Chymotrypsin	Acetylation	Amino	Active
	Oxidation (photo-)	Imidazole (1), indole (3)	
	Addition of glycine	Amino	Active
	Dinitrophenylation	Amino (phenolic)	Partially active (altered specificities)
		Imidazole	Inactive
	Methylation	Carboxyl, amino, phenolic	Decreasing activity
Crotoxin (lecithinase)	Acetylation	Amino	Inactive
	Esterification	Carboxyl	Inactive
	Sulfation	Hydroxyl	Inactive
	Iodination	Phenolic	Inactive
	Coupling, etc.	Phenolic, imidazole, etc.	Inactive
Lysozyme (egg white)	Acetylation	Amino	Inactive
	Esterification	Carboxyl	Inactive
	Sulfation	Hydroxyl	Inactive
	Iodination	Imidazole (all = 1), phenol	Partially reversibly inactive
	Coupling	Imidazole (all = 1), phenol	Partially inactive
	Oxidation (photo-)	Imidazole (all = 1), indole (1.2)	Partially inactive
	Guanidination	Amino	Active
Pepsin	Acetylation	Amino	Active
	Deamination	Amino	Active
	Iodination	Phenolic	Decreasing activity
	Acetylation	Phenolic	Decreasing activity
	Mustard gas	Carboxyl	Decreasing activity
	Reduction	Disulfide (1)	Active
		Sulfide (1)	Inactive
	Oxidation (ultra-violet)	Phenol	Inactive
Phosphatase (alkaline)	Acetylation	Amino	Active
Ribonuclease	Esterification	Carboxyl	Inactive
	Guanidination	-Amino	Active
	Deamination	-Amino	Active
	Reduction	Disulfide (2)	Active
		Disulfide (all = 4)	Inactive
	Oxidation (performic)	Disulfide	Inactive
	Oxidation (photo-)	Imidazole (1)	Inactive

* From Frankel-Conrat (292).
Note: Numbers in parentheses refer to the number of amino-acid residues.

Nucleic Acids. A most interesting and significant development in DNA biosynthesis is that concerned with the presence of glucose and cellobiose conjugated with DNA in definite ratios in the T-even series of bacteriophage (578, 532, 1). The sugars are coupled as α or β glycosides to the hydroxyl groups of hydroxymethylcytosine and are found in the ratios shown in Table 23. It has been established that following bacterial infection with phage, glucosylating enzymes characteristic for the particular phage used are produced. Further, the specific glucosylations take place after synthesis of DNA, and only with the double-helix form as substrate. These results provide convincing evidence for *in vivo* existence of the double-helix structure, but in the present connection they demonstrate the function of DNA as a substrate in enzyme-catalyzed alterations other than degradation. Other potential varieties of possible reactions such as transnucleotidation, methylation of cytosine, or other base derivations are not known at present.

Among the various classes of RNA reactions at the macromolecular level (after primary synthesis) that may be of biological significance are the removal and addition of terminal nucleotides in sRNA. Of the twenty or more sRNA molecular species (at least one for each amino acid) all apparently carry the sequence A—C—C— at one end and most have —G at the other end of the chains of around 100 nucleotides.

Table 23. Distribution of Hydroxymethylcytosine Nucleotides in DNA from Bacteriophages T2, T4, and T6 °

Phage	Digluco-sylated dHMP	Monogluco-sylated dHMP	Nongluco-sylates dHMP	Recovery of HMC of DNA as Mono-nucleo-tide
	(% of Total dHMP)			%
T2	6	69	25	97
	6	70	24	96
	5	67	28	99
T4	0.5	100	0.5	93
T6	72	3	25	90
	72	4	24	88

* From Lehman and Pratt (578).

The A—C—C— portion can be removed and reformed enzymatically. The genetic question here, then, is whether the complete functional chains including A—C—C are designated directly by direct or indirect gene action, or whether only nonidentical portions are designated. If the latter is true, as seems reasonable, then the most direct gene product in terms of totally designated primary sequence differs by at least three nucleotides (about 3%) from the product that is recognized as a functional agent for control of metabolism.

Proteins. There are a number of good examples among the proteins which illustrate processes of alteration which occur after formation of primary amino-acid sequences. A case in which a good deal of detail is known (see p. 256) is that of chymotrypsin. This proteolytic enzyme is derived from the zymogen chymotrypsinogen through specific actions of proteolysis. Activation can occur slowly through the action of trypsin to give α-chymotrypsin or by a fast sequence involving autocatalysis through the steps:

(1)   Chymotrypsinogen $\xrightarrow{\text{Trypsin}}$ π Chymotrypsin

(2) π Chymotrypsin $\xrightarrow{\pi \text{ Chymotrypsin}}$ δ Chymotrypsin

Structural relations of products are shown diagrammatically in Figure 6.5 where the solid lines indicate peptide chains and dotted lines cystine —S—S— tertiary cross linkages. Here it may be presumed that the single primary chain of chymotrypsinogen is that designated by DNA through mRNA, but this is not a functional enzyme. Full activity is achieved, however, by the splitting of one peptide bond (Ileu—Arg) giving two chains linked by cystine. The autocatalytic action of π-chymotrypsin on itself simply shortens one chain and, as shown, α-chymotrypsin is produced by a second break and peptide loss to give three separate chains bound together. Peptide A contains 13 residues, B 180, and C about 50. The active center with ·Asp·Ser·Gly is in B, but it is clear that the break between A and B is essential to provide an enzymatically active conformation, and there is evidence that a histidine residue at least is also required at the active site. (See Linderstrom-Lang (619) for further details.)

Although different in details, conversions of trypsinogen to trypsin, pepsinogen to pepsin (619), and fibrinogen to fibrin—in blood clotting (893)—all involve the same principle, that of limited and specific proteolysis of nonfunctional protein to entities with recognizable and measurable functions. In these cases, then, part of the synthesis of

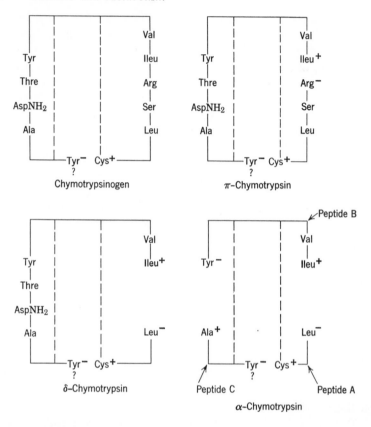

Figure 6.5. Hydrolytic steps involved in the activation of chymotrypsinogen. From Linderstrøm-Lang and Schellman (619).

specific enzymes occurs after designation of primary sequences with participation of other enzymes and products or intermediates (auto-catalysis), and all components and reactions participate individually and collectively in metabolic control.

With increasing recognition of the participation of macromolecules as substrates in reactions which alter or yield biological specificity, it should be profitable to give serious consideration to other possible kinds of reactions that might be significant in this regard. In proteins there are numerous possibilities which include amidation and deamida-tion of carboxyl groups in side chains, other substitutions such as alkylation or acylation of side-chain hydroxyl or amino groups, actual changes in side chains such as one-carbon transfers to interconvert glycine and serine, and alteration of primary chains by transpeptidation. These and similar phenomena remain to be discovered, though a recent unconfirmed report (1137) suggests occurrence of deamidation after protein synthesis.

One other aspect of protein synthesis that is receiving attention and should be clarified before too long is the function of intermediate-sized peptides. Such substances have been found in a great variety of tissues (687) and, in some cases, as nucleotide derivatives equivalent to activated amino acids (1078). These substances may be intermediates in protein synthesis, and whether derived from stepwise synthesis or protein turnover they may have an important function in regulation. Such things as the trypsin-inhibitor peptide (619) and peptide hormones (27) provide suggestive examples.

Some Effects of Mutation

Regardless of details concerning the exact nature of genes and the various aspects of the problems of how gene products interact and affect metabolic processes, a very large amount of information has been acquired in recent years which demonstrates clearly that mutation can result in a deficiency of a specific enzyme or a change in a specific protein. Deficiencies are frequently alleviated by nutritional means, and such mutants have proved invaluable in studies of metabolic pathways and patterns (Chapter 7). Loss of an enzyme or other specific product, however, provides only a correlation between mutation and a specific result and does not usually present a basis for studies of the mechanisms of gene action. Thus mutations which give rise to a change are more suitable for the latter purpose, and consideration of a number of examples is instructive as to the problems involved and the existing state of progress.

Mutations and Enzymes. One of the earliest examples of a gene-controlled deficiency of enzyme was observed in humans and described by Garrod (313). Increasing numbers of cases have been observed in higher organisms, but analysis at genetic, physiological, and enzymatic levels is usually more difficult than in microorganisms. At the same time, analyses of systems in a wide variety of species is essential to a broad understanding of the different kinds of variables and problems that one may expect to encounter. Some breadth of this nature can be attained from careful considerations of the examples that follow and others presented elsewhere for special reasons (Chapter 14, for example).

Indirect Effects. Uric acid is excreted in the urine by higher primates such as man and the chimpanzee, whereas monkeys and other mammals excrete allantoin to a greater extent than uric acid. Uric acid and allantoin are related chemically and enzymatically, as shown here:

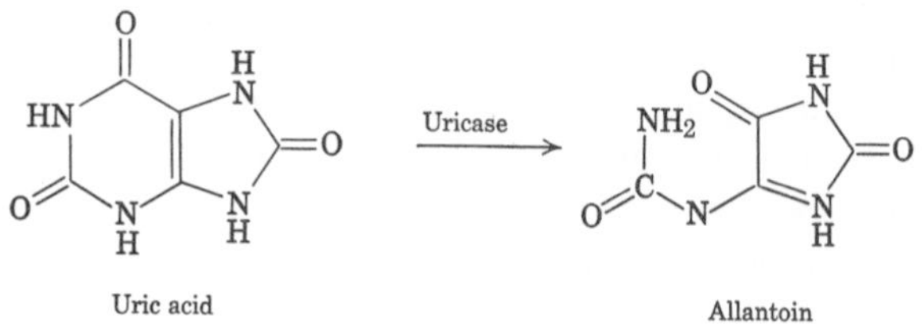

$$\text{Uric acid} \xrightarrow{\text{Uricase}} \text{Allantoin}$$

Uric acid   Allantoin

The reaction is a complex one involving several chemical steps. The enzyme has been highly purified, and there is no evidence that more than one enzyme is required. In 1916 Benedict (71) made the observation that Dalmatian dogs, unlike other dogs and carnivores in general, but like man, excrete uric acid rather than allantoin, and it thus appeared that there might be a heritable difference in dogs with respect to the presence or absence of the enzyme uricase.

A number of investigations—see Trimble and Keeler (1020)—provided evidence that the uric-acid excretion in dogs is inherited as a single recessive gene. On the biochemical side of the problem it was shown that, in actual fact, both uric acid and allantoin are excreted by all dogs but that the proportion of uric acid to allantoin is higher in the Dalmatian. With respect to uric-acid excretion, the dogs fall into two clearly separated groups, those that excrete 4–10 mg per kg of body weight per day, and those that excrete more than 28 mg per kg of body weight per day. The values are quite constant on a given diet, and intermediate animals (between 10 and 28 mg) were not found. Investigations demonstrated the occurrence of equal uricase activity in the livers of dogs of both Dalmatian and other breeds. Thus a simple direct explanation of the genetic phenomenon in the terms of uricase activity appeared to be untenable.

Much more recently Friedman and Byers (299) provided some experimental data that contribute a great deal to an understanding of the problem. It was found that the total excretion per unit of time of uric acid plus allantoin is the same for Dalmatians as for other dogs, but that, on the average, the Dalmatians excrete 11 times as much uric acid. Since uricase is found in the liver but not in the kidney, the blood was examined for its uric-acid and allantoin content. The following average figures in terms of milligrams of nitrogen per 100 ml of blood plasma were obtained:

Dalmatians	Uric acid	0.20
	Allantoin	0.23
Other dogs	Uric acid	0.11
	Allantoin	0.33

Here, as in the case of the products excreted, the totals are the same but the differences in relative proportions of the two substances are much higher in the urine. This apparent anomaly was resolved by studies of renal clearances. It was found that uric acid is passed into the glomerular filtrate of the kidneys, but in the Dalmatians it is not reabsorbed in the kidney tubules, whereas in other dogs it is reabsorbed to a considerable extent. Allantoin is not reabsorbed in the kidney of either type of dog.

It seems apparent, therefore, that the phenomenon of uric-acid excretion by Dalmatians is due primarily to a fault in the mechanism of reabsorption in the kidney and not to a fault in the production of uricase in the liver. In other dogs the reabsorbed uric acid is further converted to allantoin when carried back to the liver. There still remains one point that is not entirely clear. From the data just given on the blood composition with respect to uric acid and allantoin and from experiments by Friedman and Byers on the blood composition after tying off the ureters, it appears, superficially, that Dalmatians have a less effective uricase system than other dogs. Rimington (840) has accounted for this apparent difference by assuming that reabsorption by liver cells is faulty in Dalmatians just as has been demonstrated for kidney tubule cells. The two systems discussed are diagrammed in Figure 6.6. It is now quite clear that this single-gene difference in dogs does not necessarily have anything to do with the production of the enzyme uricase. It may have to do with the formation of structural materials directly or indirectly or with some enzymatic process concerned with the transport of uric acid through cell membranes.

These results provide a good illustration of the complexities encountered in the analysis of the results of a genetic change. Obviously, the simplest interpretation of the original observations is inadequate, and indeed the simplest interpretation of results in general is acceptable only as a basis for further experimentation.

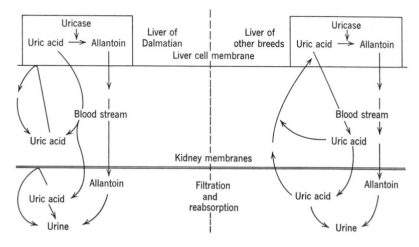

Figure 6.6. Excretion of uric acid and allantoin by Dalmatians and other dog breeds. Uric acid is not reabsorbed through the kidney membranes and possibly not through liver membranes in the Dalmatian.

In higher plants the enzyme catalase is found to be associated largely with the chloroplasts which also contain the chlorophyll of the plant. Von Euler (1042) investigated the catalase activity of a number of mutants in barley and found that catalase activity could be correlated with the degree of pigmentation of the plants. That is, normal green barley shoots contain a considerable quantity of catalase activity, whereas albino shoots contain less than one-half as much. Thus it appeared that the single dominant gene controlling the presence or absence of chlorophyll also affects the production of catalase. Both catalase and chlorophyll contain porphyrin structures, and an explanation of this phenomenon could rest on an assumption that the mutation concerns porphyrin synthesis. However, the work reported by Eyster (263) seems to provide an explanation for the catalase difference on the basis of degradation of the enzyme rather than on the basis of its formation (Figure 6.7). Eyster investigated corn seedlings from albino and yellow mutants as compared to normal green seedlings. Using one-gram samples of macerated tissue from seedlings grown in a greenhouse (exposed to light), the catalase activities in terms of O_2 in milliliters per 5 minutes were found to be: albino, 1.53; yellow, 2.83; and green, 10.14. Seedlings of all three types, when grown in the dark, gave the same value for catalase activity at a level about 25% higher than the green plant grown in the light (12 to 13). When the dark-grown seedlings were exposed to light the catalase disappeared at rates in inverse proportion to the green pigmentation. Thus it seems quite clear that the catalase level is dependent on a balance between the rate of its formation and the rate of its photochemical decomposition. The genetic control of the catalase level is quite indirect on this basis, since the chlorophyll apparently merely acts to absorb or reflect destructive radiations and thus protect the catalase.

As discussed in Chapter 2, an extremely important aspect of metabolic processes is a high degree of organization of enzymes into systems which bind the components in fixed relative positions. In such situations, mutations may be expected to affect the components individually and/or collectively, and thus indirect effects are quite likely. Examples in which it is very difficult to distinguish primary effects are found among mutants affecting plastids (835) and in those affecting mitochondrial systems in yeast (256) and *Neurospora* (695). These are also of interest in connection with cytoplasmic inheritance (Chapter 11).

In yeast it has been observed that a certain single-gene mutation results in a drastic alteration of the cytochrome system. In this strain cytochrome c is found in a twofold excess, while it is deficient in succinic-acid dehydrogenase. In the absence of oxygen the mutant behaves more or less normally, but under aerobic conditions, where the cytochrome system is needed, growth is severely limited. Two somewhat analogous mutants have been studied in *Neurospora,* and a summary of some of the characteristics of the yeast and *Neurospora* mutants is given in Table 24. Most of the differences shown represent complete or partial deficiencies of components of the oxidase system, but the presence of an abnormal cytochrome was observed in *Neurospora* strain C-117. This has absorption-spectra characteristics of cytochrome e, as shown, but it is possible that it is an altered form of another cytochrome. Numerous quantitative abnormalities in other enzymes and

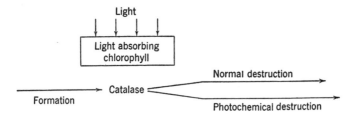

Figure 6.7. Catalase in plant seedlings.

Table 24. Alterations of Complex Enzyme Systems in *Saccharomyces* and in *Neurospora* as the Result of Gene Mutations

Strain	Succinic Dehydro- genase	Cyto- chrome b	Cyto- chrome c	Cyto- chrome a	Succinic Oxidase	Cyto- chrome
Normal yeast	+	+	+	+	+	−
"Petite" yeast segregational	−	−	++	−	−	−
Wild-type *Neurospora*	+	+	+	+	+	−
Neurospora C-115	+	+	++	−?	+?	−
Neurospora C-117	+	++	−	−	−	++

various metabolites have been observed, but if single mutational primary effects are the rule then most of these changes are indirect effects.

In view of the importance of cellular organization, mutations such as these which affect enzyme complexes are very important for study, but in no case has it yet been possible to trace the cause in terms of a mutation-induced structural change in one gene product.

Direct Effects. The three preceding examples, though taken from the older literature, are illustrative of generally occurring and important situations in which obvious and simplest interpretations of primary data were shown to be completely incorrect. Many more examples of this kind have been observed along with analyses of situations where genetic changes yield protein changes that are clearly more direct.

A variety of investigations carried to different degrees of analytical sophistication have shown mutation to result in production of enzymes with altered heat stability, electrophoretic properties and Michaelis constants—see Fincham (280). It should be kept in mind that these are positive differences based on arbitrary criteria, and it is to be expected that large numbers of changes in a protein could occur without noticeable effects. For example, substitution of leucine for valine in a primary chain might have no measurable effect on activity and be detectible only by amino-acid analysis. In any case, positive effects in terms of heat stability of an enzyme have been studied extensively for glutamic-acid dehydrogenase (*Neurospora*), pantothenate synthetase (*E. coli*), pyroline-5-carboxylate reductase (*Neurospora*), and tyrosinase (*Neurospora*) (280). Although each of these examples has special features of interest, details of one must suffice here for illustration. In this case—tyrosinase (461)—mutants studied were spontaneous and derived from nature. From a total of 12 wild-type strains of *N. crassa,* four were found to produce tyrosinases with different properties, as summarized in Table 25. Genetically, only one locus is repre-

Table 25. Properties of the Four Tyrosinases of
Neurospora crassa *

Type	Thermostability † (Half lives at 59°, min)	Electrophoretic Migration ‡ (mm per hour, on paper)
S	70	2.0
L	5	2.25
PR-15	20	1.5
Sing-2	70	1.5

* From Horowitz, Fling, Macleod, and Watanabe (461).
† In 0.1 M sodium phosphate buffer, pH 6.
‡ In 0.05 M sodium phosphate buffer, pH 6, containing 0.1% bovine serum albumin. Rates not corrected for electro-osmotic flow of the buffer.

sented and, in heterocaryons (mycelia-containing nuclei from more than one parent, p. 144), mixed tyrosinases of the parental types are produced. The enzymes from T^s and T^l have been isolated in a pure form and found to have molecular weights of about 33,000 and to contain copper but no cystine. Pure components have the same temperature-sensitivity difference characteristics as crude preparation, and thus it is likely that the four mutants yield four enzymes differing in primary amino-acid sequences, although this has not yet been established. Confirmation that the T-locus (represented by the preceding four alleles) is concerned with the primary structure of the enzyme, tyrosinase, came from studies of two other mutants, *tyr-1* and *tyr-2,* which carry mutations not linked to T or to each other. Under conditions where T strains produce large amounts of enzyme (low sulfur medium), these mutants produce very little or none. They do produce enzyme, however, in heterocaryons, and, most significantly, the type produced in this situation is that characteristic of the particular T-locus (or loci) present. The loci of *tyr-1* and *2* are therefore regulatory to the primary function of T. They may have, of course, other functions that have not yet been detected. These relations between T and *tyr* loci were further emphasized by studies on physiological aspects of enzyme production and inducibility. Tyrosinase is not required for vegetative growth of

Neurospora, but it is in some way concerned in sexual reproduction (1070), and the mutants *tyr-1* and *2* are female sterile. In the wild-type, enzyme production is favored by growth limitation, and a variety of growth inhibitors induce production in the absence of nutritional limitation. These same inhibitors induce enzyme formation in mutants *tyr-1* and *2,* again of the type characteristic of the *T*-locus present. It appears that tyrosinase synthesis occurs when more general protein synthesis ceases, with incorporation of amino acids from breakdown of other proteins and a free amino-acid pool.

In the foregoing example there appears to exist a straightforward primary gene-enzyme relation with a remarkably high frequency of spontaneous mutation to give different, but functionally similar, enzymes. Imposed on this are genetic regulatory factors which do not alter primary structures and a clear-cut distinction between structure determination and regulation function is indicated. The level at which regulation occurs (DNA, RNA, protein, etc.) cannot be designated here, as in most cases, but it is not necessarily different from the mechanism proposed by Jacob and Monod (p. 424).

A second example of gene-enzyme relations that provides additional important details is that derived from studies of the tryptophan-synthetase systems in *Neurospora* and *E. coli.* The reactions involved here, as established by Yanofsky and collaborators (188), are shown in Figure 6.8. It is probable that reaction 1 is the one of physiological significance, but the existence of 2 and 3 presents a situation which makes this example especially interesting and useful. At the time of the first studies of this mutation-induced enzyme deficiency in *Neurospora* (685), indole was believed to be an intermediate in tryptophan synthesis, and only reaction 2 was studied. This appeared to be a completely straightforward case of loss of a single enzyme as the result of a single mutation. Further, the restoration of an apparently normal enzyme by reverse mutation was demonstrated as well as actions of several modifying genes in regulation of amounts of enzyme activity (446). Thus the definition of a gene-enzyme relation, even at the level of one known enzyme activity, can be quite misleading.

Subsequent studies of the tryptophan-synthetase system, which led to the formulation of the reactions as given in Figure 6.8, were carried out using *Neurospora* and *E. coli* (85, 992). Large numbers of mutants affecting this system were obtained in both organisms, and several points concerned with similarities and differences are of special interest. In *Neurospora* there appears to be a single protein (mol. wt. about 140,000) which catalyzes all three reactions shown in Figure 6.8.

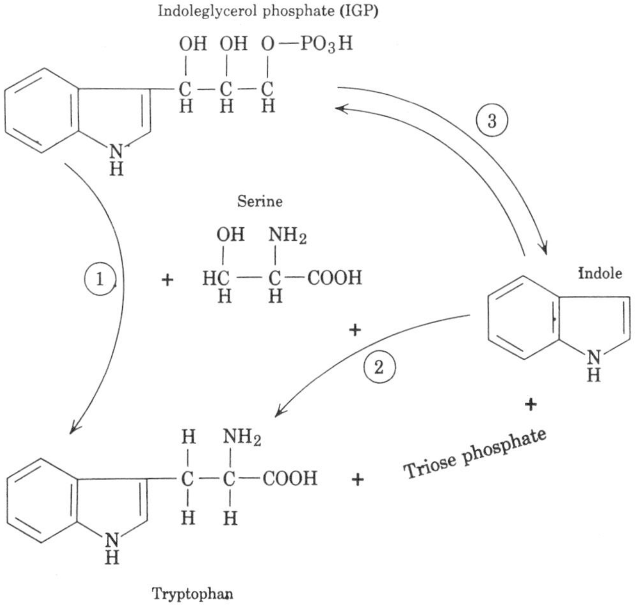

Pyridoxal phosphate required
in serine reactions

Figure 6.8. Tryptophan synthetase reactions.

On the other hand, in *E. coli,* the unit which catalyzes all three reactions is easily dissociable into two proteins, A (mol. wt. 29,500) and B (mol. wt. not determined). Protein A alone has trace activity for reaction 3 and B alone a slight activity for reaction 2, but combination to give AB is necessary for maximum rate of any of the reactions 1, 2, or 3. Thus, superficially, in *Neurospora* one gene gives rise to one enzyme (one protein), while in *E. coli* two genes give rise to two inefficient enzymes (two proteins) which combine to give one organized and enzymatically efficient system. However, genetic and biochemical analyses using many mutants in both organisms have established an interesting homology in the two systems which may lead to a less decisive conclusion. In both organisms, many mutants which cannot synthesize tryptophan because of a deficiency of fully functional tryptophan synthetase still have the capacity to carry out reactions 2 or 3 (Figure 6.8).

All of these mutants also produce material which gives a cross reaction with antibodies produced from immunization of rabbits to normal tryptophan synthetase.   For this reason they are called crossreacting material or CRM.   These substances that give partial enzyme activity and crossreact are one and the same, and thus represent altered enzymes produced under the influence of genes altered by mutation.   On an immunological basis alone *Neurospora* yields one crossreacting material, but *E. coli* yields two corresponding to proteins A and B.   However, on an enzymatic basis *Neurospora* also yields classes of altered enzymes equivalent to the A-B system of *E. coli*.   These results have been further correlated through genetic fine-structure studies, using recombination analysis as described in Chapter 8, with the demonstration that in both organisms the chromosome segment concerned with tryptophan-synthetase production can be divided into sections, one concerned with alterations affecting capacity for reaction 2 and the other reaction 3.   This information is summarized in Figure 6.9.

As shown, in *E. coli* the genetic division corresponds to production of proteins A and B and these can be defined as separate genes, or two parts of one gene, as desired.   In *Neurospora* it is not known whether the enzyme produced has a single primary chain, but for the present the entire chromosome segment is most reasonably considered as one gene.   There are here, in this comparison, interesting implications concerning evolutionary mechanisms going in either direction: toward separation to give two genes and two enzymes from one, or toward condensation of genes to give a more highly organized enzyme.   But, to return to the more specific problem at hand, genetic fine-structure studies have yielded much detail concerning mutational loci in all parts of the chromosome segments in both organisms (992).   For the present discussion only part of this information will be considered.   Further information is given in Chapter 8.

As shown at the bottom of Figure 6.9, relative positions of mutational sites for a considerable number of A mutants of *E. coli* have been established.   All of those mutants which produce altered A protein are shown in the upper half of the chart, and many of these have been isolated in a pure state for studies of amino-acid sequences in relation to the linkage map.   These various A proteins are not distinguishable immunologically nor do they differ significantly with respect to activation of B protein to enzymatic activity.   However, they do differ by other criteria, as follows.   In the first cluster (left in Figure 6.9) proteins from mutants 11, 26, 37, 41, 45, and 48 are precipitated but not inactivated at pH 4.0.   In the last cluster, A proteins of 23, 24, 27, 28, 35, 36, and 53 are all labile to heating.   Proteins from mutants in the same cluster are not identical, however, as those from 3 and 33

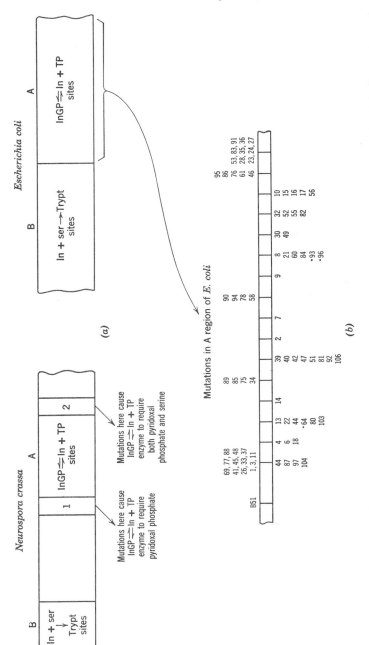

Figure 6.9. Genetic units concerned with formation of tryptophan synthetase in *N. crassa* and *E. coli*. (*a*) Comparison of subunit positions in the two organisms. (*b*) Fine structure of the A region in *E. coli*. From Suskind and Yanofsky (992); and Yanofsky, Helinski, and Maling (1126).

(cluster 1) and 46 (cluster 4) are more heat stable than normal A. A number of other distinguishing characteristics have been applied, but the information of ultimate importance is amino-acid sequence. Although such data are far from complete, some important conclusions have been reached from studies of the proteins of the distinguishable mutants 11, 33, and 3 (cluster 1), 46 (cluster 4), and 23 (cluster 5).

Considering first the question of correlation of the order of genetic linkage with relative positions of amino-acid changes in the primary peptide chain of protein A, protein samples from these mutants were subjected to specific proteolysis (p. 80) using trypsin and chymotrypsin. Resulting peptides were examined by the fingerprint technique and by separation and analysis of individual components. The general conclusion was reached by the application of these techniques that the normal, 11, and 33 strains yield different amino-acid substituents in a particular peptide from tryptic digestion, whereas 23 and 46 have abnormal amino-acid components in a totally different region of the A chain. Thus, as far as it goes, this analysis indicates a colinearity between the genetic-linkage map and the amino-acid sequence (425).

A second point of special interest derived from studies of the A protein as just described comes from a more detailed study of mutants 26 and 23 (426). As noted in Figure 6.9, these mutational loci are close, but recombination does occur. Nevertheless, a peptide (no. TP_3) from tryptic digestion shows amino-acid substitution at position 12 with glutamic acid in A46, arginine in A23, and glycine in the normal A protein. Although total sequences are not yet established, it is clear that substitution of different amino acids at the same position in a primary peptide chain can result from mutation at different sites. Interpreted in terms of DNA coding (p. 108), at least two nucleotides are required to code for one amino-acid position. In the triplet code (Table 6, p. 107) nucleotide changes of GUG (glycine) to AUG (glutamic acid in A46) or to GUC (arginine in A23) would satisfy the situation as it stands with mutation consisting of substitution of A for G and C for G in the two mutants. Recombination at the nucleotide level could thus yield normal protein (and GUG in the DNA).

Two additional developments of interest in this same system are considerations of reversion and suppression in terms of primary sequences in the A protein. Present information gives the conclusion that reversion frequently occurs by a second substitution of amino acid rather than a change back to normal. In cases of suppression (see p. 397) the interesting situation has developed in which altered protein and a small amount of apparently normal protein is produced in a suppressed mutant. One explanation of this phenomenon involves change in primary sequence at some level between the gene and the peptide in the processes of protein synthesis. Obviously, it is an important area for investigation, and fortunately much can be done with available methods.

As has already been pointed out, the gene-enzyme relation as illustrated by the tryptophan-synthetase system could have been illustrated by a number of other examples, and the reader is advised to consult current literature for further details in this very rapidly advancing field. It should be clear at this point that combinations of fine-structure genetic analysis with detailed determinations of protein structure will eventually give a much clearer picture as to what can be defined as a gene and as an enzyme. Further precise definitions are not very important at present since experimental approaches permitting definitions are clear. At the same time, the greater the variety of genetic and enzymatic systems that can be subjected to analysis, the more useful the conclusions in establishing fundamental and general principles.

In the literature there are several examples of situations where a genetic locus gives rise to an enzyme functional in different essential reactions of metabolism (320, 1048). For example, in *Neurospora,* Giles and collaborators (323) have studied mutants at the *ad-4* locus which is concerned with production of an enzyme that will catalyze two different but widely separated steps in purine synthesis. One reaction involves the splitting of succinate from 5-amino-4-imidazole-(N-succinylcarboxamide) and the other the splitting of succinate from adenylosuccinate (see p. 315). Mutation to give a thermolabile enzyme or mutation to give no enzyme followed by reversion to give an enzyme with different activities in the two reactions have been observed. Detailed analysis in terms of peptide sequence is a potential here, but in terms of mutations and metabolic control the results show clearly the potential of one locus mutating to change balances of more than one reaction.

Another kind of situation that has interesting prospects is that concerned with the production of xanthine dehydrogenase in *Drosophila* (285). In this case mutation at either of two unlinked loci (*ry* and *mal*) results in a deficiency of the enzyme. An immunologically cross-reacting substance is produced by one of the mutants and a maternal effect has been observed (328, 329). It would be possible to interpret this situation of two separated genes—one enzyme in terms of structural and regulatory action as in the tyrosinase case discussed previously, but Forrest (286) has suggested interpretation similar to that for the tryptophan-synthetase analysis in *E. coli;* that is, a complex enzyme formed from at least two primary peptides produced under the influence of genes not adjacent or even in the same chromosome. Although this and many prospective systems in higher organisms have technical limitation in analysis at the gene level, prospects at the enzyme level are good and possibilities for gaining more understanding of RNA classes of agents of metabolic control appear to be as good as in any kind of material. One need only consider the progress in studies of hemoglobin inheritance to come to this conclusion (Chapter 14).

Mutations and Structural Protein. No real distinction can be made between enzymes and structural proteins. It is true that enzyme activities provide additional and useful criteria for use in isolation and detecting mutational changes, but in organized systems enzymes are themselves structural proteins, and structural proteins must certainly be functional in dictating arrangements of organized enzyme systems. Specific amino-acid sequences are essential in both, and genetic mechanisms controlling synthesis are, surely, not fundamentally different.

Hair and silk represent extreme examples of structural proteins, while such materials as collagen and ribosomal protein are also structural proteins but have a physiological function beyond mere protection. With respect to studies of mutation and protein structure, many such components present favorable material for study. One very instructive example should be considered in some detail—that of the coat protein of tobacco mosaic virus (TMV).

As shown in Figure 2.39, page 100, TMV consists of a long rod with RNA coiled within a protein coat made of many identical units. This presents as simplified a system as has been obtained for studies of relations between nucleotide sequence and amino-acid sequence, since the RNA itself represents the genetic material and the coat protein one of its products. Replication and other functions, of course, only occur in a host cell, and little is known of the other functions and participation and influences of host metabolic systems in reproduction of the virus. In any case, infection can be produced from isolated RNA alone (914), and coat protein from new viruses is easily isolated for analysis. The coat protein (about 2130 units/virus) contains 157 amino acids (mol. wt. 17,420) and normally yields 12 peptides from tryptic digestion. Complete amino-acid sequences are known for peptides and total protein (1025, 22). The virus RNA, which gives order to the amino-acid sequence, contains about 6500 nucleotides, and if sequences of only three (see p. 107) are required for each amino acid, less than one-tenth of it is needed for coding coat protein and the rest probably has other functions.

Studies of the nucleotide-amino-acid relations have been greatly aided by use of nitrous acid as a mutagenic agent.   This is particularly useful, since it involves direct structural alterations of purines and pyrimidines in isolated RNA.   It depends on the general reaction of nitrous acid on amines to yield substitution of —OH for —NH$_2$.   Thus adenine is converted to hypoxanthine, guanine to xanthine, cytosine to uracil, and uracil remains unchanged.   In RNA which is so altered, hypoxanthine functions in base pairing as guanine, uracil as uracil, and xanthine as the original guanine from which it came (727, 1024).   Therefore, the significant changes that occur are A $\rightarrow$ G and C $\rightarrow$ U.   The total potentials, then, for specific alterations of RNA, based on the nitrous-acid reactions and the triplet code (p. 107), are those shown in Figure 6.10. Whether or not the triplet code is correct and universal, the formulation of the type shown in the figure is illustrative of the prospects of this method for investigating controlled changes in RNA base sequences. With appropriately mild conditions, it is reasonable to obtain a single change in one RNA chain.   Although other kinds of reactions of nitrous acid with the bases are not entirely ruled out, deamination is likely to be the most common.

This nitrous-acid mutation method has been applied extensively (1095, 1024, 1023) to TMV with analysis of coat proteins from

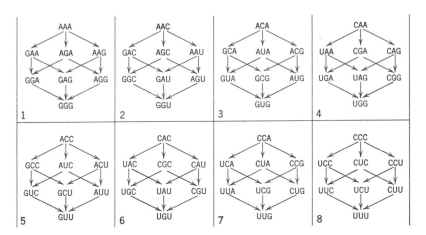

Figure 6.10.   Diagram showing 64 possible nucleotide triplets grouped in octets to show potential for interconversions. From Wittman (1096).

mutants.   An example of results obtained by Wittmann (1095) is shown in Figure 6.11.   Mutants with the Ni prefix are those derived from nitrous-acid treatment, and the others are spontaneous.   Amino-acid substitutions in various tryptic peptides (Roman numerals) are shown below the chart.   In cases where no substitution was found it is assumed that mutation affected parts of the RNA not concerned with production of coat protein.   A summary of kinds of substitution ob-

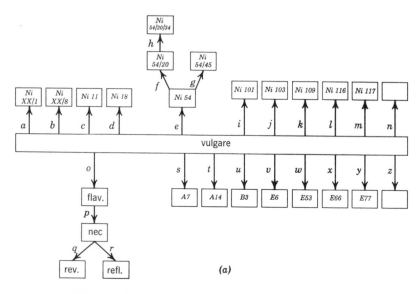

(a)

Steps $a, c, d, e, f, g, s, u, v, w, y$: no change
Steps b, h, o, z: Asp → Ala in peptide XII[1]
Steps i, j: Asp → Gly in XII
Step k: Glu → Gly in VII
Step l: Asp → Gly in IX
Step m: Ser → Phe in IV; Glu → Val in XII
Step n: Pro → Leu in XII
Step p: Ala + Phe → Val + Leu in XII; Ser → Phe in IV
Step q: Val + Leu → Asp + Phe in XII
Step r: Leu → Phe in XII
Step t: Ileu → Thr in XI
Step x: Asp → Lys in IV

(b)

Figure 6.11.   Mutants in tobacco mosaic virus.   (a) Summary of mutants obtained.   (b) Amino-acid substitution observed corresponding to the mutants shown in (a).   From Wittman (1095).

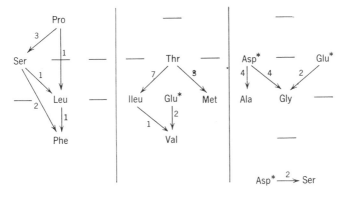

Figure 6.12. Amino-acid substitutions in the coat protein of tobacco mosaic virus. From Wittman (1096).

served from nitrous-acid treatment is given in Figure 6.12. This picture is based on the octet formulations shown in Figure 6.10 and illustrates a procedure for evaluating the validity of the triplet code in terms of amino-acid substitutions. It is clear that a complete evaluation on this basis is technically feasible, and in time it will be possible to clearly establish further fundamental aspects of the specificity relations between base sequences in polynucleotides and amino-acid sequences in polypeptide chains. Obviously, this represents a fertile field for future development in terms of finding other kinds of reactions to change base-pairing characteristics in specific ways. Alterations by change of existing bases, substitution of bases, and even total synthesis are feasible. This general approach may be expected to contribute greatly not only to the coding problem but also to increased understanding of the mechanisms of the fundamental processes by which nucleotide sequences yield expression in amino-acid sequences.

Genetic Blocks

One of the conveniences of the one gene-one enzyme hypothesis has been its use as a simple concept relating a specific genetic locus to a specific metabolic reaction. Thus, as shown in Figure 6.13*a,* and frequently in the discussions of metabolic patterns in Chapter 7, it is reasonable and useful to oversimplify and describe a "genetic block" as a mutation to a change in capacity to carry out a given metabolic

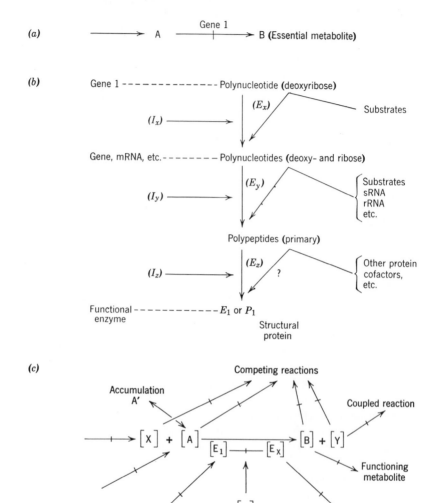

Figure 6.13. A diagrammatic representation of a genetic block. (*a*) Simplified designation. (*b*) Pattern concerned with enzyme synthesis. (*c*) Patterns concerned with enzyme function. Cross bars on arrows indicate some possible positions of change in reaction rates that can give rise to change in the rate of reaction A → B.

reaction. At the same time, this method of notation can lead to an illusion of understanding that does not exist, and it is desirable to inquire further into the character of a "genetic block," taking into account the facts and principles discussed in this and earlier chapters.

Clearly, the genetic block can mean many things in terms of the mechanisms by which it is dictated and many things with respect to its phenotypic expression, and some of the factors concerned are indicated in the diagrams in Figure 6.13a and b. Part b is concerned with the initial production of an enzyme or other specific protein and is separable into at least three major steps, each of which is subject to influences by the nature and amounts of reactants, enzymes, and inhibitors. In the first stage the size and total structure of the gene itself cannot be generalized, but assuming it to be a DNA chain with designated limits, it functions in the first step in replication of itself and in production of mRNA. Numbers and kinds of enzymes involved are not known, but included here would be polymerase reactions and any secondary reactions which alter the polymers produced (glucosylation in T-even phage, for example). The factor I_x in this step includes all inhibitors that may have influence with the extreme of total inhibition of function of gene 1, at least for mRNA production. In this case, which would be one of regulation, the lack of production of E_1 could be traced to the function of a gene giving rise to I_x rather than to gene 1. However, it should be noted that a given I_x could react with one mutant form of gene 1 and not with another and thus bring two genes into the action concerned with formation of E_1 or altered E_1. Other varieties of potential interactions will surely be evident as more details of these processes become known.

The reactions of step 2 include the possibilities of mRNA replication and degradation during function, and it is easy to see many possible sources of influence on E_1 production. These may influence only amount of enzyme, but structure control at this level is certainly not ruled out, in general, and it does occur in the case of RNA viruses. Only speculation is possible, at present, but in any case it is clear that many genes other than gene 1 may be responsible for control of at least the amount of E_1. At step 3, also, there is a broad potential for control of types related to those of the first two steps. In addition, there are special considerations concerned with the derivation of functional enzymes from primary peptide chains. Such reactions include folding processes to give proper tertiary structure, association processes when the functional E_1 is a polymer of subunits or a combination of more than one primary polypeptide chain, and enzymatic alteration of a primary protein to give a functional product as with the zymogens. Here again, there exists a potential for systematic alterations of primary structure resulting from action of genes other than gene 1 and a high probability of a wide variety of regulation activities by many genes. Thus, this incomplete scheme (Figure 6.13b) representing production of E_1 as related to gene 1 is potentially very complex in terms of possible interaction and regulations and possibly in terms of primary structure. It should be pointed out that the present evidence is consistent with a somewhat simpler concept in which all of primary structure of E_1 is determined by gene 1 with all other interactions being secondary and regulatory, but much more evidence will be necessary to establish this as a principle without exceptions.

As indicated in the foregoing discussion, the first feature of the genetic block is regulation of formation of E_1 or P_1, but this can be derived in many ways, and even if only one gene does control structure, many others can control functional presence or absence. Such a potential for indirectness is also a characteristic of the second feature of a genetic block. This is concerned with enzyme function in production of a phenotype and some aspects of this problem are diagrammed in Figure 6.13c. Here, as an example, with the genetic block expressed phenotypically as a reduced rate of some metabolite B_1, it is shown that this reduced rate can come about by a change in rate of any one of a large number of other reactions as indicated by the crossbars on the various arrows in the figure. From previous discussions in this and earlier chapters it is clear that a given reaction pattern can be somewhat simpler than shown here, but it can also be a good deal more complex. Here the rate in question is dependent on the approximately steady-state concentrations of reactants and products designated as (X), (A), (E_1), (E_x), (I), (B), and (Y). These concentrations are in turn dependent on relative rates of formation of components and diversion into side reactions. The designation $(E_1) \rightarrow (E_x)$ includes both enzyme alterations to different active forms and the concepts of sequential steps and attachments at the enzyme surface, while the coupled-reaction step is shown to indicate that the phenotype as evaluated by the amount of B may be strongly influenced by the energetics of the system and a secondary reaction of a different product (Y).

A careful consideration of the diagrams in Figure 6.13b and c will reveal oversimplification with respect to formulation of all possibilities. For example, it would be reasonable to add more steps in part b and to treat each one in the same way as the $A \rightarrow B$ reaction in part c. The important thing for emphasis, however, is not the potential detail but the principle that a genetic block, as represented in part *a* of the figure, is in itself a characteristic metabolic pattern. Furthermore, a phenotypic expression of mutation in terms of some remote metabolite depends as much on the metabolic pattern that existed before mutation as on that existing after mutation. In addition, it is unlikely that a mutant phenotype can ever be completely restored to a premutation phenotypic state by environmental changes. For example, a nutritional mutant of *Neurospora* may grow just as well as a parent wild strain if a required metabolite is provided, but the metabolic pattern of the genetic block remains. It may be altered or even largely eliminated, as in repression (see p. 419), but not restored. This is shown clearly in many cases by accumulations in mutants of metabolic intermediates or remote by-products. Often, different kinds of mutation will produce, through indirect changes of balance, accumulations or deficiencies of the same thing, and in some cases, such as in pigment formation, these may provide the major phenotype for study.

Some General Conclusions

An important conclusion that can be drawn from the more recent developments concerning mutations and the agents of metabolic control is that the prospects for coming to a real understanding of gene-enzyme relations are excellent. This is true for the following reasons. (1) It is firmly established that base-pair hydrogen bonding in nucleic acids can provide much of the chemical specificity required for gene replication. (2) A partial definition of the nature of reactants and products in the steps leading to protein synthesis has been achieved. (3) Much has been accomplished in the technology of protein structure and its relation to enzyme activity. (4) It has been demonstrated that there is a homology between linear base sequences in polynucleotides and linear amino-acid sequences in polypeptide chains. As a result of these developments, the one gene-one enzyme (or one-function) hypothesis has evolved to a more specific one gene-one primary polypeptide relation with a linear correspondence, and a tentative ratio of three nucleotides to one amino acid. Further evolution of the hypothesis is likely, but this form, like the earlier one, presents a very useful and important concept.

Developments in the specific areas discussed in this chapter have been extremely rapid, and no attempt has been made toward a complete review. The material selected is intended only to represent a number of principles useful for evaluating various experimental results and for indicating directions that may be expected in future work. At the gene level, it is clear that much additional definition is necessary, as discussed in Chapter 8. It is not a foregone conclusion that the complex structures of chromosomes as in higher organisms are functionally equivalent to the simpler system described here by the TMV system. Accepting the linear correspondence of nucleic acid and peptide, it becomes necessary to learn more about length, about terminal linkages, and about the significance of structure of genes in the other dimensions. That is, specific changes normally occur after linear polynucleotide formation (as in glucosylation of DNA) and in chromosomes secondary and tertiary structure as in nucleohistones surely represent factors in specificity in terms of mechanisms of function. Further, the kinds of chemical reactions resulting in recombination by reciprocal mechanisms and in aberrations are in need of much clarification.

At the intermediate levels between genes and polypeptides which involve s, m, and rRNA's, at least, little can be said as to the importance of secondary reactions. Here the major initial problem of isolations and determination of primary sequences is in need of solution before much indirect deductions regarding structure and function can be wholly acceptable.

At the polypeptide level several facts are clear. As illustrated by the zymogen examples, initial polypeptide products of gene action are not necessarily the direct agents of metabolic control. The subsequent changes in these cases are derived from hydrolysis, but other kinds of reactions may occur after sequence designation. Hydroxyproline perhaps is produced from proline after polypeptide formation, and amidation and deamidation of the decarboxylic amino acids may also fall in this category. It will not be surprising if many other kinds of secondary reactions are found, and primary-chain changes by transpeptidation are also likely. Present knowledge comes from relatively few cases that are favorable for study, and surely all the potential for deviations of these types have not yet been encountered. Imposed on these kinds of probable variables concerning single chains are those of polymerization, as in the lactic dehydrogenase and combinations of different proteins to give new functional activities. In such cases there is the potential for formation of longer primary chains than designated by a gene through covalent attachments of like or unlike components. For example, as a speculative situation, it seems possible that a gene could be broken in two by translocation in a chromosome, and, if each part continued to direct production of its portion of an original polypeptide, the two small pieces might function as a two-protein enzyme or become coupled after formation as a two gene-one peptide product. Obviously, what is already known provides a basis for a great variety of possibilities of this kind that may be encountered. Thus, what is important at present is not any special dogma of the gene-metabolic control relation, but the continued reevaluation of each component involved and each step involved in the greatest variety of biological material possible.

7 Metabolic Patterns

Genes exert their effects through control of metabolic processes, and an attainment of a full understanding of inheritance is therefore dependent not only on knowledge of the nature of the heritable units and the processes by which they act, but also on a complete knowledge of cellular biochemistry. An important feature of this biochemical problem is the gathering of information on the numerous reaction sequences that occur and which can be put together to give an overall metabolic pattern. As discussed in the preceding chapter, each genetic unit is related through its own special metabolic process to the production of a polypeptide chain which may itself be an enzyme and participate in the control of the rate of a single specific reaction.

At the same time, this single reaction is subject to control by a variety of factors and is best represented also by a pattern of potentially interacting components (see Figure 6.13). Thus, for a given reaction, there is a primary factor, the gene itself, but its ultimate action in terms of control may be strongly influenced at all levels, from replication to enzyme catalysis or morphology determination, by primary or indirect products of the action of other genes.

Part of this control lies in the nature of reaction sequences leading to production of substrates for a reaction and in the nature of sequences providing for utilization of reaction products. Therefore, specific knowledge of the nature of sequential reactions and interactions in converging, diverging, and cyclic patterns is the essence of any proper evaluation of a phenotype. At the present time there remain innumerable examples of phenotype inheritance which cannot yet be analyzed because of a lack of biochemical information. The number of examples in which direct gene-primary peptide relations can be studied are still few in number, and a full description of a phenotype change resulting from mutation is not possible in any known case. A mutation always alters a metabolic pattern in many ways, and, although these may result from a primary change, they are not predictable from gene structure or polypeptide structure since they depend on a balance of many reactions under the influence of many genes.

Studies in biochemistry and especially in enzymology have yielded extensive information on single reactions and series of reactions that can occur in tissues. The patterns of the citric-acid cycle (Figure 5.3, p. 231) and of sugar metabolism (Figure 5.4, p. 240) are examples. To a considerable extent these have been elucidated by studies of individual reactions using inhibitors to block specific reactions, and in these examples *in vivo* functions in the same patterns have been adequately demonstrated. In recent years this general approach to studies of reaction sequences in cells has been supplemented by the extremely useful combination of genetic and biochemical techniques. Obviously, when mutation results in the lack of capacity for production of a specific enzyme, a corresponding reaction is blocked and a product is not formed. If the product is essential for the life of the cell, then the mutation is lethal unless this product can be supplied from the external environment as a nutritional supplement. When the latter is true, the cell may survive and even grow in an apparently normal fashion. On the other hand, such a genetic block, which is commonly represented as $\rightarrow$ compound $A \xrightarrow{\text{Gene (1)}}$ compound $B \rightarrow$, may represent an extremely complex phenomenon as already discussed (Figure 6.13, p. 280). Nevertheless, mutant survival, in spite of metabolic alterations, is common, and the nutritional mutants of microorganisms have proven extremely valuable as an aid in elucidating metabolic patterns. The principle here is simple. In an organism which requires substance C, which is normally produced in the reaction sequence $\rightarrow A \xrightarrow{(1)} B \xrightarrow{(2)} C$, a mutation which blocks (by any mechanism) reaction 2 may cause accumulation of an unknown intermediate B when the mutant is grown on C. This compound *B* can then be utilized for growth of a second mutant 1 which will grow with either *B* or *C*. In this general fashion, by use of large numbers of mutants and isolation or synthesis of intermediates, it has been possible to establish extensive patterns of metabolism, especially in microorganisms. Many of these same patterns have been found to obtain in whole or in part in higher organisms as well.

Nutritional Mutants. The organisms that have been especially useful for studies of metabolic patterns through nutritional mutants are those which will grow on the simplest culture media and thus have the highest capacity for biosynthesis. One of the first and most extensively used organisms, the fungus *Neurospora crassa,* for example, grows well on a medium containing only inorganic salts, a carbon source (sucrose or other relatively simple carbon compound), and the one vitamin biotin (58, 1071). The bacterium *E. coli* requires only inorganic salts and a carbon source. In both cases the organisms have the genetic and enzymatic capacities for synthesis of all the other metabolites essential for life. Thus these apparently simple organisms are in a sense more complex than vertebrates, which depend on the synthetic capacities of plants and microorganisms for satisfaction of their complicated nutritional requirements. On the other hand, the vertebrates have additional synthetic capacities for substances not required by microorganisms. In any case, many of the fundamental systems are similar in a great variety of species, and patterns derived from studies of mutants of the fungi such as *Neurospora, Aspergillus, Ophiostoma,* yeast, and the bacteria *E. coli* and *Salmonella* have provided a great deal of information of general value.

As pointed out earlier and discussed in Chapter 6, the overall, apparently simple picture of a mutational block in metabolism can be very complex in mechanism, and prior to a consideration of some details of established metabolic patterns, some discussion of certain practical aspects of the use of nutritional mutants is desirable. With respect to the mutants themselves (Chapter 4), they are easily obtained in large numbers by various enrichment techniques (699, 591), but strong selective factors are involved. For survival of mutants it is necessary that the cell membrane be permeable to a nutrient supplied in the external environment. Thus selection methods eliminate mutants with requirements for many high-molecular-weight substances and for many substances which are strongly charged. Especially prominent in the second category are the organic phosphate esters which, in general, are not transported through cell membranes. Frequently the parent organic compounds are taken in but they cannot be used because of a lack of a special phosphorylating mechanism within the cell. Since phosphorylated intermediates are very common in biosynthetic mechanisms, this principle is a very important one in mutant selection.

At the other end of the selection scale from those which are lethal because of requirements which cannot be satisfied from the outside are those which have unstable mutations and those with partial requirements. For every usable mutant obtained in a selection experiment there are probably ten or more which fall in these categories, and often they are discarded in the selection process. Many such mutants may be especially interesting, and likely much more can be done with them than has been done so far, but some kinds present serious technical difficulties. The high reverse mutation rate of unstable mutants presents problems that can be handled properly only if means could be found to reduce the rate. This is a possibility worthy of investigation. Partial mutants, on the other hand, are of great interest but difficult to analyze when their phenotypes overlap too much with the parental character, and they grow too well to be detected as mutants. These can arise for many reasons, among which are reduction in amount of a normal enzyme activity due to inhibition from the gene level all the way to the enzyme-substrate level, and from qualitative changes in an enzyme produced. In the former situation one class of mutants that may frequently be lost is that which carries, in the parent wild strain, suppressor genes for potential mutants. These should be recoverable by further mutation. In the second category, partial mutants probably due to qualitative changes in enzymes, an important group that has received increasing attention, are the temperature mutants.

An early example from studies with *Neurospora* is that of a mutant which requires riboflavin at 35°C but behaves as wild type at 25°C (684). A partial requirement was demonstrated at 30°C. Furthermore, at 35°, where no growth was observed over a long period in the absence of riboflavin, it was shown that when a very small amount of the vitamin was supplied, growth and riboflavin synthesis occurred intermittently as illustrated in Figure 7.1. Thus the genetic block is partial even at 35° where the nutritional requirement appears to be absolute, but the physiological expression of activity at this temperature is complex. Many nutritional temperature mutants of *Neurospora,* yeast, and bacteria have been encountered, and, as discussed in Chapters 6 and 10, temperature-sensitive enzymes have been shown to result from specific gene mutations. It is worthy of note also that among the temperature mutants are many which are not nutritional, and they cannot be induced

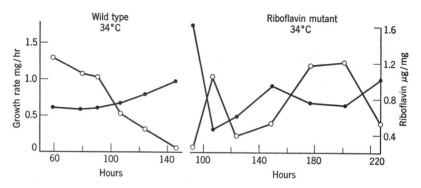

Figure 7.1. Riboflavin production by wild-type *Neurospora* and a riboflavin-requiring mutant. Growth rate is indicated by open circles and riboflavin content by solid circles. Cultures in the presence of a severely limiting concentration of the vitamin (0.015 μg/ml). From Mitchell and Houlahan (684).

to grow in the range where the mutant character is expressed. These, which were termed irreparable mutants by Leupold and Horowitz (594), would be lethals were they not temperature mutants, and they probably represent examples of mutations to deficiencies for substances that cannot be supplied from outside the cell. As such they merit reinvestigation at enzyme, protein, and nucleic-acid levels. In a recent study Edgar (245) has obtained temperature mutants representing about half the genes of bacteriophage, and without question these will be valuable in further studies of gene-protein synthesis relations.

There is probably no fundamental difference between the temperature mutants that give temperature-sensitive proteins and many other partial mutants which are not temperature-sensitive. The latter may vary in nutritional phenotype all the way from the wild-type character through different degrees of nutritional stimulation to apparent absolute requirements. Even many of these which do not grow without supplement are apparently partial mutants. For example, as demonstrated by Bonner et al. (86), three of the four tryptophan-requiring mutants listed in Table 26 showed considerable dilutions of N^{15}-labeled growth supplements (see Figure 7.10 for the metabolic pattern involved). Partial blocks in the three mutants is indicated.

From the foregoing discussion it is clear that partial genetic blocks in metabolism are very common, and when reasonably extreme such mutants are as useful as any other kind in studies of metabolic patterns. Furthermore, those which are less extreme and often discarded are likely to be particularly useful in studies of gene-enzyme relations. In a mutant with a complete block with no production of enzyme or related protein, there is no detectible gene product for structure study, but, as well demonstrated for the temperature mutants and likely for many of the weak partial-block mutants, altered enzyme is present and mutation structure-change relations can be investigated.

Another important aspect to the technique of use of nutritional mutants in studies of metabolic patterns is concerned with the metabolic changes that occur because the mutation is present. Even though a mutant may grow normally, the block exists, and reactions preceding it may go on normally under some circumstances. In this case the last substrate or intermediate before the block may accumulate in the cell and also be spilled out into the medium. This sometimes happens, but often secondary side reactions occur which may convert accumulated substances by one or more steps to materials quite unrelated to the original intermediate. These become part of the phenotype and may show as pigments, etc., but they may give little information concerning the nature of the intermediate from which they came.

As illustrated in Figure 7.8, intermediates in histidine biosynthesis are phosphate esters. But these are not taken in through the cell membrane and do not serve as nutritional supplements. Furthermore,

Table 26. Partial Blocks in Tryptophan-Requiring Mutants of *Neurospora* as Demonstrated by the Use of N^{15}-Labeled Compounds [*].

Strain Number	N^{15}-Labeled Compound Supplied	Compound Isolated	N^{15} Content of Isolated Compounds Given as % of Total Amount Isolated
39401	Anthranilic acid	Acetyltryptophan	67
7655	Anthranilic acid	Acetyltryptophan	37
		Quinolinic acid	40
10575	Indol	Acetyltryptophan	80
		Quinolinic acid	75
C-83	Tryptophan	Acetyltryptophan	98.9
		Quinolinic acid	94.3

* Bonner, Yanofsky, and Partridge (86).

though mutants do accumulate small amounts of phosphate-ester inter-mediates inside the cells, the principal accumulation is of the dephos-phorylated intermediates in the medium. These do pass through the cell membrane but in general are of no value as nutritional supplements, since they cannot be phosphorylated, and the biosynthesis occurs at the phosphate-ester level. This is a principle which probably holds in many reaction series. Another point of significance concerning accu-mulation is backup behind the block. Sometimes several intermediates are accumulated and sometimes none. Obviously, these can be func-tions of relative rates in a series and the equilibrium conditions. If the blocked reaction is one which normally provides energy coupling for several steps, then none will go on, and a precursor far back may dissi-pate in side reactions. Thus accumulation is often not straightforward and neither is nutritional supplementation with prospective intermedi-ates. Actually, it is not a simple matter to prove that a substance is an intermediate in a reaction series even if it is nutritionally satisfactory. It may undergo reaction to give an active intermediate in reverse of the situation where a true intermediate may not serve as a nutritional sup-plement. Combinations of tests are therefore necessary to establish a compound as an intermediate in a series, with the greatest confidence being derived from the demonstrations of production of the substance by a pure enzyme from an immediate precursor and the conversion of the substance by a pure enzyme to the next compound in the reaction series. This has been achieved in only a few of the situations described by the metabolic patterns presented in this chapter.

An important factor broached in the preceding discussion and having a strong influence on studies of metabolic patterns through the use of nutritional mutants is that of the nature of metabolic-control mecha-nisms. Aspects of this problem were discussed recently by Krebs (537) with fundamental parameters considered under the headings of reaction reversibility, specific enzyme inhibition, obligatory coupling, and shared cofactors. All these can yield feedback systems in which products create unfavorable conditions for further progress. Much of this occurs at the reaction-series level, and it introduces many compli-cations into analyses of reaction series. One kind of feedback of out-standing theoretical and practical interest is that represented by induc-tion and repression (see Chapter 9). In repression normal, low-molecular-weight metabolites can strongly influence the quantitative production of one or more enzymes. This may occur quite directly at the DNA-RNA level (761), and, as such, it is of considerable theo-retical interest. From the strictly practical standpoint, it has been observed many times in studies of nutritional mutants of microorganisms that good accumulation of intermediates frequently occurs only when growth is strongly limited by providing a limited amount of required product. Although there are many possible explanations for this phenomenon, it has been shown in several examples that an excess of product can repress formation of an enzyme or several enzymes in a given reaction series.

A good example is that studied by Ames and Garry (17) and summarized in Figure 7.2. Here it is shown that a histidine-requiring mutant of *Salmonella* contains only very low-activity levels of four enzymes concerned in histidine biosynthesis (see Figure 7.8) during growth in the presence of histidine, but after histidine is used up (0 hours) and growth is slow, there is a rapid and extensive increase in all four enzymes. Thus synthesis of all four enzymes is evidently repressed in the presence of histidine, whereas an unrelated enzyme, glutamic dehydrogenase, is not affected. Feedback control in which

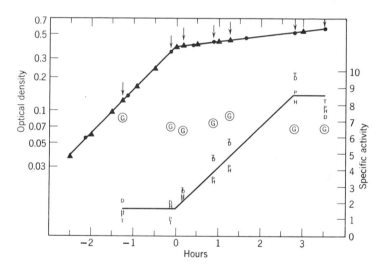

Figure 7.2. The specific activities of four histidine biosynthesis enzymes and glutamic-acid dehydrogenase during growth of a histidine-requiring mutant of *Salmonella*. 0 hours represents the time when added histidine was used up. Black circles and triangles represent separate experiments and the curve shows diminution of growth when histidine is depleted. Ⓖ represents specific activity for glutamic-acid dehydrogenase (no change). In contrast, specific activities of the histidine synthesis related enzymes D (dehydrase), T (transaminase), H (histidinol dehydrogenase), and P (phosphatase) increase together with histidine depletion. From Ames and Garry (17).

precursor accumulation is prevented but enzyme activity is not reduced in the presence of excess histidine has also been observed in histidine mutants of both *Salmonella* and *Neurospora*. This has been called feedback inhibition.

There are now many known examples of metabolic control of the kinds just described, and it is clear that a wide variety of growth conditions must always be invoked to obtain maximum information from studies of any one or a series of nutritional mutants. A mutant always differs from its parent strain, though all the differences may not be obvious under one set of conditions.

Some Metabolic Patterns

The major contribution of studies of nutritional mutants to knowledge of metabolic patterns has, of course, been concerned with biosyntheses of essential metabolites, and in many cases these have provided the primary information to direct detailed developments derived from strictly biochemical techniques. At the same time, the latter approach has yielded an enormous amount of information on reaction steps, both essential and nonessential to cell survival, and reasonably complete overall patterns can be found in recent textbooks of biochemistry and in review summaries such as the *Annual Review of Biochemistry*. Advances are rapid, and current literature should be examined thoroughly for details in any specific area. The patterns presented here are necessarily incomplete, and the literature bibliography is minimum. In general, reference is given only to reviews and a few specific recent publications from which an overall picture can be traced. In a few cases the systems are known in great detail with respect to both genetic and enzymatic characteristics. Numbered reaction steps (as step 1, 2, etc.) indicate positions of genetic blocks in metabolism, and where the system has been further defined by reasonable resolution of one enzyme involved, this is indicated by (E). Though a number without (E) may indicate a single step, in many cases there may be several steps of unknown nature. Accumulations of compounds are indicated by (A). Several of the patterns also show reactions without label. Some of these correspond to reactions that are indicated from biochemical investigations and likely will be found on a genetic basis among nutritional mutants, while many correspond to one or more reaction steps leading to substances not essential for growth. Mutational changes corresponding to this last category of reactions may well be obtainable, but special selection methods are needed to obtain the mutants.

Some general reviews pertinent to the following discussions are by Fincham (278, 280), de Serres (217), Horowitz and Owen (457), and Harris (410).

Some Established Patterns

Nitrogen Metabolism. A wide variety of microorganisms are able to make use of NO_3^-, or in some cases N_2, as a source of nitrogen for synthesis of ammonia, amino acids, and the large number of other nitrogenous cell components (735). A rather highly tentative scheme for the origin of reduced nitrogen at the NH_3 level is given in Figure 7.3. Step 1 is the only one clearly defined by a gene-enzyme relation.

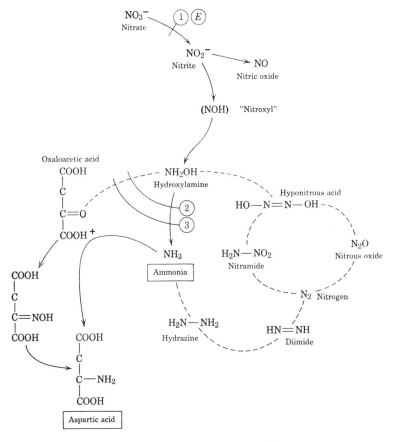

Figure 7.3. Probable pathways in the metabolism of inorganic nitrogen.

Silver and McElroy (909) studied three classes of nitrate mutants of *Neurospora,* and, of four which would utilize nitrite but not nitrate, two possessed negligible nitrate-reductase activity, one produced the enzyme only above pH 6, and the fourth was found to have a heat-labile inhibitor of the enzyme activity. Class 2 mutants were found to accumulate nitrate and hydroxylamine. These investigators proposed that these mutants are deficient in ability to form the oxime of pyridoxal from hydroxylamine rather than the oxime of oxaloacetic acid as shown in Figure 7.3. Class 3 was presumed to be concerned with reduction of the oxime to the amine. Investigations up to the present time have not provided an answer to whether nitrate reduction to NH_3 normally goes entirely at the inorganic level or through oximes such as those of pyridoxal and oxaloacetate, but the class 2 and 3 mutants of *Neurospora* should be of considerable further aid in clarification of this problem.

Some Primary Amino Acids. The amino acids—glycine, serine, alanine, aspartic acid, and glutamic acid—are all closely related in metabolism, and to the fundamental reactions of glycolysis and the citric-acid cycle (Figure 5.3, p. 231). Furthermore, as indicated in the generalized summary in Figure 7.4, these substances provide carbon-chain units for the synthesis of a very large number of important metabolites. In view of this centralized position characterized by high turnover, it is to be expected that mutants concerned with this pattern of reactions may be lethal or at least highly complicated by metabolic imbalances and multiple nutritional requirements. Nevertheless, a number of mutants involved in this system in *Neurospora, E. coli,* and *B. subtilis* have been described and studied. Those of *Neurospora* include mutants that grow if provided serine or glycine, aspartic acid, asparagine, formate, succinate, alanine, and amino nitrogen. Fincham (280) has made an especially extensive study of the amination-deficient mutants of *Neurospora*—(1) in Figure 7.4—which lack or have altered forms of glutamic dehydrogenase. This results in a nutritional requirement for one or more of several amino acids other than glutamic acid, since the reaction indicated by (1) normally provides amino nitrogen for transfer to keto acids in amino-acid synthesis. *Neurospora* mutants with reduced activities for oxaloacetic cocarboxylase (succinate requirement) and pyruvate carboxylase have been studied, and in *E. coli* a mutant deficient in condensing enzyme (α-ketoglutarate requirement) has been described. The formate mutants of *Neurospora* will utilize formaldehyde or, with low efficiency, adenine. Generally speaking, nutritional

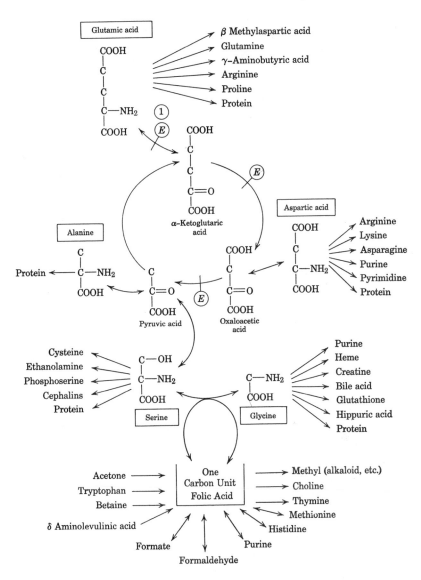

Figure 7.4. Generalized metabolic pattern for the primary amino acids that are derived most directly from the citric-acid cycle.

requirements of mutants concerned with the reactions at this primary level are partial or not fully satisfied by one metabolite, and they do not provide an adequate basis for construction of a detailed metabolic pattern as can be done in a series less involved in multiple side reactions and cyclic systems.

Isoleucine, Valine, and Leucine. As shown in Figure 7.5, the biosyntheses of the two amino acids isoleucine and valine are very intimately related to the extent that homologous reactions occur under the influence of common enzymes and common genes (1048). This is probably true in numerous microorganisms including *Neurospora, E. coli,* yeast, and *Salmonella.* In *Neurospora* and *Salmonella* some of the mutant genes occupy adjacent positions in the linkage group and in the order of reaction series. A point of some interest is illustrated by the fact that one of the *Salmonella* mutants corresponding to step 1 will grow with valine alone. This is interpreted to mean that in this mutant there is residual activity for isoleucine synthesis in a mutant enzyme. Thus it would appear that this system with double enzyme specificities would be favorable for studies of gene-enzyme structure relations. Submaximal growth of mutants corresponding to step 4 in the presence of isoleucine alone is also of interest as an illustration of a type of nutritional mutant characteristic. Here the mutants are deficient in the phenylalanine-keto (isoleucine or valine) transaminase, but valine can be produced inefficiently by a different transaminase. Alternate pathways of this kind are probably common and may account for many apparent partial mutations and a lack of mutants for particular steps in a series.

A considerable number of leucine mutants are known in a variety of microorganisms. The present evidence indicates that it is formed from α-ketoisovaleric acid through a series of steps as yet imperfectly known. Steps in the reaction patterns following leucine synthesis are also of interest in relation to formation of carotenes and steroids (161). Although the latter are not required by microorganisms, they are synthesized, and a number of *Neurospora* mutants concerned with carotenoid synthesis are known (417). These merit reexamination in view of the present knowledge of the participation of isoprenoid phosphates in the biosynthesis system.

Arginine, Proline, and Hydroxyproline. Mutants requiring arginine were among the first studied (946) with the initiation of the use of *Neurospora* in the elucidation of metabolic patterns, and they are also common in other organisms such as *Aspergillus* and *E. coli.* An ex-

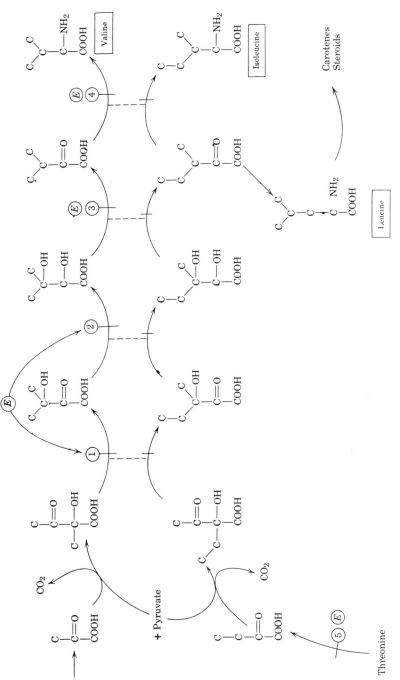

Figure 7.5. Metabolic patterns for the biosynthesis of valine, isoleucine, and leucine. Steps 1 and 2 are apparently catalyzed by the same enzyme.

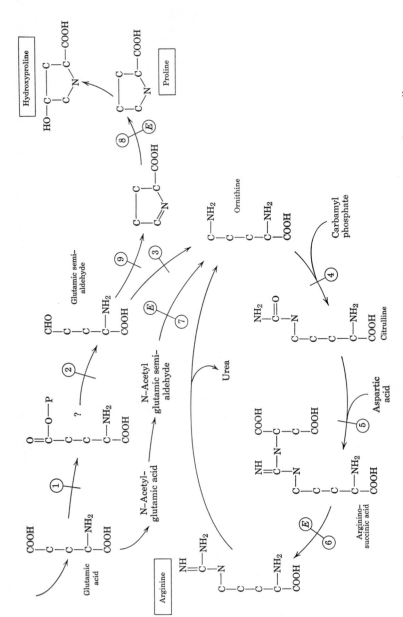

Figure 7.6.   Metabolic patterns in the biosynthesis of arginine, proline, and hydroxyproline.

tensive pattern has been established, as shown in Figure 7.6. One point of special interest concerns the two alternate series from glutamic acid to ornithine. Although several details remain to be established, it appears that the carboxyl reduction with acetylated glutamic acid is characteristic of the system in *E. coli,* while the other series obtains in *Neurospora.* Such a deviation in systems in different organisms is perhaps not surprising, but in general different organisms display more similarities than differences of this kind.

As shown in the figure, this cyclic system is linked through carbamyl phosphate and aspartic acid to the system for pyrimidine biosynthesis (see Figure 7.12, p. 316), a fact which accounts for some of the nutritional complexities and mutant-to-mutant suppressions observed earlier (470). As may be noted in the figure, a nutritional mutant for hydroxyproline has not been found. This is not surprising, since present biochemical evidence indicates that hydroxyproline is made from proline but only during protein synthesis. If so, a mutation concerned with this reaction would be lethal, since a high-molecular-weight substance would be required as a nutritional supplement. Perhaps such a mutant could be found among those with partial blocks.

Lysine. As mentioned earlier, patterns of biosynthesis of a great many essential metabolites are usually quite similar in different organisms, and the most extreme difference yet observed is that shown for lysine in Figure 7.7. The first one established is that involving α-aminoadipic acid, and this is characteristic of the fungi *Neurospora* and *Ophiostoma* (338, 874) and the yeast *Torula.* The alternative pattern with diaminopimelic acid as an intermediate is characteristic of *E. coli* and a number of other bacteria, as well as blue-green algae, green algae, and at least some higher plants (324). As shown, the latter system is intimately related, in early stages, to the biosynthesis of threonine and methionine. Accumulations of the two succinyl derivatives and diaminopimelic acid itself have been demonstrated in mutants of *E. coli,* as have three different enzyme deficiencies.

Histidine. The general pattern of histidine biosynthesis is similar in *E. coli, Neurospora,* and *Salmonella* (16, 714). A summary is shown in Figure 7.8. This system is illustrative of several important phenomena encountered in studies of patterns through use of nutritional mutants. First, histidine mutants of *Neurospora* were not found until they were selected for on a medium containing only histidine in addition to the salts, sugar, and biotin required by the wild strain. Then many were found, but none would grow on a protein hydrolysate or

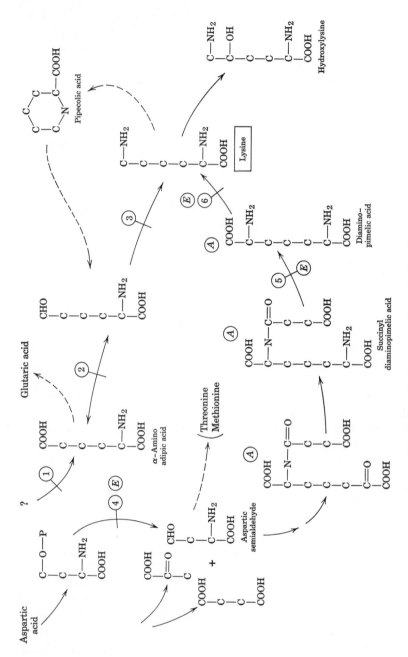

Figure 7.7.   Pathways for biosynthesis of lysine.

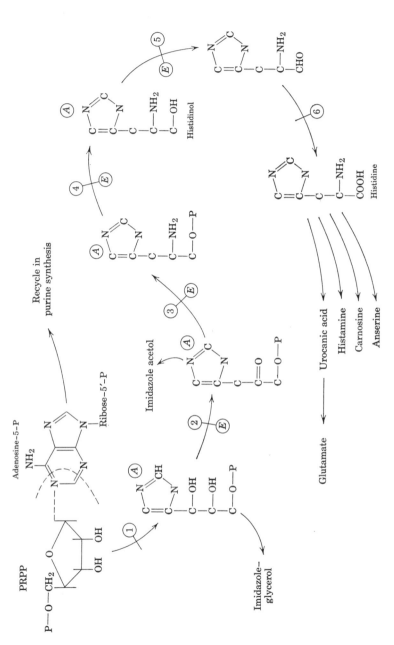

Figure 7.8. Histidine biosynthesis.

other complex medium.   All were found to be inhibited by a supplement containing a basic amino acid (as lysine, arginine, or ornithine)
plus tyrosine or any one of a variety of other neutral amino acids (387,
146).   Thus histidine mutants had been systematically selected against
by use of complex media, and probably this can happen in any system
with specific metabolites.

A second principle of great importance illustrated by this system is
concerned with the nature of the intermediate compounds.   As shown
in the figure, three of the imidazole derivatives are phosphate esters,
and these do not serve as nutritional supplements because they cannot
be taken in through the cell membrane.   On the other hand, the de-
phosphorlyated compounds, which are accumulated in mutant culture
media and thus can pass through the membrane, do not serve as nutritional supplements either; presumably they cannot be phosphorylated
in the cell.   Thus in *Neurospora* it was necessary to establish this
metabolic pattern solely on the basis of serial enzymatic conversions in
cell-free systems.   In view of the fact that a great many biosynthetic
systems of all kinds involve phosphorylated intermediates, nutritional
studies should always be supplemented with a search for phosphate
esters in extracts of mutant cells.

In *Neurospora* and yeast the genes concerned with histidine biosynthesis are distributed among linkage groups, but in *Salmonella* (411)
they are evidently at adjacent loci and located in the order of the reaction series (p. 362), and simultaneous repression of certain enzymes
in the series is exhibited, as mentioned previously.

Cysteine, Methionine, and Threonine.   A large number of nutritional
mutants concerned with the biosynthesis of the sulfur amino acids and
threonine have been obtained in *Neurospora, E. coli, Salmonella, Aspergillus, Ophiostoma, B. subtilis,* and other organisms.   Part of the total
metabolic pattern, shown in Figure 7.9 (456), has a counterpart in
mammalian tissue in that animals require methionine and can produce
cysteine by reversal of the reaction series 4, 5, and 6.   However, they
also require threonine and cannot produce cysteine from more oxidized
forms of sulfur.   This kind of difference appears to be the more common form between species.   That is, fragments of total patterns appear
in different species, but what is present is usually similar rather than
totally different as in the contrasting pathways in lysine biosynthesis.

There remains a considerable uncertainty as to details of mechanisms
of reactions in early stages of sulfate reduction with respect to the point
at which sulfur becomes attached to a carbon chain.   As shown, pre-

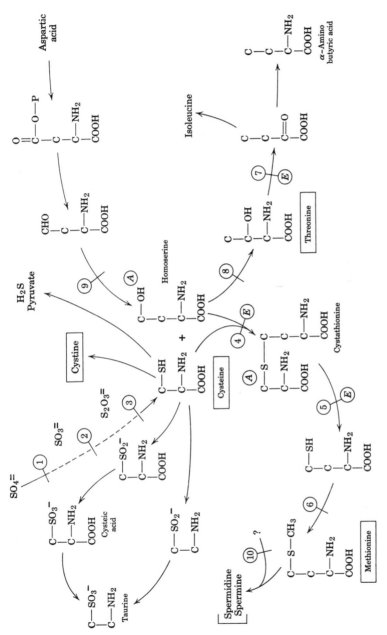

Figure 7.9.   Patterns of synthesis and metabolism of cysteine, methionine, and threonine.

cursors and products tie the system in with that of aspartic acid and the citric-acid cycle through serine and the one carbon unit of methionine as well as with isoleucine synthesis. Reaction 10 represents one concerned with the production of diaminobutane (a nutritional mutant requirement in *Aspergillus*) which reacts with methionine to give spermine and spermidine. A fairly extensive reaction series is to be expected here (997), but the products have not been found as nutritional requirements in most microorganisms. Alternate pathways or inherent suppression are possibly the cause.

Aromatic Compounds. Metabolic patterns describing the biosynthesis and metabolism of substances containing benzene rings and pyridine rings are very extensive and are summarized here in three parts in Figure 7.10*a, b,* and *c.* Although many important aromatic-metabolites are included in the system, the benzene ring of riboflavin and the pyridine ring of pyridoxine originate in other special patterns, and additional special systems may also exist. In the overall picture, microorganisms and higher plants have the capacity for ring synthesis, and animals are completely dependent on these organisms for their nutritional requirements for phenylalanine or tyrosine and tryptophan. On the other hand, the section of the pattern which describes the conversion of the tryptophan to nicotinic acid does exist in animals, as do the reaction series in part *b.*

The patterns of Figure 7.10 have been studied very extensively by use of nutritional mutants of *E. coli* (203, 917) and *Neurospora* (668) and in a number of other microorganisms and higher plants. As shown (part *a*), the six-carbon ring which comes directly from products of carbohydrate metabolism is initially alicyclic and becomes aromatic through dehydration and oxidation. The alicyclic components, such as shikimic and quinic acids, are fairly common constituents of higher plants and are accumulated by various mutants of microorganisms. Although several of the steps can be described in terms of single enzyme reactions, some, such as reaction 5, need clarification, and both plants and microorganisms produce a great variety of phenolic side products through constitutive or adaptive reactions series (955). Protocatechuic, vanillic, and gallic acids are common by-products not shown in the figure.

Reaction series shown in Figure 7.10*b* occur in animals, and that part concerned with homogentisic-acid formation is discussed further in Chapter 14. The tyrosine-to-melanin series is quite universal, giving rise to red-brown and black pigments in a great variety of organisms.

The enzyme tyrosinase indicated at step 8 is probably involved at subsequent reactions, but some are spontaneous. However, melanization is complex and probably occurs at the peptide and protein levels as well as by the series shown. This reaction series has provided the basis for the gene-enzyme relation studies described earlier (p. 269) and in Chapter 9. As shown in the scheme, the amino-acid dopa can enter into at least three reaction series including conversion to polymers in chitin synthesis (511). Mutants of insects should prove especially valuable in evaluation of this pathway and the mechanisms of control of the diversion.

The reaction pattern given in Figure 7.10*c* provides a description of a number of points of special interest. The reaction series from tryptophan through niacin cofactors, 17 through 22, is of quite universal occurrence, though there may be exceptions, and it shows a mechanism of biological production of a pyridine ring from a benzene ring. This system was first delineated by studies of *Neurospora* mutants (456) and confirmed and extended by studies with a great variety of organisms from bacteria to higher plants and animals. The reaction series preceding tryptophan does not occur in animals, but they can convert kynurenine to anthranilic acid, a reaction which provides for a cyclic system in *Neurospora* which has the capacity for carrying the entire series 1–22. Steps 9–16 have been studied most extensively in bacteria, including *E. coli, Salmonella,* and *Aerobacter*. Doy and Gibson (232) and Smith and Yanofsky (917) demonstrated the production of the unique ribulose-5-phosphate derivative of anthranilic acid which is a key intermediate in the series, and the pyruvate derivative product of step 9 has only recently been described (843). More steps, intermediates, and enzymes are to be expected in steps 9–13. The reactions (14–16) provided the basis for the extensive study of gene-protein structure relations described earlier (Chapter 6, p. 271). At step 22 the reaction mechanism is not entirely clear, and quinolinic acid may be an intermediate in the series. Subsequent steps from nicotinic acid to cofactors have received considerable attention, but since the intermediates are phosphorylated, corresponding viable nutritional mutants are not to be expected. This system illustrates the seemingly unwarranted complexity of biological biosynthetic systems. If one had been asked to predict a biological mechanism for the conversion of anthranilic acid (step 9) to hydroxy-anthranilic acid (step 20), it is unlikely that the correct answer would have been given. There are simpler ways to insert one oxygen atom. But many realities in biosynthetic mechanisms are equally improbable.

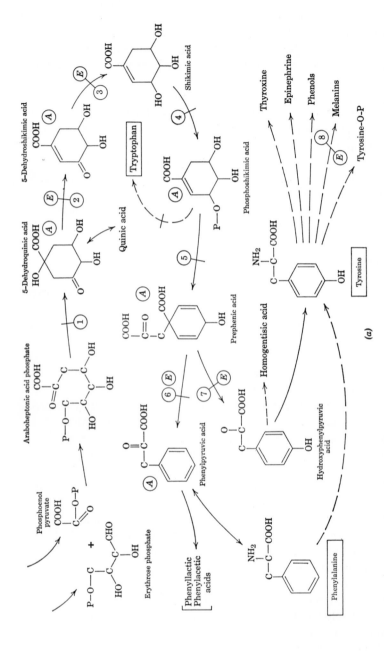

Figure 7.10. Metabolic patterns concerned with biosynthesis and metabolism of aromatic compounds. (a) Biosynthesis of tryptophan, phenylalanine, and tyrosine. (b) Metabolism of tyrosine. (c) Biosynthesis patterns for tryptophan, niacin, and insect eye pigments (brown).

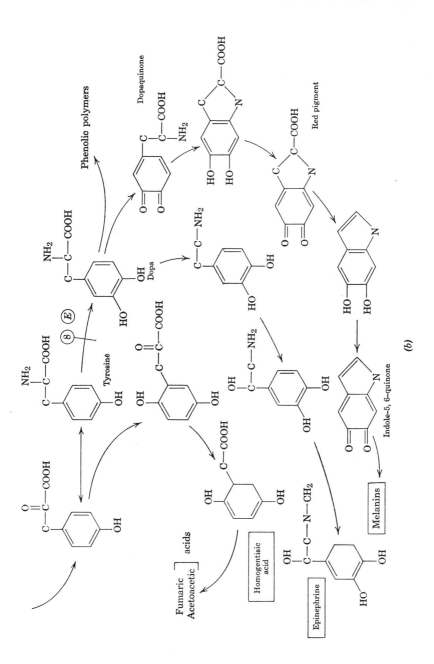

(b)

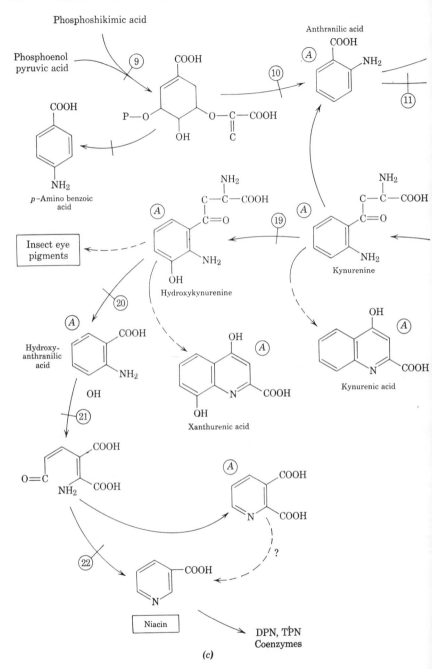

p-Amino benzoic acid

Phosphoshikimic acid

Phosphoenol pyruvic acid

Anthranilic acid

Insect eye pigments

Hydroxykynurenine

Kynurenine

Kynurenic acid

Hydroxy-anthranilic acid

Xanthurenic acid

Niacin

DPN, TPN Coenzymes

(c)

Oxidation products

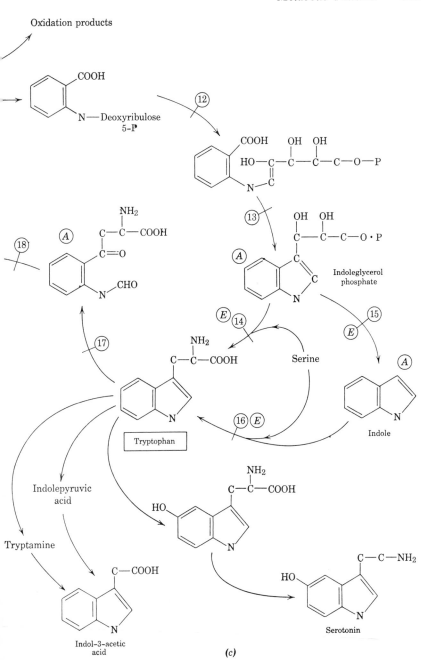

(c)

An additional system concerned with aromatic substances that is no doubt related to the pattern in Figure 7.10 is that concerned with formation of coumarins (344).

Purine Nucleotides. Mutants with nutritional requirements for the purines, adenine, hypoxanthine, and guanine are of very frequent occurrence in selection experiments using various fungi, yeasts, and bacteria. However, a large part of the metabolic pattern of synthesis shown in Figure 7.11 was established through enzyme and tracer studies (413). A major reason why mutant studies did not contribute more lies in the fact that all intermediates are phosphate esters, and the purines themselves are used as nutritional supplements only by virtue of the fact that ribose phosphate can be attached after the bases pass through cell membranes. Nucleosides can also be phosphorylated, but, in general, intermediate sugar derivatives or phosphates cannot be used. At the same time, accumulations in growth media of dephosphorylated intermediates is common, and their identification has provided a basis for placing genetic blocks in correspondence with established steps in the reaction series. For example, a series of *Neurospora* mutants have been correlated with steps 5 through 8 on the basis of accumulations of one or more precursors. Both ribosides and ribotides were found in mold extracts (78, 323). One of these mutants (step 5) and a corresponding strain in yeast present startling phenotypes in the accumulation of large quantities of a purple pigment which apparently is derived from polymerization of accumulated intermediates. As pointed out previously in connection with the patterns of synthesis for isoleucine and valine, a genetic locus can be concerned in reactions in two series, and in the purine series an interesting variation is that of one genetic locus being concerned with two different steps in the same reaction series (steps at 7). The two reactions are of the same type, but the substrates are quite different, and the case provides another favorable situation for studies of gene-protein structure relations.

Pyrimidine Nucleosides. Pyrimidine-requiring mutants arise with a high frequency among fungi and bacteria and have provided more information on the pyrimidine biosynthetic pathway than the purine mutants. As shown in Figure 7.12, several early steps in the series do not involve phosphorylated intermediates. The early observation of the accumulation of very large amounts of orotic acid by *Neurospora* mutants and the production of orotidine in the mycelium of a mutant corresponding to step 5 provided evidence for some kind of participation of these components in the biosynthesis of pyrimidines (688, 396).

It is a curious observation that the mutants excrete accumulated orotic acid into the culture medium, whereas orotidine is retained within the mold as are phosphate esters. An additional point of interest from the nutritional standpoint is that the *Neurospora* mutants utilize the free pyrimidine bases only very inefficiently in a medium containing NH_3, but just as efficiently as the nucleosides in a NO_3^- medium. Orotidine is used very inefficiently under both conditions. Thus, as with purines, the free heterocyclic substances are not normal intermediates. Furthermore, the fungus evidently contains an efficient phosphorylase for uridine and cytidine but not for orotidine. None of the *Neurospora* mutants utilize thymidine or thymine, but such strains (step 6) have been found in bacteria.

A point of considerable interest concerning this pathway of biosynthesis concerns the initial step which involves carbamyl phosphate and aspartic acid. Davis (205) has shown that one mutant (*pyr-3d*) of *Neurospora* is deficient in the enzyme, aspartic-acid transcarbamylase, whereas two other mutants (*pyr-3a* and *b*) at or near the same locus are not. The latter mutants, however, are suppressed (relieved of pyrimidine requirement) by a mutant gene *s* which alone has no obvious phenotypic effect. It has been demonstrated, however, that *s* causes a great reduction in activity and an alteration in ornithine transcarbamylase (see Figure 7.6) which also uses carbamyl phosphate as a substrate. It may be presumed that *pyr-3a* and *b* have partial genetic blocks, and they are suppressed by the partial block *s* through shunting more carbamyl phosphate into pyrimidine synthesis. Gene *s* also suppresses certain proline mutants, perhaps by shunting of the common proline-arginine precursor aspartic semialdehyde. This pattern of interactions is discussed at length in Chapter 9 (p. 399). The significant point to be made here is that all of the patterns are interrelated. This would be much more obvious if it were possible to write each reaction in a properly balanced form including all reactants and products. Thus the reaction series as presented do not show the innumerable complications that actually exist in their evaluation from studies of accumulations and nutritional studies. That much more investigation at the enzyme level will resolve many of these complications seems evident from the preceding example.

Thiamine. Although several mutants in *Neurospora, E. coli,* and yeast have been shown to require thiamin (242), the system has not been extensively exploited. The positions of genetic blocks shown in Figure 7.13 are rather tentative and based on a reaction pattern that appears

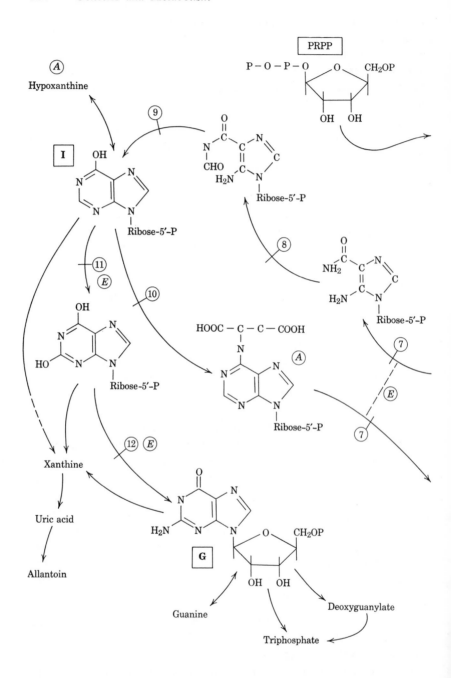

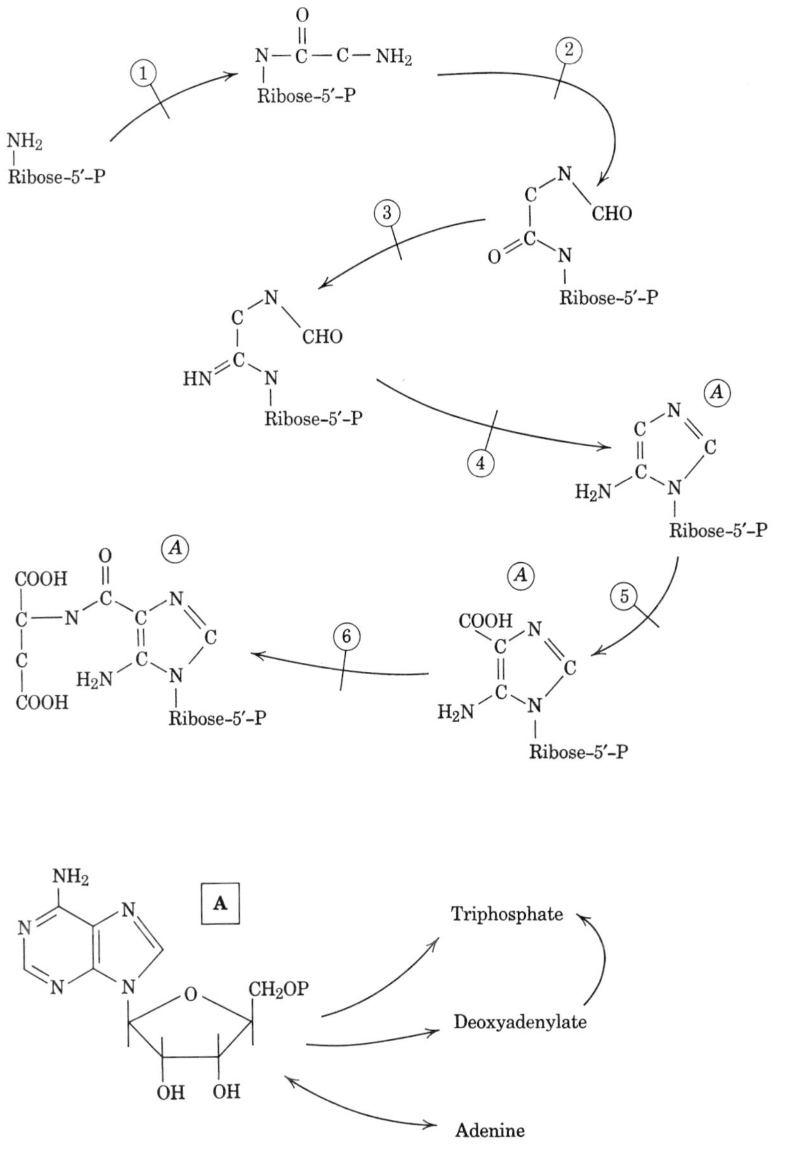

Figure 7.11. Metabolic pattern concerned with the biosynthesis of purine nucleotides.

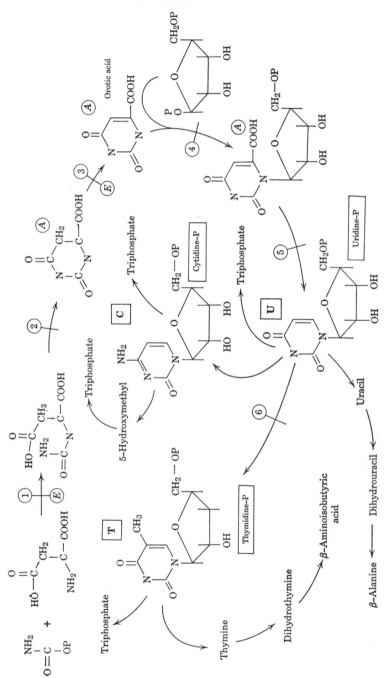

Figure 7.12.   Metabolic patterns for the biosynthesis of pyrimidine nucleotides.

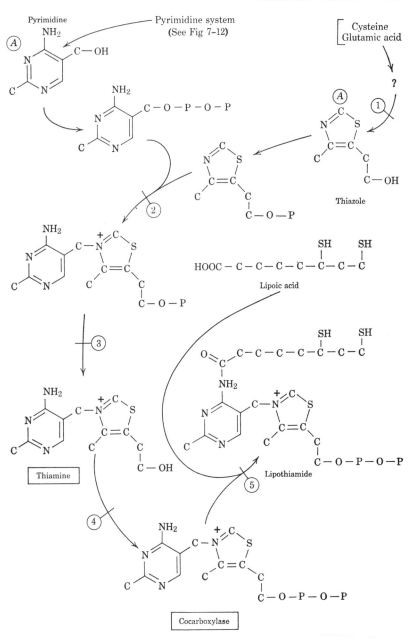

Figure 7.13. A postulated scheme for the biosynthesis of thiamin.

probable from enzyme studies. Both the pyrimidine and thiazole moieties are accumulated by a *Neurospora* mutant (step 2, presumably), but a search for accumulated phosphate esters which predominate in the system has not been reported.

Flavins and Pteridines. Relatively few mutants with nutritional requirements for riboflavin and folic acid have been encountered, and perhaps special methods of selection are needed to obtain them. As shown in Figure 7.14, the heterocyclic ring portions are derived from, or in a similar fashion to, those of purines, and again it is likely that most intermediates are phosphate esters. The ribityl-lumazine deriva-

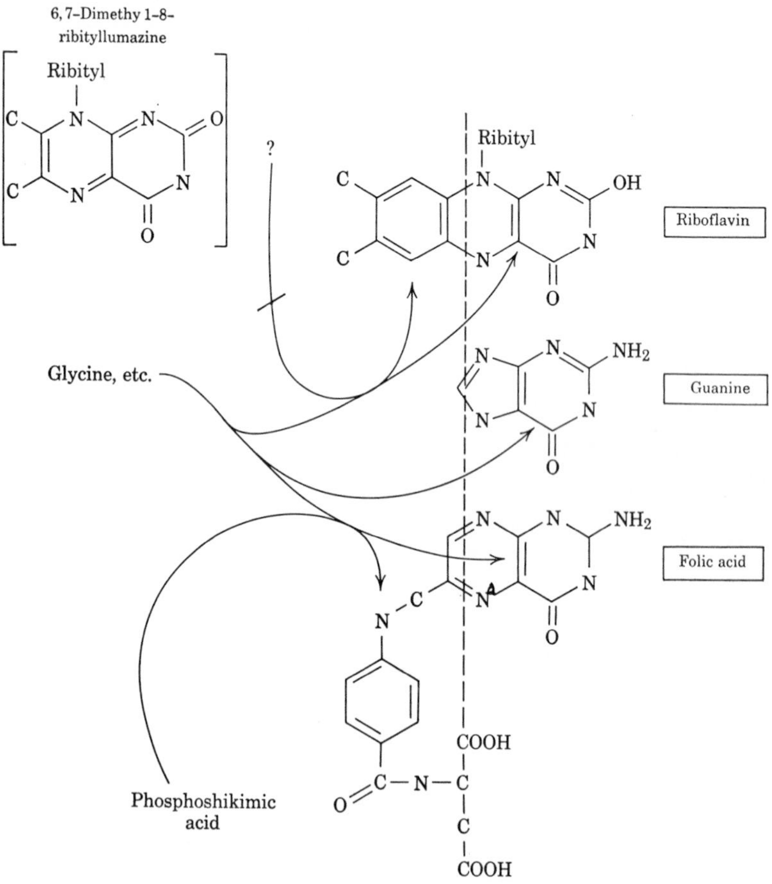

Figure 7.14. General metabolic relations among purines, flavins, and pteridines.

tive shown was isolated from a fungus which normally accumulates large quantities of riboflavin (*Erymothecium ashbyii,* a pathogen for cotton), and enzymatic conversion to riboflavin has been effected (789, 790). Thus the benzene ring of riboflavin is produced by a system different from the general scheme shown in Figure 7.10*a, b,* and *c.* On the other hand, the *p*-aminobenzoic acid portion of folic acid does arise from the general aromatic-ring syntheses pathway.

Many pteridines other than folic acid are formed, especially in insects and fishes, and further details concerning biosynthesis and inheritance are given in a later section (p. 325).

Pantothenic Acid. Mutants of *Neurospora* and *E. coli* involved in the synthesis of pantothenic acid at step 1 (Figure 7.15) have been studied extensively, particularly with respect to the relation to the enzyme pantothenate synthetase (1046, 648). A *Neurospora* mutant was shown to contain the enzyme, but apparently in an inactive form, whereas an extensive study has been made of a temperature-sensitive *E. coli* mutant which produces an altered and temperature-sensitive enzyme. Presumably, it should be possible to evaluate more of the pantothenate pathway by use of mutants, but, as shown, present evidence indicates most intermediates are phosphate esters, and partial mutants are needed for the purpose.

Inositol and Ascorbic Acid. Mutants have provided relatively little information on the biosynthesis of inositol and ascorbic acid, but both are derived fairly directly from sugar metabolism, probably by the general scheme shown in Figure 7.16. Step 1 probably represents the position of the genetic block in higher animals, which require ascorbic acid in the diet, but such a requirement has not been observed in microorganisms (122). Step 3 and a deficiency of xylulose isomerase has been studied in *Salmonella,* and step 2 refers to *Neurospora* mutants which require inositol.

Choline. Choline-requiring mutants of *Neurospora* have been useful for bioassays as well as studies of the synthetic pathway which is related to glycine and serine as shown in Figure 7.17. Horowitz (455) observed the accumulation of monomethylethanol amine by a partial mutant corresponding to step 1, and the dimethyl derivative was isolated from a mutant corresponding to step 2 (1097). Further steps in the utilization of the compounds for the formation of phospholipids (as with serine and inositol) are known on a biochemical basis, but not in relation to mutants.

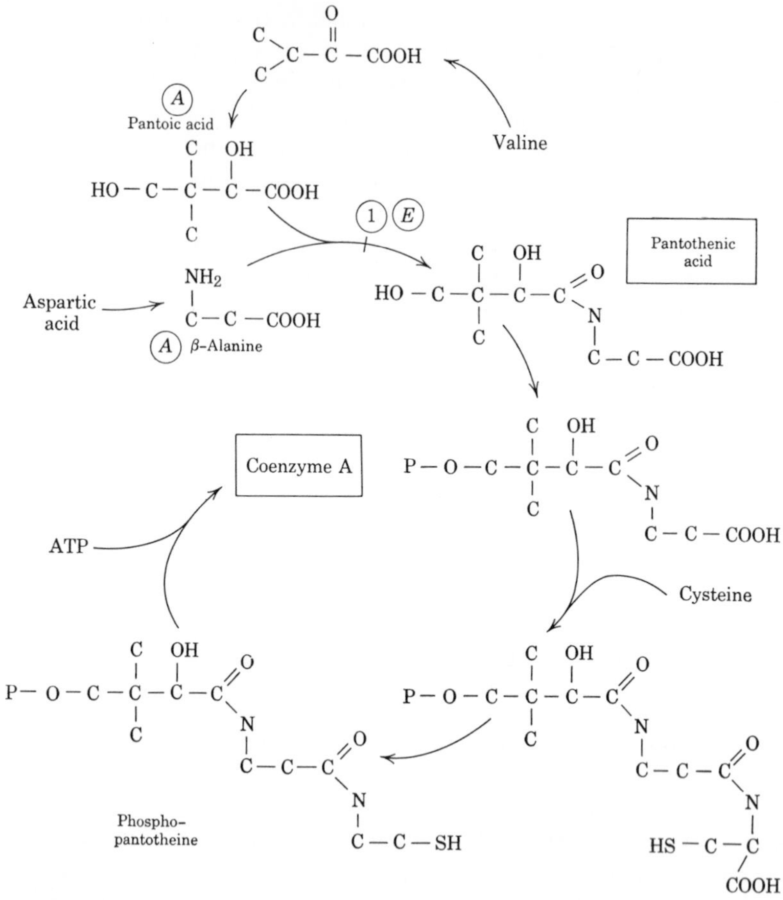

Figure 7.15. A postulated metabolic pattern for biosynthesis of pantothenic acid and coenzyme A.

Miscellaneous. A variety of investigations with many classes of substances of biological importance relate mutation and biosynthesis, but many areas remain relatively untouched. In the discussion here, a biosynthetic pattern for the vitamin niacin is given, along with aromatic substances in general (Figure 7.10*c*). Little information is available with respect to pyridoxine, biotin, and vitamin B_{12} or carnitine. Nor is it possible to give information relating mutation to the biosynthesis of the fat soluble vitamins D, E, K, and Q. Mutants with fatty-acid requirements have been studied in *Neurospora* (583), including two

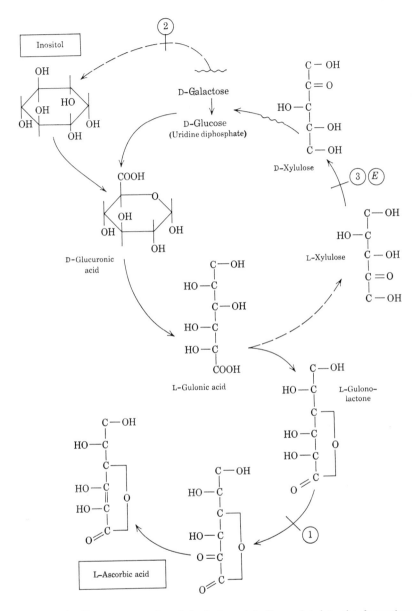

Figure 7.16. Some aspects of carbohydrate metabolism related to the formation of inositol and ascorbic acid.

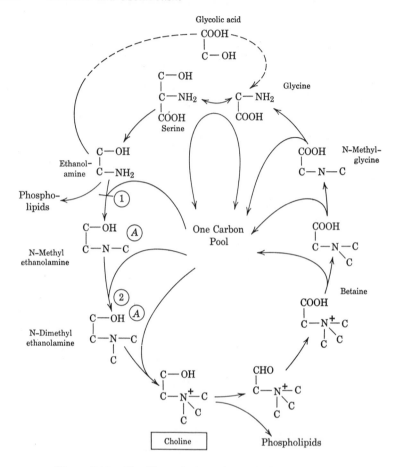

Figure 7.17. The biosynthesis and metabolism of choline.

classes, one of which uses acetate and the other long-chain unsaturated acids such as oleic, linoleic, and linolinic.

A genetic locus in *Neurospora* concerned with the production of D-amino acid oxidase has been studied by Ohnishi et al. (752). The enzyme deficiency is not lethal, nor does it cause accumulation of D-amino acids. In a similar vein, a very large amount of work has been presented concerning mutants in many organisms, from bacteria to man, having to do with carbohydrate metabolism. Perhaps the greatest attention has been given to galactose metabolism.

Some details of the biosynthesis of porphyrins, which are essential parts of chlorophyll, cytochromes, catalase, peroxidase, and hemoglobins, have been elucidated by use of mutants of the algae *Chlorella* (350). The steps concerned with later stages of chlorophyll synthesis and mutants concerned with earlier steps, which should occur in most microorganisms as well as higher forms, have not been reported. These steps involve the condensation of succinate and glycine to form δ-aminolevulinic acid and subsequent further condensation to the pyrrol-unit porphobilinogen, but all steps may occur with combinations such as with coenzyme *A,* and perhaps mutants cannot be preserved by nutritional means.

Phinney and collaborators (783) have made extensive use of dwarf mutants of maize in studies of the pathway of biosynthesis of the plant-growth hormone gibberellin. This substance counteracts dwarfism in maize, and several steps in its synthesis have been defined by a series of maize mutants. The system has considerable promise for studies of gene-enzyme relations in a higher plant when further chemical details have been elucidated. It is of interest that no gibberellin mutants have been found in the fungus *Gibberella fujikuroi* which produce considerable quantities of the hormone. Since the biologically active substances fall in the class of isoprenoid compounds, it is likely that they are derived from leucine through phosphorylated mevalonic acid and other phosphorylated intermediates, and it is possible that only partial mutants of the fungus, classed as irreparable from the mutational standpoint, would be available.

An interesting system in which mutants may become of value in elucidation of pathways concerns those of light emission in fireflies and luminous bacteria (645). A mutant of *Achromobacter* deficient in the enzyme luciferase has been reported (849), and biosynthesis of the compound luciferin is likely related to those of thiamin and aromatic substances.

The metabolism of phosphate has presented some interesting and useful situations for study. Levinthal and collaborators have very successfully exploited mutants of *E. coli* which affect alkaline phosphatase in studies of gene-protein structure relations (601). A less direct, but perhaps highly significant, mutation-phosphate relation concerns the accumulation of polyphosphate in a variety of microorganisms. Harold (408) extended earlier observations on *Neurospora* mutants by relating polyphosphate accumulation to ATP and RNA metabolism.

In general, certain specific mutants from many reaction patterns (e.g., histidine, methionine, tryptophan, and pyridoxine) accumulate quite large quantities of lower-molecular-weight polyphosphate (in the order of 20 residues) when growth is limited. Obviously, these accumulations are not directly related to the different metabolites at the biosynthesis level, since only specific mutants in a series show the characteristic, but there exists the interesting possibility that the phenomenon is related to RNA metabolism in the specific pathways of gene-to-protein synthesis.

Insect Eye Pigments. Pigment systems such as in the higher plants (Chapter 13) and animals (p. 427) have naturally received special attention in genetics since results of mutation are so easily observable. The same is true for the insect pigments, and, though the metabolic patterns involved are far from completely defined, the large amount of genetic information available, especially in *Drosophila,* and the extensive observations on the chemical basis of phenotypes and relation to development present a wealth of material for future investigations. The Hadorn group (390) has been particularly active in this area, and a general review was recently presented by Ziegler (1132). The latter includes references to the chemical aspects of the problem which are under investigation, especially in the laboratories of Butenandt (123), Viscontini (1036), and Forrest (287).

A general picture of the eye-pigment pattern is shown in Figure 7.18. Involved are problems of structure, orientation, and synthesis within the pigment granules themselves, as well as in the reaction series which provide the colored compounds. The granules contain proteins and RNA, and some evidence exists for participation of the tyrosinase system, at least in connection with the oxidation-reduction state of the environment in which the two classes of pigments (ommachromes and pteridines) are formed. A good deal is known about the ommachrome-synthesis reaction series in the early stages as shown here and in Figure 7.18. Tryptophan is degraded along the niacin pathway characteristic of many higher plants, animals, and microorganisms, but diversion occurs at hydroxykynurenine with the condensation of two molecules to form xanthommatin (shown in oxidized form). Niacin is a dietary requirement in *Drosophila,* so it is evident that one or more steps in the hydroxykynurenine-to-niacin series are normally blocked in this and related organisms. Accumulations by the mutants are indicated in Figure 7.18. In the steps following ommatins, more complex condensation products, ommins, are formed, but details of deposition of pigments within the granules and interrelations with other components

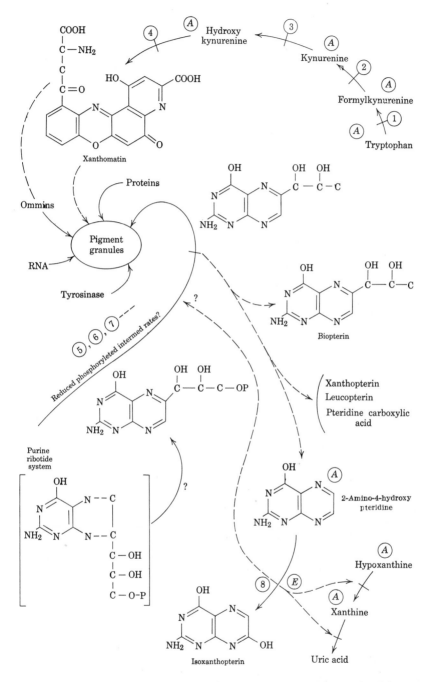

Figure 7.18. Some facts and speculations concerning the biosynthesis of insect eye pigments.

are not known.   This system provides brown pigment in *Drosophila* and in many related organisms.

The red pigments, which are found only in *Drosophila* sp., are derived from pteridines, and, as pointed out previously, the biosynthesis is related to that of riboflavin and the purine-synthesis system.   The observations of Ziegler (1132) and of Goto and Forrest (345) suggest the possibility that the intermediates are reduced pteridines and phosphate esters with a possible origin from purine and sugar phosphate, as indicated in the figure.   If so, the many oxidized pteridines observed directly by color or fluorescence and found to be accumulated by certain mutants are secondary products.   Nevertheless, a consideration of these along with enzyme studies should provide a basis for elucidation of the synthetic pathway and a further evaluation of the wealth of information that has been obtained regarding plieotropy and developmental changes in mutants having alterations in this metabolic pattern (390, 1132).   Although the prospects are very good for obtaining cases for the study of gene-enzyme relations in the pteridine-synthesis pattern, so far only one mutational enzyme-deficiency system is well known.   The deficiency observed is for the enzyme xanthine dehydrogenase, which is concerned in two reactions in purine metabolism, at least one at the secondary level in pteridine metabolism and probably at least one in the series of primary pteridine synthesis.   It is of special interest that the deficiency results from mutation at at least two different genetic loci (*ry* and *ma-1* in *Drosophila*), and in one of these mutants (*ma-1*) a protein which crossreacts immunologically with enzyme antibodies is produced (285, 328, 329, 349) (see also p. 275).

Although the primary concern here is with the metabolic patterns involved in eye-pigment synthesis, both ommachrome and pteridine reaction systems are functional in other tissues than eyes, and an important general method of study of them in *Drosophila* and related organisms is tissue transplantation.   This method, which was used by Beadle and Ephrussi (55) in early work in biochemical genetics and has been very extensively exploited by Hadorn and collaborators (389), involves transplantation of adult tissue anlage of larvae into larvae of a different genotype.   As shown in Figure 7.19, the development of the transplanted tissue in the abdomen of a host permits a direct evaluation of the interaction of tissues of different genotypes.   Nonautonomous interaction (equivalent to nutritional supplementation) by accumulated intermediates in microorganisms is shown for eye-disc transplantation in Figure 7.19 and Table 27.   Here the *v* mutant corresponds to step 1 and *cn* to step 3 in the ommachrome reactions series (Figure 7.18),

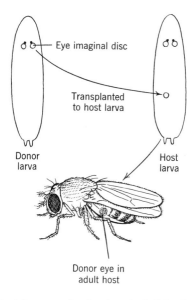

Figure 7.19. A method for transplanting imaginal disks from a fly of one geno-type to the body of a fly of another genotype. The donor eye (or other trans-planted tissue) can be disected from the body of the host for examination of its phenotype.

Table 27. Pigment Formation in *Drosophila* Eye Transplants

Donor Genotype	Host Genotype	Formation of Pigment in Donor Eye Transplant
+/+	+/+	+
v/v	+/+	+
cn/cn	+/+	+
+/+	*v/v* or *cn/cn*	+
v/v	*cn/cn*	+
cn/cn	*v/v*	−
cn/cn	*cn/cn*	−
v/v	*v/v*	−
cn/cn or *v/v*	*st/st* or *cd/cd*	+
st/st or *cd/cd*	+/+ or *v/v* or *cn/cn*	−

and the wild-type eye colors result from accumulations of diffusible intermediates; for example, a *cn/cn* host accumulates kynurenine which diffuses into a *v/v* eye transplant and is used to give a normal pigmentation in the *v/v* eye. Although most of the mutants in the pteridine series are autonomous, Hadorn and Graf (392) have presented data showing an influence on the formation of red pigment and isoxanthopterin in *ry* hosts by transplants of wild eye discs, Malpighian tubes, or fat body.

As just mentioned, the reaction systems which give rise to eye pigments are by no means existent only in eye tissue, nor are these systems characteristic of *Drosophila* alone except for the final red-eye character. The various intermediates in secondary products of both pigment systems are formed and metabolized in many tissues where at least some probably have other special functions. From extensive studies of pteridine distribution, it is clear that metabolic patterns in different tissues are at least quantitatively different and related to environment. For example, male *Drosophila* have a much higher content of isoxanthopterin than females due to a large accumulation in testes and some in eye tissue (393). However, transplant of testes or eye discs into females results in much reduced levels of isoxanthopterin in the transplanted tissue. The chemical basis for this kind of control of pattern has not yet been established.

Although there are drastic tissue differences within a species, striking homologies exist, nevertheless, among species with respect to both wild-type and mutant characteristics. This is especially evident in the ommachrome-mutant reaction series, and a summary of information on equivalent mutations in several species is shown in Table 28. Data on accumulations are by no means complete, and, as noted, mutant symbols which were derived from eye-color comparisons are not necessarily comparable in different species. In general, feeding, injection, or diffusion from transplants of compounds following a genetic block will yield normal eye colors up to but not following the hydroxykynurenine level.

General Comments

It has been mentioned earlier, but should be emphasized again, that the more or less separate metabolic patterns that have been presented are incomplete in several ways. Each reaction step has usually been treated in its simplest form without regard to the details of what ac-

Table 28. Mutations in Insects Affecting Ommochrome Synthesis

Reaction	Mutant	Organism	Accumulation
Tryptophan			
	v	D. melanogaster	tryp
	v^{40d}	D. virilis	
	cd	D. virilis	
	green	Musca domestica	tryp
	white eye	Periplaneta americana	tryp
	snow	Apis mellifica	tryp
	v	D. pseudoobscura	
Formylkynurenine			
	a	Ephestia kühniella	tryp
	v	D. affinis, D. simulans	
	v^{48a}	D. virilis	
Kynurenine			
	cn	D. melanogaster, D. affinis	kyn, kynurenic acid
	or	D. pseudoobscura	
	st	D. virilis	kyn
	candida	Phryne fenestralis	
	yellow	Phormia regina	kyn
	ivory	Apis mellifica	kyn
	white-1	Bombyx mori	kyn
	o-series	Habrobracon juglandis	
	white	Calliphora erythrocephala	
	ra	Plodia interpunctella	
Hydroxykynurenine			
	st, cd	D. melanogaster	
	cn	D. virilis	
	white-2	Bombyx mori	OH-kyn
	g	Plodia interpunctella	kyn, OH-kyn
	pallida	Phryne fenestralis	
	br	Ephestia kühniella	kyn

Ommin
Ommatin
Brown   pigments

tually enters into constitution of a genetic block. As discussed previously (Chapter 6, 280), each step may represent a complex system in itself. Further, in most cases all reactants and all products are not shown, and, if they were, in cases where they are known, the different patterns would be much more intimately tied together to yield a huge network of potentially interacting reactions. This concept is particularly important in any understanding of the full effects of mutation. In many cases a slight shift in balance of a reaction network may cause a phenotypic change quite far removed from the primary reaction change, and only an extensive knowledge of a large part of the metabolic patterns would permit tracing the change to the primary lesion.

Further efforts in extending information on metabolic patterns are also important in other ways. An obvious one is that they provide further systems for studies of gene-protein structure relations. It is already evident from the relatively few that are under investigation (Chapter 6) that there are several kinds of gene expression in terms of enzyme structure and function. For example, a simple enzyme may be a single direct product, an enzyme may be a complex of protein derived from the action of more than one gene, or an enzyme may come from a more complex protein by secondary degradation reactions. One can imagine a number of other possible parameters to the problem, and a true definition can come only from detailed studies of many more situations. Then there is the closely related matter of control; the sequential order of events, which is a fundamental characteristic of living things from cells to higher organisms, is intimately bound in cause and effect with changing metabolic patterns. Thus problems of differentiation and development are intimately bound to those of potential biochemical capacities.

With respect to directions of effort for the future in this important area, it is clear that the general approach described here may be expected to be profitable for a long time to come. Much more can be expected from nutritional mutants, but an increased emphasis on an enzymological approach would seem to be especially profitable. The latter is evident from the fact that now it is well established that many important reaction-series intermediates are phosphate esters or other metabolites which cannot be supplied from outside a cell. Thus enzymological studies of lethal or semilethal mutants should yield much new information. In addition, mutants must exist in very large numbers with slightly altered but functional enzymes which are not easily distinguishable from wild types by direct nutritional tests (see isoalleles, Chapter 8). These, too, would be most useful for studies of metabolic patterns and gene-enzyme relations, but selection for such mutants presents difficulties. Nevertheless, it seems likely that much more use can be made of extreme environmental conditions to aid expression in partial mutants. In microorganisms, temperature and pH alterations as well as nutritional excesses, deficiencies, and inhibitors are generally useful, as would be expected from the properties of enzymes and the general nature of genetic blocks (Chapter 6). Some of these conditions are applicable in higher organisms, and an additional criterion, useful in any case where regular differentiation and development is easily observable, is that of mutant expression at special stages. Examples are to be found in higher organisms (394) as well as at the virus level (245).

8 Genetic Units of Structure and Function

In physics the search for units of matter led to the discovery of atoms and the elementary particles which constitute them. In chemistry the units are molecules and aggregations of molecules. In biology we are faced with somewhat more complexity. The organism can be broken down structurally to molecules and atoms, but, as anyone knows, an organism is considerably more than an aggregation of molecules. Above the level of molecules and macromolecules are found such structural and functional units as mitochondria and ribosomes which function as units within the framework of the cell.

The results of Mendel's discoveries and those of the cytologists of the nineteenth century and the geneticists of the twentieth century led to the "theory of the gene" (709) which established the gene not only as the unit of heredity, but as a physical unit in a linear array with other genes on the chromosome. This physical entity, the gene, was a unit in three senses. (1) It was the unit of structure, since it was assumed that no crossing over or other types of exchanges occurred within it. (2) It was a unit of function. (Indeed, in 1941 Wright stated that the gene was *the* physiological unit and that it actually superseded the cell in this regard, being the ultimate significant unit of function.) (3) It was considered as a unit of mutation.

No sooner was the theory of the gene as the genetic unit of structure and function generally accepted and considered established by the majority of geneticists than the seeds of its destruction began to be sown. It is now evident that the theory is in need of considerable modification, and it will be one of the functions of this chapter to consider some of the possible modifications that are in order. However, like most theories in science, the new can best be understood against the background of the old. The newer aspects of structure and function will be introduced, therefore, after certain of the classical aspects, not all of which can or should be discarded, are first considered.

The analysis of the meaning and consequences of classical theory can best be accomplished by considering the nature of allelism. This is basic to our understanding of what we mean by gene.

Allelism

Allelic genes produce phenotypic differences but can be shown by breeding tests to occupy equivalent loci on homologous chromosomes. This is a definition based on experimental evidence and refers back in its origin to the concept of the Mendelian allelomorphic factors or contrasting characters which segregate at the time of reproduction.

Proof of allelism between genes is obtainable only by means of breeding tests with sexually reproducing organisms. If the organism is a diploid, these tests require first that strains be produced by inbreeding which are homozygous for the genes suspected of allelism. The homozygotes of different genetic constitution are then crossed and the F_1 hybrids inbred. The second generation (F_2) is examined for the phenotypic ratios which should not deviate significantly from a $1:2:1$ ratio (for incomplete dominance) or a $3:1$ ratio (for complete dominance of one of the genes), if the genes are allelic. A significant deviation from these ratios shows that the genes are nonallelic, and that the character is due to two or more nonallelic genes, or that some type of chromosomal aberration is involved. In haploid organisms such as *Neurospora* the expected ratio from a cross between strains carrying different alleles is $1:1$.

To appreciate the difference between the inheritance of a character due to a single gene or a pair of alleles (monogenic inheritance), and one due to two or more nonallelic genes (digenic, etc., inheritance), it is necessary to understand the mechanics of certain types of breeding tests and their consequences. For example, consider the inheritance of white eyes in a strain of *Drosophila melanogaster* which breeds true for this character. The mutant white-eyes strain is crossed, following the usual procedure, to a wild-type strain known to be homozygous for the normal red eye color. The resultant F_1 is composed only of flies with red eyes, showing that the mutant character is recessive to normal. The F_1 hybrids are then inbred and the F_2 phenotypic ratio determined to be 9 wild type:3 brown-eyed:3 bright red-eyed:1 white-eyed. This is the ratio expected from two pairs of genes segregating independently and demonstrates that white eyes are determined here by the simultaneous presence of two mutant genes. Further test crosses of the mutant types to genotypically known mutant strains make it possible to determine the identity of the mutant genes involved. It will be assumed in this example that they are scarlet (*st*) and brown (*bw*). The genotypic formula of the white-eyes strain may then be written as *st/st, bw/bw* and the original test crosses described with the appropriate symbols as shown in Figure 8.1.

The gene *st* has been discussed previously (p. 329) as blocking the formation of the brown-pigment component of the normal colored eye. It is also known that the gene *bw* blocks the formation of the red and yellow eye pigments which together with the brown pigment give the normal eye pigmentation of the wild-type fly. With this information the F_2 becomes intelligible. Wild-type flies have both components due to the presence of both st^+ and bw^+; brown-eyed flies are st^+bw, and bright red-eyed flies have no brown pigment, being $st\ bw^+$. With the failure of production of both pigments in $st\ bw$ flies, white eyes result.

$$\text{Parents} \qquad \frac{st^+\ bw^+}{st^+\ bw^+} \quad \times \quad \frac{st\ bw}{st\ bw}$$

$$\text{Wild type} \qquad \text{White eyes}$$

$$F_1 \quad \text{All} \quad \frac{st^+\ bw^+}{st\ bw}$$

F_2 Ratio of:

$$9 \ \substack{\text{Wild} \\ \text{type}} \left[\begin{array}{c} \dfrac{st^+\ bw^+}{st\ bw} \\[2mm] \dfrac{st\ bw^+}{st^+\ bw^+} \\[2mm] \dfrac{st^+\ bw}{st^+\ bw^+} \\[2mm] \dfrac{st^+\ bw^+}{st^+\ bw^+} \end{array} \right] :3 \quad \substack{\text{Brown} \\ \text{eye}} \left[\begin{array}{c} \dfrac{st\ bw}{st^+\ bw} \\[2mm] \dfrac{st^+\ bw}{st^+\ bw} \end{array} \right] :3 \quad \substack{\text{Bright} \\ \text{red} \\ \text{eye}} \left[\begin{array}{c} \dfrac{st\ bw^+}{st\ bw^+} \\[2mm] \dfrac{st\ bw}{st\ bw^+} \end{array} \right] :1 \quad \substack{\text{White} \\ \text{eye}} \left\{ \begin{array}{c} \dfrac{st\ bw}{st\ bw} \end{array} \right.$$

Figure 8.1. A demonstration of digenic inheritance involving the character white eye in *Drosophila melanogaster*.

If *bw* and *st* were closely linked instead of being on separate chromosomes (the second and third, respectively), the ratio obtained in the F_2 would have been different. The recombination classes *st bw+* and *st+bw* would be proportionately lower in number, and the ratio would approach 3 wild type:1 white-eyed, which is characteristic of monogenic inheritance. The closer the two genes are linked, the closer the approach to the 3:1 ratio. Complete linkage, that is, no crossing over between *st* and *bw,* would result in a simple 3:1 ratio, and the conclusion would be that only one gene is involved in producing white eyes. Thus, in another situation, if a white-eyed strain of the genotype *w/w* (see p. 329) is tested, the F_2 ratio is 3 wild type: 1 white-eyed with no intermediate eye colors among the progeny. The genotypic formula of the two white-eyed strains should be written $\dfrac{w\ st^+\ bw^+}{w\ st^+\ bw^+}$ for the monogenic strain, and $\dfrac{w^+\ st\ bw}{w^+\ st\ bw}$ for the digenic, to be complete.

Many gene loci are known by only two alleles, one being the "wild type," which is standard and usually dominant, and the other a recessive "mutant" gene. Frequently, however, *multiple allelic series* occur; these series of seemingly allelic genes produce anywhere from 3 on up to 20 or even 80 distinguishable phenotypes, for example, the genes determining some of the cellular antigens in cattle (976). Multiple alleles can be tested only two at a time in test crosses. If three or more of the "alleles" are found in the same individual, it means that some are nonallelic.

Phenotypic Effects of Allelic Substitutions. One of the earliest attempts to explain genetic differences in phenotype was that of Bateson and Punnet (46) who postulated the "presence-absence theory," stating that the dominant phenotype is due to the presence of a gene whereas the recessive phenotype results from its absence. This theory became inadequate when multiple-allelic series and the phenomenon of dominant-mutant phenotype caused by a heterozygous deficiency were discovered. However, it still persists in a modified form to explain situations in which a recessive allele when homozygous produces none of the type effect produced by the dominant. For example, in the inheritance of pigmentation, the albino, or colorless animal or plant, may be assumed to be due to an absence of a gene determining pigmentation, or the presence of an allele which is inactive in forming pigment. It is possible to show that either of these cases may hold.

The recessive mutant characters: acheate, yellow (719), roughest[2] (*rst*[2]) (255), and white (760) in *Drosophila melanogaster* may be produced either by homozygous visible deficiencies, or by gene mutations. The deficiencies can be shown by cytological analysis of the salivary-gland chromosomes to include the loci of the + genes which mutate to the mutant condition. Thus a deficiency in the X-chromosome at the white-eye locus, or a point mutation ($w^+ \to w$) will, in either case, result in recessive white eyes showing no trace of reddish eye pigment. The gene mutation can be differentiated from what otherwise might be a minute, invisible deficiency by showing that reverse mutations to an "active" allele are possible as described in Chapter 4 (p. 172).

Since similar phenotypic changes may be attained by either means, it is necessary only to modify the original presence-absence theory by adding the possibility of the genes becoming inactive. Genes which appear to be completely inactive with respect to the production of type effect are frequently termed *amorphs* (718). The term will be used here in its widest sense; that is, an amorph may be either an allele determining absence of type effect or the absence of a gene, for it is quite impossible to distinguish between the two conditions in many organisms. It is probable, however, that most amorphs are not deficiencies, for deficiencies are usually lethal homozygous; the ones just described are rare exceptions, and, therefore, inactivity is probably more frequent than absence.

It must be emphasized that the term "inactivity" as applied to a gene is a description of a phenotypic effect, and although it implies that the gene is inactive per se, there is no way of proving this except when deficiencies are involved. The apparent inactivation may involve only one aspect of the gene's function, and it may theoretically still be active in producing other effects in the organism. The allelic genes a^k and a both produce a recessive, colorless larval skin in the moth *Ephestia kühniella*, but a^k/a^k adults have brown eyes and a/a adults, red eyes. The wild-type a^+/a^+ larvae are pigmented and the adult eyes are black (135). Therefore a^k and a are both amorphs with respect to larval pigmentation, but not with respect to pigmentation of the adult eye.

In contrast to the amorph, the gene without apparent activity toward producing type effect, there exist recessive genes producing type effect characteristic of the standard, + allele, but less of it. These, termed *hypomorphs* by Muller (718), appear to produce a quantitatively different phenotype from the + allele. Certain alleles in the *albino* series of the guinea pig can be used as examples of hypomorph effect, since substitutions at this locus cause quantitative measurable changes in melanin pigmentation in the hairs. (See p. 427 et seq. for a complete description of this series and its effects in different guinea-pig coat-color strains.) The five known alleles, C, c^k, c^d, c^r, c^a, produce different relative amounts of pigment in the hairs of pink-eyed sepia guinea pigs as shown in Table 29. It will be observed that there are three general levels: zero pigmentation in $c^a c^a$ animals, full pigmentation in the presence of dominant C, homozygous or heterozygous, and intermediate levels in the presence of c^k, c^d, c^r, homozygous, and in various heterozygous combinations with one another. The allele, c^a, is an amorph, and the intermediates are clearly hypomorphs, causing the production of less pigment than the dominant standard, C.

One explanation of hypomorph and amorph activity in the guinea pig might be that they are acting against the standard (intense pigmentation) by inhibiting the production of type effect—the inhibition of the hypomorphs being incomplete, that of the amorph complete when homozygous. This interpretation requires that the C series of genes is

Table 29. Relative Amounts of Melanin Pigment in the Hair of Pink-Eyed Sepia Guinea Pigs *

Genotype	Per Cent Melanin Pigment	
$C-$	100	Full pink-eyed sepia
$c^k c^k$	75	
$c^k c^d$	65	
$c^k c^r$	45	
$c^k c^a$	40	
$c^d c^d$	40	Intermediate sepia
$c^d c^r$	25	
$c^d c^a$	20	
$c^r c^r$	15	
$c^r c^a$	5	
$c^a c^a$	0	No melanin

* Wright (1111, 1112); Wright and Braddock (1114).

not involved in the production of melanin in a positive way, but negatively by controlling the production of an inhibitor. Thus the complete dominant, *C,* would be an allele which inhibits the action of the other alleles and permits pigment production while the hypomorphs and the amorph, c^a, the allele with most inhibitory activity, inhibit pigment production. The data in Table 29 are not inconsistent with this interpretation. An alternative explanation would be that the hypomorphs are acting in the direction of standard.

A test to distinguish between these alternatives is not feasible with the guinea pig, but it is possible in *Drosophila,* for in this organism the chromosomes may be duplicated and the duplications easily recognized. Schultz (885) has described an example in *melanogaster* involving the fourth chromosome gene, *shaven (sv)*, which is recessive and reduces the number of bristles on the body. The fourth chromosome in *melanogaster* is very short relative to the rest of the chromosomes, and changes in its dosage do not have so drastic an effect on the phenotype as do dosage changes in other chromosomes (see Chapter 9, p. 438 et seq.). The following effects are noted when the number of shaven genes is increased by adding fourth chromosomes in an otherwise diploid fly: $sv < sv/sv < sv/sv/sv < sv/+ = +/+$. The symbol here indicates that as the number of *sv* genes is increased the phenotype approaches the wild-type condition more closely by the addition of bristles. Hence the most extreme shaven phenotype with fewest bristles is obtained in the presence of but one *sv* gene. From this it is difficult to avoid the conclusion that the shaven gene is acting in the direction of the standard allele, sv^+, but to a lesser degree. As might be expected, it is possible to add too many *sv* genes and obtain a fly with more bristles than wild type. Thus flies of the genotype *sv/sv/sv/sv* have extra bristles.

An allele can be assigned to the hypomorph category with certainty only if it fulfills the conditions just stated as determined by duplication and deficiency tests. These tests are practical only in a very few organisms. However, since many of the genes in *Drosophila,* and in maize and other plants, give an increasing effect toward type with increase in dosage, it may be assumed that the occurrence of hypomorphic recessives is relatively common in all organisms.

Occasionally it is found that certain genes produce a type effect when homozygous or in combination with an allelic amorph, but behave aberrantly as heterozygotes with other alleles. The cubitus interruptus (*ci*) allelic series in *D. melanogaster* located on chromosome 4 includes several good examples of this phenomenon (966, 967, 969). The mutant alleles in this series cause an interruption of the fourth vein (L4) in the wings (Figure 8.2). The phenotypic expression is modified by the temperature with the most extreme mutant effect expressed at low temperatures around 14°C. In the experiments to be described, the degree of mutant expression was determined by the extent of interruption of the vein and the penetrance (the percentage of individuals observed of a particular genotype which show a mutant phenotype). All gradations between the extremes of intact fourth vein and its complete absence occur as a result of substituting mutant alleles in different combinations. For convenience in measurement of gene effects the four arbitrary standards illustrated in Figure 8.2 were defined, and mutant types were grouped in classes around these.

The effects of the three alleles, ci^+, ci, and ci^w, are as follows. The allele ci gives a more complete vein with increasing dosage,

$$+/+ > ci/ci/ci > ci/ci > ci/o$$

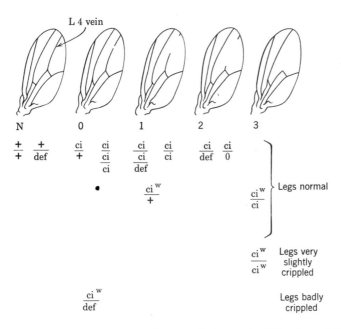

Figure 8.2. The effects of genes ci^+, ci, and ci^w on the L4 vein and legs of *Drosophila melanogaster*. The deficiency is *def M-4*. It is a lethal when homozygous and may include loci other than ci; o indicates the absence of one of the fourth chromosomes. After Stern (967).

This is precisely what we would expect from a hypomorph, but here the similarity ends, for when ci is in combination with its $+$ allele, a greater deviation from the wild type results than is produced by $+$ heterozygous with a deficiency which includes the ci locus:

$$+/+ \; > \; +/def \; > \; ci/+$$

Thus, although both ci and $+$ act positively toward the production of an intact vein, $+$ is haplo-insufficient, and ci "competes" with $+$ when they are in combination. A more extreme example is provided by the ci^w allele (cubitus interruptus of Wallace), which is partially dominant over $+$ and causes abnormal venation with increasing dosage:

$$+/+ \; > \; +/def \; > \; ci^w/def \; > \; ci^w/+ \; > \; ci^w/ci^w$$

From this it would appear that ci^w is acting away from type effect and "competes" strongly with the $+$ allele so that the vein is even more reduced in $+/ci^w$ than in ci^w/def flies. A similar competition is shown in ci/ci^w flies. The subtractive effect of ci^w on type is accompanied by some capacity to produce type since one ci^w approaches normality. The other genes in the genome with effects on this character must not be overlooked, however, for the indications are that the remainder of the deficient fourth chromosome is active in producing type effect too, in both ci and ci^w flies, as shown in Figure 8.2.

Besides interrupting the wing vein, ci^w produces a crippled condition of the legs which in extreme cases may be so serious as to prevent the emergence of the adult from the pupa case. The most extreme crippling results in ci^w/def flies which, it will be recalled, have almost normal wings. The other combinations of ci^w have the following relative phenotypic effects on the legs:

$$ci^w/def \; < \; ci^w/ci^w \; < \; ci^w/+ \; = \; +/+$$

The $+$ allele is thus completely dominant to ci^w with respect to leg development, and since two ci^w produce more type effect than one ci^w, this allele is a hypomorph with respect to leg development.

Since the ci alleles show an inhibition or competitive effect on type manifested particularly in compounds with $+$ and one another, they cannot be classed either as hypomorphs which always show, by definition, an additive effect when present with a dominant, or as amorphs, which by definition show no effects.

Alleles which give the peculiar qualitative and quantitative effects described for ci and ci^w are by no means uncommon. A whole series of *A* alleles in maize has been carefully analyzed by Laughnan (556, 557, 558, 559) for phenotypic effects on pigment production and found to present the same difficulties of interpretation. Indeed, it is quite possible that most alleles are neither purely hypomorphs nor amorphs, but exhibit some degree of subtractive effect in certain combinations; sometimes this is measurable, sometimes it is not.

Biochemical explanations of the action of the various types of alleles just discussed may be made, if it is assumed that the effect on the phenotype by substitution of alleles can be extrapolated to the level of primary gene function. Hypomorphic alleles may then be assumed to produce a primary gene product directly involved in the production of type effect, which is only quantitatively different from the gene product of the normal, dominant allele but limited in amount so that full type effect does not result. The amorphs may be assumed to produce no gene product or an amount so small as to have no effect on the phenotype. Actually, it is highly improbable that either one of these explanations is correct. It is more probable that both hypomorphs and amorphs produce just as much gene product in the form of RNA as the wild type and just as much protein in the cytoplasm. But the protein produced by the mutant alleles is either not as efficient in its role of catalysis, etc., or it is completely inactive. Hence, rather than assume *quantitative* differences, it is more in agreement with recent findings such as discussed in Chapter 6 to assume *qualitative* differences in the products of different alleles. Such qualitative differences may give quite unexpected results in heterozygotes, as in the cubitus interruptus example described earlier, which we will probably be unable to interpret satisfactorily until more is known about the mechanism of protein synthesis and the interactions of proteins in the cytoplasm.

Dominance. In the usual allelic series one allele will exhibit dominance over all others such that in heterozygotes with hypomorphs, amorphs, or deficiencies it will produce essentially the same phenotype as when homozygous. If the dosage of this "haplo-sufficient" allele is increased beyond the diplo condition, there will usually be no change in phenotype. Between the extremes of complete dominance and complete recessiveness there exist varying degrees of incomplete dominance as exhibited by hypomorphs. The two phenomena, complete and incomplete dominance, appear to differ only in degree, and it will be on this assumption that they will be discussed here.

It is a relatively simple matter to make a model to explain dominance in terms of gene action. If it be assumed that the lower alleles are hypomorphs or amorphs, the explanation becomes clearly one of degree of activity toward the production of the normal phenotype. The dominant allele is not only the most active, but it is overactive in the sense that when present once it produces sufficient or more than sufficient gene product to convert some substrate to a product necessary for the production of the normal phenotype. Any increase in its dosage may be expected to provide for more gene product, but since limits are set on the amount of product that may be active in producing type effect in the cellular environment, there will be no change in phenotype. Therefore excess gene product will be ineffective. If the hypomorphic alleles produce a less efficient gene product than the dominant even when homozygous, they will obviously act as recessives, and when heterozygous with other hypomorphs they will be expected to show intermediate effects with no dominance of one allele over another. A diagram of this model is presented in Figure 8.3. The quantitative data given for the guinea pig C series are in rough agreement with this interpretation, as are the phenotypic expressions of heterozygous combinations of many other allelic series. The threshold effect which

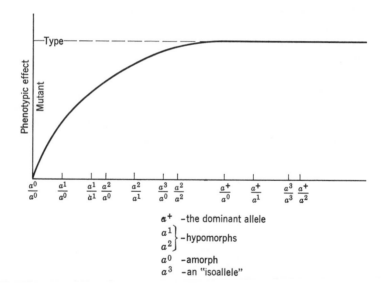

Figure 8.3. Dosage in terms of concentration of gene product and its effect on the phenotype. A graph illustrating the hypothesis that an increase in the number of hypomorphic genes causes an increase in gene product with a concomitant effect on the phenotype. Any increase of gene product beyond a certain threshold value has no further effect on the phenotype for reasons discussed in the text.

prevents further change in phenotype with increase in gene-product concentration may be explained in terms of an assumed relationship between enzyme and substrate concentration, an explanation developed by Wright (1108, 1109).

It will be recalled from Chapter 5 in the discussion on enzyme kinetics that under certain conditions of enzyme excess relative to substrate a further increase in enzyme concentration will have little or no effect on the rate of formation of the product (see Figure 5.2a and b). If it is assumed that the dominant gene when present once allows for the production of a concentration of enzyme which is always in great excess of its substrate, then a further increase in enzyme (by increasing the gene dosage perhaps) will not be expected to increase the rate of product formation. The only factor which theoretically will affect the rate of product formation under these circumstances is a change in substrate concentration. Application of enzyme kinetic theory can also be made to explain the results with hypomorphs. If it is assumed that a hypomorph allows for the production of an inefficient enzyme so that the substrate concentration is effectively in relative excess (see Figure 5.2b), then any change in enzyme concentration resulting from change in gene dosage will be expected to affect the rate of the product formation and hence the phenotype.

The concept that dominance of a gene is a manifestation of its superior activity as compared to its recessive alleles has its roots in the presence-absence hypothesis. More recently supplementary observations have been added that intermediate alleles exist which show intermediate degrees of dominance.

Isoalleles. The foregoing explanation of dominance may be used to predict the existence of alleles which, when homozygous, produce full type effect, but when heterozygous with hypomorphs or amorphs produce a mutant effect. By reference to Figure 8.3 it will be seen that if an allele, a^3, were assumed to produce the concentration of gene product indicated when homozygous, the phenotype would be normal, but mutant when a^3 is hemizygous or in heterozygotes with a^0 or a^1. Therefore a^3 would be separable from a^+ only in combinations with lower alleles in the series. The further possibility exists that any number of a^+ alleles may occur which have more potency than the a^+ in Figure 8.3. These may be termed *hypermorphs* (718). They would theoretically not be distinguishable from one another or the a^+ allele, unless the shape of the dosage-effect curve were altered in such a way that they manifest themselves as alleles of a different potency.

Actual examples which fit the description of these hypothetical situations exist in what are known as isoalleles. These are so named because they are to be distinguished from one another, phenotypically, only when in certain heterozygous combinations, or when the gene-modifier background or the external environment is changed. The ci series in *D. melanogaster* has been shown by Stern and Schaeffer (969) to include at least three isoallelic ci^+ genes, $+^{ci}$, $+^2$, and $+^3$. The differentiation between these three alleles is based on their different phenotypic expressions at two temperatures, 14°C and 26°C, and their different expressions when in heterozygous combinations with an M-4 deficiency, ci or ci^w. When homozygous, each of the isoalleles, $+^{ci}$, $+^2$, and $+^3$, causes the development of a complete L4 vein, except for $+^3$ at 14°C, in which a few flies with type O venation appear (circa 2.5%). The really significant differences appear in the heterozygotes with ci and ci^w at the two different temperatures. The heterozygote $+^3/ci$ produces a much higher percentage of wings with an interrupted L4 vein than $+^{ci}/ci$ and $+^2/ci$. In heterozygous combination with ci^w, none develop to the adult at 14°, but at 26° the three isoalleles are clearly different from one another on the basis of their effects on the L4 vein— $+^{ci}$ is most dominant to ci^w, while $+^3$ is actually recessive and $+^2$ is intermediate.

The action of modifying genes on the expression of these isoalleles must be taken into consideration, for, as will be discussed in Chapter 9, the expression of probably any gene can be changed by changing its genetic background. Therefore it is of great importance particularly when investigating isoalleles to make the genetic background of each as uniform as possible in order to insure that the relatively slight phenotypic differences which distinguish them are not due to different modifiers. In the ci case this was approached by making the stocks isogenic for the X, second, and third chromosomes, but not for the fourth on which the ci locus is found. The possibility exists, therefore, that modifiers close to the ci locus are the cause of the difference in expression noted, and that really only one $+$ allele is present. It can be said only that, within the limits of discrimination set by the genetic techniques, there exist three isoalleles in the ci series. This qualification applies, of course, to any allele, even those not considered as isoalleles.

Another example of wild-type isoalleles in *Drosophila melanogaster* has been found by Muller (719) and Green (361) by somewhat similar methods. Two wild-type stocks, Canton and Oregon, which are phenotypically inseparable were shown to differ at the white locus. When Canton ($+^c$) and Oregon ($+^o$) are combined with the white amorph, *w*, in triploids, $w/w/+^c$ and $w/w/+^o$, a distinct difference is noted in the eye colors (362). A somewhat lesser difference is noted if the diploid heterozygotes $w/+^c$ and $w/+^o$ are compared. Since the white locus is known to be compound, as discussed hereafter, Green attempted to determine which part of the locus was different and found that they both differed in the right segment in the area occupied by wch and *sp-w* (see Figure 8.9). They appear to be identical in the region to the left of this.

Isoalleles may be differentiated not only on the basis of their phenotypic expression under different environmental and genetic conditions, but by the effects of mutagens on them. Thus Timofeeff-Ressovsky (1008) identified two different wild-type alleles at the white locus, *w*, in *Drosophila melanogaster*. One, found in the American stocks and designated as w^{+A}, was twice as sensitive to x-rays as the other, w^{+R}, found in a Russian stock. Lefevre (577), working with a different sex-linked gene, *y* (yellow), found that the same situation exists with respect to this locus. It is probable that many isoalleles exist which differ only in mutability but are otherwise identical in their phenotypic effect.

Every gene probably has the potentiality of mutating to a large number of different states and being transformed into an amorph, or a completely wild-type dominant, or hypomorphs of a number of different potencies. The possibilities may be graded in a spectrum, from the amorphic condition on up. Those close together in the spectrum may be considered as the isoalleles, and the differences in phenotypic expression of these must frequently be magnified by the alteration of the gene's background in order for them to be recognized. On the other hand, relative position in the spectrum may be so dependent on external factors that members of isoallelic groups may appear to exchange positions, depending on the controlling circumstances.

A glimpse into the great complexity of the actual situation is afforded by the results of Spencer (935) obtained in his study of the isoalleles at the bobbed locus in *Drosophila hydei*. The bobbed genes in *hydei* reduce the size of the bristles just as does the bobbed gene previously described for *melanogaster*. Spencer found that the bobbed alleles procured from different populations of *hydei* produce a complex series of bobbed phenotypes. It was recognized that the differences in phenotype could be due to a single type mutant allele acted on by modifiers, but results obtained from the appropriate crosses indicate that a multiple allelic series is involved. The alleles detected may be graded in a series according to phenotypic effect in homozygotes from extreme lethal

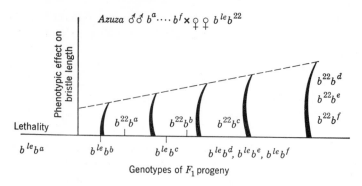

Figure 8.4. A demonstration of isoalleles at the bobbed locus in *Drosophila hydei*. After Spencer (935).

types to those which give normal phenotypes homozygous, but which produce a bobbed phenotype when heterozygous with intermediate bobbed alleles. An example which indicates the high incidence of these isoalleles in the population is presented in Figure 8.4. Shown here are the results from a cross involving 12 phenotypically wild $\delta\ \delta$, chosen at random and crossed to a tester stock made up of Azusa (b^{22})/Azusa bobbed lethal (b^{le}) $\,\varphi\ \varphi$.

Analysis of Genetic Material by Recombination

In the previous section alleles have been considered primarily from the functional aspect. Implicit in the discussion was the hypothesis that alleles are genes that always segregate from the heterozygotes in a one-to-one ratio, and that they usually show some relation in their effect on the phenotype. The relationship of alleles through function is on the whole well supported and has led to the point of view that allelic genes are merely different forms of the same gene. On the other hand, the conclusion that alleles, as defined functionally, never show recombination is not supported by the experimental data. It is the purpose of this section to describe and explain some of these data and discuss their significance.

The possibility of the existence of genetic units with similar or related functions occurring closely linked together on the chromosomes has been postulated on many occasions (896, 3, 818) to explain the peculiar phenotypic manifestations of some allelic series. Such "subgenes" or "pseudoalleles," as they have been variously described, would be expected to be inherited as a unit with only extremely rare separation by crossing over. If mutation occurred in one of them, the "allele" so produced would be considered as allelic to the whole group of units rather than only the subunit from which it was derived, unless it could be separated from the group and recognized.

Pseudoallelism in Drosophila.   Until relatively recently the existence of such subunits, or the compound nature of certain loci, was in the realm of speculation.   In 1940, however, Oliver (754) obtained a small number of unexpected wild-type *Drosophila melanogaster* from a cross involving two lozenge alleles, lz^g (glossy), and lz^s (spectacle), which could best be explained by assuming that the lozenge "locus" is compound.   The lozenge alleles are recessive sex-linked genes which affect the phenotype in a number of ways: alterations in the amount of eye pigment, changes in the eye facets, and reduction of female fertility associated with the absence of spermathecae (756, 170, 171).   Heterozygotes between the mutant alleles show the lozenge effects on the phenotype, but the phenotype is not always intermediate.   For example, in certain heterozygotes such as lz^g/lz^s, the flies are considerably more viable and fertile than the homozygotes, lz^g/lz^g or lz^s/lz^s (757, 165).   Thus a complementary effect is demonstrated rather than an intermediate quantitative effect which would be the expected result from the interaction of hypomorphs as discussed in the previous sections.

Oliver (755) found that, in a cross between heterozygous lozenge females lz^g/lz^s and either lz^g or lz^s males, some wild-type males resulted which were genotypically lz^+.   This could be explained on the basis of unequal crossing over as demonstrated for the *Bar* mutant by Sturtevant (985) which would mean that two lz alleles were caused to be located on one chromosome, whereas the complementary type produced by the unequal crossing over would produce a normal phenotype.   On the other hand, it could be the result of equal crossing over between two closely linked genes which have similar effects on the phenotype.   Thus, if the "glossy" chromosome were genotypically $lz^g\ lz^{s+}$ and the "spectacle," $lz^{g+}\ lz^s$, and a crossover occurred between them, then the genotypes $lz^g\ lz^s$ and $lz^{g+}\ lz^{s+}$ would be expected.

The latter explanation is the correct one in this example, as established by the work of Green and Green (365, 366), who showed convincingly the lozenge "gene" consists of at least three pseudoallelic loci which are separable by crossing over.   Their map for these three loci is given in Figure 8.5.   Other alleles have been located which appear to be "true" alleles at the indicated loci in the sense that they do not show crossing over except with those at the other loci. Heterozygotes between the lozenge alleles in the same pseudoallelic group always give a mutant phenotype, which is as expected, since they are presumably "true" alleles. The same is true for pseudoallelic lozenge alleles when they are in the *trans* relationship, for example:

$$\frac{lz^g\ +}{+\ lz^{4b}}$$

but not when they are in the *cis* configuration, viz.:

$$\frac{lz^g\ lz^{4b}}{+\ +}$$

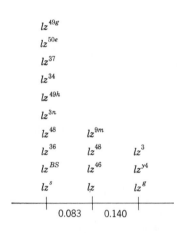

Figure 8.5. The three lozenge pseudoallelic loci according to Green and Green (366). Some of the known "true alleles" are shown for each locus. The figures refer to the map units separating the loci.

The *cis* condition always gives the wild-type phenotype, whereas the *trans* is mutant. Those alleles at the same locus always give mutant phenotypes when heterozygous. Two exceptions have been found which are of considerable interest: lz^{49h} and lz^{50e} both give an approximately wild phenotype in the *trans* configuration with all others, those at the same locus with them and those that are pseudoallelic to them. For this reason they are said to *complement* with the other alleles. Complementation is discussed in detail later in this chapter.

The question naturally arises at this juncture as to why $\dfrac{lz^{46}\ +}{+\ lz^g}$ flies, for example, are mutant in phenotype, whereas $\dfrac{lz^{46}\ lz^g}{+\ +}$ flies are normal. Since these are recessive genes, both genotypes would be expected to be normal. But, other than the fact that heterozygotes with alleles on different chromosomes are more fecund than the homozygotes, lozenge phenotypic effects are expressed unaltered in the heterozygotes. The reasons for this peculiar type of "position effect" are not known. It has been noted not only in the lozenge series but in a considerable number of other allelic series in *Drosophila*. The number continues to grow larger,

Table 30. Examples of Pseudoallelic Loci in *Drosophila melanogaster*

Locus Designation	Chromosome	Number of Groups	Reference
lozenge	X	3	755, 366
star-asteroid	II	2	606
stubble	III	2	609
vermilion	X	2	358
forked	X	2	359
white	X	4	362, 610, 631, 500, 501
singed	X	3	440
garnet	X	3	439, 163, 164
bithorax	III	5	609, 611, 612
dumpy	II	7	130
notch	X	7	1065, 1066

and it is becoming increasingly evident that pseudoallelic loci may be the rule rather than the exception. A partial list of pseudoallelic loci in *D. melanogaster* is given in Table 30. Because of the great importance of the studies that have been made on some of these groups, two representative series, the bithorax and white, are considered further hereafter to give the reader some comprehension of the significance of the findings in this area to the study of genetics in general.

The bithorax series of pseudoalleles studied by Lewis (609, 611, 612) is of unusual interest because it affects the development of a set of related morphological characters in *Drosophila* and allows an analysis of their ontogenetic and phylogenetic relationships via the changes manifested by the various combinations of alleles in the series. At least five loci have been identified with this group (Figure 8.6). In general, *Ubx, bx, bxd,* and *pbx* have closely related effects. They cause the metathoracic segment of *D. melanogaster* to resemble the mesothorax, whereas both *pbx* and *bxd* affect the posterior metathorax. In addition, *bxd* causes the first abdominal segment to develop in the direction of the anterior metathoracic segment. The effects of these mutant genes can be somewhat bizarre, especially in double mutants. For example, a fly carrying both *bx* and *pbx* in the double-homozygous condition will have the complete metathorax resembling the mesothorax. This results, among other things, in the development of the halteres into wings. Hence a four-winged dipteran is produced.

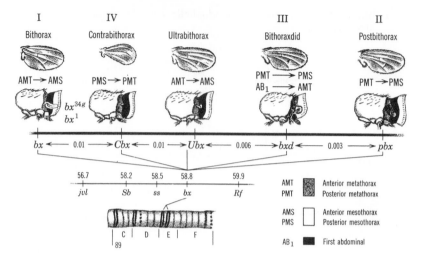

Figure 8.6. Bithorax in *Drosophila melanogaster*. The manifold effects of each pseudoallelic locus on the wing, haltere, thorax, legs, and abdomen are shown. From Carlson (130).

Cbx also acts on the thorax, but opposite to the way in which the others act, since it causes the mesothorax to resemble the more posterior metathorax. It is obviously closely related, however, to the others in its effects.

These genes are all located on chromosome 3 and occupy a very short segment. Figure 8.6 shows their approximate location on the chromosome, their order, and their distance apart as determined by recombination analysis.

Ubx and *Cbx* are dominant genes. *Ubx* is lethal when it is present homozygous and terminates development in the larval stage. When it is present in the heterozygous state, it has very slight effects, except when combined with certain of its pseudoalleles. All the other pseudoalleles are recessive and exhibit mutant phenotypic effects when homozygous, or in heterozygous combination with their other mutant alleles or certain of the pseudoalleles.

The phenotypic results of combining the pseudoalleles in pairs are given in Table 31. The *"cis"* and *"trans"* in the table refer to the two possible ways in which a pair of pseudoalleles can be combined, as already described for the lozenge alleles.

As can be seen from the table, all *cis* combinations are wild type, except for those examples in which one or both of the genes are dominant. In two combinations in which *Cbx* is involved, that is, *bx³* and *Cbx*, and *Cbx* and *Ubx,* the type IV phenotype is suppressed to a considerable extent in the *cis* arrangements. This is a rather exceptional situation, recorded, so far as known, elsewhere only for the Stubble-stubbloid case described by Lewis (609) in which stubbloid acts as a complete suppressor of the dominant Stubble phenotype in the *cis* arrangement.

When the genes are in the *trans* arrangement, it will be seen that *bx³* complements with *bxd* and *pbx* to give a wild type; that is, *bx³* does not act phenotypically like an allele of either of these genes. The same is true of *bxd* and *Cbx*. Whether in the *cis* or *trans* condition, the phenotype is the same. However, in all other cases presented in the table,

Table 31. Effects of Bithorax-Bithoraxoid Pseudoalleles *

Mutant Genes	Type of Hetero-zygote	I	II	III	IV
bx³ and *bxd*	cis	0	0	0	0
	trans	0	0	0	0
bx³ and *pbx*	cis	0	0	0	0
	trans	0	0	0	0
bxd and *pbx*	cis	0	0	0	0
	trans	0	+++	0	0
bx³ and *Ubx*	cis	+	0	0	0
	trans	+++	+	0	0
Ubx and *bxd*	cis	+	0	0	0
	trans	+	+++	+++	0
Ubx and *pbx*	cis	+	0	0	0
	trans	+	++++	0	0
bx³ and *Cbx*	cis	0	0	0	++
	trans	0 to +	0	0	++++
Cbx and *bxd*	cis	0	0	0	++++
	trans	0	0	0	++++
Cbx and *pbx*	cis	0	0	0	++++
	trans	0	0 to +	0	++++
Cbx and *Ubx*	cis	+	0	0	+
	trans	++	+	0	+++

Deviation from Wild Phenotype † spans columns I–IV.

* From Lewis (612).
† See Figure 8.6 for phenotypic descriptions of I, II, III, and IV.

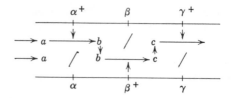

Figure 8.7. An hypothesis to explain the differences obtained in *cis* and *trans* arrangement of the bithorax pseudoalleles. *a, b, c, d* indicate substrates acted on by products of genes $\alpha+$, $\beta+$, and $\gamma+$. The diagonal lines indicate blocks. *b* and *c* are assumed to diffuse.

the *trans* configuration gives a mutant phenotype. This may be described as the *cis-trans* position effect. It is basically identical to the situation already described for lozenge and is typical for all of the pseudoallelic groups of genes listed in Table 30.

Why some pseudoalleles show a mutant phenotype in the *trans* configuration while others do not, but show complementation to give the wild phenotype, cannot be answered satisfactorily at present using the data obtainable from *Drosophila* alone, but additional experiments carried out by Lewis may throw some light on this question.

Lewis (611) first made the following assumptions. (1) The genes in this series are all related in function insofar as they act together in the production of a particular phenotype. (2) They are arranged together in a linear array because they are involved in a sequence of reactions. (3) Any interruption or disruption of this close proximity and array will result in inefficiency and a mutant phenotype. (4) The reactions proceed in close proximity to the chromosomes. From Figure 8.7 it will be seen that by making these assumptions one can explain the difference in phenotype between the *cis* and the *trans*. To test this hypothesis further, Lewis assumed that somatic pairing (which occurs in the Diptera) enables the chromosomes to collaborate reasonably efficiently, but that this efficiency would be considerably reduced if pairing in the bithorax region were reduced by a rearrangement. By producing rearrangements in the third chromosomes of such a nature that the bithorax region was moved to another position on the chromosome and then incorporating this into a heterozygote, Lewis was able to show that the combination R* $(bx^{34e} +)/+$ *Ubx* had a more drastic mutant phenotype than $bx^{34e} +/+$ *Ubx*. If the same rearrangement is present in both chromosomes, R $(bx^{34e} +)/$R $(+ Ubx)$, then the

* R indicates that what follows in parentheses is included in the rearrangement.

mutant phenotype is the same as if a rearrangement is not present. The mutant phenotype can also be produced by transposing the segment containing bx^+, Cbx^+, Ubx^+ to the left arm of the third chromosome, leaving bxd^+ and pbx^+ in their original position. When homozygous, this transposition gives an extreme type II and III phenotype even though theoretically all the genes are wild type.

The hypothesis that interference in pairing in structural heterozygotes reduces the transport of gene products from one homologue to another is an attractive one, but unfortunately no further data have been forthcoming to support it. In fact, it is difficult to conceive how the hypothesis can be put to further test at the present time.

On the other hand, one does not need to resort to the hypothesis that exchange takes place between the products of homologous chromosomes when they are paired or even unpaired. The bulk of evidence in general does not indicate an active exchange of products between chromosomes (see the discussion on enzymes produced in heterozygotes and heterocaryons later in this chapter). It may be necessary in the first place to keep certain segments of the chromosome together because by virtue of their organization they cause the production of a certain type of cytoplasmic organization. From a *trans*-type heterozygous double mutant one expects two types of organization, and a requirement that the substrates diffuse between them (see Figure 8.8). A transposition, such as that involving bx^+, Cbx^+, and Ubx^+, in this light would be expected to produce a considerable derangement of the bithorax locus product.

Perhaps no locus known to geneticists in any animal or plant has received the attention accorded the white locus in *Drosophila melano-*

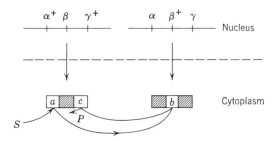

Figure 8.8. Hypothetical explanation of the *cis-trans* effect based on the assumption that the pseudoallelic loci are involved in the synthesis of organized particles in the cytoplasm. Substrate S is converted to product P by three steps involving two different kinds of particles in the cytoplasm.

gaster. One of the first mutants discovered in *melanogaster* by Morgan, white eye (*w*), gives the locus its name and is the amorph in the series. For many years new alleles were described at the white locus, and it became the ideal to use as an example to describe what was meant by gene, gene action, and multiple-allelic series. The intermediate alleles causing various amounts of pigment formation in the eyes were useful in demonstrating hypomorphic action and also widely used in back mutation tests to wild type.

In 1952 two workers simultaneously showed that the white locus is like the lozenge locus (610, 390) by demonstrating recombinations between certain members of the white series. The further work of Judd (500) and Green (361) established the existence of at least five subloci or recombination sites within the white locus. Each of these with the exception of Brownex (w^{Bwx}) has a series of "true" alleles, as can be seen in Figure 8.9. It is quite possible that more than five recombination sites will be found to exist.

Two peculiarities of the white-locus pseudoalleles make a study of them particularly rewarding and interesting. The first is that no com-

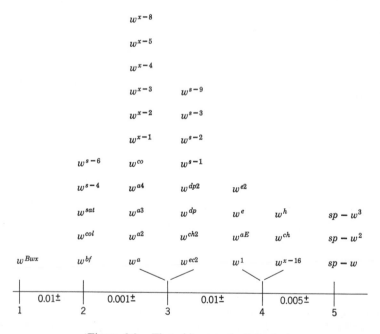

Figure 8.9. The white pseudoallelic loci.

plementation seems to occur between the members in different recombinational groups, as well as within groups, except between the spotted whites, *sp-w,* and the others. Second, unequal crossing over seems to occur regularly within the groups as determined by the appearance of progeny with a white phenotype from certain heterozygotes with intermediate alleles such as w^a/w^{a4}, w^{Bwx}/w^a, w^a/w^{bf}, w^e/w^h. Since the white "recombinants" (w^r) do not show recombination with any of the other members of the series (500), and $w^r/sp\text{-}w$ females give a phenotype like $w^{def}/sp\text{-}w$ (363), the conclusion is that they are small deficiencies in the white region resulting from unequal crossing over. By carrying out a cross with attached-X flies in which w^a and w^{bf} were present heterozygous and in *trans* relationship, Judd (501) was able to demonstrate that the expected complementary exchange products, a duplication and a deficiency for a portion of the white locus, were produced.

The demonstration of unequal crossing over at the white locus is of general interest, because it indicates that the various pseudoalleles may be related in ways other than functional. Lewis (609) and Stephens (964, 965) have suggested that pseudoalleles have arisen through duplication from a common ancestor and that during the evolutionary process some of the duplicates changed in function to give the pseudoallelic complex. If this were so, it would be expected that even after some evolutionary change the divergent duplicates may still have enough homology to pair. This asymmetrical pairing would lead to unequal crossing over, as shown in Figure 8.10, and lead to some quite unexpected results.

The white pseudoalleles may be differentiated to some extent functionally, as well as by recombination, by combining them with zeste (z), a mutation located about 0.5 units to the left of the white locus. Female flies homozygous for z have yellowish eyes (309). If homozygous z/z flies are also heterozygous for certain of the w alleles, the zeste phenotype is suppressed. It has been found by Green (361) that all w alleles in group 4 (Figure 8.9) and the *sp-w* alleles suppress zeste, but the others do not. It is in fact possible that zeste should be considered part of the white locus, not only because of the effect of some of the white alleles, but because heterozygous z/z^+ flies, also heterozygous for a small white duplication, w^+/w^{dup}, give a mutant phenotype (501). This is true whether the genotype is $z^+ \ w^{dup}/z \ w^+$ or $z \ w^{dup}/z^+ \ w^+$. In both cases the flies have reddish-brown mottled eyes. Hence the combination $z \ w^{dup}$ acts as a dominant "gene."

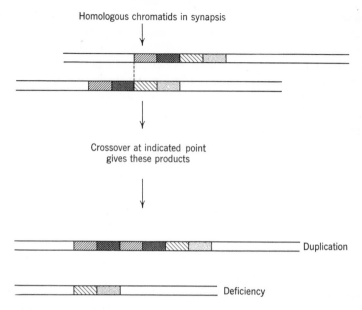

Figure 8.10. The results of unequal crossing over. Exchanges between non-homologous regions result in duplications and deficiencies.

Some Definitions. It is necessary at this point in the discussion to introduce some new terms and redefine some old ones.

The term *gene* has been used in the classical sense up to this point, but in what follows it will be given a less restricted definition. It will apply to that segment of genetic material which determines the linear order of nucleotides in an RNA participating in ordering a unit or subunit of biological function, for example, an enzyme or a chain of hemoglobin. When the protein is not known, the term will denote that segment which appears to control a single function in the production of the phenotype. To be considered synonymous with gene in what follows is the term *locus*. Previously, locus was merely the point on the map which the gene occupied. Current usage, however, now gives it material reality and makes it synonymous with gene. This is perhaps unfortunate, but there is no advantage in insisting on archaic meanings which all words acquire as the language evolves. *Allele* will be used here in the same sense as previously, except that recognition must be given to the change in definition of gene which no longer includes the requisite that exchanges cannot occur within it. The term allele is considered obsolete by some authors (799), but it will not be so considered here provided recognition is made of its change in definition.

Mutational site will refer to a point within a gene where a change (mutation) has occurred. Since a gene may mutate at a number of different points within its length, it may have many mutational sites. This, it will be recognized, causes difficulties with use of the term allele, for if we have two genes derived from the same parental gene but mutant at different sites are they allelic or not? This difficulty can be easily resolved by naming those which are mutant at the same site as *identical* alleles or *homoalleles,* and those which are mutant at different sites as *nonidentical* alleles or *heteroalleles.* Homoalleles are also called *homologous* alleles or *noncomplementary* alleles by some authors, and *heteroalleles* are referred to as *nonhomologous* or *complementary* alleles. The use of these latter terms, complementary and noncomplementary, in this connection is to be discouraged, however, as it leads to more than the usual amount of ambiguity found in expositions on this general subject, as is evident in the following.

Other terms will be introduced as the need arises, but the ones just given will form the basic vocabulary of the remainder of this chapter.

Recombination in Phages. The same criteria of recombination and linkage cannot be used for phages as are used in the higher organisms with meiosis, because a phage "cross" is not analogous to a cross in plants and animals. When a cross is made between dihybrids of a plant or animal such as $\dfrac{a\ b^+}{a^+b} \times \dfrac{a\ b^+}{a^+b}$, we generally assume that each of the recombinant gametes produced, that is, $a\ b$ and a^+b^+, is the result of a crossover event at or just prior to the time of meiosis in a single spore mother or gonial cell. Hence we can state with some confidence that the degree of linkage or map distance can be determined by scoring the number of recombinant chromosomes and determining the fraction of the total number of chromosomes retrieved in the gametes. Other things being equal, we can take this fraction, expressed as a percentage, to be a function of the actual physical distance between the markers.

As pointed out in Chapter 3, a phage cross must be looked upon as an experiment in population genetics. What is controllable in a phage cross is the input ratio of genotypes, that is, the ratio of $a^+\ b$ to $a\ b^+$, for example, and what is measurable are the frequencies of the genotypes after the production of infectious phages. The input ratio can be controlled, but the processes which proceed after infection and during replication and reassortment can neither be closely controlled nor

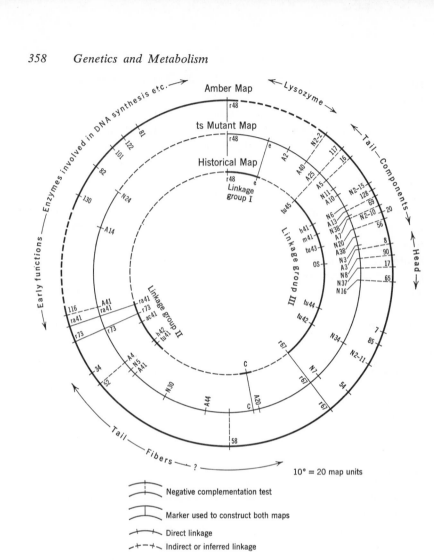

Figure 8.11. Circular linkage map for bacteriophage T4D. Functions of various regions indicated on the periphery. From Edgar (245).

directly observed. Such factors as the number of "matings," and the number of phage particles involved in each mating, determine, among other unknown ones, what types and ratios of recombinants are to be obtained. Thus it is quite apparent that the problem facing the phage geneticist is quite a formidable one. In order to learn anything about the basic aspects of recombination in phage, he must analyze second hand a complex population phenomenon hidden from him within a bacterial cell. Attempts to do this have been made, particularly by Visconti and Delbrück (1035), Hershey (431), and Bresch (103). The results of the analyses emphasize the complexity of the problem and show that linkage relationships in phage should be defined free of assumptions concerning the mechanism of mating and recombination (981). Linkage data for markers in phage, therefore, should not be considered equivalent to linkage data obtained from the eucaryotic organisms. What the phage data mean we cannot say at the present time with any degree of assurance, but some of them are sufficiently interesting to be described and may have a bearing on the structure of genetic material in general (957).

A genetic cross in phage is made by infecting a bacterium with phage of the desired genotypes simultaneously and then scoring the types of progeny by using them to infect indicator cells. Two, three, or more markers may be employed for infection simultaneously. By scoring the genotypes of the progeny, linkage maps have been constructed which are interpreted to mean that the markers are linearly arranged, that the map distances are roughly additive, and that there is one linkage group in phages T2 and T4 (51, 74, 981). The single chromosome indicated by the linkage data appears to be circular just as in *Escherichia coli*. The circular map for T4D is shown in Figure 8.11.

Difficulty is experienced when attempts are made to determine map distances quantitatively in phage rather than just order of elements because of the occurrence of high negative interference. In the higher organisms positive interference is the rule. If the distances between three markers, as established by two-point test crosses, are determined, and a three-point cross is made as illustrated in Figure 8.12, the expected frequency of double crossovers is invariably (unless the markers are far apart) greater than the observed frequency. This is generally called positive interference and expressed as a coincidence value derived by dividing the observed frequency of doubles by the calculated expected frequency. If interference is positive the coincidence value is, of course, less than one, and if it is zero, it is one. With the T-even

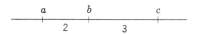

Figure 8.12. Explanation of positive and negative interference. Cross $ab+c \times a+bc+$. Expected doubles: abc and $a+b+c+$. *Calculated* expected frequency: $0.02 \times 0.03 = 0.06\%$. If *observed* frequency < 0.06, then $0/c < 1$ and interference is positive. If observed frequency > 0.06, then $0/c > 1$ and interference is negative.

phage, however, coincidence values greater than one are found ranging up to 31 (159). This is called negative interference. Ordinarily, in eucaryotes the closer together three markers are, the greater the positive interference; but in phage the opposite holds, the frequency of multiple recombination such as doubles is inversely related to the map interval. These data are difficult to correlate with those from organisms in which standard crossing over occurs. In these it is assumed that two crossovers will not occur simultaneously close together easily, because they interfere with one another, hence positive interference. If we take the negative interference data from phage at face value, then we must assume that an exchange at one point actually induces or encourages another exchange close by.

Benzer (74, 75) has carried out a comprehensive analysis of the plaque-type *r* mutants of phage T4. Many hundreds of *r* mutants have been isolated and found to fall into several categories as determined by their plaque morphology and ability to reproduce in a series of different strains of bacteria. The phenotypic effects are of two types, plaque type, which is either wild type or *r* type, and inability to develop in a host. By testing each *r* mutant on a series of three host strains, *E. coli* B, S, and K, it was found that the *r* mutants fell into three groups, *r*I, *r*II, and *r*III, as shown in Table 32. It is, of course, quite possible that more than three groups exist, but this is improbable (247). The number detected depends to a large extent on the number of different bacterial host strains tested with the phage.

By making crosses between *r* mutants, both within and between the *r* groups, it has been established that they all fall within what would appear to be a limited segment of the phage "chromosome" arranged linearly in the order *r*II, *r*I, *r*III. The mutants used to arrive at this

Table 32.   Plaque   Morphology of
Phage T4 Strains on Different Host
Strains *

Phage Strain	Bacterial Host Strain		
	B	S	K
Wild	Wild	Wild	Wild
*r*I	*r*	*r*	*r*
*r*II	*r*	Wild	. . .
*r*III	*r*	Wild	Wild

* From Benzer (75).

map were of two general types: (1) those that always gave r^+ "recombinants" when crossed and (2) those that did not give r^+ progeny when crossed to some of the first type. Benzer (75) considered the first category as representing "point" mutations and the second type as anomalous mutants with deletions in their genetic material. If two r mutants did not produce r^+ recombinants when crossed, it was considered either that they both carried deletions which were identical or overlapping, or that one was a deletion and the other a point mutation within the map area covered by the deletion.

Aside from establishing by the use of recombination data that the different r mutants are not distributed along the map at random, but are grouped together according to apparent function, Benzer also established that recombination occurs extensively within each of the three groups. That is, the rII, rI, and rIII regions may each be broken up into subregions by recombination. The rII group of mutants particularly has been thoroughly analyzed. The exceedingly extensive data which have been accumulated by Benzer for the rII mutants are all consistent with the conclusion that they are in linear array on the phage "chromosome" (74). Over 300 mutational sites have been identified in the rII region, and it has been calculated by Benzer (76) that 120 more may yet be expected to be found.

The resolving power of recombination analysis with bacteriophage is very great, because of the large number of progeny phages that can be tested for phenotype relatively easily and quickly. Hence the "fine structure" of the genetic material can be studied. Phages have so far demonstrated one drawback, however, with regard to this, which is that the phenotypic effects of mutational changes are not easily studied. This makes it difficult to relate the linkage findings to anything concrete at the protein level. However, the fine structure analyses have provided sufficient data to indicate the possibility that exchanges may occur between any pair of nucleotides in a DNA chain. This is in itself a major step forward in our understanding of recombination and, most important, of the structure of the genetic material.

As a result of a very extensive analysis of the entire genome of phage T4D by recombination and complementation (a technique of analysis described hereafter), it has been possible to show that genes having similar functions are located contiguously along the chromosome (246) as shown in Figure 8.11. This close correlation between function and linkage is quite characteristic of bacteria, too, as will now be described.

Genetic Analysis in Bacteria. The recombinational analysis of linked mutational sites in bacteria has been done primarily in *Salmonella typhi-*

murium and *Escherichia coli* by transduction, and by transformation in pneumococcus (*Diplococcus pneumoniae*).

Transductional analysis of linkage in bacteria is based on the observation that the vector phage can carry only a limited amount of the donor host genetic material to the recipient. Presumably, the host material transported is carried as a single short segment and is expected to carry two markers from the host only if they are closely linked. Hence simultaneous transduction of two markers can be used as a measure of linkage. The higher the frequency of simultaneous transduction, the closer together two markers are assumed to be. This technique can also be used to order closely linked markers by making three-point test crosses (214). It should be noted, however, that only a limited amount of linkage data can be derived by this method, for markers too widely spaced to be incorporated into a single transducing particle cannot be ordered with respect to one another.

By means of transductional analysis Demerec, Hartman, and associates have been able to map several regions of the *Salmonella* chromosome. They have been particularly successful with mutants which have requirements for histidine, tryptophan, and isoleucine. Some of the results obtained by Hartman and associates (411, 412) are summarized in Figure 8.13. Several interesting facts are evident. First, as in phages, many mutational sites are found to exist in a relatively short segment of the map. Second, the mutational sites with similar or identical phenotypic effects are clustered together. Thus all those sites which affect the production of the enzyme histidinol dehydrogenase are clustered together as shown in Figure 8.13. They are not distinguishable phenotypically, but they are separable by recombination. Demerec considers clusters of sites with identical or similar function of this type as loci or genes. Hence there is a gene for each of the specific functions indicated in Figure 8.13, but none of these genes can be delimited by recombination data alone. The criterion for establishing such an entity here is purely functional.

The third fact of great interest is that a close correlation between the location of a gene locus in the linear map and its function in a biosynthetic pathway is found in most of the pathways investigated in *Salmonella*. For example, Figure 8.13 shows that the genes controlling the enzymes for the biosynthesis of histidine are arranged in approximately the same order as the reactions catalyzed by the enzymes (412). The same seems to be true for the gene-enzyme relationship in tryptophan (212) and isoleucine-valine biosynthesis (326, 1048), among others. It has been found by using transduction techniques that galactose utili-

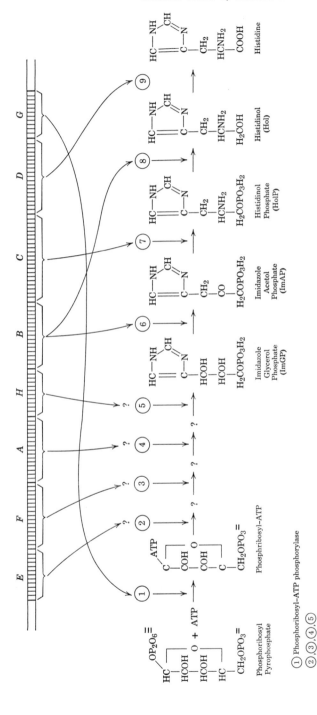

Figure 8.13. Relationship between the genetic organization of the histidine-controlling region in *Salmonella typhimurium* and the sequence of reactions controlled by the histidine genes.

zation (568) in *E. coli* and tryptophan biosynthesis in *Bacillus subtilis* (20) are similarly controlled by gene loci arranged in clustered sequences. The more that is learned about the linkage and fine structure of the genetic units in the coliform bacteria, the more evident it becomes that the "assembly line" relationship between genes and enzymatic reactions which they control is general. The phenomenon may be general throughout the bacteria. It is not general in eucaryotic organisms such as *Neurospora* and *Aspergillus*, but, even in these, isolated examples occur which cannot be assumed to be solely chance association.

Perhaps one of the most thorough analyses of a correlation between a single-enzyme activity and the structure of the genetic material which controls it has been made in the tryptophan synthetase system in *E. coli* by Yanofsky and his coworkers. This work has been described in Chapter 6, where it was pointed out that tryptophan synthetase can be resolved into two distinct components, the A and B proteins. Neither one of these is active alone in what is considered to be the physiologically significant reaction, the reaction of indoleglycerolphosphate with serine to form tryptophan, but together they are active. By transductional analysis it has been possible to show that the mutational sites affecting the B protein are clustered, as are those affecting the A protein, and, further, the B protein sites are adjacent to the A sites (1121). The close apparent physical association of two genetic elements, each concerned with the synthesis of part of an enzyme, may well be a general phenomenon. The pseudoallelic loci in *Drosophila* may well represent similar situations. Practically nothing is known about the genetic control of enzyme formation in *Drosophila* at present, but it would seem that investigation of it in the light of what is known about the genetics of *Drosophila* should prove quite rewarding.

An alternative method for studying recombination within a small genetic segment in bacteria is transformation. Here, presumably, a small piece of DNA is transferred from a donor to a recipient and incorporated by the recipient into its genetic structure. It is apparently a similar process to transduction, but a phage vector is not involved. The same reasoning can be applied, however, with respect to simultaneous incorporation of two markers being an indication of their close linkage. This method has been applied by Hotchkiss and Evans (466, 467) to an analysis of sulfonamide resistance in pneumococcus (*Diplococcus pneumoniae*). It was discovered that sulfonamide resistant strains could be classified into seven different categories of degree of sulfonamide resistance, ranging from a strain that required 1200 μg/ml

Table 33. Effects of Sulfanilamide on *Pneumococcus*
Strains *

Strain	Sulfanilamide Level for ½ Maximum Growth, μg/ml
Wild	5
a	20
d	80
b	15
ad	400
db	300
ab	70
adb	1200

* Hotchkiss and Evans (467).

of sulfonamide for half-normal growth to one which required only 15 μg/ml. (The growth of the wild-type sensitive strain is depressed one-half by the presence of 5 μg/ml.) (See Table 33.)

When the highly resistant strain is used as the donor in transformation to wild type, recipients are produced which may have either the high degree of resistance of the donor or one of the lesser degrees of resistance down to the least resistance, with one exception. These results indicate that the most resistant strain carries a complex locus which may be broken up and only portions of it incorporated into the recipient genome. This was verified by making a large number of transformations between all the different types and showing that the most resistant strain could be reconstituted from the less resistant strains by making the proper transformations. Figure 8.14 diagrams the types of transformations made using the symbols *a, d,* and *b* to indicate the subunits of the resistance locus. It is considered that the most resistant strain is *adb,* while the others have only two or one of these factors. The phenotypic effect increases with the addition of factors, but some combinations are more effective than others as shown in Table 33. Because the *ab* strain cannot be obtained by transformation of wild type with *adb* (Figure 8.14), it is assumed that *d* is located between *a* and *b*. Supporting this conclusion, in addition, is the finding that *ab* cannot be transferred simultaneously to *d* cells.

Hotchkiss and Evans (467) interpret their data to mean that the *adb*

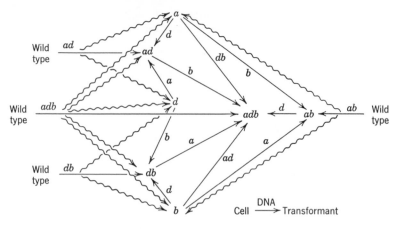

Figure 8.14.   Scheme showing relations between sulfonamide resistance genotypes regularly demonstrable by transformation.   Conversions which have not been demonstrated or indicated in the figure include the simultaneous transfer of *a* and *b* from donor *ab* to cell *d* or of the same two without *d* from *abd* to wild type.   Additional conversions which have been demonstrated but are not included in the figure are: linked displacement of *a* and *b* from strains *a* and *b* by donor *d*, linked displacement of *d* with introduction of *a* or *b*, and conversion of *d* to wild type.   From Hotchkiss and Evans (466).

locus controls the building of an enzyme which incorporates *p*-amino-benzoic acid into folic acid, which in turn becomes the functional coenzyme.   This enzyme is sensitive to sulfanilamide in the wild type, but a change in this locus at *a, d,* or *b* alters the enzyme and confers upon it less sensitivity.   The relationship visualized between the locus and the protein is diagrammed in Figure 8.15.

Recombination in Fungi.   The ascomycete fungi are ideal for the study of recombination, because the products of a single mating and meiotic event are analyzable by the simple procedure of separating the asco-spores of a single tetrad.   This makes it possible to demonstrate that reciprocal products are formed as a result of a single exchange, as well as to demonstrate that they are not formed on occasion—a fact which is just as significant.

As indicated in Chapter 3, recombination can be considered to be the result of breakage and reunion of two homologous chromatids as in crossing over, or, as the result of synthesis of new genetic material partly along one and partly along another of alternative "homologous" templates, as in "copy choice" (Figure 3.4).   While the copy choice hypothesis fits some of the phage and bacterial recombination data, the crossing over explanation fits best with what is known about recombination in fungi with certain important exceptions which will be discussed hereafter.

The analysis of tetrads, in *Neurospora* in particular, demonstrates rather conclusively that crossing over occurs in the four-strand stage. At what time it occurs is not certain, but it would obviously have to occur either in the first prophase of meiosis or earlier in the interphase nucleus. This point is of some significance because one would like to know if, in organisms like the fungi, replication and recombination occur simultaneously or are separate events. If, as seems likely, replication always occurs in interphase, then the demonstration that recombination does, too, would require a new evaluation of the current interpretation of the mechanism of crossing over which in turn would require a reevaluation of certain of the data obtained from recombination studies.

An analysis of linkage relationships and recombination in the three principal fungi used in genetic research, *Neurospora, Aspergillus,* and certain yeasts, reveals that there is a superficial resemblance to that found in *Drosophila.* Functional units are scattered among the various chromosomes with no apparent high degree of organization as found in *Salmonella.* But, on the other hand, when "allelic" mutants are crossed, wild-type segregants may occur.

When a cross is made between two functionally closely related mutants which are also closely linked, wild-type recombinations will

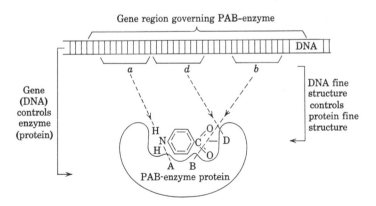

Figure 8.15. Postulated relationships between sulfanilamide locus and protein. From Hotchkiss and Evans (467).

occur if the mutants involve changes at different mutational sites. If
such a cross is made with closely linked outside markers present to act
as indicators of exchange in the segment of the chromosome under
scrutiny, a considerable amount of information can be obtained. As
an example, consider the results found by M. B. Mitchell (689, 690)
in a cross between two pyridoxineless mutants, *pdx* and *pdxp*. Phenotypically, these two mutants are separable because *pdxp* is a pH mutant,
while the other is not. If *pdxp* is crossed to *pdx* with the markers *co*
(colonial) and *pyr* (pyrimidine) present on either side of the pyridoxine
locus, about 0.2% recombinants are obtained which are neither *pdx*
nor *pdxp*, and these prototrophic strains are not always clearly the
result of a recombination between *pyr* and *co*. From a cross, pyr^+pdxp
$co \times pyr \ pdx \ co^+$, Mitchell (689) obtained pyridoxine independent
segregants which on further testing proved to be of four genotypes:
(1) + [++] *co*, (2) *pyr* [++] *co*, (3) + [++] +, and (4) *pyr*
[++] + (in which [++] designates $pdx^+ \ pdxp^+$). These are the
expected segregants, as shown in Figure 8.16, from a cross of this type.
However, if *pdxp* is to the right of *pdx*, then the *pyr* (++) *co* segregants should predominate, and if to the left the + (++) + segregants
predominate, for these would be the result of single crossovers between
pdx and *pdxp*. Since *pyr* and *co* are only 4 to 5 units apart and *pdx*
is only 0.5 to 1 units from *pyr*, segregants resulting from double crossovers (+ [++] *co* and *pyr* [++] +) should account for only about
1–4% of the $pxd^+ \ pxdp^+$ segregants. Triples, of course, should occur
even more infrequently at about 0.04%. Yet it was found that in a
sample of 22 $pdxp^+ \ pdx^+$ segregants, five were + (++) *co*, seven
pyr (++) *co*, seven + (++) +, and thirteen *pyr* (++) +. Clearly
these data are not in agreement with the standard interpretation of the
origin of recombinants by crossing over, and the *pdxp* and *pdx* loci cannot be ordered sequentially between *pyr* and *co* by means of them.

Complete asci from this same cross were obtained in which all spores
germinated. The genotypes of the segregants from four asci in which

Figure 8.16.   Expected segregants from a cross between *pdxp* and *pdx* with the
indicated outside markers.

Table 34. Asci Obtained from *pdxp* × *pdx* **Cross** *

Spore Pairs	Ascus			
	1	2	3	4
1	+ *pdxp co*	*pyr pdx* +	+ (++) +	+ *pdxp co*
2	+ (++) *co*	*pyr pdx* +	+ *pdxp co*	+ (++) *co*
3	*pyr pdxp* +	+ (++) *co*	*pyr pdx* +	*pyr pdx* +
4	*pyr pdx* +	+ *pdxp co*	*pyr* (++) *co*	*pyr pdx* +

* M. B. Mitchell (690).

pdx⁺ *pdxp*⁺ segregants appeared are given in Table 34. The occurrence of *pdxp* and *pdx* in the segregants and the absence of *pdxp pdx* double mutants was verified by making the appropriate crosses to the known *pdx* and *pdxp* parent strains. It is evident, then, that the wild-type segregants did not arise by standard crossing over, but by some other process such as mutation or by a copy-choice mechanism. Whatever the mechanism, the phenomenon has been given the name "gene conversion" (615). To reproduce the phenomenon it is not necessary to cross two closely linked mutants as just described. When *pdx* is crossed to *pdx*⁺, a small number of tetrads is produced (in one experiment, 2 out of 246) which contain 3:1 ratios of *pdx*⁺ to *pdx* (690, 691). The conversion occurs only when a *pdx* mutant is crossed to wild type or a closely linked pyridoxineless mutant, and never when the mutants are selfed.

Gene conversion as determined by tetrad analysis has been observed in other *Neurospora* mutants, in cysteineless mutants by Stadler (949), and in pantothenicless mutants by Case and Giles (134). It has also been established to occur in yeast by Lindegren (615), Roman (850), and Leupold (593).

One of the peculiarities of gene conversion is that it is generally associated with recombinations between the converted locus and the outside markers. These recombinations are different from conversion, however, since ordinarily they are reciprocal. Hence, as a general rule, gene conversion seems to be accompanied by a crossover. This amounts to two separate events (949), although attempts have been made to formulate single-event explanations (294).

Gene conversion does not necessarily always occur whenever closely linked and functionally related markers are involved in a cross. Results have been reported from *Neurospora* crosses in which wild types have been produced along with the expected reciprocals. Case and Giles (133) have shown by a tetrad analysis of two *pan-2* mutants, *B5* and *B3,* that several kinds of tetrads with wild-type spores may be obtained when *B5* is crossed to *B3:* some containing *B5*+ *B3*+ with the reciprocal *B5 B3* double mutants, and others containing no reciprocals, and different ratios of wild-type to mutant spores. Thus it is apparent that either standard crossing over or gene conversion may occur at a given locus.

Because of the low number of expected wild types and the low germination rate of ascospores obtained in many crosses in which closely linked markers are obtained, it is almost impossible to find complete tetrads with wild-type spores. In these cases, it is necessary to forego tetrad analysis and resort to a random ascospore-plating technique which enables one to detect only the wild-type recombinants (741). Although this method eliminates the possibility of finding reciprocals, if they occur, it does enable one to determine the incidence of crossovers between outside markers and to use this to determine the position of the mutational sites being investigated relative to one another and the outside markers as shown in Figure 8.17. By these means mutational sites at the *pan-2* (134) and an isoleucine-valine locus (1050, 526) in *Neurospora* have been mapped in somewhat the same fashion that the *r*II locus in bacteriophage T4 has been mapped. This procedure does not always work, however, for one does not always obtain

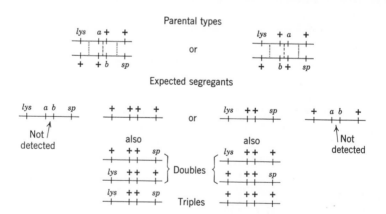

Figure 8.17. The types of recombinants expected from the crosses indicated, depending on the order of mutant genes.

Table 35. Examples of Some Aberrant Segregation in *Neurospora crassa*

Locus	Cross	Genotypes of Wild-Type Progeny				Reference
		Noncrossover		Crossover		
		P_1	P_2	R_1	R_2	
ad-3A	$\dfrac{\textit{hist-2 ad-3A(2) nic-2}}{\times \\ + \textit{ad-3A(4)} +}$	34	35	17	18	216
pan-2	$\dfrac{\textit{ylo pan-2(10)} +}{\times \\ + \textit{pan-2(25) tryp}}$	22	20	33	26	133
am	$\dfrac{+ \textit{am}^2 \textit{inos}}{\times \\ \textit{sp am}^3 +}$	7	9	0	8	768
hist	$\dfrac{+ \textit{hist(2) inos}}{\times \\ \textit{iv hist(1)} +}$	11	17	8	13	294
cys	$\dfrac{\textit{un cys-c} +}{\times \\ + \textit{cys-t ylo}}$	68	26	18	25	949

data which show a preponderance of single exchanges accompanied by the assumed doubles and triples in lower frequency. One example of these types of results has already been given in connection with the *pdxp* and *pdx* cross.

This is a puzzling phenomenon which may or may not be related to gene conversion. It occurs so frequently when crosses are made involving heteroalleles that it may be considered a definite limitation on the method of recombination analysis of genetic structure. It has been reported for practically all of the loci in *Neurospora crassa* in which a number of heteroalleles have been tested for their production of wild-type recombinants in crosses. Samples of some of the data which have been obtained from a number of crosses involving different loci in *Neurospora crassa* are shown in Table 35. Similar results have been obtained by Pritchard (808, 809) from crosses involving the *ad-8* heteroalleles of *Aspergillus nidulans*.

In Table 35 the wild-type progeny which carry the parental type outside markers are indicated as P_1 and P_2. Those which show a recombination between the outside markers are indicated as R_1 and R_2 to indicate crossovers for regions 1 and 2 respectively. In all examples the outside markers are close enough together to expect lower frequencies of recombination between the markers than are indicated. If the data are to be interpreted in terms of standard crossing over, they lead to the conclusion that high negative interference such as has been described for phages previously (p. 359) is operating. Furthermore, if gene conversion is used as the working hypothesis, then it must be assumed that it is generally accompanied by a recombination between the outside markers. No data exist at present to make a decision on this matter. It can only be noted that considerable speculation has been made centering around the concept of limited effective pairing of chromosomes at the time of exchange. This hypothesis, mainly the result of Pritchard's (807, 808, 809) attempt to interpret similar data from *Aspergillus* crosses, states that chromosomes synapse at only limited segments along their length and that exchanges or "switches" occur only at these regions of "effective pairing." The closer two markers are together the more likely they are to be included in a single region of effective pairing. If it is assumed that many switches may occur within a region of effective pairing, then high negative interference should be expected. An even number of switches, one of which occurs between two heteroalleles, should result in parental-type marker distribution for the outside markers, whereas an odd number should result in an outside "crossover" recombinant.

High negative interference has not been reported to occur in any of the many crosses which have been made between the various pseudoalleles studied in *Drosophila*. It may, however, occur in the fourth, or dot, chromosome of *melanogaster,* according to Sturtevant (987).

Whatever the mechanism or mechanisms of recombination operating in fungi, recombination analysis combined with functional analysis has led to several interesting findings. The tryptophan locus (*td*) in *Neurospora crassa* which controls the synthesis of the enzyme-tryptophan synthetase has already been discussed in Chapter 6. A genetic analysis has been made of this locus by Bonner and coworkers (85), the results of which indicate clearly that it is complex, consisting of many mutational sites (see Figure 6.9). The significance of these findings is not only that a number of different sites occur at this locus, but that different alterations are produced depending on which sites mutate. Further, the sites are not distributed haphazardly; specific types of alterations appear to be produced by changes at sites close together. Thus recombination analysis demonstrates that the genetic material, in this example, has a highly organized structure dedicated to the production of a specific protein.

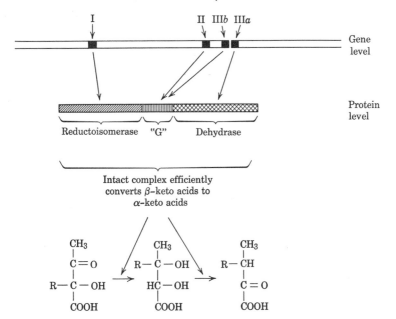

Figure 8.18. Hypothetical relationship between isoleucine-valine genes in *Neurospora crassa* and the conversion of the α-keto acid precursors to α-keto acids. See text.

Several other investigations give some indication of the degree to which genetic material is organized in *Neurospora*. The enzymes involved in two steps in the biogenesis of isoleucine and valine are controlled by two closely linked loci (1050), as shown in Figure 8.18. A locus controlling the production of the enzyme, reductoisomerase, is closely linked to a cluster of sites which appear to control dihydroxy-acid dehydrase activity. Since the reductoisomerase acts in the step just preceding the action of the dehydrase, this relationship is similar to that found in *Salmonella* on a somewhat more extensive scale. Located between these two mutant groups, designated as I and IIIa in Figure 8.18, are a number of sites designated as II which produce mutant strains that are deficient in neither the reductoisomerase nor dehydrase yet require isoleucine and valine for growth. It has been postulated that these sites are located in the area which controls the production of a protein which holds the two active enzymes together to form a single complex functioning *in vivo*. Also included in the same area as the IIIa mutants are the IIIb mutants, one of which is similar to the II mutants in its properties, and another which resembles the I mutants because it shows a deficiency in reductoisomerase rather than dehydrase. Such evidence as this would indicate that the whole segment extending from I to III, comprising about four map units, should be considered as a functional whole. Also found to be related in function and position are the genes controlling certain steps in the biosynthesis of the aromatic amino acids (373) in *Neurospora*.

The gross organization pattern of the genetic material in the fungi is by no means as obvious as it is in such bacteria as *Salmonella* where there seems to be a selective advantage to keeping genes arranged in the order in which they presumably act. A remnant of this possibly primitive organization may still exist in the fungi in such examples as part of the isoleucine-valine pathway and the aromatic amino-acid pathway, but it is by no means as general as it seems to be in the bacteria. This does not mean that the genetic material of fungi is less highly organized than that found in the bacteria, however. The organization may be simply more complex and more difficult to understand.

Complementation Analysis

The study of the fine structure of genetic material by recombination analysis has proved effective in showing that the genetic material is complex, and a structural gene is difficult to define. But, like all methods, recombination analysis has its limitations, among them being the occurrence of gene conversion which makes it impossible to determine linear order of mutational sites. Fortunately, another method of analysis exists which has proved to be a correspondingly powerful tool, although it, too, has certain limitations, and the results obtained by it are not completely compatible with those obtained by recombination analysis.

One of the expected results of two allelic mutant genes in heterozygous combination is that the heterozygote will be different from either homozygote. It has been recognized for many years that the heterozygote will generally be intermediate in phenotype, but examples are known in which the heterozygote is more mutant in phenotype than either homozygote, or at the other extreme the heterozygote is wild type or at least approximates it. The condition in which heterozygotes are inclined to be phenotypically like the wild types is called *complementation,* for the alleles obviously complement one another in the production of a wild phenotype.

If two genes are far apart and quite different in function, it is expected, of course, that their recessive mutant alleles in heterozygous combination will give a wild phenotype, whether in *cis*, $\frac{a\ b}{+\ +}$, or *trans*, $\frac{a\ +}{+\ b}$, arrangement. If, however, two mutant genetic elements are apparently related in function as evidenced by their homology, then they should be expected to always give the mutant condition in the *trans* condition and a wild phenotype in the *cis*. That they do in many of the examples of *Drosophila* pseudoalleles discussed previously is in agreement with this expectation. Two deficient elements, each on a different chromosome, should result in a mutant phenotype, whereas, if they are on the same chromosome, the normal homologue should result in a wild-type product. The fact that this does not always happen even in *Drosophila* is of extreme interest.

Complementation in Microorganisms. Complementation has been extensively studied in microorganisms, particularly in fungi and to a more limited extent in bacteria and phages. Benzer (75) has succeeded in demonstrating that complementation occurs between certain *r*II mutants of phage T4. This is made possible by the fact that the *E. coli* strain K-12S will not support the multiplication of *r*II mutants, although the phage particles may infect and kill the host cells. If two *r*II mutants with different mutational sites, as determined by recombination, are used to infect K-12S simultaneously, this is comparable to performing a *trans* test, since each mutant site is on a different "chromosome." Under these conditions the particles may or may not multiply within the host cell. If they do, they are said to complement. A *cis* test is performed by infecting the cell with a double mutant and a wild type. Both types of particles multiply under these conditions.

Benzer found that by performing the *cis-trans* test with a large number of different *r*II mutants he could classify them into two groups which he designated as *cistrons* A and B. Mutants within each cistron do not complement, but do with members of the other cistron. Recombination analysis shows that the members of group A are in one part of the linkage map and those in B in an adjacent region. Hence there is a degree of correlation between the recombination and the complementation data. The cistron, being a region within which complementation does not occur, is considered to be a functional unit by Benzer.

Edgar and his associates (245) have been able by the complementation test in T4 to determine that at least 40 cistrons or genes exist. Each cistron appears to have a specific function and is closely linked to neighbors with similar functions as described earlier.

The use of the term cistron has become so widespread that further comment is required. Currently, many workers, particularly those in microbial genetics, use the term synonymously with gene. This usage is improper, if one insists on the strict adherence to the original definition, which is that the cistron is a complementation unit. Actually, few workers insist on this and use the term without discrimination, the result being that gene and cistron are rapidly becoming different words for the same thing. Obviously, this is to be discouraged, and for this reason the term cistron is used infrequently in this book. When it is used, it is applied to complementation and nothing else.

A phenomenon termed *abortive transduction* has been described in *Salmonella* which seems to be a manifestation of complementation (213). If a mutant strain of *Salmonella* receives by transduction a chromosome segment from a donor which is wild type, or a nonallelic mutant, the recipient may receive and incorporate into its genome a wild-type segment in place of its deficient segment, and hence become genotypically and phenotypically wild type. This is referred to as *complete transduction,* and the cell so produced gives rise to only wild-type progeny which produce large colonies on minimal medium. Occasionally, however, in transduction experiments minute or microcolonies appear in addition to the expected larger colonies. These contain cells which apparently arise from progenitors which receive a fragment carrying the wild-type genotype for the region in which they are mutant, but the fragments are not incorporated and do not duplicate. The cell carrying a fragment is capable of maintaining itself, however, and dividing, because the incorporated fragment complements with the nonidentical host genome. When it divides, it produces a daughter without the fragment and one with it. The minute colony is produced by successive divisions of this type but is composed of deficient daughters capable of only slight growth on minimal medium and a single cell carrying the unincorporated fragment. Ordinarily, in *Salmonella,* abortive transduction, or complementation, is evident only when transduction occurs between cells which carry nonallelic mutations. Mutants which are mutant at different sites or the same sites within the same gene or locus do not generally demonstrate it (411). Significant exceptions to this rule are known, however, in *Salmonella.* For example, the *D* gene involved in histidine biosynthesis which controls the production of histidinol dehydrogenase has at least three cistrons (18). The *B* gene apparently forms a single protein which catalyzes two reactions but has at least four cistrons (1440). Examples such as these make it doubtful that the term cistron should have continued wide usage.

By far the most extensive studies on complementation have been carried out with fungi. Yeast (850), *Aspergillus* (798), and *Neurospora* have been the organisms most extensively involved, and the largest number of different loci have been investigated in *Neurospora crassa*. Mutants with requirements for tryptophan (144), adenine (1103), arginine (147), methionine (728), isoleucine and valine (1050), pantothenic acid (133), amino nitrogen (282), pyrimidine (693, 1105), and histidine (1057) have been found to complement, even though the mutations in each case are closely linked and would ordinarily be considered to be allelic, since they seem to produce the same types of blocks. Perhaps the most instructive are those examples in which neither mutant produces a detectible enzyme, but the heterocaryon does. Since no evidence exists in *Neurospora* for nuclear fusion, and a consequent possible mitotic recombination with the production of normal chromosomes, one is forced to the conclusion that some type of interaction must occur in the cytoplasm which results in the formation of active enzyme. The production of active enzyme in *Neurospora* heterocaryons formed from strains carrying allelic mutations has been found for mutants deficient in glutamic dehydrogenase (282), adenylosuccinase (323), and tryptophan synthetase (543). In none of these, however, has the enzyme activity been found to equal the wild-type activity. Generally, the activity of the heterocaryon enzyme does not exceed 25 to 30% of the wild-type activity.

The *ad-4* mutants of *Neurospora crassa* are deficient in the enzyme adenylosuccinase which catalyzes the reaction-splitting fumarate from adenosine-monophosphate succinate (Figure 7.11, p. 315). More than 100 of the *ad-4* mutants of independent origin have been tested two at a time for complementation in heterocaryons on minimal medium. About 50% of those tested complemented with at least one other allelic mutant (1103). An example of the type of complementation pattern found is diagrammed in Figure 8.19. In this diagram overlapping lines indicate that no complementation has been observed, and when lines do not overlap complementation occurs. By this means complementation data may be reduced to a simple diagram known as a complementation map. It will be noted that some of the mutants complement with a considerable number of others, while some complement with only one or two. If each line is visualized as a segment of some linear array (such as a polynucleotide or polypeptide would be), then it can be postulated that the linear dimensions of the line indicates the segment in which a "mutant" condition or lesion has been formed. The segments represented by the lines may be fitted into a linear array. For example, A,

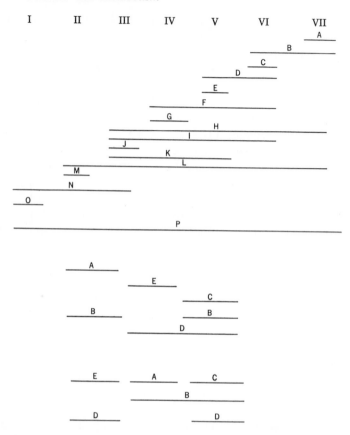

Figure 8.19. Complementation pattern for some of the *ad-4* mutants in *Neurospora crassa*. See text.

C, and E may be considered to exist in this order rather than A, E, C or E, A, C because of the overlap of B with both A and C, and the overlap of D with E, C, and B, but not A. If the order were A, E, C, then it would require an interrupted line to represent B, or if E, A, C, an interrupted line would be required to represent D. The order A, B, C, D, E permits us to represent each with an uninterrupted line. By continuing this reasoning, the linear array for all of the mutants has been deduced as shown in Figure 8.19. From this it can be hypothesized that there is a segment having to do with adenylosuccinase activity that may have lesions at a number of different sites which are in a linear array. Occasionally, a mutant is found that does not fit a linear mutation map as a single line, but must be represented by two or more lines. There has been a general tendency to ignore such mutants, but it is evident that they should be considered of some significance in any attempts to interpret the meaning of complementation.

Following the nomenclature of Benzer (75), one may consider each of the overlap areas in Figure 8.19 a cistron, and thus seven cistrons are indicated by the Roman numerals. Of interest, in addition, is the fact that many of the complementing mutants overlap several cistrons while 50% or more overlap all, since they do not complement. The question is raised, then, are these short-to-long deletions, or are they the result of multiple-point mutations? This question has been partially answered by inducing reverse mutations among these types of mutants back to the wild-type phenotype. The revertants have been demonstrated to produce anywhere from 3 to 100% of the wild-type activity of adenylosuccinase and to grow on minimal medium without adenine supplementation (1104). This fact alone eliminates the explanation that the overlapping mutants are all the result of deletions. The possibility that they are the result of multiple mutations at the *ad-4* locus is not entirely eliminated but seems unlikely.

Enzyme Activity in Complementing Heterocaryons. All mutants at the *ad-4* locus that complement do not by any means form complementing heterocaryons equally as readily, nor do the heterocaryons always grow as well as the wild-type strain. By comparing the adenylosuccinase activities, and the time required for heterocaryon formation of a number of different mutant combinations, it was found that the most widely separated mutants on the complementation map form heterocaryons the most readily and have the highest adenylosuccinase activities (1103). Thus mutants with lesions indicated in adjacent regions or cistrons on the map may take up to ten or more days to form heterocaryons and produce less than 10% of wild-type adenylosuccinase activity, whereas those widely separated may show complementation uniformly within two days and produce 20–25% of adenylosuccinase activity.

The adenylosuccinase produced by the *ad-4* heterocaryons which grow on minimal medium has been investigated by Partridge (766) to determine whether it is different from that produced by the wild type. He found that the enzyme obtained in a partially purified state from different heterocaryons showed quite different properties in two out of four examined. The differences detected were in thermostability, metal inhibition (Zn^{++} and Cu^{++}), urea inhibition, and pH at half maximum velocity. Studies with crude extracts from other heterocaryons resulted in essentially the same findings: some produce an enzyme similar in physical properties to the wild type, others do not. The glutamic dehydrogenase of two heterocaryons of mutants at the *am* locus (amination deficient) has been found to be different, in somewhat the same fashion, from the wild-type enzyme (279). The mutants am^1, am^2, and am^3 form complementing heterocaryons in the combinations $am^1 + am^2$ and $am^1 + am^3$. The homocaryotic mutants produce no detectible glutamic dehydrogenase, but $am^1 + am^2$ has about 10% and $am^1 + am^3$ about 20 to 25% the activity of the wild type. Fincham (279) purified the heterocaryon enzymes twenty- to thirtyfold and found that they both had different properties from the wild type and from one another. The $am^1 + am^2$ enzyme is thermally activated when the temperature is raised between 20° and 35°, and is very unstable at 60°C. The $am^1 + am^3$ enzyme has a very low affinity for glutamate and is also more thermolabile than wild type. These results are extremely interesting and quite pertinent to an analysis of complementation for they indicate that the enzyme synthesized by the heterocaryon may be quite different from that produced by the wild type even though still capable of acting as a catalyst in the same reaction.

Actually, it is not necessary to form heterocaryons between complementing *ad-4* mutants to produce active adenylosuccinase. Woodward (1102) has demonstrated that merely mixing fresh homogenates from different mutants, which are capable of complementing, results in the appearance of the activity. In order to obtain this restoration the homogenates from the two different strains must be mixed immediately after disruption of the mycelium. Mixtures of homogenates from strains that do not complement *in vivo* do not produce active enzyme, whereas all mixtures of homogenates from strains that do complement show some enzyme activity. A correlation exists between the enzyme activity recovered from a homogenate mixture and the distance between the mutant contributors on the complementation map. Hence *in vitro* and *in vivo* results correlate very well.

Complementation and Genetic Maps.   Although it has been possible to demonstrate that prototrophs arise at a low rate from crosses between some of the *ad-4* mutants, it has not been possible to obtain sufficient data to order them. However, certain pantothenicless mutants (the *pan-2* locus) which have been investigated by Case and Giles (134) have yielded more easily to genetic analysis. A detailed complementation map has been made and compared to a genetic map for the same mutants. A degree of correspondence between the two maps was found.

In general, the mutants that complement most readily are spaced farthest apart on the genetic map, whereas those close together on the genetic map either do not complement or complement less readily. The degree of correspondence is not perfect, however, since some exceptions were found. A similar partial agreement between a complementation pattern and the genetic map has been found for a group of isoleucine-valine mutants by Wagner, Kiritani, and Bergquist (1050). A group of adenineless mutants at the *ad-3* locus are also interesting in this regard. This locus controls a different enzyme than the *ad-4* locus already discussed and is located on a different chromosome. Genetic analysis shows that it consists of at least two subgroups, *ad-3A* and *ad-3B*. Crosses of *3A* to *3B* mutants produce some recombinants which are wild type. In all the tetrads that have been analyzed the reciprocal double mutants also occur (322). Thus the evidence indicates that standard crossing over takes place. Crosses between the mutants within the *3A* group also produce wild-type recombinants, but at a lower frequency, and exhibit a complete lack of asymmetry in distribution of the segregant, with respect to the outside markers. Thus it is impossible to order the mutational sites within the *3A* region. The complementation pattern correlated with this may be of significance, because while *3A* mutants complement readily with *3B* mutants, complementation has not been observed between sites within the groups (215, 216). Hence, here complementing mutants show standard crossing over, but non-complementing mutants do not.

Complementation between two mutants is ordinarily interpreted to mean that they must be mutant at different sites. In this event, crosses between complementing mutants should always be expected to produce recombinants in the form of wild types. This, however, is not always the case, nor do mutants which recombine always complement (133, 216, 1105). Actually, negative results from complementation tests cannot always be considered conclusive, because it is known that incompatibility factors can inhibit heterocaryon formation in *Neurospora* in strains that would complement were these factors absent (312). However, heterocaryon formation can be forced by the use of other markers (216), and when complementation does not occur under these circumstances it is assumed that noncomplementation is demonstrated.

Pseudowild Types in Neurospora. When mutants of *Neurospora* which have mutational sites within the same locus are crossed, it is frequently observed that two types of spores are produced which result in cultures with a wild-type phenotype. One type is a true wild type as can be demonstrated by crossing to a standard wild type, for only wild-type progeny are obtained. The other type is not, because, on being crossed to the standard wild type, both wild-type and mutant spores with the original mutant phenotype are obtained. These latter are called pseudowild types (696). They are assumed to be the result not of recombination but of complementation. It is hypothesized that at the times of meiosis disomic nuclei are produced, some of which carry chromosomes with both mutations (787). If these are capable of complementation, then a strain will arise which will have wild-type characteristics. Presumably, the disomic breaks down in subsequent mitoses, and a simple heterocaryon is formed. Obviously, when it is crossed it can contribute only mutant genes. This interpretation is supported by the observation that only mutant strains that complement form pseudowild types (134). Further evidence is needed before this can be considered a valid generalization, however.

Complementation in Higher Organisms. When two pseudoallelic loci are tested for phenotypic effect in the *trans* position in *Drosophila,* it may be considered a complementation test analogous to heterocaryon tests in the fungi. It is of interest, therefore, to consider whether any relationship exists between complementation and the relative positions of different pseudoalleles in *Drosophila.* As already pointed out, the white alleles, except for white spotted (*w-sp*), do not complement, but some of the other pseudoallelic series do. The recessive bithorax pseudoalleles occur in the order *bx, bxd, pbx,* and as will be seen from Table 30, *bx* complements with *bxd* and *pbx,* but *pbx* does not complement with *bxd.* A pattern may be detected here insofar as *bx* is separated from *bxd* and *pbx* by about 0.03 units, but *bxd* is immediately adjacent to *pbx.* Hence it may be postulated that widely separated loci complement, whereas those close together show the *cis-trans* position effect. Unfortunately, the matter is not as simple as this. In another pseudoallelic series, the dumpy group investigated by Carlson (131), the complementation pattern appears to bear no relationship to the arrangement of the different loci. Some of the pertinent data relative to recombination and complementation at the dumpy locus are shown in Figure 8.20.

The members of the dumpy series of alleles are recessive and have three major phenotypic effects: (1) a truncation of the wings (*o*); (2) thoracic modifications with disturbed bristle patterns (*v*); and (3) a homozygous lethal effect (*l*). When combined in heterozygotes, they may show either a lethal effect, the *o* or *v* effect, a combination *ov*

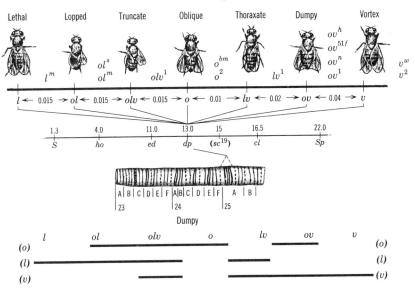

Figure 8.20. The effects of the various dumpy pseudoalleles in *Drosophila melanogaster*. The complementation pattern is shown at the bottom of the figure. From Carlson (130).

effect, or they may complement and produce flies with normal phenotype. Seven subloci exist which are designated in the Figure 8.20 as *l, ol, olv, o, lv, ov,* and *v*, each with its own specific phenotypic effect. Those designated as *l, ol, olv,* or *lv* are lethal homozygous but in compounds with *o, ov,* or *v* express the particular phenotype which they hold in common. For example, *olv/o* has an *o* phenotype, *olv/ov,* an *ov* phenotype, etc., or the phenotype is wild. Six different combinations complement as shown in Figure 8.20, *l/o, l/ov, l/v, ol/v, o/lv,* and *o/v,* but a linear complementation map cannot be constructed, and no correlation seems to exist between the distance apart on the map and the ability to complement.

Complementation between heteroalleles or subloci is by no means unknown in the higher plants such as cotton (965) or in mammals such as the mouse (237, 907). The work since 1955 in this field has primarily been with microorganisms, but it should be recognized that the phenomenon had been known for many years prior to that year in diploid organisms. Whenever one examines the vast reservoir of genetic material in natural populations such as in *Drosophila* (935, 361) and in the house mouse (72), one finds many isoalleles which when tested in different combinations reveal that many genes exist which may be more beneficial to the population in the heterozygous than in the homozygous condition (222). Complementation should not be looked upon, therefore, as a curiosity but as an important aspect of gene expression.

Molecular Explanations of Complementation. In conclusion, it may be stated that complementation studies in a variety of organisms have proved to be quite successful in uncovering a number of interesting facts such as: (1) there is a rough correlation between the genetic map and the complementation map in a number of them, but little correlation in others; and (2) the enzyme formed by a heterocaryon need not be identical to the wild-type enzyme (indeed, it may always be different). As is usual, more mysteries are created than are solved. Chief among these is the question of how an active protein is formed in a heterocaryon without recombination at the nuclear level. Many hypotheses have been advanced to explain how this may occur in the cytoplasm after each gene produces its primary product, which we may assume to be messenger RNA. These all break down roughly to two primary explanations which are not necessarily antithetical or exclusive.

First, it has been postulated that recombination at the RNA level or the protein level may occur, with the result that from two deficient molecules a molecule is formed that has at least some of the properties of the wild type (1104, 1047). This is analogous to obtaining a wild type from a cross between two mutants.

The second hypothesis is based on the observation that enzymes may ordinarily only be active as polymers of basic monomeric structures. For example, glutamic dehydrogenase is active in the tetrameric state, and not as a monomer (1016); functional hemoglobin is made up of two of each of two types of molecules (p. 615). If this is true for functional protein in general, it is possible that complementation is the result of a functional polymer being formed from two or more different defective monomers, each compensating for the other's defect (279, 101).

Adequate experimental evidence is not available to prove either of these hypotheses. Actually, it is highly probable that a reasonable explanation of complementation will not be forthcoming until a more complete understanding of the mechanism of protein synthesis in the cytoplasm is at hand.

A second problem which is raised by the complementation data is its significance with regard to our attempts to understand genetic units.

Obviously, complementation at a single locus has some significance with respect to gene structure. If all mutational sites within a single locus complemented and also recombined, the analysis and correlation might be quite simple. But this is not the case. Furthermore, in several complementation tests of a large number of mutations affecting the same locus, at least 50% or more of the mutants have not been found to complement with any of the other mutants. Yet these recombine readily with the others and act in crossing experiments like single-site mutations. In addition, some complementing mutants cannot be fitted into a linear-complementation map as a single site; they require two separate linear arrays to indicate their pattern. Such results as these indicate that the problem is not a simple one. Complementation data may be telling us something about the structure of the protein under control of the locus, but much more must be known before we can decipher the message.

Hybrid Substances

Before the discovery of active enzymes in the heterocaryons of *Neurospora* made up of parental strains which could not produce them in active form, it was known that the heterozygotes of diploids occasionally produced proteins which were different from those produced by the homozygous parents. These are known as hybrid substances, and they were first discovered in a cross between two species of dove, *Columba livea* and *Columba guinea*. *Columba livea* produces a cellular antigen, C, while *C. guinea* produces an antigen, C′, distinguishable from C by antibody tests. The hybrid between the two species produces an antigen CC′ which is different from either C or C′ as determined by tests with antisera (678). A similar situation has been found in rabbits except that an interspecific hybrid is not involved. The rabbit has a series of allelic blood-group factor genes designated as Hg^A, Hg^D and Hg^F. In general, heterozygotes for any combination save one have both antigens as expected. However, Hg^A/Hg^D heterozygotes, in addition to producing antigens A and D, produce a hybrid substance, I, different serologically from either parent antigen (175). Certain lozenge mutants of *Drosophila melanogaster* have been investigated for antigenic differences, and it has been found that heterozygotes carrying both the lz^{46} and lz^{BS} alleles produce a different antigen than either parent produces. This occurs whether the mutant alleles are in the *cis* or *trans* relationship (165).

After being first recognized solely at the serological level, hybrid substances were quickly found in protein fractions from various sources by other means. By the use of starch-gel electrophoresis, it has been possible to distinguish hybrid substances in both man and corn. The haptoglobins, hemoglobin-binding proteins in the blood serum of man, have been found to occur as two basic types, 1-1 and 2-2, presumably inherited through a pair of alleles Hp^1 and Hp^2. (See Chapter 14, p. 602). In some individuals, however, who are assumed to be heterozygous (Hp^1/Hp^2), haptoglobins are formed which do not migrate electrophoretically at the same rate as in the assumed homozygotes (Figure 14.6) (920). This occurrence of a hybrid substance(s) in the haptoglobins has been verified by ultracentrifugation. Homozygotes show either a 6 S or 10 S sedimentation rate in the haptoglobin fraction, but in the heterozygotes the major component has a sedimentation rate of about 8 S (64).

Corn esterases which are active on an α-naphthylacetate substrate have been investigated by Schwartz (1210). Starch-gel electrophoresis of the endosperm from an inbred population of corn showed that three general types of esterase occurred, designated as fast (F), normal (N), and slow (S). Genetic analysis indicates that each type is determined by one of a set of three alleles, E^F, E^N, and E^S. Homozygotes produce their respective expected esterases, but the heterozygotes E^F/E^N, E^F/E^S, and E^N/E^S produce esterases which move electrophoretically at rates intermediate to those produced by the homozygotes.

Hybrid substances such as those just discussed which appear only in heterozygotes may well be related to the substances with enzymatic activity that appear as the result of heterocaryon formation between closely linked mutations affecting the same enzyme. The two may be, in fact, the result of basically the same mechanism. Hence anything learned by studying complementation in heterocaryons may be of value in understanding hybrid-substance formation in heterozygotes in diploids and vice versa. For example, one striking observation by Smithies (920) on the human haptoglobins is the change in pattern obtained from the haptoglobin of the heterozygote after treatment with mercaptoethanol and urea before electrophoresis. The pattern obtained is similar to the homozygotes insofar as both homozygote patterns are present. This may well mean that the haptoglobins are polymers and that the formation of the hybrid substance is the result of the polymerization of two kinds of monomers. When the polymers are broken down by the treatment with urea (which presumably breaks hydrogen bonds), parental-type molecules are reconstituted. This interpretation agrees, it will be noted, with one of the two hypotheses mentioned (p. 384) to explain complementation in heterocaryons. On the other hand, Schwartz (888) has found evidence that the hybrid esterases formed in the corn plant are synthesized *as* hybrids and do not form from the dimerization of preformed monomers. Here again, it is evident that a single explanation is not going to fit all occurrences of hybrid molecules formed in heterozygotes or heterocaryons, but that further insight into the mechanism of protein synthesis is needed.

General Considerations and Conclusions

The foregoing discussion serves to indicate that the present status of what constitutes a genetic unit is in a state of confusion. It is quite clear that at present a gene may be defined as that segment of DNA which controls the linear sequence of bases in a specific messenger RNA. The RNA in turn determines a specific polypeptide which becomes a protein. Any change in the sequence of the bases within the DNA may be reflected in a changed protein. A change in base sequence is a mutation and is recognized as a "mutational site" which may be distinguished from other different mutational sites by recombination analysis. That is, two different sites should produce a wild type by recombination. Sites which do not do so may be identical (or homo-allelic), overlapping, or deletions rather than simple alteration of sequence without loss of base pairs. In any event, a mutation within the segment of DNA controlling the amino-acid sequence of a particular polypeptide should be reflected in that polypeptide and no others. Thus it should be possible to determine the *limits of a gene* by plotting mutational sites. This method has, of course, been employed since the beginnings of genetics. It has attained a high degree of sophistication in the fine-structure analysis of phage, bacteria, and fungi and has resulted in the realization that there is probably a colinearity between the sequence of bases in DNA and amino-acid sequences in polypeptides.

Essentially consistent results have been obtained by this method, and there can be no doubt of its importance as an analytical tool in defining genetic units by their *structure* in terms of DNA, provided the polypeptides involved are known. It also serves to define the gene as a *functional* unit in its determination of a specific polypeptide. But function must also be considered from the aspect of the functional units of protein which act as enzymes, carriers of oxygen, structural elements, and so forth. The one gene-one function hypothesis had this point of view implicit in it. When we look at the problem from this vantage, it is immediately evident that more than one gene may be involved in the synthesis of a functional protein. Thus genes *A* and *B* of *E. coli* each produces a different protein. These, when combined, give the active enzyme which is tryptophan synthetase. The genes do not even have to be closely linked. For example, the α and β genes determining the α and β proteins which make up hemoglobin are apparently widely separated in the human genome (see Chapter 14). To the qualification that more than one gene may be involved in the formation of a single enzyme must be added the further one that enzymes may operate *in vivo* only in complexes such as in mitochondria or smaller units. Thus we must bear in mind a series of units, the complex, the enzyme, the polypeptide, and recognize that the current aphorism should be "one gene-one polypeptide unit," rather than one gene-one function which might lead to misunderstanding.

As to the question of what constitutes allelism, this has a relatively simple answer in light of the foregoing. Allelic genes are those concerned with the formation of the same polypeptide. Heteroalleles should produce polypeptides with partially different amino-acid sequences, whereas homoalleles may be considered as those producing identical amino-acid sequences. One can only be certain that two genes are homoallelic under this definition after making a complete sequence analysis. Absence of recombination does not necessarily mean that two alleles are identical. Indeed, it will be recalled that the "true" alleles in the lozenge series of *Drosophila melanogaster* do not recombine, but they produce different phenotypes which probably mean that they produce different polypeptides. Of course, it should be recognized that a fine-structure analysis is not feasible in *Drosophila* to the same degree that it is feasible in microorganisms, and it is quite possible that the different "true" alleles do in fact represent different mutational sites so close together that they cannot be resolved by observing a relatively small number of meiotic products.

Other difficulties are raised when one considers pseudoalleles such as found in *Drosophila*. What is the relationship, for example, of pseudoalleles to the genes in microorganisms? At present there exists too wide a gap in our knowledge to even begin to answer this question. It is quite possible that a rather fundamental difference exists in the genetic organization of higher plants and animals as compared to bacteria and phages. Fungi may well represent an intermediate condition between these two groups, for they show certain characteristics found in each group. For example, they not only demonstrate orthodox crossing over, but also gene conversion and high negative interference, phenomena which may be related to events which occur in bacteria and phages. These considerations raise the possibility that genetic units may be different in different groups of organisms. This is not meant to imply that all organisms do not maintain the ultimate determination of specificity of protein in nucleic acids, but rather that the organization of the nucleic acids together with their attendant protein in the chromosomes may be different in different groups. It must be recognized that our present knowledge of the chemical organization of chromosomes is quite deficient.

The smallest conceivable subunit of a gene is the nucleotide pair. Theoretically, a mutation may consist of a change in a single nucleotide pair out of many thousands of pairs in a single gene. Hence a mutational unit, or *muton* (75), may have very small dimensions. It is also possible that exchanges resulting in recombination may occur between any two nucleotide pairs, and the recombinational unit, or *recon* (75), may also consist of a single nucleotide pair. However, the alternative possibility exists that a recon may consist of a small to large number of contiguous nucleotide pairs within which exchange is prevented by protein associated with the DNA, for example. This may be the explanation for the apparent lack of exchanges *within* the pseudoalleles of *Drosophila*. These considerations must be borne in mind when attempts are made to compare genetic units, as defined by recombination, in different organisms (799).

As has already been pointed out, considerable difficulty is met with when attempts are made to correlate the data from complementation studies with recombination data. These difficulties are highly significant and must not be ignored. On the other hand, it should also be recognized that complementation occurs readily between different genes, and not so readily, but generally spotted, or not at all, between allelic genes. This fact is also significant. It is easy to explain why complementation should occur between nonallelic genes, but why does it sometimes occur between allelic genes and sometimes not? The apparent conflicts in the complementation data and in the relation of complementation maps to recombination maps will probably not be resolved until we understand more about the formation of hybrid proteins in heterozygotes, and functional enzymes in heterocaryons. It is very possible that the two phenomena are closely related. It is doubtful that further complementation studies per se will clear up the conflicts. What is almost certainly needed is a better understanding of what takes place in the cytoplasm during the synthesis of polypeptides, enzymes, and the various types of complexes in the presence of the ribosomes and messenger RNA. In addition, some of the earlier steps leading up to the production of RNA are not by any means understood, and it is quite possible that events which occur in these stages in heterocaryons and heterozygotes are reflected in unexpected types of protein synthesis.

As an aid to an understanding of some possible generalizations, the diagram presented in Figure 8.21 may be useful. Much of this is

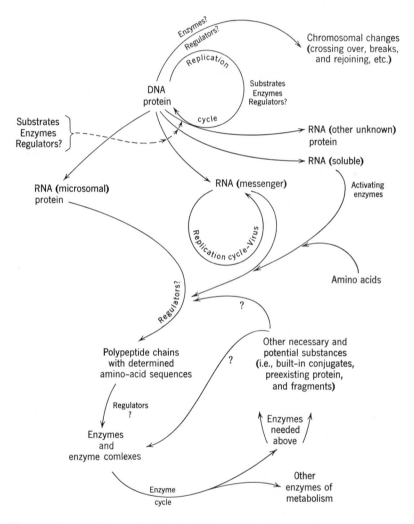

Figure 8.21. A diagram showing the interrelations of DNA, RNA, and protein.

hypothetical, but it has a basis in some important facts (pp. 105, 411). At the outset, it should be noted that in the process of ordering amino acids in sequence by genetic material there exist many complex chemical reactions. Each of these is subject to a variety of controlling factors, as illustrated in Figure 6.13, p. 280, and, in addition, to limitations and control by structural organization and compartmentalization.

In such a scheme of things, the varied aspects of allelism are expected to have their origins in changes in the base sequence of nucleic acids. DNA changes may be always primary, except for RNA viruses, but independent RNA replication in other systems with apparent DNA control is not ruled out, nor are the mechanisms of synthesis of the different kinds of RNA at all understood whether under DNA control or not.

From the diagram it can be seen that a large number of possible foci of interaction and regulation occur between the initial synthesis of the different kinds of RNA and the formation of a specific enzyme. In a heterozygote carrying a deficiency the situation may be simple, with no production of specific RNA at the deficient locus and complete dominance of the RNA produced by the normal locus. However, even here one must consider the influence of a deficiency on normal DNA replication, and possible abnormal regulatory action by the lack of RNA production at the deficiency site. This may be important in cases in which genes are closely linked and show a sequential linear relationship. It may also be important in position effects.

As for interactions between alleles not involving deficiencies, it should be evident that in a complex situation such as that depicted in Figure 8.21 many possibilities exist for the immediate and subsequent products of two different DNA's interacting with the production of a protein or proteins which would not be produced if each were acting alone.

9 Gene Interaction and Balance

Whenever breeding tests demonstrate that the inheritance of a particular character is dependent on more than one known gene, the phenotype is defined as a manifestation of gene interaction. This definition holds whether the genes are allelic or nonallelic. Since allelic interactions have already been discussed, it will be the purpose here to consider nonallelic interactions.

By using the term gene interaction we do no more than imply that some type of interaction occurs in the vast unknown between the genes and final phenotype which reflects on the activities of two or more genes. This interaction may exist at any level, directly between the genes themselves, at the level of action of their primary products, at some very superficial level in the cytoplasm, or it may represent an interaction between cells or organs many steps removed from the genes. To be able to specify the level at which an interaction occurs and the nature of the interaction is one of the chief objectives of the study of gene interaction. This requires only a superficial knowledge of genetics, but a rather complete knowledge of the biochemistry and physiology of the organism.

The production of any given aspect of the phenotype is dependent on the whole complex of genes making up the genome. Obviously, it is quite impossible to approach the study of gene interaction by considering the totality of interactions, but much insight can be gained by considering those simple situations in which variations in a characteristic are dependent on the segregation of two nonallelic genes. This is called digenic inheritance.

Digenic Characters

Listed in Table 36 are a number of different types of digenic character inheritance and the phenotypic ratios which identify them. These ratios are determined by assuming independent assortment of the gene pairs considered, and the cited ratios are obtained only under this con-

Table 36. Ratios Expected in Different Types of Digenic Character Inheritance

	Standard or Wild Type		Mutant	F_2 Ratio Diploid	F_1 Ratio Haploid
Complementary	a^+b^+		ab a^+b ab^+	9:7	1:3
Duplicates	a^+b^+ a^+b ab^+		ab	15:1	3:1
Suppressors					
Recessive suppressor of a recessive mutant gene	ab a^+b a^+b^+	(b is the suppressor)	ab^+	13:3	3:1
Dominant suppressor of a dominant mutant gene	CD C^+D C^+D^+	(D is the suppressor)	CD^+	13:3	. . .
Recessive suppressor of a dominant mutant gene	Cb C^+b C^+b^+	(b is the suppressor)	Cb^+	7:9	. . .
Dominant suppressor of a recessive gene	aD a^+D a^+D^+	(D is the suppressor)	aD^+	15:1	. . .

Note: a and b, recessive genes; a^+ and b^+, respective normal alleles; C and D dominant genes; C^+ and D^+ respective recessive normal alleles.

dition.   Linkage of the interacting genes, as is frequently the case, would alter these ratios.   Ratios for both diploid and haploid meiotic organisms are given, except where dominance is involved.

Complementary Genes.   One of the commonest types of gene interaction involves nonallelic genes which give the same mutant phenotype. They are called complementary genes or mimic genes.   In order for the wild type or standard phenotype to be expressed, all complementary genes must be represented by their dominant wild-type alleles; the failure of any one or all will produce the mutant phenotype.   Many examples of this type of interaction have already been discussed; cf. the *v, cn, st,* and *cd* genes in *Drosophila,* the different groups of genes which bring about the requirement for the same compounds in *Neurospora,* and many of the genes involved in flavonoid-pigment formation in plants described in Chapter 13.

Two complementary genes segregating independently will give a phenotypic ratio of 9 wild type to 7 mutant, unless single recessives can be distinguished phenotypically from one another and the double recessive, with the result that a $9:3:3:1$ ratio is obtained.   In haploids such as *Neurospora* the result of a cross between two phenotypically similar mutant strains each carrying a different mutant gene will be a ratio of 1 wild type to 3 mutant.

The usual biochemical interpretation of the action of complementary genes is that they are involved in a reaction sequence.   Thus, if *a* and *b* are complementary in the wild type, the following model of the action of the normal alleles may be made:

$$S \xrightarrow{\ a^+\ } I \xrightarrow{\ b^+\ } P$$

where *S* is the substrate, *I* an intermediate, and *P* is the product required for the expression of the wild phenotype.   If either *a* or *b* or both were present to block its production, the mutant phenotype would result. However, alternative explanations such as

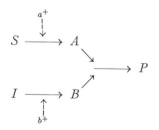

may also be applied to many examples, and it is to be recognized that these models as well as those given hereafter have heuristic value only, unless accompanied by biochemical data which confirms them.

Duplicate Genes. Duplicate genes have been found in a number of plants. These are recessive, nonallelic genes which when present together in the homozygous condition produce a digenic mutant phenotype. Duplicate genes are known in maize, especially among those affecting chlorophyll production, and in tobacco, wheat, and cotton (251, 169, 891). Duplicate-gene inheritance is recognized by a 15:1 ratio of wild type to mutant in the F_2, if there is independent assortment.

The relatively common occurrence of duplicate and even triplicate and quadruplicate genes in plants is probably the result of the fact that many plants are hybrid polyploids. Origin from polyploids would be expected to result in the duplication of genes of like function, each parent contributing a member of a duplicate set as shown in Figure 9.1. The phenomenon of two mutant genes being required to produce a mu-

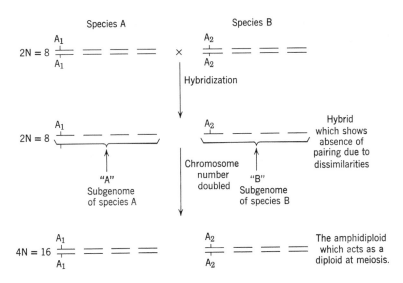

A_1 and A_2 are genes with identical functions. If they exist as amorphs, then, in the amphidiploid, only the genotype $\frac{a_1}{a_1} \frac{a_2}{a_2}$ will show the mutant character.

Figure 9.1. The origin of an amphidiploid tetraploid by hybridization and doubling.

tant phenotype may then be the result of duplication of action rather than gene interaction. In animals polyploidy is an uncommon phenomenon (1073) which must have played a very minor role in evolution, and, in view of this, it is not surprising to find that duplicate genes of this type are practically unknown among them.

Although duplicate-gene inheritance is not strictly an example of gene interaction, it is evident that genes with related functions showing interactions may have arisen from duplicates during the evolutionary process. Among the various species of cultivated cotton, for example, which are known to be amphidiploids, there are a number of peculiar types of inheritance of mutant characters best explained by assuming a divergence of function of originally duplicate genes.

The amphidiploid cotton species, *Gossypium hirsutum* and *G. barbadense,* each has a mutant phenotype inherited as a single-gene difference involving a shortening of the flower branches to produce flower clusters. The mutant is known as cluster in *hirsutum,* and short branch in *barbadense.* Cytologically, *hirsutum* and *barbadense* are similar. Both appear to have been derived originally by hybridization of a diploid species with 13 pairs of chromosomes of the "D" type and a diploid with 13 pairs of "A" chromosomes (961, 964). The amphidiploids, therefore, have a chromosome constitution of 26 A + 26 D and would be assumed to carry a number of duplicate sets of genes. This assumption is in part supported by the presence of the cluster mutant gene in the D chromosome set or subgenome of *hirsutum* and short branch in the A subgenome of *barbadense.* However, the assumption that cluster, cl_1, is a true duplicate of short branch, cl_2, because they both produce the same phenotype, is not borne out by the results of breeding experiments. If cluster *hirsutum,* cl_1cl_1, is crossed to short-branch *barbadense,* cl_2cl_2, the F_1 progeny are all normal, whereas one would expect only mutants if they were duplicate genes. These results can best be explained by assuming the genotypes of cluster to be cl_1cl_1 Cl_2Cl_2, and short branch to be $Cl_1Cl_1cl_2cl_2$, and that the normal phenotype is produced by the genotype $Cl_1Cl_1Cl_2Cl_2$ (or $Cl_1cl_1Cl_2cl_2$) only. Hence, although cl_1 and cl_2 appear to have arisen as duplicates as a consequence of amphidiploidy, their respective normal alleles do not act as duplicates. As pointed out by Stephens (965), cl_1 and cl_2 act as complementary genes, each producing the same phenotype when one is present as the mutant and the other represented by the normal alleles. The inference may be made that, although originally identical in function, they have diverged in evolution in much the same manner as discussed for pseudoalleles in Chapter 8.

A continued evolution of this sort would be expected to result in the amphidiploid becoming essentially a diploid no longer carrying duplicate genes. This process is apparently well on its way in the amphidiploid cottons. If leaf-shape genes from A and D natural diploids are transferred into a natural amphidiploid, those that come from the D species act as alleles of the amphidiploid leaf-shape locus, whereas those that come from the A diploid act as duplicates in the amphidiploid (837).

It is evident that at least two possible means for the origin of new genes are indicated: (1) through the divergence of repeats in tandem on the same chromosome, and (2) through the divergence of duplicates resulting from hybridization. It should be emphasized that these processes of divergence have not been proved to occur in any example studied. Nonetheless, an appreciation of the possible existence of such a mechanism and the study of the biochemical relationships of genes apparently so related may well lead to a better understanding of gene action and interaction in metabolism, for present-day genotypes are the result of long evolutionary development.

Suppressor Genes. The geneticist working with homozygous or homocaryotic mutant strains is occasionally surprised to find normal phenotypes appearing in his supposedly pure-breeding populations. Sometimes these "reversions" to type can be shown to be due to reverse mutation, but frequently they are the result of a mutation at another locus which modifies the mutant phenotype toward wild type. Genes which cause a wild phenotype despite the presence of a nonallelic mutant gene are called *suppressors*. A suppressor may have no other apparent effect than the suppression of the mutant phenotype in whole or part, but, on the other hand, it may also have a mutant effect of its own. Many examples of suppressors are known, and it will be of value to consider some of them in detail.

Suppressor genes were first recognized in *Drosophila melanogaster,* but have since been found in most other organisms used in genetic work including microorganisms and even phage. In Table 37 are listed some of the suppressors found in *Drosophila melanogaster.* It will be noted that a suppressor may be dominant (*Su-S* and *Su-ss*) or recessive, as in *su-s* the suppressor of sable, and that it may suppress the mutant phenotype given by a dominant (*Su-S* and su^2-*Hw*) or recessive mutant gene. A number of them are obviously not very specific for the character suppressed, particularly su^2-*Hw,* su^3-*s,* and su^{82}-*v pr.* Just how specific the others are is difficult to decide, since it is not always clear from the

Table 37. Some Suppressors Found in *Drosophila melanogaster*

Suppressor Symbol	Character Suppressed	Reference
su^2-Hw	Hairy wing (Hw) scute-1 (sc^1) cut-6 (ct^6) forked (f) bithorax-3 (bx^3) bithoraxoid (bxd) cubitus interruptus (ci^2) yellow wing color of y^2	607
su^B-pr	purple (pr)	886
su-s ⎫ su^2-s ⎬ * su^3-s ⎭	sable (s) vermilion (v) and sable (s) speck (sp), sable and vermilion	886
su^{S2}-v pr	vermilion and purple (v and pr)	108
Su-S	Star (S)	710
Su-ss	spineless (ss)	710

* Alleles or pseudoalleles; also called suppressor of vermilion.

descriptions given in the literature whether they have been tested for interaction with other mutant genes. The otherwise nonspecific su^2-Hw shows a peculiar specificity toward the scute series of alleles, for it suppresses only scute-1 (sc^1) and not the others of the series. It shows a similar specificity toward the cut, bithorax, cubitus interruptus, and yellow allelic series. The suppressors of sable and vermilion have no visible effect alone and therefore can be recognized only when combined with the mutant genes they suppress. When present alone, the suppressors of purple (su^B-pr) and hairy wing give an altered phenotype not related in appearance to the phenotype suppressed, while Su-S and Su-ss give wild phenotypes when heterozygous, but are lethal homozygous.

In *Neurospora* suppressors have been described which relieve the requirement for pyrimidine, proline, methionine, inositol (692, 319,

321), acetate (584, 980), pantothenic acid (1049), and tryptophan and related compounds (415, 1119) in mutants requiring these substances.

The suppressor of certain pyrimidine- and proline-requiring mutants (*s* or *su-pyr*) is of particular interest because of its proven interaction with a number of nonallelic mutant genes. Table 38 summarizes the effects of this gene which alone has no observable effect on the phenotype. When in combination with any one of four nonallelic genes, *pyr-3a, prol-2, prol-3,* or *prol-4,* it completely relieves the requirements of these mutants so that the double mutants can grow on minimal medium. The suppression of the proline requirement is complete, and the double mutant is otherwise phenotypically identical to wild type. But the double mutant, *pyr-3a, s* is not identical with wild type, for although it grows on minimal in the absence of pyrimidine, it is inhibited by the indicated related compounds. These compounds do not inhibit the growth of wild type significantly. The same inhibition and its relief by lysine is exhibited by the triple mutant *pyr-3a, prol-2, s.*

This suppressor has other peculiarities, among them being the results of its combination with certain ornithine and lysine mutants. When either of these is combined with *su-pyr,* the double mutant still retains its requirement for the respective amino acid, but it cannot use pre-

Table 38. The Effects of *su* and *pyr* in Combination with Certain Pyrimidine and Proline Mutants [*]

Suppressed Mutants	Phenotypic Characteristics of Mutants Alone	Phenotypic Characteristics of Suppressed Mutants
pyr-3a *pyr-3b*	Require pyrimidine	Grows on minimal like wild type, but inhibited by ornithine, citrulline, and arginine. Inhibitions relieved by lysine or pyrimidine.
prol-2 *prol-3* *prol-4*	Grow slowly on minimal, stimulated by proline, ornithine, citrulline, or arginine	Identical to wild type

[*] Data of Mitchell and Mitchell (692).

cursors of the amino acid which satisfy the requirement when *s* is not present.

The *pyr-3* locus is a complex one. The mutant gene *pyr-3d* is about 0.02 units from the *pyr-3a* site, and has quite different characteristics. Strains carrying *pyr-3d* require pyrimidine, but they are not suppressed by *s*, and, furthermore, they are lacking in the enzyme aspartic-carbamyl transferase which catalyzes the formation of ureidosuccinate from aspartate and carbamyl phosphate (Figure 9.2). This enzyme is not missing from the *pyr-3a* and *3b* mutants which must therefore have blocks at some other step leading to the formation of pyrimidine.

The suppressor (*s*) of *pyr-3a* has been investigated for its enzyme activity by Davis (204) and found to be fifteen- to one hundredfold lower in ornithine-transcarbamylase activity than wild type, or *pyr-3a*, both of which have similar activities. This enzyme catalyzes the formation of citrulline from carbamyl phosphate and ornithine, and hence controls arginine biosynthesis, for arginine is formed from citrulline (Figure 9.2). The suppressor strain apparently has enough of the enzyme to produce arginine for normal growth despite the low level of the enzyme.

It should be possible to explain the action of *s* as a suppressor on the basis of its low ornithine-transcarbamylase activity. Reference to

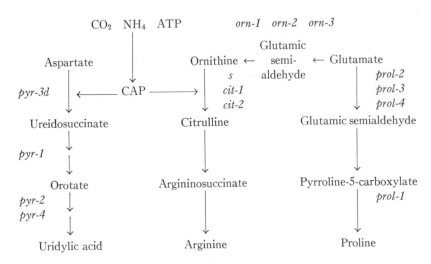

Figure 9.2. Relationship of the pyrimidine, arginine, and proline pathways in *Neurospora*. Mutants are indicated where they appear to be blocked. Most of the reactions shown are reversible. From Davis (205).

Figure 9.2 shows that the step controlled by this enzyme is at a point where three pathways are closely related. Both uridylic acid and arginine require carbamyl phosphate as a precursor, and glutamic semialdehyde is required for both arginine and proline biosynthesis. If glutamic semialdehyde is synthesized in limited amounts from glutamic acid, it may be postulated that a reduction in the enzyme activity leading to the formation of proline might cause a channeling of available glutamic semialdehyde to arginine biosynthesis only. In other words, when either *prol-2,3,* or *4* are present, the proline pathway cannot compete effectively with the arginine pathway. Actually, the steps at which these proline mutants are blocked are not known, but it seems reasonable to assume that the genes they represent act early in the sequence leading to proline, because in addition to proline, ornithine, citrulline, or arginine will also satisfy these mutants' requirements for growth. The action of the *s* gene can be explained, then, by assuming that since it reduces the rate of utilization of ornithine, adequate glutamic-semialdehyde remains to go to the formation of proline (205, 1039). The *s, prol-2,3,* or *4* combinations should not be affected by the presence of exogenous arginine, and, as stated in Table 38, they are not.

The suppression of *pyr-3a* mutants by *s* may have a somewhat similar explanation, because carbamyl phosphate is common to both the pyrimidine and arginine pathways. Thus, if it is assumed that *pyr-3a*$^+$ controls the synthesis of carbamyl phosphate, then a reduction in carbamyl phosphate in the mutant form may leave only sufficient precursor for arginine biosynthesis. As in the proline explanation, a balance may be reestablished by reducing the demands on the limited supply of carbamyl phosphate by the addition of *su-pyr* whereupon both arginine and pyrimidine are formed in sufficient quantities for normal growth. The inhibition of the suppressed mutant *s, pyr-3a* by arginine may be explained by assuming that excess endogenous arginine causes a repression of the pathway to pyrimidine at some point common to both pathways (205).

Perhaps the most thorough analysis of suppressor-gene activity has been made in the *td* mutants of *Neurospora crassa* and *Escherichia coli*. These mutants are all deficient in tryptophan-synthetase activity for reaction 1 in Figure 6.9 and therefore require tryptophan. Some of them produce crossreacting material or CRM as described on page 271.

Among the *Neurospora td* mutants producing CRM, some have been found which are suppressed by nonallelic suppressor genes. The suppressed mutants produce active tryptophan synthetase. A number of different suppressors have been found which may be distinguished on the basis of their effects on the different *td* alleles. A summary of some of the data obtained by Yanofsky and Bonner (1122) is given in Table 39 for three of the suppressor genes isolated by them. Two significant observations have arisen from data such as these. First, those *Neurospora td* mutants which do not produce CRM appear to be unsuppressible, and second, there appears to be a definite allele-specific relationship among suppressor genes. For example, td_1 does not produce CRM, and a suppressor has not been found for it. (However, it is also true that some CRM-producing strains do not have known suppressors.) With respect to allele specificity, it will be noted from the table that *su-2* suppresses only td_2 and not td_6, whereas td_2 and td_6 are both suppressed by *su-6*. The third suppressor, *su-3*, does not overlap with the other two in its specificity. Allele specificity of this type has also been well established in *Drosophila* and should be recognized as a general characteristic of suppressor genes.

When the suppressed mutants were examined for tryptophan-synthetase activity, this enzyme activity was found to be present, but at lower levels than in the wild type (1120). Such a finding leads one to the initial conclusion that suppressors are merely duplicate genes which are able to produce the enzyme in place of the inoperative *td* allele. However, the fact that suppressors have different specificities for suppressing the different alleles at the *td* locus makes this conclusion untenable, for if they were duplicates all should suppress all *td* alleles.

Table 39. **Suppressor Gene Specificity in *Neurospora crassa* ***

Mutant Allele	CRM	Suppressor		
		su-2	*su-3*	*su-6*
td_1	−	−	−	−
td_2	+	+	−	+
td_6	+	−	−	+
td_3	+	−	+	−
td_{24}	+	−	+	−

* From Suskind (990).

Note: Suppression of *td*, and presence of CRM is indicated by a + sign.

The more likely assumption is that enzyme activity is produced in the double mutant as a result of some type of interaction. Support for this conclusion is to be found in the work of Suskind and coworkers (991) who studied the suppressor of td_{24}. This strain is a temperature mutant and grows in the absence of tryptophan at temperatures above 30°C. The suppressed strain *su-24, td*$_{24}$ is also temperature sensitive, but to a lesser degree, since it gives good, although not wild type, growth at 30° (1120). If the tryptophan synthetase of the td_{24} mutant grown above 30°C is examined and compared to that extracted from the suppressed mutant, they appear to be similar insofar as both of them are tenfold more sensitive to Zn^{++} than the wild-type tryptophan synthetase (991). This result would indicate that the td_{24} mutant produces an altered tryptophan synthetase whose sensitivity to Zn^{++} makes it incapable of being active at temperatures below 30°C in the presence of this and perhaps other ions in the cell. The suppressor, *su-td*$_{24}$, may be assumed to be merely reducing the concentration of free Zn^{++} and other ions in the cell to a level which permits the otherwise unchanged td_{24} tryptophan synthetase to be active.

As has been pointed out in previous discussions, the absence of a specific enzyme activity in the presence of a mutant gene does not mean that the gene is not directing the formation of a protein. The evidence is, in fact, that more often than not a protein is formed which is related to the active enzyme but does not have enzyme activity itself under the conditions present in the cell. It would follow from this that an alteration of the internal cellular environment might enable the altered protein to have activity as an enzyme. Many suppressors may be acting to alter the cellular environment so that an inactive protein may become active. If this is so, then it should be expected that some will be quite allele-specific, for each different mutant allele should be modifying the protein in a different way. To make each mutant protein active should require a different specific type of change in the environment and hence a different suppressor. In addition, one might also expect the same mutant allele to be suppressed by a number of different nonallelic suppressors which all have similar effects (as, for example, complementary genes) on the cellular environment. This possibility is supported by the observation that at least six nonallelic suppressor genes suppress td_2 in *Neurospora*.

Another example of suppressor activity which may be the result of the relief of inhibition, but in a somewhat different fashion, is found in the studies on the acetate mutants of *Neurospora* (584, 980). The acetate mutants, *ac,* investigated by Strauss and Pierog (980) are unable to oxidize pyruvate to form acetaldehyde.   The acetate requirement is presumably met by its formation from acetaldehyde.   However, growth is not rapid under these conditions because acetaldehyde, or some other substance formed from it, is inhibitory.   This inhibition is enhanced by glucose which goes directly to acetaldehyde by the Embden-Meyerhof pathway.   Two other mutant genes, *sp* and *car,* suppress this inhibition when combined with *ac* by reducing the carboxylase activity and hence also the rate of acetaldehyde formation. The net result is enhanced growth of the double mutants *ac,sp* and *car,sp*.   This represents partial suppression by relief of inhibition of an alternate pathway.

While the analyses of the *td* and *pyr-3a* suppressors in *Neurospora* lead to rather simple possible explanations of suppressor action, not all suppressors act either by relief of inhibition or competition for limited substrate.   Studies of suppressors of the *td* mutants of *Escherichia coli* show that the suppression phenomenon may have a far more complex and esoteric explanation in many cases.

The *td,* or tryptophan-synthetase-deficient, mutants of *Escherichia coli* are similar in many ways to the *Neurospora* mutants, but they show significant differences with respect to their actions.   The *E. coli td* mutants which do not produce CRM are, unlike equivalent *Neurospora* mutants, suppressible.   Furthermore, the tryptophan synthetase of *E. coli* is easily resolved into two protein subunits A and B (1124, 1126) which are controlled by adjacent, but genetically separable, segments of genetic material as described on page 272.   Hence two types of *td* mutants are distinguishable, *A* mutants which affect only the A component, and *B* mutants which affect only the B component (see pp. 272 and 364).   The suppressors that have been found are specific for either *A* or *B* mutants, but not both (1121, 1126, 1123).

In Table 40 three *A*-type mutants are listed, each of which produces an A-type CRM (a protein which crossreacts with the antibody against normal A protein).   The modified A protein from each of these mutants has activity for the reaction, indole + serine → tryptophan when in the presence of normal B, as shown in the table, but none has activity for the reaction involving indoleglycerol phosphate.   Suppressors have been detected for all three of these mutations, and the activities of tryptophan synthetases produced by each of the suppressed mutants have been determined.   It will be noted from the table that all three suppressed mutants produce an A protein which is capable of catalyzing the synthesis of tryptophan from indoleglycerol phosphate and serine in addition to the reaction involving indole.   These findings lead to the

Table 40. Activity of A Proteins from Suppressed and Unsuppressed *td* **Mutants of** *Escherichia coli* *

Strain	A Protein Activity in Presence of Normal B Protein		$\dfrac{\text{InGP} \rightarrow \text{Tryp}}{\text{In} \rightarrow \text{Tryp}}$ Per Cent	Property of A in Mutant
	In → Tryp	InGP → Tryp		
Wild type	2.5	1	40	
A36	22	0	. . .	Heat labile
A36su	6	0.25	4.2	
A3	35	0	. . .	Heat stable
A3su	44	0.19	0.44	
A11	31	0	. . .	Acid precip-
A11su	55	0.48	0.87	itable

* From Yanofsky et al. (1126).

question of whether the suppressed mutants are producing a single A protein with the specificity of wild type, but somewhat less active, or a mixture of two A proteins, one the CRM of the unsuppressed mutant and a second with wild-type activity. The answer to this was found by examining the stability of the activity for the indole to tryptophan conversion. It will be noted from the table that the A proteins of the unsuppressed mutants have different stabilities. It was found by checking these stabilities in the suppressed mutants that the indole to tryptophan activity had, in each case, the same stability as the A protein in the corresponding unsuppressed mutant. In all three cases, however, the activity for indole-glycerol phosphate to tryptophan was inactivated at a different rate, one which corresponded to the inactivation of wild-type A protein. Complete inactivation of the unstable A from the *A36su* strain by prolonged heating left an A protein that had a ratio of the two enzymatic activities corresponding to the wild-type ratio (as shown in the last column of Table 40).

These results definitely show that the suppressed mutants form two A proteins. These have been separated by chromatography. Similar results have been obtained with B proteins formed in B-suppressed mutants. When *A* and *B* mutants which do not form CRM in the unsuppressed state are combined with suppressors, they produce A and B proteins with the properties of wild type.

The appearance of two different but presumably related proteins in suppressed mutants, as in the examples just cited, presents great difficulty in interpretation of the function of the suppressors, since the possibility of their being simple duplicates seems to be eliminated because they show allelic specificity. It has been suggested that suppressors of this type permit the formation of an effective amino-acid peptide sequence which can substitute for the damaged portion of a defective protein and endow it with effectiveness as an enzyme (1125). In this sense the suppressor would be a partial duplicate of the gene it suppresses. Another possibility which has been suggested is that the suppressor mutation affects the activity of an activating enzyme for attaching a specific amino acid to soluble or transfer RNA. This might result in the wrong amino acid being inserted into a protein at the time of its assembly. Such "mistakes" might lead to the formation of functional molecules as well as nonfunctional ones (1126).

All of the examples of suppressor action discussed in the foregoing have been concerned with suppressors outside the limits of the genes suppressed, that is, with nonallelic suppressors. It has become evident, however, that intragenic or internal suppressors are also possible. These have been discovered in bacteriophage T4 in connection with reversion studies on *h*III and *tu45* mutants which have heat sensitivities several to a hundred times higher than that of wild type at 45°C. Some of the heat-stable revertants have been analyzed genetically, and it has been found that the suppressors map within the mutant gene they suppress (492). Hence we have a situation in which a mutation at one site is rectified by a second mutation at a different site within the same gene. It is quite possible, of course, that internal suppression of this type may occur in cellular organisms. Its detection in these would be somewhat more difficult because of the difficulty of mapping within short segments of genetic material. The existence of internal suppressors in phage, however, brings to the fore the interesting possibility that a mutation at one site causing an amino substitution in a protein giving it mutant properties may be alleviated by a second amino-acid substitution somewhere else in the chain of the same protein. Hence we may imagine that the protein cannot fold properly after the first mutation, but can after a second.

The characteristics of suppressors are complex and diverse, so much so that even to summarize them presents difficulties. However, several characteristics are outstanding and should be reiterated. (1) A suppressor may suppress the mutant action of a number of nonallelic genes seemingly unrelated in function. (2) The action of a suppressor may be quite specific for only one of the alleles in a series of allelic genes.

(3) The phenotype of a suppressed mutant may superficially resemble the wild type, but in reality be quite different when examined closely. With respect to suppressors being duplicate genes, some of them may be, but certainly not all are for the reasons just stated, and because some of them may actually be deletions as, for example, the suppressor of Star in *Drosophila melanogaster* which seems to be included within a deficiency (606).

As for generalizations about the mode of action of suppressors, none can be made except the obvious one that suppressors apparently act in many different ways. From the examples discussed here it is evident that some suppressors may act by relieving an inhibition, some by opening up an alternate pathway, some by relieving competitive demands for a common substrate in short supply, and some by causing the production of an active enzyme by as yet unknown mechanisms. Explanations for the synthesis of active enzymes by mutants carrying suppressors present somewhat the same problem as trying to explain enzyme production by complementing heterocaryons and heterozygotes as discussed in Chapter 8. It is quite evident that the action of suppressors will not be completely understood until more is known about protein synthesis and its regulation. It should also be obvious that the study of suppressor action is one of the most important approaches to the analysis of gene action. Organisms after all may be looked upon as having systems of genes which "suppress" one another. If one mutates, another which "suppresses" it must also mutate in order to reestablish conditions that enable the system to survive.

Modifiers in General

To describe a gene as a modifier is to state that it interacts with another gene and modifies the phenotype produced by that gene. The modifier may or may not have an observed effect by itself. All the examples of gene interaction described earlier involve modifiers which drastically change the phenotype to the wild-type or mutant condition. Many intermediate conditions exist which lie between the extreme examples of suppressors and complementary genes on the one hand, and genes which seemingly produce their phenotypic effect completely independently of all other genes, on the other hand. Most of these intermediate types are best described by the general term modifiers, although here again certain distinctions can be made on the basis of observed effects on the phenotype.

Modifiers may intensify or enhance the phenotypic effect of other genes. Thus the *B* gene in maize is a dominant intensifier of anthocyan pigmentation (250), causing the production of more pigment in those plants which have been determined to have pigment by other anthocyan genes. In *Drosophila* there are intensifiers of certain mutant phenotypes, such as, for example, the dominant enhancers of Minute (*E-M (3)g*) and Star (*E-S*) and the recessive enhancer of Notch (*e-N*[8]) (108). A mutual enhancement is found to occur between two recessive mutant genes, abbreviated (*abb*) and shrunk (*shr*) (108). Flies homozygous for either of these alone show only a relatively weak mutant phenotype and a considerable overlap with wild type. However, the double mutants show complete penetrance for reduced bristle size and shrunken body. An enhancer may have effects other than causing a more drastic mutant phenotype, as demonstrated by the suppressor of purple (*su*[B]-*pr*) which, in addition to its suppressor action, also enhances hairy wing.

Studies on the inheritance of coat color in mammals have shown that in most species there are genes which cause a dilution of the pigmentation of the hair causing black to become a shade of gray, or brown to become tan, and so on (138, 139, 396). These *dilution genes* are recessive and appear to reduce the amount of pigment present quantitatively without having any other marked visible phenotypic effect. If the main color genes determine the animal to be albino, the effect of substitutions at the diluting-gene loci will, of course, not be recognized. The dominant alleles of dilution genes can be called intensifiers or enhancers.

Related to dilution genes in phenotypic effect, but more drastic in their action, are the *inhibitors*. These are usually dominant genes which prevent the expression of a character such as pigmentation even though all genes determining pigmentation are present in the active form. The dominant genes causing absence of pigment in the chicken (474) and horse (138, 139), and the rabbit, dog, and cat (396), are examples of pigment inhibitors. A dominant inhibitor would be expected to give a ratio of 3 standard to 13 mutant in the F_2.

The terms inhibitor or epistatic are best not used in connection with recessive mutant alleles. These genes may best be considered as complementary with the gene which they appear to inhibit. An example of this latter type would be the white-eye gene, *w,* in *Drosophila,* which produces white eyes when homozygous whatever the state of the other eye-color genes.

A study of any character in a population comprised of one species, all members of which are classified as the wild phenotype, will always reveal a degree of variability in the expression of the character. Part of the variability can be ascribed to the environment, but genetic tests will usually demonstrate that it is primarily inherent. If the character is one such as length of a part, weight, degree of pigmentation, or the yield of a crop plant which can be determined quantitatively, it will usually be found that the variations fit a binomial distribution. Crossing experiments between variants will give results showing that the character is controlled by a number of genes each of which contributes toward the production of the character, but each having a relatively small effect alone. For example, one can breed for extreme shortness or extreme tallness in an animal or plant by constant inbreeding of short or tall. This will usually result in shorts and talls which breed fairly close to type, showing that a high degree of homozygosity has been attained. If now the true breeding extremes are crossed, the hybrids will be intermediate, none being either extremely tall or short. Crossing the hybrids will give an F_2 with a wide spread of variation distributed again from the extremely short to the extremely long with most of the offspring intermediate—a pattern of distribution which fits the binomial curve. The genes responsible for this type of continuous character variation are referred to as multiple or quantitative factors.

Quantitative inheritance is extremely important to the plant or animal breeder whose main concern is the "improvement" of particular strains or breeds. To improve a plant or animal, the breeder selects certain individuals possessing the desirable characteristics and uses these for breeding stock. In this way he may, slowly, over a period of generations, attain the desired degree of improvement presumably by selecting for a number of genes which, when together in a single individual, have an additive effect in producing a superior strain. Whether, in actual fact, the interaction of these factors is additive, multiplicative, or subtractive has never been established and will not be until the physiological and chemical approach is used successfully in the analysis of modifier effects. Obviously, modifiers may do many things at the biochemical level, from suppressing the rate at which some reactions occur to enhancing the rates of others, and it is illogical to assume that all have an "additive" effect other than at the superficial phenotypic level.

The genes involved in quantitative inheritance must be considered as modifiers of one another, since they act together to produce a particular phenotype. They must also be considered as modifiers of the "main" genes which produce relatively drastic changes in the phenotype, for a mutant character may show the same degree of variability as the wild type or standard. It would follow from this that a main gene produces its phenotypic effect not only by its own activity but subject to the activity of modifiers with assumed slight effects. Proof that this is so is to be found in the results of the vast number of genetic experiments dealing with quantitative variation of characters. Some of the best examples are to be found in the results of crosses between various strains and species of cotton. Silow (908) has described the effect of crossing the Asiatic cotton, *Gossypium anomalum*, with a leaf type determined by the leaf-shape allele L^A to another Asiatic species, *G. arboreum*. By repeated backcross to the *G. arboreum* parent it was possible to transfer the *anomalum* L^A allele to an essentially *arboreum* background of genes. As will be seen from Figure 9.3 the L^A allele produces a quite different phenotype in *anomalum* than it does in *arboreum*. The gene has not mutated, but its phenotypic expression has been altered by changing its environment of modifying genes.

Modifying effects have also been noted on the genes determining nutritional requirements in *Neurospora*, as described for the suppressor of pyrimidine. Another striking example of the modification of a nutritional mutant phenotype is the effect of modifying genes on the ex-

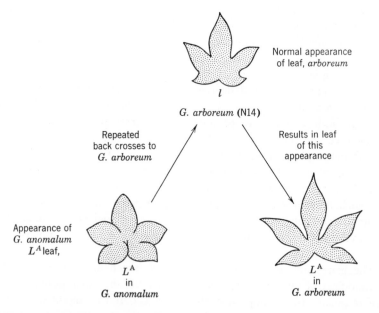

Figure 9.3. The effect of transferring gene L^A from *Gossypium anomalum* to an essentially *G. arboreum* background of genes. After Silow (908).

Table 41. The Effect of Modifying Genes on the Phenotypic Expression of Two Mutants of *Neurospora* Stimulated by Tryptophan [*]

	C-86	*int*	39401	*nic*
Phenylalanine	++	+ to trace	Trace	Trace
Anthranilic acid	++	++	Trace	Trace
Indol	++	++	++	Trace
Tryptophan	++	++	++	Trace
Kynurenine	++	++	++	++
3-OH kynurenine	++	++	++	++
3-OH anthranilic acid	++	++	++	++
Nicotinic acid	++	++	++	++

[*] From Haskins and Mitchell (415).

pression of the mutant strains *C86* and *39401* (415). These are both tryptophanless mutants whose requirements are satisfied by a number of other related compounds in addition to tryptophan as shown in Table 41. At least six modifying genes produce in various combinations with *C86* and *39401* intermediate phenotypes with altered requirements such as the segregants *int* and *nic* indicated in Table 41. The effect of the modifiers is to move the apparent position of the "genetic blocks" by modifying the action of the gene which appears to cause the block. It is obvious that under proper conditions of observation these would not be considered "modifiers of small effect" but, rather, main genes with large effects. A possible explanation of the type of interactions exhibited here may be found in the discussion in the next section.

Controlling Genes

In classical genetics it was generally thought that genes always did something in a positive direction. When the gene-enzyme relationship became evident, it was assumed that the "something" positive was the production of an enzyme. Failure of a gene to act was explained by its inability to produce a protein, or its production of an altered protein with no enzyme activity. Both of these states of inactivity, it was assumed, were attained by gene mutation. Certain types of dominant inhibitor genes were recognized, as discussed in the previous section, but these were generally interpreted as working at the cytoplasmic level and producing a protein product which catalyzed the synthesis of some nonprotein inhibitor of the protein produced by some other gene. It was not until the decade 1950–1960 that it began to be realized that genetic units might exist which modify the functioning of genes producing functional proteins. Since 1960 work done principally with bacteria and maize has made it clear that there may be at least two types of genetic elements, *structural genes* and *controlling genes*. Structural genes are those which carry the codes for specific types of protein biosynthesis via messenger RNA. Controlling genes, on the other hand, are postulated to modify the activities of the structural genes by stopping protein synthesis entirely or otherwise regulating its rate.

The realization of the possible existence of controlling genes comes from three sources: first, the studies in enzyme adaptation or induction in microorganisms; second, from studies on repression of synthesis of enzymes by compounds which are products of the enzyme's activities; and, third, from the discovery of controlling elements in maize. The evidence from these three sources will be discussed in that order.

Enzyme Induction. It has long been known that many microorganisms will not produce a given specific enzyme in quantity unless an inducing substance, generally a small organic molecule, is present. This substance is called the inducer, and ordinarily it is the substrate of the enzyme it induces. However, the inducer may only resemble the natural substrate and not be affected by it. Furthermore, the inducer may induce the formation of several enzymes sequentially involved in its metabolism (954, 702).

Although enzyme induction is a common phenomenon probably to be found in all organisms, most of the work on it has been done in microorganisms for technical reasons. Specifically, the experiments which have yielded the most fruitful results in the analysis of induction have involved the studies of the induction of β-galactosidase in *Escherichia coli* principally by Monod and his associates (484).

The hydrolysis of lactose to glucose and galactose is carried out in *E. coli* by β-galactosidase. This enzyme has been crystallized and shown to be active exclusively for β-galactosides in which the galactose is unsubstituted. It is apparently the only β-galactosidase made by *E. coli,* for mutations at the *lac* locus, which controls its production, give rise to strains which are unable to grow on lactose as the sole carbon source.

Two other enzymes are associated with β-galactosidase. One of them, β-galactoside permease, is necessary for the entrance of β-galactosides into the cell and their concentration there. Without permease, *E. coli* cells cannot hydrolyze β-galactosides even though they possess active β-galactosidase. The second enzyme, β-galactoside acetylase, is characterized by the ability to acetylate β-thiogalactosides to form the 6-acetyl thiogalactoside in the presence of coenzyme A. This is almost certainly not its natural function in the cell, but the enzyme is considered to be of significance, because its synthesis is controlled by the *lac* locus just as is the galactosidase and permease, and its appearance is induced by the same compounds that induce the galactosidase.

Wild-type *E. coli* grown in the absence of β-galactosides produces a barely detectible amount of β-galactosidase. However, if grown in the presence of β-galactosides the bacterium may produce up to 10,000 times more enzyme. The inductive effect is the appearance of this activity in the presence of the inducer. This tremendous increase in activity may have several explanations. First, it may be the result of selection of "mutant" strains of *E. coli* which are always present in small numbers and capable of producing β-galactosidase in contradistinction to the wild type which is unable to do so. This possibility can be discounted by showing that enzyme production begins about three minutes after the addition of the inducer (764), and then proceeds to increase at a rate proportional to the increase in total protein of the culture. (Figure 9.4). The presence of the inducer is necessary for the continued increase in enzyme activity, for if it is removed the increase ceases as abruptly as it begins. These observations make it quite evident that induction of β-galactosidase is not an enzymatic, genotypic adaptation in a population. The second possibility is that the increase in activity is the result of the uncovering, activation, or conversion of a preexisting protein. This has been eliminated by demonstrating (1) that prior to induction there is no protein present which has the antigenic properties of β-galactosidase (177), and (2) that induced β-galactosidase does not derive any of its sulfur and carbon from preexisting protein in the induced cells (447). The elimination of these two possibilities leaves the possibility that the appearance of enzyme activity is the result of *de novo* synthesis of a new protein, β-galactosidase, in response to the presence of the inducer. This may be assumed to be applicable to enzyme induction in general. The studies on other induced enzymes all indicate that the inducer causes the onset of biosynthesis of a protein, or an increase in its rate of biosynthesis, from amino acids.

Of particular significance is the fact that the inducers need not be

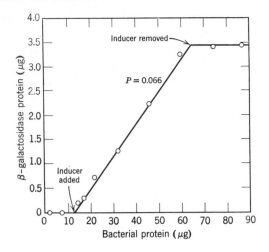

Figure 9.4. Kinetics of induced enzyme synthesis. Differential plot expressing accumulation of β-galactosidase as a function of increase of mass of cells in a growing culture of *E. coli*. Since abscissa and ordinates are expressed in the same units (micrograms of protein) the slope of the straight line gives galactosidase as the fraction (P) of total protein synthesized in the presence of inducer. After Cohn (176).

actual substrates, that is, capable of being hydrolyzed. Such inducers are called gratuitous inducers. As shown in Table 42, β-glucosides and galactose are not effective as inducers, but the remaining compounds listed, with the exception of β-phenylthiogalactoside, are effective in varying degrees. All the β-thiogalactosides with inducing ability are, however, gratuitous, as is melibiose, an α-galactoside, insofar as they induce the enzyme but are not hydrolyzed by it. Melibiose does not even have a measurable affinity for the enzyme. A remarkable fact brought out by these data is that the presumed natural substrate, lactose, has both a low inductive value and a low affinity for β-galactosidase. A second important fact is that all these inducers, insofar as they have been tested, induce the appearance of the β-galactoside acetylase as well as the β-galactosidase. In addition, they also induce the appearance of the permease.

It should be recognized that these observations raise some extremely important questions, chief among which relates to the gene-enzyme relationship. If we accept it as established that a specific protein will not be synthesized except in the presence of a gene which dictates its structure, we must assume that wild-type *E. coli* carries the structural

Table 42. Induction of Galactosidase and Galactoside-Transacetylase by Various Galactosides *

Compound		Concentrations	β-galactosidase Induction Value	V	$1/K_m$	Galactoside Transacetylase Induction Value	V/K_m
β-D-thiogalactosides	(isopropyl)	$10^{-4}M$	100	0	140	100	80
	(methyl)	$10^{-4}M$	78	0	7	74	30
		$10^{-5}M$	7.5	…	…	10	…
	(phenyl)	$10^{-3}M$	0.1	0	100	1	100
	(phenylethyl)	$10^{-3}M$	5	0	10,000	3	…
β-D-galactosides	(lactose)	$10^{-3}M$	17	30	14	12	35
	(phenyl)	$10^{-3}M$	15	100	100	11	…
α-D-galactoside	(melibiose)	$10^{-3}M$	35	0	0.1	37	1
β-D-glucoside	(phenyl)	$10^{-3}M$	0.1	0	0	1	50
	(galactose)	$10^{-3}M$	0.1	…	4	1	1

* From Jacob and Monod (484).
Note: Induction value refers to specific activity of enzyme produced in presence of indicated galactoside. Values are given in per cent of induction obtained with isopropylthiogalactoside at $10^{-4}M$. V refers to maximal substrate activity of each compound in per cent of activity obtained with phenylgalactosidase. $1/K_m$ expresses affinity of each compound with respect to galactosidase.

gene for β-galactosidase.   If this is so, why, then, does this gene not express itself except in the presence of a specific inducer?   At least two general explanations may be advanced to account for this phenomenon.   First, it may be hypothesized that the inducer acts as some sort of template which triggers the synthesis of the enzyme because it is, or resembles, the substrate for the enzyme.   Or, second, it is possible that the enzyme is not synthesized, because its active formation is inhibited by some agent.   In this case the inducer may itself be viewed as an inhibitor which inhibits the inhibiting agent.   This latter possibility may appear overly complex, but it has a considerable amount of evidence in its support, whereas the evidence does not support the first hypothesis.   The fact that an inducer may show an insignificant affinity for the enzyme is not consonant with the inducer playing the role of a template—neither is the fact that a substance may show affinity for the enzyme, but have no inducing ability.   On the other hand, the quick action of the inducer in starting enzyme biosynthesis, and the almost immediate cessation of this activity when it is removed, point to its being some sort of inhibitor.

Several genes have been distinguished in the *lac* region of *E. coli.* Among these are: (1) the gene z^+, which in the mutant conditions results in the loss of ability to synthesize active β-galactosidase in the presence or absence of inducer; (2) the gene y^+, which in the mutant condition results in loss of capacity to synthesize the β-galactoside

Table 43.   **Phenotypes of Various Genotypes of the *lac* Locus in *E. coli***

Strain	Genotype	Noninduced Enzymes *			Induced enzymes		
		a	b	c	a	b	c
1	$z^+y^+i^+$	−	−	−	+	+	+
2	$z^-y^+i^+$	−	−	−	−	+	+
3	$z^+y^-i^+$	−	−	−	+	−	+
4	$z^-y^-i^+$	−	−	−	−	−	+
5	$z^+y^+i^-$	+	+	+	+	+	+
6	$z^+y^-i^-$	+	−	+	+	−	+
7	$z^-y^+i^-$	−	+	+	−	+	+
8	$z^-y^-i^-$	−	−	±	−	−	±

* a: β-galactosidase; b: permease; c: acetylase.

permease in the presence or absence of inducer; and (3) the gene i^+, which causes changes in the influence of inducers on the synthesis of β-galactosidase. Many i mutants synthesize large amounts of the enzyme in the absence of inducer, and are therefore termed *constitutive*. Table 43 shows the expected phenotypes of *E. coli* strains having various combinations of these genes. It will be noted that i controls only the *inducibility* of the enzymes while z and y control the ability to synthesize them, or not, depending on the conditions. Thus strains numbered 5 through 8 in Table 43 are constitutive for all, two, or one of the enzymes, depending on the state of z and y. (Mutants unable to synthesize the acetylase have also been obtained, but the genetics of these mutants have not been defined.) All three of these genes have been shown by recombination analysis to be closely linked in the *lac* region of the *coli* chromosome. Each locus consists of a number of different mutational sites. Because of their complete control over the ability to produce active enzyme, or not, z and y are considered *structural genes*. The validity of this description is made certain for z, at least, by the observation that certain z^- mutants produce crossreacting material (CRM) which reacts with antibody specific to normal β-galactosidase, but the CRM itself has no activity as enzyme. Furthermore, the production of CRM in these mutants is induced by the same inducers that cause the biosynthesis of β-galactosidase in cells with i^+ present.

A number of different types of experiments (the results of one is illustrated in Figure 9.5) indicate that i^+ produces some substance which inhibits the production of the enzymes in the presence of z^+ and y^+ (484, 485). It has been suggested that the constitutive condition is the result of the absence of this inhibiting substance in i^- cells. This hypothetical inhibiting substance has been called *repressor* substance by Jacob and Monod (484, 485). The action of the inducer is to reverse the repression of enzyme synthesis by repressor.

The nature of the postulated repressor substance is not known. Some evidence exists that it is not a protein (763, 484), because the inhibition of protein synthesis does not seem to inhibit repressor synthesis. However, it is possible that the repressor is not a substance per se, but some sort of mechanism regulating some still unknown step or steps in protein synthesis.

The i locus may be described as a modifier gene. In their action, certain of its forms show the same characteristics as inhibitor genes and suppressor genes described in this and the preceding chapter. Jacob and Monod (484) consider that the i gene is sufficiently distinguished

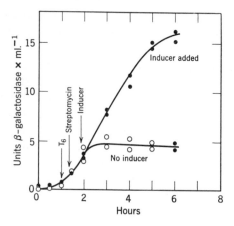

Figure 9.5. Synthesis of β-galactosidase by merozygotes formed by conjugation between inducible, galactosidase-positive males and constitutive, galactosidase-negative females. Male (*Hfr i⁻z⁺T6Sˢ*) and female (*F⁻i⁻z⁻T6ʳSʳ*) bacteria grown in a synthetic medium containing glycerol as carbon source are mixed in the same medium (time 0) in the absence of inducer. In such a cross, the first zygotes which receive the *lac* region from the males are formed from the twentieth minute. The rate of enzyme synthesis is determined from enzyme activity measurement on the whole population, to which streptomycin and phage T6 are added at times indicated by arrows to block further formation of recombinants and induction of the male parents. It may be seen that in the absence of inducer enzyme synthesis stops about 60 to 80 minutes after penetration of the first *z⁺i⁺* segment, but is resumed by addition of inducer. From Pardee et al. (762).

by its action that it should be classified as a controlling gene which has no function other than to regulate the activity of *z⁺*, *y⁺* and possibly other genes at the *lac* locus in their production of specific proteins. That this is their sole function remains to be demonstrated. However, present evidence certainly justifies the hypothesis that not all genes directly control the specificity of proteins by dictating their primary structure through RNA.

A number of other types of inducible enzyme systems have been found in which there is a mutation to the constitutive condition, and hence are similar to the *i⁻* state in *E. coli*. Some examples of these are the penicillinase system of *Bacillus cereus* (528), and the amylomaltase (173) and galactose systems of *E. coli* (124, 507).

Actually, there seems to be another kind of genetic element at the *lac* locus which converts the system to the constitutive condition by mutation. It may be distinguished from the regulator just described because its mutant forms interact differently. For example, the gene o^+ has the mutant form, o^c, which determines that the cell carrying it shall be constitutive. It is, however, different from the regulator gene, i^-, which also gives constitutivity, because o^c is dominant to o^+, whereas i^- is recessive to i^+. Like i, o affects all three enzymes of the lactose system equally, but it affects them in a somewhat different way than i. Thus, in the heterogenotes

$$\frac{o^+z^+y^+}{o^cz^-y^-} \quad \text{and} \quad \frac{o^+z^-y^-}{o^cz^+y^+}$$
$$\text{(trans)} \qquad\qquad \text{(cis)}$$

o^c produces the *constitutive* condition only when it is *cis* to z^+y^+. The *trans* relationship is inducible. Thus z^+ and y^+ attached to an o^+ are repressor sensitive, or inducible, despite the presence of the o^c allele in the partial diploid; o^c is active only when on the *same* chromosome with the active forms of the structural genes. This is not true of the i alleles.

In addition to becoming insensitive to repressor, it might also be expected that the o^+ gene could mutate to a condition such that it is completely inoperative, and none of the structural genes controlled by it is active. Such a mutant allele has been found. Designated as o^0, it can be shown to be *recessive*. Thus, heterogenotes of the constitution

$$\frac{i^+o^0z^+y^+}{i^-o^+z^+y^+}$$

are inducible, and the haploids, $i^+o^0z^+y^+$, produce none of the enzymes expected in the presence of z^+y^+ even with inducer.

The *lac* locus is obviously a complex one containing four regions separable by their function, $i, z, y,$ and o. This region has been mapped and the four genes indicated found to lie immediately adjacent to one another in the order indicated in Figure 9.6.

Enzyme Repression. It has been observed in many enzyme systems that the end product may cause a repression of the *synthesis* of the enzymes involved in its own production. This phenomenon is called enzyme repression, and is distinctly different from *inhibition* of already-formed enzyme by the end product in a process sometimes called *feedback inhibition*. The two processes give somewhat the same result, that is, a reduction in amount of end product formed, but the mechanisms are quite different.

Like enzyme induction, enzyme repression has been studied primarily

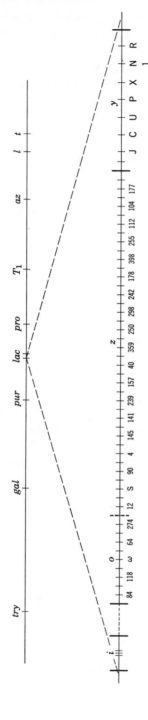

Figure 9.6. Genetic map of the *lac* region of *E. coli*. The upper line locates the *lac* region on the *coli* chromosome. The lower line shows the fine structure of the *lac* region and its division into the functional regions *i, o, z,* and *y*. After Jacob and Monod (484).

in the bacteria. An example, the enzyme system involved in the biosynthesis of histidine in *Salmonella typhimurium,* has already been briefly considered on page 362. As shown in Figure 8.13 genetic and biochemical evidence indicate that eight genes and nine enzymes are involved in the conversion of 5-phosphoribosyl-1-phosphate to histidine (411, 19, 622). The genes, designated as *E, F, A, H, B, C, D,* and *G* in Figure 8.13, are closely linked in the order indicated. This order corresponds to the order of reactions from 5-phosphoribosyl-1-pyrophosphate to histidine except that *G* controls the first enzyme in the series, and the others, starting with *E,* the reactions leading to histidine (Figure 8.13). It has been shown that the biosynthesis of each of the enzymes

> pyrophosphorylase
> imidazoleglycerolphosphate dehydrase
> imidazoleacetolphosphate transaminase
> histidinol phosphate phosphatase
> histidinol dehydrogenase

is repressed simultaneously and to the same degree in the presence of histidine. The repression was not complete under the conditions used, but the differences between the amounts of enzymes produced in the repressed and unrepressed states ranged up to ten- to fifteenfold. The other four enzymes in the series may also be repressed, but not enough was known of them to test them. However, the coordinate repression of the five known enzymes demonstrates a phenomenon of considerable significance in metabolic control (17). It has been suggested that coordinate repression acts at the genic level. The reason for this hypothesis is that the genes are in a cluster on the chromosome, and, therefore, possibly under the control of a genetic element associated with them which responds directly or indirectly to the presence of histidine by slowing down the rate of biosynthesis of the enzymes. Evidence for such a controlling gene is the existence of several mutants which involve deletions at the *G* locus. These mutants lack *all* enzymes for histidine biosynthesis despite the fact that the structural genes for them can be shown to be present. Point mutations at the *G* locus affect only the pyrophosphorlyase, but when part or all of it is deleted the widespread consequences would indicate that the *G* locus also includes a controlling element similar to those described for the *lac* locus in *E. coli.* Viewed from this aspect the *G* deletions would correspond to the o^0 mutation in *E. coli.*

A repression similar to histidine repression has been found for the arginine biosynthetic enzymes in *E. coli*. Seven genes have been located on the *coli* chromosome, each of which causes an arginine requirement when it mutates (Figure 9.7). These genes are associated with the seven enzymes controlling the synthesis of arginine from glutamic acid (Figure 7.6) as structural genes. It has been found that certain strains (K-12 and W) of wild-type *E. coli* show repression for six of the enzymes in the presence of exogenous arginine (1038, 343). The repression by arginine is not coordinate because the enzymes are not affected to a similar degree, but the effect is noted on all six. Not all strains of *E. coli* show arginine repression, however. Indeed, some show an *increase* in the enzymes in the pathway in the presence of arginine (342). Nonrepressible mutants of strains K-12 and W have been obtained which produce all six enzymes in high amounts in the presence of arginine. Genetic analysis of these mutants shows that a

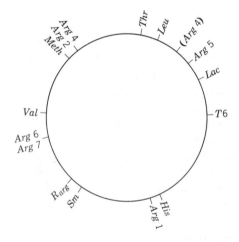

Time of entry of *Arg* 5 = 22–23 min.
 "   "   "   " *Arg* 4 = 27–28 & 34–35 min.
 "   "   "   " *Arg* 2 = 36–37 min.
 "   "   "   " *Arg* 6 or 7 = 55–59 min.
 "   "   "   " *Arg* 1 = 68 min.
For reference
 Lac enters at 18 min.
 His "   " 70 min.

Figure 9.7. Map of some of the genes involved in arginine biosynthesis in *E. coli* K-12. The order and distance was determined by time of entry during conjugation (see Chapter 4). After Gorini, Gunderin, and Burger (343).

gene designated as *R arg,* and located between *val* and *sm* (Figure 9.7), is responsible for the change from repressibility to nonrepressibility. When in the *R arg+* state the cells are repressible, and in the mutant *R arg−* state nonrepressible. The evidence indicates no change in the structure of the enzymes or in the structural genes producing them, but only a change in the rate of synthesis of the enzymes when *R arg+* mutates. It would seem from these data that the existence of a controlling gene for arginine biosynthesis seems fairly well established.

Controlling genes have also been found in *Neurospora crassa* which involve the synthesis of tyrosinase (460). It will be recalled from Chapter 6 that Horowitz and his associates have demonstrated that the *T* gene in *Neurospora* directs the synthesis of tyrosine as a structural gene, since changes at this locus cause apparent changes in the structure of tyrosinase, as described in Chapter 6. Two other genes, *ty-1* and *ty-2,* also affect the synthesis of tyrosinase. These are linked to neither *T* nor each other. Both *ty-1* and *ty-2* exist in mutant forms which produce little or no tyrosinase on sulfur-deficient medium, although the wild type produces abundant tyrosinase under these conditions. When grown in sulfur-deficient medium, the *ty* mutants do not produce a protein which crossreacts with antibody to the wild-type tyrosinase. This absence of crossreacting material indicates a virtually complete absence of a protein even similar to tyrosinase. However, if a heterocaryon is formed containing *ty-1* and *ty-2,* the *T* gene is expressed, and the tyrosinase formed is that which is expected from the particular *T* allele present (460). Homocaryons carrying either *ty-1* or *ty-2* produce large amounts of tyrosinase in the presence of tyrosine, dopa, phenylalanine, and tryptophan in the D-stereoisomeric forms. The natural L-isomers have no activity as inducers. The synthesis of tyrosinase by heterocaryons composed of *ty-1 + ty-2,* or in the presence of inducers in homocaryons, makes it quite evident that *ty-1* and *ty-2* are controlling genes of some sort that interact with the structural gene, *T.* The fact that the + alleles of *ty-1* and *ty-2* are active in the heterocaryons indicates that some sort of cytoplasmic action is involved. Furthermore, it should be noted that these genes are quite different from the *i* genes of *E. coli,* because the allele that confers inducibility is *recessive* to the constitutive condition in *Neurospora.*

Induction and Repression. The phenomena of enzyme induction and repression were discovered quite independently, but when the mechanism of induction and its attendant genetic factors began to be understood, it was quickly recognized that the two might be closely related. Indeed, models may easily be constructed which incorporate most of the known facts about both phenomena with considerable agreement and consistency into a unified theory. One such model as devised by Jacob and Monod (485) is given in Figure 9.8.

This model assumes that there are two general types of genetic elements: controlling genes and structural genes. The controlling genes in turn exist as two types: *regulators* and *operators*. Regulators may or may not be closely associated with the genes they regulate, but in any event they act through the medium of a repressor substance. The repressor acts specifically on an operator causing it to inhibit the action of the structural genes under its control so that no protein is formed. The operator is always close to the structural genes it controls and the whole unit is called an *operon*. Enzyme induction and repression are considered to be different aspects of the same phenomenon. It is assumed that repressor substances react with certain small molecules specific for each type of repressor. In inducible systems the inducer interacts with the repressor and prevents it from affecting the activity of the operator. As a result, the structural genes are active. In repressible systems the repressor can only be active after it has reacted with the repressing substance which may be called corepressor. The

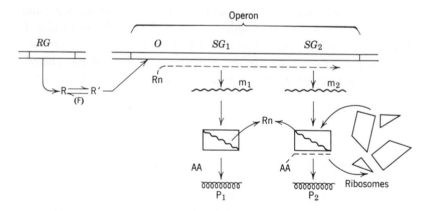

Figure 9.8. The Jacob-Monod model for the regulation of enzyme synthesis. RG = regulator gene; R = repressor converted to R^1 in presence of effector F (inducing a repressing metabolite); O = operator gene; SG_1 and SG_2 = structural genes; rn = ribonucleotides, m_1 and m_2, messengers made by SG_1 and SG_2; aa = amino acids; P_1P_2 = proteins made under control of SG_1 and SG_2 via specific messenger RNA. From Jacob and Monod (485).

repressor-corepressor complex prevents the operator from acting and no proteins controlled by the operon are formed.

In order to account for the sudden cessation of synthesis of an enzyme when an inducer is removed or a corepressor added, it is necessary to assume that something highly short-lived or labile exists in the protein-making machinery. As discussed previously, the primary action of a gene is to make messenger RNA which in turn acts as a protein template on, or in, a ribosome in the cytoplasm. Jacob and Monod make the assumption that messenger RNA is labile and acts as a template only once, whereupon it must be replenished from the nucleus. The corollary to this is that while a protein is being synthesized actively the structural gene behind it must also be active. Protein synthesis, then, can only occur in the presence of active nuclei. A considerable body of evidence exists to support the contention that active protein and RNA synthesis requires a nucleus or genic material, and that there is a labile RNA fraction in the cytoplasm which must be continually resupplied from the nucleus. This evidence comes primarily from studies on bacteria, and *Amoeba* and other protozoa. A universal dependence of protein biosynthesis upon the nucleus is, however, not established. In fact, it has been shown that in the alga, *Acetabularia,* both RNA and protein synthesis continue for long periods (up to two to three weeks) after the removal of the nucleus (94). Synthesis does eventually stop, but these results do not support the contention that all protein is synthesized via highly labile messenger RNA constantly renewed from the nucleus. The overriding importance of the nucleus in protein synthesis cannot be denied, however, for the specific type of protein synthesized must, so far as can be determined at present, be dictated by a genetic element. The only question really at issue is how independent can the biosynthetic mechanism for protein synthesis *in vivo* become from active gene participation once initial specificity has been dictated?

Whatever the answer to this question, it is abundantly clear that in microorganisms, at least, there are built-in mechanisms for the control of protein synthesis. These mechanisms are in turn under the control of elements which have the properties of genes insofar as they duplicate with the genome and can mutate to a stable new condition. One can reason teleologically that the control mechanisms exist for the specific purpose of regulating the cellular economy, that is, the prevention of protein synthesis above what is needed for carrying out the minimum requirements of the cell. It should be evident that if a cell were to concentrate too much of its efforts on the uncontrolled synthesis of just a few proteins it might not be able to synthesize other proteins just as necessary for its maintenance. Thus, from this point of view, it is quite logical to assert the existence of controlling genes which interact with structural genes for the purpose of maintaining a balanced metabolism.

Controlling Genes in Maize. In Chapter 4 the controlling elements discovered in maize by Rhoades and McClintock were discussed primarily with respect to their effect on the "mutation rate." As has been pointed out by McClintock (641), a more realistic view of them would be to consider them in the light of the work just described in bacteria as controlling genes of the regulator and operator type. Thus, in the *Ds-Ac* system, *Ds* may be considered an operator and *Ac* a regulator. The *Dt* gene of Rhoades may also be considered a regulator.

The basis for making such an analogy can best be appreciated by considering the Suppressor-mutator systems of McClintock (640). Three independent loci, A_1, A_2, and *Wx,* were found to be affected by this system. The *A* loci are associated with anthocyanin formation in the kernels and plant, and *Wx* controls the formation of amylose in the pollen grains and endosperm. A number of changes involving these loci were found which could be explained by assuming the insertion of an "operator" at the loci of the structural genes. Thus, from A_1 three "mutant" forms, a_1^{m-1}, a_1^{m-2}, and a_1^{m-5}, were derived which could be differentiated from A_1, and one another, on the basis of the amount of pigment produced in the kernels and plant tissues. A second controlling element, the Suppressor, *Spm,* has been described which regulates the activity of the modified alleles such as a_1^{m-1}, a_1^{m-2}, and a_1^{m-5}. When *Spm* is present, and it need not be adjacent to the A_1 locus to be active, it suppresses pigment production by the operator associated states, a_1^{m-1}, etc., so that no pigment is produced except in spots or patches. In the presence of active *Spm* two general classes of effects may be obtained. First, no pigment is produced unless *Spm* enters an inactive phase, or is lost by somatic transposition or meiotic segregation. There is no alteration of the *A* locus in the nature of a gene mutation. The structural gene present merely expresses itself because a regulator gene, *Spm,* becomes inactive or is lost and presumably no longer affects the operator at the *A* locus. The second type of effect is also brought about in the presence of *Spm,* but by changes in the operator-locus association rather than loss or change in the regulator. Two main consequences may result: production of a stable mutant, or production of a new operator-associated state.

The *Spm* factor itself may undergo changes of a number of different types such as (1) time of its transposition during development, (2) weakening of its ability to effect mutation at the A locus, and (3) alternation between active and inactive phases in a cyclic fashion.

Although much remains to be explained about the *Spm* element in maize, it is quite clear that it resembles the regulator gene described for bacteria because it seems able to turn off the activity of apparent structural genes through the medium of its effect on associated operators. The existence of an operator is indicated by the fact that there are two different states at the A_1, A_2, and Wx loci—a state of sensitivity to *Spm* and a state of insensitivity. It is assumed that when the operators are present at the loci, sensitivity to *Spm* is found; otherwise in plants carrying *Spm* without operator there is no repression of gene action, that is, the synthesis of some enzyme is carried out.

Groups of Interacting Genes with Large Effects on Specific Characters

An insight into the complexities of gene interaction is perhaps best afforded by considering some examples of the effects of groups of genes on a specific character such as pigmentation. The inheritance of mammalian coat or skin color has been a favored subject for the investigation of gene interaction. The results have provided adequate evidence for the great complexity of gene interaction and, in addition, have brought to light the existence of a number of interesting phenomena which are undoubtedly significant manifestations of genes acting in groups. One of these investigations will be considered in some detail, and attempts will be made to analyze some of the results in terms of the simpler examples already cited.

Differences in Melanin Pigment in the Guinea Pig. One of the most elegant and thorough investigations of pigment inheritance to be found in the literature is that conducted by Wright (1112) and his associates over a period of more than 40 years on the melanin pigments of the guinea pig. The results have demonstrated that melanin in animals is produced by the interaction of a large number of genes.

Melanin is the primary pigment found in the vertebrate integument and the integumentary derivatives such as hair, feathers, and scales. Chemically, melanin is a highly intractable substance, because it is insoluble in most solvents and it ordinarily is combined with protein. However, there is no doubt that tyrosine provides the primary precursor compound for vertebrate melanin. By a series of reactions tyrosine is converted to indole-5,6-quinone which then polymerizes in an as yet undetermined manner (662). The polymerized indole-quinone is generally attached to a protein matrix, the whole forming a granule, the melanin granule. Melanin granules are produced by specialized cells, the melanocytes, which are derived from the neural-crest ectoderm in the vertebrates. The granules are produced within the cytoplasm of the melanocytes by an as yet unknown process, but apparently the protein matrix is formed independently, the melanochrome synthesized in the presence of the enzyme, tyrosinase, and then combined with the protein. After their formation, the granules may be extruded from the melanocytes and even enter other cells. Melanin granules coloring the hair of mammals or feathers of birds are deposited there by the melanocytes as the hair or feather structure is formed in the follicles. Hair color, shade, and intensity are determined by a number of factors such as the actual color of the melanin, the concentration of granules, the amount of melanin per granule, and the size and shape of the granules. Two types of melanins, based on color, are formed in mammals: the dark melanins, or eumelanins, and the yellow-orange red melanins, or phaeomelanins.

A considerable number of factors, genetic and otherwise, are involved in the determination of coat color in the guinea pig. Some of these are listed in Table 44. The major factors are considered first.

The gene E determines the presence of eumelanin; its recessive allele, e, gives a phaeomelanic animal. The allele, e^p, produces the tortoise pattern of a eumelanic background with phaeomelanic patches. Two qualitatively different eumelanins are produced in eumelanic animals, brown in bb animals, and sepia in B animals. Both sepia and brown pigment may be reduced in intensity in the presence of the recessive allele of the gene P. The yellow, phaeomelanic pigment is affected quantitatively by the gene F which is incompletely dominant. FF animals may be intense yellow, with Ff and ff genotypes, in that order, having correspondingly less yellow pigment. F has no effect on eumelanic animals in the presence of P, but $ffpE$ animals have no eumelanin and may have a small amount of phaeomelanin.

The intensity of the pigmentation produced by E, P, F, and B and their respective alleles is determined by a series of five albino alleles, C, c^k, c^d, c^r, and c^a. C is in most cases completely dominant over the lower alleles and gives the most intense pigmentation. Animals homozygous for c^a are albinos with pink eyes. The other alleles give intermediate intensities.

Despite the lack of knowledge of melanin chemistry, it has been

Table 44. Major Factors and Modifiers Determining Coat Color in the Guinea Pig [*]

Major Factors

E,e^p,e: Eumelanic, tortoise, or phaeomelanic
C,c^k,c^d,c^r,c^a: Albino series of quantitative amount of pigment
P,p^r,p: Dilution of eumelanic pigment
B,b: Sepia or brown eumelanin
F,f: Reduction in phaeomelanin

Major Modifiers

A,A^r,a: Agouti or self-colored
S,s: Spotting
Si,si: Silvering
Dm,dm: Diminution
Gr,gr: Grizzling

Other Modifying Factors

Age of dam
Age of individual
Amount of androgen
Temperature
Region of skin
Minor modifying genes; number not determined

---

[*] From Wright (1113).

found possible to make quantitative estimates of the amount of melanin in the hairs of various colors and shades of guinea pigs by gravimetric and colorimetric methods after extraction of the melanin with suitable solvents (423, 866). In addition, Wright (1111, 1112) has estimated the relative amounts of pigment in the different phenotypes by using a series of standard skins. The estimates obtained by the colorimetric method agree quite closely with those determined by means of the empirical grades (1114) so that a fair degree of confidence may be placed in the estimates. Table 45 gives the most recent data obtained by using the empirical grades as reported by Wright and includes

Table 45. Quantitative Estimate of Amount of Melanin Pigment in the Hair of Guinea Pigs at Birth, and Effects of Some of the Factors on Eye Color*

| | Eumelanin | | | | Phaeomelanin | | | Eye Color | | |
| | Sepia | | Brown | | Yellow and Red | | | Ec and Ff without Effect | | |
	EPB	EpB	EPb	Epb	Epf	eFF	eff	PB	Pb	P
$C-$	100	20	50	15	7	100	34	Black	Brown	Pink
$c^k c^k$	93	15	49	12	0	37	4	Black	Brown	Pink
$c^k c^d$	90	13	50	12	0	42	6	Black	Brown	Pink
$c^k c^r$	96	9	49	12	0	18	—	Black	Brown	Pink
$c^k c^a$	77	8	38	8	0	18	0+	Black	Brown	Pink
$c^d c^d$	66	8	42	10	0	39	4	Black	Brown	Pink
$c^d c^r$	76	5	48	6	0	17	0+	Black	Brown	Pink
$c^d c^a$	40	4	33	5	0	17	0+	Black	Brown	Pink
$c^r c^r$	87	3	48	3	0	0	0	Dark red	Dark Brown red	Pink
$c^r c^a$	44	1	34	1	0	0	0	Light red	Light brown red	Pink
$c^a c^a$	0	0	0	0	0	0	0	Pink	Pink	Pink

* Data from Wright (1111, 1112, 1114).

descriptions of the eye colors produced by the various genotypes. The estimates for eumelanin are given on the basis of the most intense sepias taken as 100%. The phaeomelanins are given in percentage of the most intense phaemelanic phenotypes.

The C alleles express themselves differently in the eumelanic and phaeomelanic animals. In the browns and sepias there are five discernible alleles, which are listed here in increasing order of effect, c^a, c^r, c^d, c^k, and C. The phaeomelanic, ee animals, however, give evidence for the existence of only three alleles, designated as c^{ra}, c^{kd}, and C; for $c^a = c^r$, and $c^k = c^d$ in effect. The only C alleles which reduce eye melanin are c^a and c^r. All animals homozygous for $c^a c^a$ are pink-eyed albinos regardless of the residual genotype. When c^r is present homozygous or heterozygous with c^a, the pigment of the eye is reduced sufficiently for the red color of the hemoglobin in the iris to be evident. Within the eumelanic animals, the dark-eyed sepias and browns (EPB and EPb) show peculiar effects with the different combinations of the C alleles. There are a series of "waves" of effect when the data are plotted as they are in Figure 9.9. These undulations are not evident in the pink-eyed sepias (ppB) and browns ($ppbb$). If the quantitative effect of the C alleles on sepia in EPB animals is compared to their effect on the yellow pigment in eFF and eFf animals, a correspondence will be noted between the dilution in the sepias in the presence of c^d, c^a, and $c^d c^d$ and the intensification of the yellow pigment. On the other hand, the combinations $c^d c^a$, $c^d c^r$, $c^k c^a$, $c^k c^r$, and $c^d c^d$, $c^k c^d$, $c^k c^k$ give quantitatively the same amount of phaeomelanic pigment, but there is an increase in eumelanin in their presence.

The genes p and b both reduce the concentration of eumelanin. In addition to a quantitative reduction, b also produces a qualitatively different pigment, brown, instead of sepia. There seems to be no effect of p on eumelanin other than that it reduces both brown and sepia and eliminates eumelanin completely in the presence of ff. The eye color is changed by both P and B. In pp animals the eyes are always pink. The eyes are brown in Pbb animals unless, as just noted, $c^a c^a$ is also present.

Wright's interpretation of these data with respect to hair color is given in Figure 9.10. This scheme proposes that the genes of the $ECPFB$ group each act at a different step in pigment granule formation, but not in a simple complementary fashion. The E series is assumed to act at the level of melanocyte differentiation in eumelanin- or phaeomelanin-producing cells. The C series genes act on both phaeomelanic and eumelanic directed processes. But since both of these processes

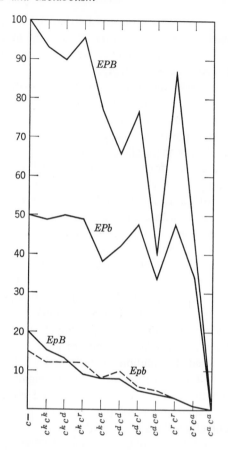

Figure 9.9. Per cent of melanin in different genotypes of guinea pigs.

are assumed to depend on some limited factor or substrate, there is a competition between the two processes. The C alleles are assumed to produce products (enzymes) of different specificity with respect to the phaeomelanic and eumelanic processes. C acts on both equally well; c^k shows greater specificity for the eumelanic process and c^d for the phaeomelanic. Neither c^r nor c^a is active for the phaeomelanic process, but c^r is active for the eumelanic process. The end result of the action of the active C alleles is hypothesized to be a eumelanic-enzyme precursor when C, c^k, c^d, or c^r are present, and a phaeomelanic-enzyme precursor when C, c^k, or c^d are present. Both precursors are presumably produced when C, c^k, or c^d are present. The enzyme precursors are

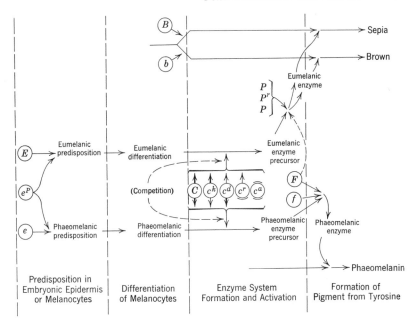

Figure 9.10. Wright's explanation for the action of the different kinds of factors controlling melanin formation in the hair of guinea pigs. From Wright (1113).

then assumed to be acted upon by *P* and *F,* the eumelanic precursor by *P,* and the phaeomelanic by *F.* The result is the formation of a eumelanic and a phaeomelanic enzyme, each of which acts upon an appropriate substrate to form eumelanin and phaeomelanin respectively. The *B* gene is assumed to act at the level of substrate preparation such that *B* action leads to sepia melanin and *b* to brown melanin. The original melanin precursor substrate for all melanin is considered to be tyrosine.

This interpretation at first sight seems to be unnecessarily complex until it is recognized that the quantitative data in Table 45 are quite well explained by it. It is obvious that the alleles c^r and c^a are not active toward phaeomelanin production, but the rest of the *C* alleles are. If, now, *EPB* and *EPb* animals are considered, it will be seen that homozygous *C*, c^k, C^d, and c^r have much the same capacity for the quantitative production of melanin. Only $c^d c^d$ animals are low with respect to sepia-melanin production. This is explained by assuming that all of these genes produce sufficient enzyme to make enough eumelanic- and phaeomelanic-enzyme precursor to maintain a high concentration of eumelanic and phaeomelanic enzyme. This concentration is maintained at a high enough level so that the enzymes are always in excess with respect to their substrates. The $c^d c^d$ homozygote is low relative to the rest, because it is more active for the phaeomelanic process as shown by Table 45. The c^r allele is highly active, both in the homozygous and in the heterozygous condition, for eumelanic production even though it is presumably a low-grade hypomorph, because it is completely inactive for the phaeomelanic process, and hence there is no competition for its product. The other combinations are explained in a similar fashion. The pink-eyed eumelanics (*EpBFF* and *EpbFF*) have reduced eumelanin but also show, much more so than the dark-eyed eumelanics, the effects of substitution at the *C* locus. Actually, if these differences were not revealed in the pink-eyed sepias and browns, it would not be possible to distinguish five alleles. Why are these alleles expressed when *P* is recessive, but not when dominant? The explanation may again be based on enzyme-substrate relationships. When *P* is present the enzyme produced by this locus is in excess relative to its substrate, the eumelanic-enzyme precursor. All precursor produced is readily transformed to eumelanic enzyme, and the eumelanic enzyme is in excess with respect to *its* substrate. However, when *pp* is present only small amounts, or no enzyme, are formed by this locus. If none, the enzyme formed by the *F* locus is assumed to be weakly active for the eumelanic precursor. In either event, only small amounts of eumelanic enzyme are formed, and the eumelanic substrate is now in excess relative to its enzyme. Under these conditions one expects change in the rate of production of product when there are small changes in enzyme concentration. Such changes would occur by substitutions at the *C* locus.

The actions of the major genes just described are modified by a number of modifier genes which are fairly well characterized with respect to their segregation in inheritance. The most important of these are listed in Table 44 under major modifiers. The agouti gene, *A,* and its allele, *A^y,* cause the hairs of eumelanic animals to develop with a subterminal band of phaeomelanin. This means that when the hair starts its development the melanocytes incorporate eumelanin, but after part of the hair is formed only phaeomelanin is formed for a period and incorporation switches back to eumelanin again. Whether the melanocytes stop synthesizing eumelanin in the follicles during the period when only phaeomelanin is formed is not known. Agouti animals have only yellow hairs on the belly. Just as mysterious in its functioning is the spotting gene *S*. The genes *Si* and *Dm* have rather drastic effects on coat color, especially when both are homozygous recessive. Animals of the genotype *si si dm dm* have complete absence of pigment. These "silver-white" animals are also anemic and have low viability. The males always appear to be sterile, and the females demonstrate low fertility. Substitution of *Dm* and *Si* causes an increase in the amount of pigment present, the effect being cumulative and roughly proportional to the number of *Dm* and *Si* genes present. The silvering caused in *si si Dm*− and *Si si Dm*− animals is manifested as a blotching and sprinkling of white and a slight dilution in the unsilvered areas. When *dm* is present in *Si Si* or *Si si* animals it causes a slight diminution of the pigment. Both gene loci have independent effects, but their recessives act synergistically. The effects of the recessives, particularly that of *dm,* is most marked with the lower hypomorphs of *C.*

The grizzled gene, *gr,* has no effect at birth, but as the animals grow older a progressive whitening of the hairs on their backs occurs.

In addition to these modifiers which have been well characterized, there exists a host of others which cause somewhat less drastic modification of the coat color.

The activity of an enzyme presumed to be of importance in the production of melanin in the guinea pig has been investigated by W. L. Russell (866), Ginsburg (325), and Foster (288) in an attempt to correlate enzyme activity with observed phenotypic effects. The problem was approached by assuming that when a black compound is formed in the presence of tyrosine or dopa (3,4-dihydroxyphenylalanine), an oxidation product of tyrosine, it is evidence for the presence of tyrosinase. This is an assumption which has considerable experimental evidence in its favor, but there is no direct evidence for it in the guinea pig, other than that in the presence of dopa a black pigment is formed by those tissues, and their extracts, which are most active in producing natural melanin. The pigment formed from externally introduced dopa is black even in what seem to be pure phaeomelanic guinea pigs. The results of testing extracts and tissue slices of guinea-pig skin from a large number of different genotypes for dopa-oxidizing activity by the enzyme, dopa oxidase, as determined by the appearance of a black color, were quite clear in indicating that substitutions at the *P* and *B* loci have little or no effect and that *E, C,* and *F* have definite effects. The activity of the *C* alleles in *both* the eumelanic and phaeomelanic animals is restricted to four levels corresponding to the effects of three alleles, *C,* c^{kd}, and c^{ra}. These correspond to the same three alleles which are expressed in phaeomelanic *ee* animals. Substitution of *f* for *F* has a similar reducing effect on oxidase activity in both pigment types. The conclusion has been drawn, therefore, that the enzyme measured by its activity for dopa oxidation is concerned primarily with phaeomelanin production (325, 866). The dopa-oxidase activity of the active melanin-producing tissues of the house mouse has also been tested, with essentially the same results. A correspondence of dopa-oxidase activity with the levels of phaeomelanic concentration is indicated, but little or no correspondence with changes in eumelanic pigment (865) were noted.

Another method for determining tyrosinase activity takes advantage of the fact that oxygen is taken up as tyrosine is oxidized, and therefore the reaction rate can be determined by noting the uptake of oxygen manometrically. When this is done with guinea-pig skin homogenates, a rough correlation with the result just described is obtained. The significant differences are that the replacement of $P-$ by pp reduces the amount of O_2 uptake about 25 to 50%, whereas no significant change is noted when the appearance of melanin is determined turbidometrically. In addition, no change in O_2 uptake is noted when ff substitutes for $F-$ which is the reverse of what is found by turbidity measurements (288).

These results with the guinea pig reveal quite strikingly that a supposedly simple characteristic such as pigmentation may have a complicated genesis which involves the interaction of numerous genes. Only a few of these genes are generally recognized by the geneticist, for he is limited to the visible mutations affecting coat color. Work on inherited differences in pigmentation has been done in a number of other vertebrates which supplements and confirms a great many of the conclusions drawn from the guinea-pig results.

Differences in Melanin Pigment in Other Vertebrates. The nature of pigmentation in the house mouse has been investigated in great detail by E. S. Russell (860–862) with respect to four characteristics: (1) nature of granule color; (2) degree of pigmentation; (3) size of granules; and (4) clumping of granules. All four of these are of significance in determining the final coat color. The effects of gene substitution on these different aspects of pigmentation are quite striking and can be effectively studied by observing the condition of the hair follicles in section.

In general it was found that the C series seems to affect only the degree of pigmentation by changing the number of granules or reducing their size, and that the P locus (corresponding in effect to P in the guinea pig) produced a similar reduction in volume, but by means of changing the shape of the granules to shred-like particles, as well as reducing their size. Qualitative pigment changes in the granules were obvious in substitutions at the B (brown) and the A^y (agouti) loci. Both of these seem to control the type of pigment present, black or fuscous to brown in the B to b substitutions, and eumelanins to phaeo-melanin in the A^y to a changes. The change produced by B in the quality of the granular pigment is accompanied by a change in shape and total volume of the pigment. The so-called dilution gene, d, in the mouse, which has no known counterpart in the guinea pig, appears from visual impressions of the coat color to reduce the total volume of pigment in both eumelanic and phaeomelanic animals when it re-places dominant D which gives no dilution. However, histological ex-amination reveals that substitution of D by d does not reduce the volume of pigment (actually, it appears to increase it), but that the apparent dilution is caused by a clumping of granules. The melanocytes present in dd animals can be quite clearly distinguished from those present in $D-$ animals. The dilute animals have melanocytes with fewer and thinner dendritic processes than the nondilute animals. The gene, leaden (ln), gives approximately the same phenotype when homozygous recessive as dd (660).

These observations are of considerable significance in the analysis of coat-color differences. They make it clear that a chemical approach to the problem, even provided the chemistry of the melanins were un-derstood, must be accompanied by morphological and developmental considerations.

This is also evident from the analysis of the albino axolotl, a neotenic salamander. The albino condition occurs in the presence of the reces-sive gene d. DD or Dd axolotls are heavily pigmented with melanin, but dd animals are not; only longitudinal streaks of pigment may appear in the middorsal region above the spinal cord. It can easily be demon-strated that the albino produces melanocytes from the neural-crest re-gion and that these cells are indeed capable of producing melanin, if they are explanted to an inorganic salt solution or to an animal which carries the D gene (196). It would appear, then, that the dd animals are albino because their melanin-forming cells do not migrate into the surrounding integument from the neural crest. In support of this ex-planation are the chimera experiments of Church (168) and the trans-plantation results of Dalton (195). If the anterior portion of an albino embryo is fused with the posterior portion of a black embryo before the time of migration of the melanoblasts from the neural crest, the resultant adult chimera is white in the anterior and black in the posterior portions. No signs of migration of melanocytes from the posterior $D-$ portion

into *dd* skin are evident.   When a genotypically black anterior is joined
to a white posterior, a slight amount of melanocyte migration occurs
from the anterior into the immediately adjoining posterior portion.
Only dermal melanocytes migrate; however, the epidermal melanocytes
do not.   These results show that the conjoining of *D*— and *dd* parts
does not cause any considerable effect in either of them.   The possibility that the *dd* animals have a completely defective pituitary is eliminated, because the normal pituitary in an anterior *D* half has little effect
on the pigmentation of an albino posterior.   On the other hand, an
albino pituitary has no effect on a *D*— posterior.   This is of some significance, because it is known that hypophysectimized embryos do not
produce pigment.   The slight effect of the anterior black part on the
posterior white may be due, however, to the *D* pituitary being slightly
more effective than a *dd* pituitary.   Some indication that this may be
true is obtained when a pituitary from a *D*— animal is substituted for
the pituitary of an albino.   The albino still remains largely unpigmented,
but its melanocytes are larger and darker than in control albinos (197).
However, the bulk of the evidence indicates that the albinos are largely
unpigmented, because their melanocytes do not migrate.

Genic Balance

An organism is the resultant of thousands of genes acting in concert.
Its phenotype results from the blended, balanced, and synchronized
activities of the products of all genes in the genome.   On this point
of view all genes interact with one another through their products and
are therefore modifiers of the activities of one another.   In order for the
modifying system to act harmoniously to produce a functional organism,
the system must be in balance—this is *genic balance*.   It is necessary,
in other words, not only to have a full complement of genes, but a
complement of genes present in the proper dosage relations necessary
to produce a balanced system.

In a sense the concept of genic balance receives its support from all
experiments which show the dependence of a specific character upon
more than a single gene, but the idea is perhaps best brought out by
considering the normal process of determination of sex in *Drosophila*.
In this species, as in nearly all organisms with sexual dimorphism, sex
is determined by the chromosomes.   There is a mechanism which acts
to produce an approximately equal distribution of sexes, and in *Drosophila* it appears to be associated with a balance between the genes on

the autosomes versus those on the X-chromosome. Bridges (107) has described the effects of changing the normal chromosome complement of *Drosophila melanogaster* (two X's and two pairs of each autosomal type in the female, and XY with two pairs of each autosomal type in the male) on sex. Some of the combinations and the results are given in Table 46. In this table, 1A means that each of the autosomes is present once, 2A each twice, and so on. It will be noticed that females are produced whenever the ratio of X-chromosomes to autosomes in each of the homologous groups is 1:1. Exceptional females, the so-called "super females," result from a ratio of 3X:2A. If the number of X-chromosomes is reduced relative to the number in each group of autosomes, normal males (1X:2A, 2X:4A) and "super males" (1X:3A) result. Individuals with both male and female secondary sex characteristics, the so-called "intersexes," arise as a result of ratios intermediate between 1X:2A and 1X:1A. Thus 3X:4A and 2X:3A individuals are intersexes. The Y-chromosome normally present in the male apparently has no role to play in sex determination for 2XY:2A females are typical females and XO:2A individuals are definitely male in appearance, although sterile.

The nature of the ratios, in which a female is always produced when the X-chromosomes are equal to or greater in number than the autosomes in the homologous group, would indicate a preponderance of female-determining capacity in the X-chromosome. This could be due to a single gene or a group of genes on the X. Extensive tests by means of special techniques involving duplication of small segments of the X

Table 46. The Effect of Changing the Ratio of X-Chromosomes to Autosomes on the Determination of Sex in *Drosophila melanogaster*

Super Female	3X:2A
Female	4X:4A
Female	3X:3A
Female	2X:2A
Intersex	3X:4A
Intersex	2X:3A
Male	1X:2A
Male	2X:4A
Super male	1X:3A

in males and intersexes have proved that there is no one gene on the X or even a group of genes close together which will cause a male or intersex to become female (223, 663, 772, 785). The multiple female-determining genes are scattered almost at random, and they express themselves cumulatively in the female direction in conjunction with genes on the autosomes which may be considered to be acting in the direction of maleness. But attempts to shift female flies in the direction of maleness by adding segments of autosomes 2 and 3 have been as unsuccessful as the attempts to make the shift to femaleness by adding small portions of the X (786).

Even though under normal conditions the sex of *Drosophila* can be changed only by manipulating blocks of genes on the autosomes and X-chromosome to establish a new balance, this does not mean that sex cannot be changed by mutation of a single gene. A number of auto-somal gene mutations have been described which convert diploid (2X: 2A) females into sterile males or intersexes in *D. melanogaster* (348, 986), *D. virilis* (567, 973), and *D. subobscura* (944). Thus, despite the fact that a large number of genes are known to be involved in the determination of sex, a single gene change may cause a complete or partial reversal of sex. This may seem to be a paradox, but in reality it is precisely what one would expect in view of the preceding discussion in this chapter. Among the many genes acting together to produce sex organs and other secondary sexual characteristics, there are those which must be present in certain relative frequencies in order for one or the other sex to be produced. These latter may be acting against one another, one set in the male direction, the other in the female di-rection, as the ratios of X-chromosomes to autosomes would indicate. It is therefore likely that a mutation of any one of them by reducing or increasing the dosage activity sufficiently toward producing one sex would cause a reversal of sex, even though all other genes are unaffected.

Support for the concept of genic balance is by no means confined to the results of analysis of sex determination in *Drosophila*. Investi-gations of sex determination of different animal species such as the gypsy moth, *Lymantria* (334), the guppy fish, *Lebistes* (1088), the wasp, *Habrobracon* (1074), and even a number of plants (1068) have resulted in data consonant with the idea of genic balance.

A considerable amount of evidence is available, even from man, which shows that similar though not identical balances are necessary for normal development. As shown in Table 47 a 3X:2A individual is a female. This is in agreement with what is known in *Drosophila*. However, 2X:1Y, 3X:1Y, and 4X:1Y individuals are all males,

Table 47. Various Abnormal Chromosome Constitutions in Man

Constitution	Phenotypic Effect	Reference
3X:2A	Female, generally with deficient mentality	921
X:2A	Females, Turner's Syndrome, gonads vestigial	379
2X:1Y:2A 2X:2Y:2A	Males, Kleinfelter's Syndrome, with various anatomic derangements	921
3X:1Y:2A 4X:1Y:2A	Males, with microorchidism and deficient mentality	921
XY:2A + extra 21 2X:2A + extra 21	Mongolism, mental retardation	586
XY:2A + extra 22	Sturge-Weber Syndrome, mental retardation	421

Note: The normal male is 1X:1Y:2A and female 2X:2A.

albeit abnormal males with varying degrees of microorchidism and mental deficiency. The sex-determining mechanism seems to be somewhat different from *Drosophila* insofar as the Y-chromosome seems to have a definite male-determining role. Thus, no matter what the ratio of X-chromosomes to autosomes is, the individual is a female, unless a Y-chromosome is present, in which case it is a male. The role of the Y-chromosome as a male-determining chromosome apparently is also true for the mouse, because in this mammal XXY:2A individuals are males (864). However, a proper balance between the X, Y, and autosomes is necessary in man for a normal individual to develop. This is evident from the data in Table 47. Frequently, anatomical abnormalities are accompanied by mental retardation. In fact, a small, but significant fraction of the individuals inhabiting mental institutions have abnormal chromosome ratios.

By utilizing the technique of duplicating or reducing various sections of the genome, it is possible to show that the genic-balance mechanism acts for characters other than sex. In some plant species individuals frequently arise with one or more extra chromosomes in addition to the normal complement, a condition related to the aberrant chromosomal complements discussed in the preceding paragraphs in connection with sex determination. The general term applied to this condition is *aneuploidy*. Figure 9.11 illustrates some aneuploid types. Trisomic aneuploids ($2n + 1$, $2n + 1 + 1$, etc.) are usually readily distinguishable by phenotype from the normal diploid or the almost normal triploids or tetraploids. The addition of extra chromosomes in both the Jimson weed, *Datura* (40), and tobacco, *Nicotiana sylvestris* (340), has a distinct effect on the phenotype. The phenotypic alterations produced depend on the chromosome duplicated and the size of the duplicated piece, if it be only a portion of a chromosome. They may vary in type from changes in overall size of the plant to shape of leaves and flowers, and in a number of physiological characters such as time of flowering, and so on. Extreme divergence is noted particularly in haploids of *Datura* which are disomic. As shown in Figure 9.12, the haploid is smaller than the normal diploid, and the addition of a single chromosome does not carry it toward the diploid phenotype, but to an extreme degree in the opposite direction.

Abnormal development reflected primarily in changes in fertility and viability is also obtained in *D. melanogaster* when sections of chromosomes are duplicated (120, 771). This is true for duplication of parts of both the sex chromosomes and autosomes, although changes are more drastic when certain parts of the autosomes are duplicated than when the sex chromosomes are duplicated. This indicates an intrinsic tolerance toward duplication on the part of the X associated with its role

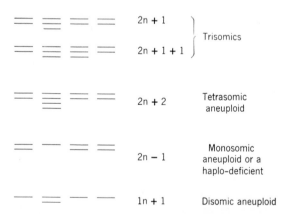

Figure 9.11.   Some types of aneuploids.

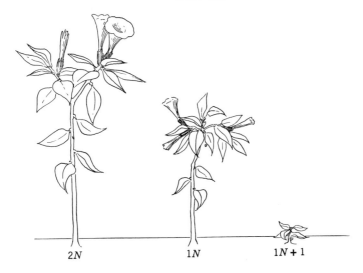

Figure 9.12. A comparison of diploid, haploid, and disomic Datura plants. Copied from a photograph published in the *Journal of Heredity* after Satina, Blakeslee, and Avery (877).

in sex determination, a fact no doubt related to its being regularly haploid in males and diploid in females. Abnormal conditions also result in man when autosomes are duplicated either in whole or in part. The condition called Mongolism or Mongolian idiocy occurs in those individuals who are trisomic for chromosome number 21. Mongoloids show only minor morphological deviations from the norm, but they are all deficient mentally. A number of other pathological conditions are caused by trisomy in chromosomes other than 21 (Table 47), so that it is apparent that genic unbalance due to unbalanced duplication is an important source of abnormalities in man.

Crosses between different species frequently result in the appearance of sterile progeny, or progeny which are strikingly different from either of the parent forms. Sterility in species hybrids has been especially studied in *Drosophila* species—see Patterson and Stone (770). Gross changes in morphology in *Drosophila* species hybrids have been reported in crosses between *D. athabasca* and *D. azteca*. The hybrids in this case may be giants or dwarfs compared to the parents, depending on the type of cross made (988). Similar morphological changes also result from plant interspecific crosses such as in mosses and flowering plants. Some of these results are to be attributed to nondisjunction resulting in aneuploids, but for the most part the abnormal hybrids are regular diploids or polyploids and not comparable to the aneuploids just discussed.   Nonetheless, the general explanation is the same: the combining of two somewhat dissimilar complements of genes into one individual results in an unbalanced genetic system which cannot function in a normal fashion.

On the other hand, hybrids resulting from crosses between unrelated strains frequently show a marked enhancement of a particular character or group of characters which are of adaptive value to the organism or of economic value to the breeder.   This phenomenon is termed *heterosis* or *hybrid vigor,* and it is the direct result of achieving a high degree of heterozygosity in a hybrid.   A number of hypotheses are extant concerning the explanation of this phenomenon which is a manifestation of genic balance.

Although it is generally recognized that heterosis is a complex phenomenon with a variety of causes, two explanations in particular seem to fit the known data best (191, 192).   These are both based on the fact that heterosis arises from heterozygosity.   One of them states that heterosis is the result of dominant, "beneficial" alleles masking the effects of recessive "deleterious" alleles in the hybrid.   It assumes that in producing the parents of the hybrid by inbreeding, certain recessive genes with deleterious effects are made homozygous, and their beneficial dominant alleles thereby eliminated.   The chances are slight that the same sets of genes will be made homozygous for deleterious recessives in any two independent strains by inbreeding, and hence when these are crossed many of the recessives will be covered by dominants in the hybrid.

A second hypothesis is based on the assumption that heterozygotes resulting from two alleles, for example, a_1 and a_2, in combination may be more "vigorous" than the respective homozygotes, a_1/a_1 and a_2/a_2. This is actually more than an assumption, for it has already been noted that in certain allelic series certain heterozygous combinations will produce phenotypes which are unexpected on the basis of their action when homozygous.   The action of pseudoalleles in heterozygotes will also be recalled.   Furthermore, many extreme cases of complementary action of what appear superficially to be allelic genes have been recognized in which genes giving a distinct mutant phenotype when homozygous produce a normal phenotype characteristic of the normal dominant allele when heterozygous.   The maize mutant gene yg_2, when homozygous, results in seedlings with a yellow-green color, but in combination with a mutant allele, pale yellow (py), the heterozygote is normal green.

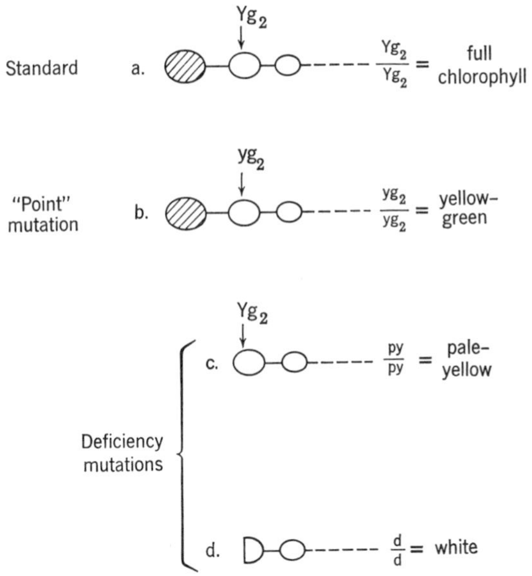

Figure 9.13. A point mutation and deficiencies in maize which produce recessive mutant phenotypes. After McClintock (637). The phenotype yellow-green is produced by a point mutation. The phenotypes pale-yellow and white are produced by deficiencies at the tip of the chromosome. Only white involves the loss of the Yg_2 locus.

The explanation for this unexpected interaction between yg_2 and py seems to be that py is associated with a deficiency (Figure 9.13) which does not include the yg_2 locus. Since the py-bearing chromosome carries the normal allele, Yg_2, and the yg_2 chromosome is not deficient, the two complement one another and the phenotype of the heterozygote is normal.

Other hypotheses may, of course, be formulated to explain heterosis in its varied aspects. From what is known about gene interaction, the possible hypotheses to explain specific cases are almost limitless in number. But the basic principle of all of these explanations will of necessity be that different gene combinations from the same restricted pool of genes will give many different phenotypes—many more than would be expected by considering the action of each gene independently without reference to the results of its interactions.

Although experimental data of a biochemical nature necessary to establish a biochemical basis for phenomena such as heterosis are lack-

ing at present for diploid plants and animals, certain experimental results from *Neurospora* may be useful in the future for aiding in the
explanation of these phenomena in the higher forms (252, 253, 829).
For example, the possibility of inhibition being the cause of certain
partial or complete genetic blocks has already been considered (Figure
6.13*c*).   It is apparent that the relief of such an inhibition by reduction
of the inhibitory agent in heterozygotes may well be one biochemical
explanation for heterosis.   Emerson (253) has presented biochemical
models based on known situations in *Neurospora* mutants which serve
to illustrate further some of the biochemical possibilities.

Modifiers of Allelic Expression and Interaction

The relations between alleles, particularly their dominance, can be
modified by alterations in the genetic background.   The vestigial gene
vg in *Drosophila* is an excellent example.   The mutant alleles at the *vg*
locus cause modifications of the wing size and shape.   It has been shown
that the dominance of vg^+ over *vg* and other mutant alleles (vg^{no2}, vg^{nw},
etc.) can be greatly reduced by (1) an allele at the cut locus, ci^{do-vg}
(335), and (2) a number of different minutes, $M(2)l^2$, $M(1)n,$ and
$M(3)w$ (357).   Some data showing the effects of the minutes are given
in Table 48.   The vestigial alleles, like the *ci* alleles, have incomplete
penetrance.   Dominance modification by the minutes and ci^{do-vg} consists
in increasing the penetrance in heterozygotes with $\pm$ and also in increasing the intensity with which the mutant phenotype is expressed.   These
modifiers may therefore be considered as dominance modifiers of the
$+$ alleles as well as enhancers of the mutant alleles.   The flies which

Table 48.   The Effect of Certain Minutes on the Dominance of vg^+ over
Its Recessive Mutant Alleles *

$vg/+$	Per Cent Normal Flies	$vg^{nw}/+$	Per Cent Normal Flies	vg^{no2}	Per Cent Normal Flies
Alone	99.2	Alone	97.78	Alone	95.3
$M(2)l^2$	68.8	$M(2)l^2$	12.10	$M(2)l^2$	5.4
$M(1)n$	27.6	...	...	$M(1)n$	...
$M(3)w$	12.4	$M(3)w$	0.00	$M(3)w$	0.0

* From Green (357).

are homozygous for $ci^{do\text{-}vg}$ are wild type unless in combination with mutant *vg* alleles. Minute flies have reduced bristles and a longer life cycle primarily due to prolongation of the larval period.

Certain translocations involving the fourth chromosome of *Drosophila melanogaster* result in the weakening of the dominance of the wild-type allele of cubitus interruptus. Thus, although ci^+/ci flies are quite normal (see p. 339), flies heterozygous for a translocation R $(ci+)ci$ show a mutant phenotype. This is called the Dubinin effect (234, 235, 967), and is probably related to the phenomenon of variegated-position effect discussed in Chapter 4. The effect is partially suppressed by the presence of Y-chromosomes (367). This is of some interest because the addition of Y-chromosomes to ci/ci flies does not cause a suppression of the mutant phenotype.

Allelic expression is affected in other ways by modifying genes besides the modification of dominance. The number of alleles in an allelic series which may be recognized is subject to the condition of other genes. This has been illustrated in the previous discussion of the guinea-pig C series of alleles in which it was shown that animals with phaeomelanic pigment (eF) are affected differently by substitutions of those alleles than are the eumelanic (EF or Ef) animals. In the phaeomelanic animals only three C alleles are recognizable, whereas five are necessary to explain the variations in eumelanic pigment. It will also be recognized that suppressor genes prevent the recognition of allelic changes of the genes they interact with when they are present in such a state as to cause the suppression phenomenon. On the other hand, the gene suppressed when in the normal state will mask certain allelic changes of the suppressor gene. From these and other examples it is abundantly clear that not only the number of recognizable alleles of a gene is modified by the condition of other genes, but the measurement of the absolute mutation rate of a gene to other allelic states is considerably handicapped by the possibility of many of the alleles not being capable of expressing a different phenotype from the gene from which they were derived.

General Considerations and Conclusions

As more and more is learned about the relations between DNA, RNA, and protein, and the protein-synthesizing machinery is better understood, more and more possibilities for interactions become evident. In the previous chapter we touched on the possible explanations for interactions of alleles but could come to no firm conclusions because the status of our knowledge is presumably not adequate. In the case of interactions of nonallelic genes, we are again forced to assume somewhat tentative explanations, but some of the more superficial types of interactions seem to have ready explanations.

If the interactions of the product of an enzyme-controlled reaction are considered, it is evident that such a product can be (1) stimulatory to the actions of one or more other enzymes, (2) inhibitory to other enzymes, or (3) affect the protein-synthesizing machinery of one or more other enzymes by acting, for example, as a corepressor or inducer. Thus the enzyme controlling the formation of the product can strongly influence the activities and synthesis of other enzymes.

Direct protein-to-protein interactions may also be very important. In this category may be considered competition for a common substrate between two or more different enzymes. As has been indicated, if the substrate is limited in supply the number of products formed from it will depend not only on the number of enzymes acting on it, but on their ability to compete. It is also conceivable that enzymes with quite different specificities and acting on different substrates will interact directly and either enhance or inhibit the activities of one or more of an interacting set. For example, if two or more enzymes act *in vivo* as a complex, the failure of one of them to attach properly in the complex, because it has been modified by mutation, may cause the entire complex to be inactive. This may, in fact, be one explanation for some types of controlling genes discussed in this chapter. Also, in this connection, should be considered the possibility that some proteins associated with enzymes may have no enzyme activity per se but act as cementing materials to hold complexes together (1045). Failure of these to function properly may result in failure of, or modification of, function of an entire complex *in vivo* (1050). Thus it is important to recognize that structural proteins may be as important in the scheme of things as the proteins which act as catalysts.

When one considers the complex series of interrelated reactions between DNA and a finished, functional protein, a large number of other possible interactions are evident. Indeed, working out these interactions may be one of the chief ways of learning more about the pathways, for, as is evident from Figure 8.21, the relation between the DNA molecules and the RNA and protein molecules they control is by no means a one-way relationship. Protein as enzyme is essential for the synthesis of DNA and RNA. Hence a mutation at a locus drastically affecting one of the enzymes controlling DNA synthesis would have the effect of bringing an end to DNA duplication. Alterations of enzymes involved in RNA synthesis would also have obvious drastic general effects. Additionally, alterations affecting the various types of soluble RNA's or the associated activating enzymes would have the effect of impeding or preventing the incorporation of specific amino acids into protein. This could result in the complete breakdown of protein synthesis, or the synthesis of many nonfunctional proteins.

In addition to these drastic effects resulting from interactions of defective genes in the protein synthesizing pathway, other less drastic ones can be visualized. For example, some nucleic acids may act as regulators. The repressor substances mentioned in connection with enzyme induction and repression may be nucleic acids whose specific role is to regulate the activity of structural genes. As has already been postulated, mutations affecting them may affect whole blocks of genes involved in the synthesis of specific end products.

The net result of considering all these various possibilities is to bring to the foreground for the emphasis it deserves the fact that a living cell is a highly balanced system. This, indeed, may serve as a summary of this entire chapter. The nucleus and the cytoplasm are not living systems. But the proper nucleus in the proper cytoplasm is a living system, because it is capable of growth and reproduction. The various parts are so interdependent that the slightest change in one may have effects all through the system.

10 Environmental Modification of Phenotype

Phenotypic changes incident to mutation have been the major concern of the preceding chapters. Genetic control of the phenotype is, however, only one aspect of the production of the phenotype. Just as gene changes modify the internal environment of the cell and hence the phenotype, so does the external environment. The essential difference is that environmental influences on the phenotypes are not inherited unless they cause genetic material to mutate.

The influence of the environment on the phenotype has been the cause of considerable misunderstanding on the part of many nonbiologists and biologists alike. Most of this misunderstanding may be summed up in the "nature-nurture" question. Which is more important in the determination of the characteristics of an organism, its heredity or its environment? The answer is that both are important. An organism is essentially a segment of the environment which exists at the expense of *its* external environment and is subject to the conditions of this environment. If it is not adapted to the environment, it will die. If one grants that this is true, one is still left with the question of the relative roles of heredity and environment in determining the phenotype. The answer to this question is not simple, and most of this chapter is an attempt to answer it.

A second important question, not unrelated to the first, is *what role does the environment play in determining what the genotype is to be?* Actually, it is probably best to answer this question first for it has the simplest answer, and an understanding of it makes it easier to understand the environment-phenotype relationship.

Inherited and Noninherited Adaptation

During the course of its life span an organism changes. This period of change as a single organism terminates either in death or cell division. The changes that occur are comprised of development through the embryonic stages (if any) to maturity and then senescence. These are characteristics of life. Another characteristic of life is the ability to adapt to the environment. The more responsive an organism is in making adjustments to its environment, the longer it will survive. The question is, what controls this change which is in essence a response to an environmental stimulus?

The answer to this question is really quite simple. The ability to adapt to an environment is always inherited, but the phenotypic state resulting from the adaptation may or may not be inherited; adaptation may therefore result in either inherited or noninherited changes. Inheritable adaptation is a population phenomenon. It is the result of the response of a population of organisms of the same species to the environment guided by natural selection and resulting in the establishment of genotypes which produce the best-fitted phenotypes. The significant factor is that *change in genotype* always accompanies the phenotypic shift toward a state better adapted to the environment. For this reason, it may be called *genotypic adaptation.* Obviously, it can be expressed only in an actively reproducing population to allow for selection to act upon new genotypes which arise by chance mutation, or by new chance combinations of genes produced by sexual reproduction.

Noninheritable, *phenotypic adaptation* is the response of the *individual* to the environment with no accompanying changes in the genotype. It can therefore hardly be the result of mutation and selection of the genetic elements within the nucleus.

Ordinarily, the distinction between the two meanings of adaptation is easily resolved. When large organisms such as corn plants or fruit flies are being dealt with, the question of the origin of individual phenotypic changes can easily be answered by breeding experiments. But a closer scrutiny of the distinction becomes mandatory when dealing with microorganisms, since populations of independent single cells or nuclei produce the observed phenotype as a unit. Unless the experimenter is able to observe and breed a single cell, it becomes necessary for him to establish the type of adaptation resulting in phenotypic change by methods of analysis which may be quite circuitous.

If the microorganism is one which can be carried through a sexual cycle, variants which appear can be tested by crossing them to the original, unadapted strain. An example of this direct approach may be found in the adaptation of a pantothenic acid-requiring mutant of *Saccharomyces cerevisiae* described originally by Lindegren and Lindegren (616). If cells of this strain are inoculated into a medium completely deficient in pantothenic acid, they may remain viable for a month or more and accomplish so little growth that no visible turbidity appears in the tubes (824). During this period, however, some cells develop the capacity to synthesize pantothenic acid, multiply, and within a few days cause a definite turbidity. Genetic analysis of crosses between the adapted cells and the original unadapted strains shows that the spores resulting from the cross segregate in a one-to-one ratio of pantothenate-dependent to pantothenate-independent, proving that the adaptation is due to mutation to ability to synthesize this compound and subsequent selection in a pantothenate-deficient medium. All the pantothenate-independent cells that have been tested by Raut (824) have also been determined to be different from the original strains of *S. cerevisiae* which are able to synthesize pantothenic acid. The reversion mutations therefore were probably suppressors of the pantothenicless gene in the unadapted strain.

Several conclusions may be derived from this analysis which emphasize the importance of considering the origin of adaptation by mutation and selection in any experiments concerned with the phenotypic change in microorganisms. First, cells of some organisms may remain viable for long periods with only slight growth in a hostile environment —in the example given, a deficient medium. Second, given the *time* provided by this extended viability, the chances of mutations providing for synthesis of the absent and required compound are increased. Third, the mutation may not necessarily be a simple reverse mutation to an identical gene type present in all synthesizers, but may be of the suppressor type. Thus chances for a mutation to adaptation to a deficiency are theoretically increased by the possibility of mutation at a number of different loci all resulting in essentially the same effect.

It is also to be noted that genotypic variation may arise by mutation in a population, and the phenotypic effects may be masked by the phenotype of the original dominant genotype which may be best fitted to the environment.   The variants, if they are nutritional mutants, for example, can easily maintain themselves by having their nutritional requirements satisfied by the strains which produce the required compounds and excrete them into the medium.   A comparable condition is well illustrated by fungi such as *Neurospora* which, being essentially coenocytic, can exist as stable heterocaryons with a population of diverse nuclei sustaining one another's capacities symbiotically.   If the environment changes, the relationship between the different genotypes may change by selection and cause a shift in the phenotype, an event which would be termed an adaptation.

These observations make it clear that despite the random nature of mutation and the rareness with which it may occur for any particular gene in the direction of fitness to the environment, adaptation in populations of cells or nuclei, as in *Neurospora,* by mutation and selection, is to be regarded as a highly probable explanation whenever a shift in phenotype is noted.

When dealing with the asexual strains of bacteria frequently used in adaptation studies, the technique of breeding cannot be applied and other methods must be relied on to distinguish between phenotypic and genotypic adaptation.   Some evidence that an observed adaptation in these forms is of the purely phenotypic variety may be had if the adaptation is a rapid one, that is, within a span of a few hours, for one would not expect adaptable variants to increase in numbers sufficient to be recognized in so short a time.   However, such a method is of no value if the adaptation is a slow one.

A second type of evidence which may point to phenotypic adaptation is a positive capacity to deadapt to the original phenotypic condition in the presence of the original environment.   This again, however, is not critical evidence that phenotypic adaptation and not mutation has occurred.   An example taken from the work of Ryan and Schneider (869, 870) will make this clear.   These workers have shown that a strain of *E. coli* which requires histidine readily adapts to growth on a medium deficient in histidine, and once adapted just as readily deadapts to the original requirement when cultured in the presence of histidine.   The use of refined plating techniques has shown that the adaptation to histidine independence on histidine-free medium is due to mutation and selection, and that the mutation rate among the mutants back to the original dependent condition is sufficiently high to provide for the presence of enough of them to account for deadaptation.   There appears to be a definite inhibition of growth of the histidine-independent cells when in the presence of those dependent on this compound.   This unexpected relationship leads to very rapid growth of the histidine-dependent mutants on histidine-containing media and to the illusion of deadaptation of the purely phenotypic variety.

A short period for adaptation and capacity to deadapt are both *prima facie* evidence for phenotypic adaptation, but neither constitutes critical proof. They must be supplemented by additional tests which are designed to detect adaptive capacity in the absence of factors which might select for or against the changed phenotype. Such a test has been devised by Luria and Delbruck (627) and used successfully by them and numerous others to distinguish between mutation and simple adaptation in microorganisms. The test is so designed as to detect the origin of adaptable strains in the absence of the adapting substrate, an event which, if it occurs, can only be assumed to be the result of mutation.

Perhaps the most direct method of elucidating the cause of adaptation when dealing with microorganisms is to observe whether phenotypic change occurs in a nonproliferating culture. Such changes as then arise (for example, the appearance of a new enzyme activity in the presence of a new substrate) cannot be due to mutation unless all cells in the population mutate to the same condition—a highly unlikely occurrence. This method cannot always be applied since many true phenotypic adaptations in microorganisms will not occur unless the culture is growing. In this case the other methods just described must be resorted to.

To summarize, the environment brings about changes in genotype primarily by selection. Environmental influences on the mutation rate are for the most part random (but see discussion in Chapter 4) and lead to a random assortment of mutations which are selected for. The environment itself, as far as we can determine at present, does not cause specific gene changes which will produce the best adapted phenotypes as a direct response.

Modifying the "Normal" Phenotype

The "normal" phenotype is that obtained under "normal" environmental conditions with the "normal" or wild-type genotype. Modifications of the environment may cause a departure from what is considered "normal" with one set of conditions to what is considered "normal" under the new conditions. The extent of phenotypic change is dependent, of course, on two factors, the degree of environmental change and the response of the organism. This response is determined by the genotype, and different responses to the same types of environmental change will be elicited with different genotypes, as will be discussed in the succeeding section.

Ordinarily, the responses of an organism to the usual environmental changes under the control of the observer, such as light, temperature, and food supply, are familiar to the observer as being usual for the organism. There may be changes in growth rate, size, pigmentation, and so on, or if the environmental change is drastic, death may ensue. Frequently, these changes (except for death) are reversible. When the environment is returned to its original state, the phenotype accordingly reverts. However, by the employment of special techniques, the phenotype of the organism may be so changed that it resembles specific mutant phenotypes which, under normal environmental conditions, can be obtained only by a mutant genotype. Mutant phenotypes so produced in genotypically wild-type organisms are called *phenocopies*.

Goldschmidt (336) produced a number of phenocopies in wild-type *Drosophila melanogaster* with temperature shocks of 35° to 37°C for varying intervals of time at different stages of the larval period. Many of the adults developing from the heat-treated larvae resemble to a remarkable degree certain known inherited mutant phenotypes. Some of the phenocopies obtained by Goldschmidt are listed in Table 49.

Heat is not the only factor which can induce phenocopies. By means of sublethal doses of cyanide, silver salts, quinone, and derivatives of quinone given the larvae in the medium, Rapoport (822*a*) was able to induce noninherited phenotypic modifications in the adults which again were strikingly similar to inherited mutant phenotypes known to

Table 49. Some Examples of Phenocopies Obtained in *D. melanogaster* by Treatment with High Temperatures[*]

Phenotype Induced	Developmental Period Treated, Days	Temperature of Treatment, °C	Exposure Time, Hours	Per Cent of Phenocopies
Scalloped	4½–5½	35	12–24	70
Curly	6–7	35–37	18–24	76
Spread	5½	35	18–24	91
Trident	7	35–37	6–24	82

[*] Goldschmidt (336).

be dependent on single-gene mutations.   Phenocopies in *D. melanogaster* of the mutant lethal meander (*lme*) have been obtained by Schmid (880) by starvation of the larvae during a certain period, and Gloor (330) has obtained a phenocopy of the mutant condition, tetraptera, by treating the embryos with ether.   The production of phenocopies is by no means confined to insects.   Landauer (547–549) has induced phenocopies of a number of different types in the fowl by injecting insulin and various other agents into the eggs.   What appear to be nutritional phenocopies have been obtained by Zamenhof (1130) in bacteria by means of heat treatment of dried spores.

Genetically determined mutant phenotypes may also be modified in the wild-type direction.   Many examples of this have already been given in the descriptions of the responses of nutritional mutants to the compounds required, and in the modification of eye color in mutants of *D. melanogaster* by transplantation of normal tissues or the addition of an active compound to the food medium.   Partial starvation of the homozygous vermilion larvae of *D. melanogaster,* which ordinarily do not produce brown pigment, results in the synthesis of considerable amounts of this pigment and hence a wild phenotype (59).

These examples serve to emphasize the tremendous plasticity of the phenotype and show that it must be defined in terms of both genotype and environment.   Further evidence for this important generalization comes from a consideration of certain peculiar mutant conditions which are extremely sensitive to environmental factors.

Temperature-"Sensitive" Alleles

The degree of pigmentation of many mammals is considerably affected by the temperature conditions under which they develop and live. Familiar examples of this are certain species of rabbits, weasels, and others, which change from dark to light pelage with the advent of cool weather.   The mechanism involved is undoubtedly complicated and related to changes in endocrine activity during different seasons of the year.   Generally, a change of temperature for a short period has little or no effect on the pelage of most mammals, but gradual and prolonged changes bring about the condition alluded to.   However, the domestic rabbit has a color variety known as the "sooty" or Himalayan strain which, unlike the other varieties, is extremely sensitive to temperature conditions.   When raised at "room" temperature (circa 22°C), the Himalayan rabbit develops a white pelage with melanic pigmentation in the extremities; the ears, forepaws, tail, and nose darken, while the rest of the body remains albino (Figure 10.1). The degree of pigmentation is variable in these parts unless the temperature is kept relatively constant. When the skin in any part of the body is cooled to temperatures under 34°C, melanin production results in that part. If the temperature is raised above this in any part, including the extremities, the hairs developing from that part are free of pigment (199). Thus the characteristic Himalayan pattern resulting when the rabbit develops under ordinary temperature conditions is a consequence of the temperature in the extremities. It is lower in these than in the other parts of the body, which, having a better blood supply, are normally kept above 34°C. A heteronomous condition exists under control of the temperature, rather than an autonomous pattern-controlling process which is the more usual situation in mammalian pigmentation.

The Himalayan strain differs from the other color varieties by a single gene, a^n, an allele of the albino series. This series, quite comparable to the C series of the guinea pig and mouse previously discussed, includes four alleles; A, full pigment; a^{chi}, chinchilla; a^n, Himalayan; and a, albino. Only animals homozygous for a^n, or heterozygous for a^n/a, respond to temperature changes.

Enzyme experiments with extracts of Himalayan rabbit skin indicate that there are at least two phases or reactions involved in the production of the melanic pigment (199). The first phase proceeds under anaerobic conditions and only if the temperature is below 34°C. A temperature of 25°C is optimal. This phase

Figure 10.1. The phenotypic appearance of Himalayan rabbits under different temperature conditions. (*a*) A rabbit raised at a temperature above 30°C. (*b*) A rabbit raised at a temperature about 25°C. (*c*) A rabbit which has had the left flank artificially cooled at a temperature below 25°C. After Danneel (199).

may be associated with the production of an enzyme which converts precursors into melanin in the second phase. The second phase results in the melanin production and requires oxygen, as might be expected. According to Danneel (199), it is the first, or anaerobic, phase which is different in the Himalayan strains insofar as it is temperature-sensitive to a degree not obtainable in extracts from the other strains. There seems to be a direct relation between the a^n allele and the production of the enzyme, since active extracts from a^n/a^n animals are about 30% more active than those from a^n/a animals.

A considerable number of the nutritional mutants of *Neurospora* are sensitive to changes in temperature. They require the addition of growth factors to the minimal medium at certain temperatures but not at others. Figure 7.1 gives some data for a temperature-sensitive riboflavin mutant which show that in the absence of riboflavin the mutant growth approximates that of wild type under 25°C, but above this temperature growth is sharply reduced to zero at about 28°C.

The presumed allelic series of pyrimidine mutants of *Neurospora* described by Houlahan and Mitchell (469) includes two temperature-sensitive alleles. Strains 37815 and 67602 both require a preformed source of pyrimidine at a temperature of 35°C, but at 25°C 37815 grows without pyrimidine, and 67602 requires much less than at 35°C. An allelic strain, 37301, requires pyrimidine at the same concentration for growth at both temperatures.

Nutritional mutants with a definite temperature dependence have also been found in bacteria and yeast. These demonstrate essentially the same characteristics as the *Neurospora* mutants discussed earlier. Even phage mutants may demonstrate a temperature sensitivity. Wild-type T4D phage form plaques at temperatures as high as 42°C, but the temperature mutants, *ts,* are unable to do so, although they are quite able to duplicate and lyse at 25°C (248).

Most temperature-sensitive nutritional mutants probably produce an enzyme which differs from the wild type in heat stability. For example, a pantothenic acid-requiring mutant of *E. coli* requires pantothenic acid only at temperatures above 30°C. An analysis of the pantothenic acid-synthesizing system in this strain shows that it is quite unstable at temperatures above 25°C (648). Heat lability of proteins is certainly to be expected as a result of mutation, since changes in amino-acid sequence may be expected to lead to heat denaturation at lower temperatures than in the wild type. Hence a certain fraction of mutants of any organism should be expected to demonstrate a temperature sensitivity because of modified protein.

Perhaps one of the most detailed and careful investigations into differences in heat inactivation of enzymes from different strains has been made in *Neurospora* by Horowitz and Fling and associates (459). Four wild-type strains of *Neurospora* collected from various parts of the world have been investigated for their tyrosinase activities (see p. 269). All produce tyrosinase, but all four tyrosinases are different, as determined by electrophoresis and heat-inactivation studies. The heat inactivation data for tyrosinases from each of the four strains are given in Table 25. Two strains, T^s and T^{Sing-2}, show the same rate of inactivation at 59°C, but their tyrosinases have different electrophoretic mobilities. Genetic analysis of the strains shows that the tyrosinase differences are allelic.

Many of the known mutant genes in *Drosophila* have a very variable expression. This is particularly true if the character has only partial penetrance, as in the cubitus interruptus example mentioned previously (p. 338). Failure to control the temperature rigidly during the course of experiments involving the *ci* series results in great variations in phenotypic ratios from one experiment to the next. The vestigial series of alleles (p. 446) also has members which have a temperature-labile phenotypic expression. Harnly (405, 406, 407) has conducted extensive investigations into the effects of temperature on the expression of the alleles *vg* and *vg^p* (pennant). Figure 10.2 describes the results obtained by raising flies of the genotypes *vg^p/vg^p*, *vg^p/vg*, and *vg/vg* at temperatures ranging from 16° to 32°C. The effect of each genotype was determined by measuring the wing area. The expression of homozygous pennant closely parallels that of wild type in the reduction of wing area over the temperature range indicated in the figure, and in this sense *vg^p* is not any more temperature-sensitive than *vg^+*. But when *vg^p* is heterozygous with *vg,* a marked temperature effect is demonstrated. Homozygous *vg* is modified only at temperatures above 29°C under which conditions *vg/vg* flies approach the wild phenotype. The range is a narrow one, however, for above 32°C the homozygous *vg* flies do not develop. A corresponding lethal effect of temperature in a range which is not lethal for wild type is also noted in *vg^p/vg^p* flies. These will not develop at temperatures above 30°C. Similar extreme temperature-modifying effects have been noted in *Drosophila* for the Bar mutants (429) and the white allele, blood (*w^{bl}*) (258) (see Figure 10.2) and the zeste mutants (66).

These observations and experimental results clearly establish the great differences in sensitivity to environmental change shown in the phenotypic expression of different genes. Extreme sensitivity is, fur-

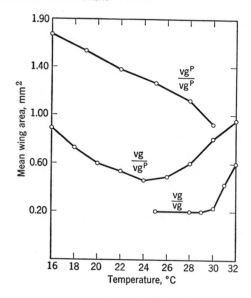

Figure 10.2. The effect of temperature on the area of the wings in vg^P and vg *Drosophila melanogaster.* Data for males only. After Harnly and Harnly (407).

thermore, not necessarily a general property of groups of allelic genes related by their effect on the phenotype, but one of particular genes within each allelic series.

Serotype Transformation in *Paramecium aurelia*

When paramecia are injected into a rabbit, the rabbit reacts with the production of antisera which immobilize the strain used as the antigen source. The active antigens are ciliary in origin and cause a paralysis of ciliary motion (812–815). Sonneborn (924, 925) and Beale (60) were able, by the establishment of homozygous stocks of *Paramecium aurelia,* to demonstrate that a given homozygous strain produces with few exceptions (657) only one type of ciliary antigen at a time. The particular antigen produced depends on the conditions of culture and the previous history of the stock. For example, Sonneborn has shown that stock 51 of *P. aurelia,* variety 4, which arose from a single homozygous individual, exhibits any one of a number of antigenic types, or *serotypes,* designated as A, B, C, D, E, G, H, and S. A culture of any one of these serotypes may maintain its antigenic specificity through many vegetative fissions (or cell divisions) and through conjugation, provided the growth conditions are maintained relatively constant and optimal for the particular serotype with respect to temperature and nutrition. Serotypes A, B, and D of stock 51 were maintained for more than four years at 26°C with no change (926). However, by a number of different environment modifications such as (1) change in temperature, (2) change in nutrition, (3) subjecting the serotype to sublethal doses of its specific antiserum, (4) exposure to ultraviolet light (927), and (5) treatment with certain chemicals (37) it is possible to induce *transformations in serotype*. Thus serotype A can be transformed to B by lowering the temperature to about 19°C. Once established, serotype B is constant provided the proper conditions of culture are maintained, but the transformation is reversible, for B may be induced to change back to A. Breeding experiments have demonstrated conclusively that no genic changes accompany the transformations, but apparently paramecia of the same genotype can be of any one of the seven serotypes just enumerated.

Four homozygous stocks of *P. aurelia* variety 1, collected from different parts of North America, have been intensively studied by Beale (60) with a view to extending the original observations of Sonneborn on variety 4 already partially described. The variety 1 stocks, 90, 60, 41, and 61, have each been marked with three different loci identified phenotypically by the different serotypes S, G, and D. The particular antigenic conditions demonstrated by a culture of any one of these stocks depend on the temperature. At 18°C the serotype of a population is predominantly S; at 25°C, G; and at 29° to 36°C, serotype D (Figure 10.3). The transformations are for the most part easily reversible but occur only after a number of fissions have taken place at the new temperature. In stock 90, serotype G is transformed to D

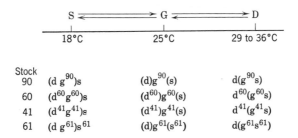

Figure 10.3. The serotypes produced at different temperatures in *Paramecium aurelia*. After Beale (61).

when the temperature is raised to 29°C, but it is complete only after 50 fissions (ten days). The transformation itself is relatively sudden, taking place within the space of two fissions. Prior to this, the sero-type is G. During the short transition the cells react to G and D anti-gens, but when transformation is completed only D antigens are effec-tive. As demonstrated by Sonneborn with variety 4, the serotypes are mutually exclusive; both cannot be maintained simultaneously except during the short period of shifting from one stable antigenic state to another. Serotype D can be caused to revert back to G in stock 90 by lowering the temperature to 25°C. Transformation is complete after 11 fissions (three days).

The corresponding sterotypes of the four homozygous stocks are designated by identical letters because they induce heterologous anti-sera which crossreact in varying degrees with each other's antigens. Hence the S antigens, 90S, 60S, 41S, and 61S, all induce antisera which crossreact with the S serotypes of all four stocks. Indeed, the S anti-gens of stocks 90, 60, and 41 cannot be serotypically distinguished. The antibody of 61S crossreacts, but only at low dilutions. The D antigens of stocks 90 and 61 appear to be identical, those of 60 and 41 dissimilar but capable of crossreacting with one another's antibodies. The G antigens show only weak crossreactions among the four stocks and are hence presumed to be all four different.

As the result of breeding tests in which recombinations were obtained in crosses between different stocks and in which heterologous antigen production segregated according to Mendelian expectations, Beale has postulated the existence of three genes in each stock. Each gene is concerned with the production of a different antigen. Thus stock 61 is genotypically d^{61}/d^{61}, g^{61}/g^{61}, s^{61}/s^{61}. The serotype expressed depends on the conditions just expressed. In Figure 10.3 the genotypes for the four stocks are summarized. Heterologous antigens are designated by allelic genes because of the results obtained from crosses. The paren-thesis enclose those genes which are presumably not expressed at the temperature indicated. Nearly all possible recombinations have been obtained from crosses between the different stocks. Those with re-combinations demonstrate the same ability to change serotype, but the specific serotypic shift is different from that found in the original homo-zygous stocks. For example, the F_2 animals from a cross between 90G × 60G produce recombinations of the genotype $\dfrac{d^{60}g^{90}s}{d^{60}g^{90}s}$. At 25°C these show the serotype 90G, but at 29°+ the serotype is 60D, not 90G.

Further results from crosses between stocks illustrate the effects obtained in heterozygotes, as shown in Figure 10.4. If 90G is crossed to 60G (60), the F_1 hybrids after several fissions at 25°C demonstrate both serotypes, 90G and 60G. It is interesting that the effect of the immobilizing antisera—90G and 60G—is less marked on the heterozygotes than on the respective homozygous parents. This fact may be interpreted as meaning that the same amount of antigen of general type G is produced in the heterozygotes as in the homozygotes, but less of each specific type. Here again is another piece of evidence for the generalization that the amount of gene product is related to gene dosage. By causing the heterozygotes to undergo autogamy, a kind of internal sexual reproduction (see Chapter 3) which results in homozygosity,

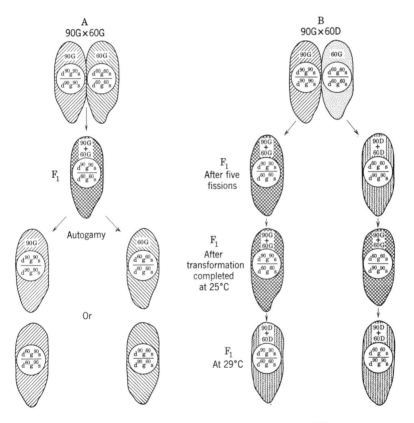

Figure 10.4. The results of crosses between paramecia of different serotypes. After Beale (61).

animals with either 90G or 60G serotypes, but not both, are obtained. It is, incidentally, in these exautogamous individuals (F_2) that recombinations will be detected.

Since 50 or more fissions may be required to transform serotypes, crosses may be made between individuals of different stocks and serotypes, as illustrated in Figure 10.4, with a cross between 90G and 60D at 25°C. The heterozygotes produced here show, as expected, both 90 and 60 stock traits, but the initial individuals derived from the original conjugant with G antigen are 90G + 60G, whereas those from the other parent are 90D + 60D. However, after a more prolonged period at 25°C, the transformation is completed and only the G serotype is expressed in all individuals. This observation is of great interest, for it demonstrates the principle that the phenotypic expression of a gene depends on the immediate environment. Gene d^{60} removed from the 60D cytoplasm and transferred into 90G cytoplasm is not expressed, but g^{60} is. Gene d^{60} expresses itself only when the temperature is raised to 29°C (Figure 10.4, bottom row).

Some of the different types of ciliary antigens have been isolated from stock 51 and have been shown to have distinctive characteristics which enable one to differentiate them by diffusion on agar gel, solubility in ammonium sulfate, electrophoretic mobility, and sensitivity to trypsin (812–814). The antigens are proteins, and the evidence is quite conclusive that antigens A, B, and D, which can be differentiated electrophoretically, differ in some peptides, as demonstrated after trypsin digestion, while having many other peptides in common. This is good evidence that they are related, because they have some amino-acid sequences in common, but also that they differ in amino-acid sequences in other parts of their chains. These findings are of extreme interest because they show that a number of nonallelic genes, that is, those controlling antigens A, B, and D, produce similar proteins. Which protein is produced at any given time depends on the conditions.

The work with the serotypes in *Paramecium* makes it quite evident that even so important an aspect of the phenotype as protein structure can be dictated by the environment. These changes are brought about without gene mutation. In addition, this work shows that, although the cytoplasm (i.e., protein in the cilia) is subject to change by the environment, there are limits to the degree and kind of changes which are dictated by the nuclear genes. That is, a strain carrying antigen D60 will not produce antigen D41 under any conditions unless, of course, there is a gene mutation.

The Sensitive Period

The phenocopies obtained by Goldschmidt, as described on page 455, were induced by the application of temperature shocks of different duration and at different times during the larval periods, as shown in Table 49. The time of application and the duration of application of the shocks for the maximum production of the different phenocopies seem to differ. For example, a miniature phenocopy is produced optimally by treating old larvae, whereas scalloped phenocopies are best induced by treatment of somewhat younger larvae. This would indicate a *sensitive* period in development, at which time a particular abnormality is most easily produced by interfering with a specific phase of development. Changing the time of application of the shock would be expected to interfere with a different phase of development and hence produce a different phenocopy.

A specific sensitive period is particularly evident in the influence of temperature on the phenotypic expression of vestigial. The modification of the vestigial phenotype in vg/vg flies caused by temperatures of 30° to 32°C is most extreme when larvae otherwise raised at optimal temperatures (circa 25°C) are subjected to elevated temperatures at the beginning of the third larval instar about 64 hours after hatching (405, 407). Temperature shocks applied prior to the third instar have no effect on the expression of vg. Hence the third instar is a sensitive period with respect to the expression of vg.

The white allele, w^{bl}, also has a temperature-sensitive phenotypic expression with a definite time of induction (258). Homozygous w^{bl} flies have pale brown eyes at 30°C and deep red-purple eyes at 17°C. Both the brown and red pigment components are affected (Figure 10.5*a*). The temperature-sensitive period occurs during the pupal stage about 40 to 48 hours after the onset of pupation (Figure 10.5*b*). Treating larvae of any stage, or young or old pupae, at low or high temperature has no effect on the eye color of flies raised at 25°C.

Sensitivity at a particular stage of development is not restricted to temperature effects. As previously mentioned, treatment of the embryo of *Drosophila* with ether produces a tetraptera phenocopy, that is, the flies develop four wings, a second pair developing from the halteres, as in the case of bithoraxoid discussed in Chapter 8. The sensitive period is in the first six hours of larval development, both for ether and for elevated temperature (Figure 10.6). Treatment with either

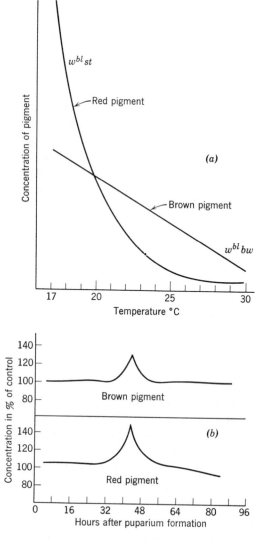

Figure 10.5. The effect of temperature on the production of red and brown pigment in *Drosophila melanogaster,* homozygous for w^{bl}. After Ephrussi and Herold (258). (*a*) The concentration of pigments in adult blood flies raised at indicated temperatures. (*b*) The concentration of pigments in adult blood flies raised at 25°C but subjected for 8-hour periods to a temperature of 17°. Note that the sensitive period for pigment production is between the fortieth and eightieth hour of pupal development.

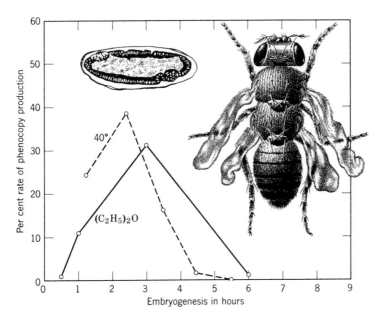

Figure 10.6. Sensitive phases in the production of the tetraptera phenocopy in *Drosophila melanogaster* by means of ether and heat shock. The solid curve gives the percentage rate of phenocopy production after ether treatment at different embryonic stages. The broken curve gives the percentage rate of phenocopy production after heat shock. Top left: section through an embryo in the blastoderm stage (aged 3½ hours, i.e., just before the end of the sensitive phase). Top right: tetraptera phenocopy; as the fly had to be dissected out of the puparium, the wings have not unfolded. From Hadorn (391).

of these agents at any stage after three hours does not produce this phenocopy. A similar result is obtained with birds. Injection of insulin into the egg of a chick which is one-to-two days old results in a rumpless phenocopy, but if injected at age three-to-six days, short upper beak or micromelia result.

The occurrence of sensitive periods in the development of multicellular, differentiated organisms, during which certain phenotypic modifications are most easily produced, is precisely the same phenomenon as that observed by the experimental embryologist who effects changes in the final form of an organism by altering the course of development of the embryo with chemicals, transplantations, and physical injury. Various types of abnormalities may be produced, depending on the stage of development subjected to treatment. The genetic observations, however, add to the embryological observations in showing that the degree of sensitivity of phenotypic response to stimuli during certain periods of development is markedly altered by the genotype.

Goldschmidt (336) has attempted to equate the observed effect of the environment to produce phenotypic alterations with the action of genes which appear to control the particular phenotypes. In connection with his analysis of phenocopies he concluded that ". . . the processes underlying the formation of phenocopies are the same as those set in motion by the mutant genes." This is an elaboration of his "rate concept" which states that the role of genes is the regulation of rates of reactions and hence development. There is no doubt that genes, like temperature, control the rates of reaction through their undoubted effects on enzyme production and activity, but it is obviously not justifiable to conclude that the processes involved in the modification of phenotype by environment are identical to those produced by gene mutation.

The Genotype as the Reaction Norm

The essential fact brought to the foreground by the experiments on the effects of environment on the phenotypic expression of different genotypes is that genes respond differently to environmental stimuli. This observation is contained within the concept of the genotype as the *reaction norm* (1099) which states that the modification of expression of the genotype is set within limits characteristic of the genotype. Figure 10.7 describes this concept diagrammatically. The genes a and a^1 may be considered as alleles, each with a different range of action and not overlapping in their phenotypic effect. The precise phenotypic effect which results from either will be determined by the environmental conditions, but under no known conditions will the phenotypes overlap. Actual examples to fit these conditions may be found in the many allelic genes which are autonomous in their expression and show no overlap of phenotype regardless of the environmental conditions. On the other hand, gene a_2 overlaps with a_1 in its expression range, and under certain environmental conditions their phenotypic expression will be identical. Isoalleles such as ci^{+2} and ci^{+3} of the cubitus interruptus series (p. 344) may be considered as genes with overlapping reaction norms, as should any mutant genes which under certain environmental conditions overlap in phenotypic expression with their wild-type alleles.

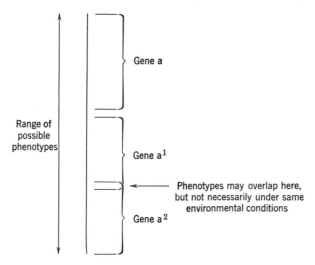

Figure 10.7. The limits of possible phenotypic change in the presence of three hypothetical allelic genes, two of which overlap in their expression under certain environmental conditions.

The reaction-norm concept is of fundamental importance to an understanding of what is meant by "gene action." The gene never acts alone, but in an environment—directly in the internal-cell environment, more indirectly in the external environment. Modification of the internal environment, either by outside forces or the action of other genes, may be expected to change the expression of any specific gene within the limits characteristic of that gene's ability to act. The genotype sets the limits of possible phenotypes.

A living system can be looked on as having two fundamental components. First, it consists of a coordinated system of genes which may be considered as the inflexible, or conservative, element. Second, it consists of an adaptive system which is subject both to the genetic system and to the environment. The adaptive system is flexible, whereas the genetic component is relatively inflexible. Both are dependent on each other. The range of response of the adaptive system to the environment, and the manner of its response, is gene-controlled, but the adaptive system in turn insures the survival and reproduction of the genetic system for the succeeding generations.

11 The Continuity of Cellular Organization

It is an important experimental fact that a particular chromosomal genetic constitution can result in more than one stable phenotype. This is evident in processes of differentiation and development where a single cell ultimately yields many kinds of cells which are very different in morphology and metabolism but apparently alike in chromosomal gene structure and functional potential (see Chapter 12). Such differences may be justifiably ascribed to nuclear differentiation resulting from completion of the cellular control cycle by action of extrachromosomal components back on the systems which designate primary synthesis.

The mechanisms by which such fundamental control systems work are not known and therefore it is quite arbitrary to separate this kind of extrachromosomal influence from a second category of phenomena which may be described as extrachromosomal inheritance or, perhaps more commonly, as cytoplasmic inheritance. This terminology, of course, implies the existence of primary structure-designating systems which are not parts of chromosomes and may undergo independent replication in the nucleoplasm or cytoplasm. For illustration, the most extreme situation which exists and could be classified in this way is that of a cytoplasmic-bacterial symbiont which has a strong influence on the phenotype of the host cell in which it reproduces. In such a case it is obvious that there are two genetic systems, one living within the other, and at division of the host cell "inheritance" of the bacteria is cytoplasmic and dependent on the manner and amount of host cytoplasm distributed to the daughter cells. Although interactions of the systems in this situation may be very complex, there is no conceptual difficulty involved as to the nature of inheritance. This is true also if the invader is a virus instead of a bacterium even though the virus is a complete parasite totally unable to undergo replication outside of a host cell. But what if the replicable unit is a normal component organelle or macromolecule of the host cell? In such a case independent genetic systems are not so obvious and the complete dictatorship of chromosomal order over all of primary sequence determination does not obtain. If this is true then the phenomenon of extrachromosomal inheritance is indeed one of great importance. Some pertinent examples and considerations are reviewed in the discussions which follow, but a final solution has not been reached. It is clear that chromosomal and nonchromosomal inheritance are intimately related and it is unlikely that either will be fully understood without the other.

Organization Continuity

Although much detail remains to be understood it is virtually certain that chromosomal characters are inherited by separation of nucleic-acid components after replication. A single copy of each genetic molecule may be sufficient to give chromosomal genetic continuity to an additional new cell. But such a complement of molecules would be useless alone. Cell division involves more than distribution of DNA; it involves also the division of the nucleus and division of the cytoplasm to give a functional system. Extrachromosomal material always goes along with the established genetic substance in its highly specialized process of essentially equal segregation, but the degree of organization and tendency toward equal distribution is of a different order of magnitude. That is, the distribution of cytoplasm at cell division can be very unequal without impairment of genetic and physiological capabilities of daughter cells. This could be taken to mean that all of the genetic potential of the cell is chromosomal, but such a conclusion is not justifiable. An alternative to the insurance afforded by the chromosomal system is a system in which individual genetic units are replicated extensively in the cytoplasm to an extent which provides a high probability that at least one unit will always segregate even with very unequal divisions of the cytoplasm. Such a mechanism seems more reasonable for formative systems than the highly organized chromosomal mechanisms, but there is no proof for its existence.

Cytoplasmic Segregations. As discussed previously, especially in Chapter 2, cells contain a great variety of highly organized inclusions. The most obvious, the nucleus, may be represented only once, but many others such as mitochondria, microsomes, membranous units of various kinds, and likely innumerable loosely organized macromolecules are represented many times and likely are essentially always distributed to daughter cells. Actually, there seems to be, in some cases at least, special mechanisms for distribution of cytoplasmic components. For example, mitochondria, which are nearly universally essential in metabolism, undergo regular segregations in some species. Wilson (1085) observed that in centrurid scorpions the mitochondria form a closed ring at the onset of spermatogenesis. This appears in the primary spermatocytes just outside the spindle and generally parallel to it. In the first metaphase the ring is elongated in a plane parallel to the long axis of the spindle. The ellipse so formed then breaks at the ends, producing two mitochondrial rods lying parallel. Each of these then breaks transversely and equally at telophase, giving each secondary spermatocyte two mitochondria. In the second division an equal transverse division again occurs, and therefore four spermatids each with two mitochondria are formed.

The spermatogonial mitotic divisions unfortunately could not be followed as easily as the meiotic divisions, but it is clear, nonetheless, that the mitochondria form the ring only in spermatogenesis. During mitotic divisions the mitochondria show a bipolar segregation so that each daughter cell receives a share of them. The origin of new mitochondria during the course of continued mitosis is, however, not understood, though presumably it may be by transverse division as in spermatogenesis.

In the noncentrurid scorpions, the ring is not formed during spermatogenesis, but an equivalent equal distribution of mitochondria is nonetheless attained. For example, the scorpion, *Opisthacanthus,* show 24 mitochondria in its primary spermatocytes. These are generally divided equally in the first division and again in the second division so that four spermatids each with six mitochondria are regularly formed from each primary spermatocyte. Out of about 500 cases examined by

Wilson, 76% demonstrated four spermatids with six mitochondria, 17% with five, and 7% with seven. Wilson interpreted these data as evidence for a mechanism insuring an approximately equal distribution, but he did not presume that anything more than a random assortment is indicated. Although nondisjunction of chromosomes during meiosis occurs to a much lesser degree than is exhibited here by unequal distribution of mitochondria in scorpions, it is possible that Wilson tended to belittle the significance of his own observations.

Since the mitochondria distributed to each spermatid are incorporated into the cytoplasm of the mature spermatozoa, and are therefore presumably passed on to the next generation, it is quite possible that mitochondria do have a genetic continuity in the scorpions. Similar observations on the regularity of mitochondrial distribution during meiosis have been made in a number of insects (793) and in the protozoan *Spirostomum* (291). The phenomenon may be somewhat more widespread than heretofore thought. Perhaps better methods of demonstrating mitochondria in dividing cells will make it possible to study this important mechanism to better advantage in the future and to determine whether it is a general phenomenon or merely a natural oddity occurring in certain arthropods and Protozoa.

The evidence for the continuity of chloroplastids in the green plants is somewhat better than that for mitochondria. But even here the crucial observations needed for proof of genetic continuity are wanting. In the lower green plants such as the algae, the chloroplastids can be seen to divide and the division products distributed among the daughter cells resulting from mitosis. In certain of the Chrysophyceae such as *Rhizochrysis* and *Myxochrysis* the chloroplasts do not always divide as rapidly as the remainder of the cell, in which event cells are produced without plastids (767). This loss of chloroplasts during cell division may account for the fact that many algal groups have colorless counterparts which are identical in appearance to the green forms, and differ only in the absence of chlorophyll and hence their mode of nutrition. The loss of chloroplasts might also be caused, of course, by gene mutations, as has been amply demonstrated in the higher plants.

The disappearance of chlorophyll and chloroplastids has been studied experimentally in various species of *Euglena*. Most *Euglena* species and indeed many other algae which are not obligate autotrophs lose their chlorophyll pigment when cultured in the dark but quickly regain it when returned to the light. *Euglena gracilis* has been shown by Baker (42) to lose all signs of chloroplastid structure with the disappearance of chlorophyll. The chloroplastids might therefore be considered to arise *de novo* in colorless individuals in the presence of light. On the other hand, Lwoff and Dusi (624) working with *Euglena mesnili* found that the complete loss of chloroplastids in this species in the course of dark growth resulted in colorless strains which were completely incapable of producing chlorophyll or chloroplasts on being returned to the light. Furthermore, the treatment of *Euglena gracilis* with sublethal doses of streptomycin rids the cells of their chloroplasts even in the light. The colorless strains so obtained are completely incapable of regenerating chloroplasts (811). Heat treatments appear to act in a fashion similar to streptomycin, but another antibiotic, furadantin, yields colorless *Euglena* by a seemingly different mechanism (633). Streptomycin effects on *Chlamydomonas* are discussed in detail subsequently.

Observations such as the foregoing suggest that at least some cytoplasmic units cannot be reproduced in the absence of preexisting units and that when they are present in only small numbers there may be special mechanisms for segregation in daughter cells. This applies also to components such as centrioles (623, 1029, 1083), but there is little or no evidence pertaining to most cytoplasmic complexes or to individual macromolecules essential to cytoplasmic function and usually present in large numbers. Replicability among nonchromosomal components is therefore not ruled out as the basis for important genetic systems. As will become obvious in subsequent discussions, such systems are not easy to recognize, and stability is to be expected only under special circumstances.

Recognition of Extrachromosomal Inheritance. As discussed in Chapter 3 studies of mutant character segregations through sexual reproduction in various kinds of systems have permitted the development of rules of order in the localization of genetic units on chromosomes. In contrast there are no known rules to describe the location and properties of extrachromosomal units of inheritance. However, best recognition is also through sexual reproduction and two general situations exist in which expression can be observed. The first is that in which the male gamete in a cross supplies little cytoplasm to the zygote while the female supplies most of it. In this situation if the female carries a cytoplasmic genetic determinant then maternal inheritance is exhibited with all progeny assuming the phenotype of the female parent. Occasional passage through the male may occur since the gametes are never entirely devoid of cytoplasm. Of the examples to be considered in this chapter these are the general characteristics of the systems in the higher plants, *Drosophila,* and the fungi. In these cases inheritance through the sexual cycle can be only a matter of segregation of units in small or large amounts of cytoplasms.   The mode of inheritance in bacteria and paramecia, however, is not so easily analyzed, since in both systems cytoplasmic bridges form in the sexual processes and considerable exchange of cytoplasm is possible.   Furthermore, in yeast and *Chlamydomonas* zygotes are formed by cell fusion and yet cytoplasmic inheritance is exhibited.   That is, the mutant character does not segregate and all of the progeny assume the phenotype of one of the parents even though both parents supply about the same amount of cytoplasm to the zygote.   Some aspect of this behavior is also noted in the fungi in heterocaryons.   Obviously, in this last system a character must be dominant to be expressed, but perhaps in the first system where a cytoplasmic unit might be physically eliminated in formation and function of a male gamete a less extreme expression might be retained.

These are the general operational principles by which cytoplasmic genetic determinants can be recognized, and it is obvious that strong expression and selection is demanded by the existing methods of detection.   Consider for example the contrasting situation with respect to probable numbers of genetic units per cell in chromosomal and extrachromosomal inheritance.   In the former there are one (haploid) or two (diploid) units per cell, whereas in the latter the number can be very large and variable.   Thus with a small constant number the expression of a change by mutation is usually assured, but with large and variable numbers expression of a mutation would only appear slowly and when the mutant form had some kind of selective value.   Without selection it might be detectible in terms of heterogeneity of primary structure of products, but no case has yet been studied at this level.   Also, with no selection among large numbers and with relatively small genetic units (e.g., stable and replicable messenger RNA), it is not expected that even the cytoplasm limitation in male gametes would be sufficient to exclude either normal or mutant units.   These points are of course speculative since there is little concrete information on the nature and number of genetic units involved in many cases of cytoplasmic inheritance, but they are presented here for consideration in connection with the factual information that follows.

Examples of Extrachromosomal Inheritance

Examples of cytoplasmic inheritance in higher plants were reported as early as 1909 by Bauer (47) and by Correns (185).   The latter investigator was one of three who rediscovered the works of Mendel, and he evidently accepted the existence of two systems of heredity, chromosomal and cytoplasmic. In intervening years a large number of additional examples have been described (in the order of a hundred) in a great variety of organisms. In recent years Nanney (733, 734) has propounded the thesis of the existence of "genetic" and "epigenetic" factors with "genetic" being concerned with designation of primary structural order (chromosomal) and "epigenetic" factors being concerned with determination of higher-order organization including stable alternate metabolic patterns existing under differentiation and development. This is a broader concept which emphasizes the important point of interdependence of chromosomal and cytoplasmic factors and it is useful as such, indicating a need for further integrating studies of developmental processes and cytoplasmic inheritance.

In the sections that follow some details of selected examples are given, first of genetic aspects of cytoplasmic inheritance and then of some biochemical aspects of phenotypes produced. In general, the examples are among those currently under investigation and for which there seem to be reasonable prospects for obtaining information on specific mechanisms involved.

Genetic Characteristics. The examples considered here are first those from higher plants, *Drosophila,* and fungi, in which there is a great difference in cytoplasm contribution by male and female gametes, and then those in which the sexual cycle involves cytoplasmic bridges or fusion as in algae, yeast, and bacteria.

A general method that has been used for delineation of cytoplasmic inheritance is that of interspecific crosses with repeated backcrosses to minimize chromosomal differences. Michaelis (671, 672) has presented extensive data from studies of the willow herb *Epilobium.* This is a small flowering plant of the family *Onagraceae* which is normally diploid, develops air-borne seeds, and has pollen grains that produce fertilization without transfer of any large amount of cytoplasm from the paternal parent. This last point is particularly important since it permits a measure of inheritance based almost entirely on the properties of the mother cell when the gene complements of the two parents are alike. In an exemplary experiment *Epilobium luteum* was crossed to *Epilobium hirsutum* through 25 generations over a period of 23 years through the following scheme:

$$[(\textit{luteum } ♀ \times \textit{hirsutum } ♂) \times \textit{hirsutum } ♂] \times \textit{hirsutum } ♀$$
$$\text{for 25 generations (strain } lh^{25})$$

In this way the genome of *hirsutum* was gradually introduced into the mother cell of *luteum* to provide a strain with the total cell characteristics very similar to *luteum* but with gene constitution of *hirsutum*. Reciprocal crosses were then made as follows:

(1) $\qquad\qquad\qquad lh^{25}$ ♀ × *hirsutum* ♂

(2) $\qquad\qquad\qquad$ *hirsutum* ♀ × lh^{25} ♂

The male and female parents in both of these crosses carry essentially the same genome, but they differ by whatever contribution is made by the cytoplasm of the maternal parent. The phenotypic expressions of these cytoplasmic differences were then evaluated by comparing the progeny from the reciprocal crosses.

Examinations of progeny of a large number of different lines which were developed as described yielded evidence for cytoplasmic inheritance of a great variety of phenotypic characteristics. These include differences in lethality, sterility, plant anatomy, flower color, heterosis, cytoplasm viscosity, permeability, sensitivity to poisons, fungus resistance, reaction to temperature and light, enzyme activities, and concentrations of metabolites. The variety of changes observed here thus encompasses nearly the same phenotypic expressions that can be observed as the result of mutation of nuclear genes. The differences observed vary in degree among the progeny from one cross, and the ranges of expression sometimes overlap considerably among the progeny from reciprocal crosses, but this is also true of characters altered by chromosomal mutation, although in the latter case nonoverlapping characters are usually selected for study. Some examples of variations in size and character of plants derived from reciprocal crosses among different races of *Epilobium* are shown in Figure 11.1.

Interspecies crosses in *Oenothera* have also provided good examples of the cytoplasmic inheritance in higher plants (830). Here reciprocal crosses can be examined directly since segregations of chromosomes at meiosis is not random as in *Epilobium*. They are associated in complexes that move as units in the first anaphase division. *Oenothera odorata* carries the two chromosome complexes *v* and *I* whereas *Oenothera Berteriana* carries the two complexes *B* and *l*. Each complex can be identified cytologically, and crosses between these two species give rise to the new combinations *B v, B I, l v,* and *l I*. Results from reciprocal crosses describing the effects of the maternal cytoplasm on plants carrying these new combinations of chromosome complexes are summarized in Table 50.   As shown, the hybrid products of these crosses have growth and viability characteristics that are as dependent on which species served as the maternal parent as on the genome.   Subsequent selfing of a viable hybrid usually gives rise to plants with the same characteristics as the hybrid, and after several generations a back cross to restore the normal chromosome complexes will give normal plants.   Slow and heritable changes have been observed on continued self-crossing of one of the hybrids.

The origin of the cytologically inherited differences found through interspecific crosses in *Epilobium, Oenothera,* and other organisms is obscure.   They may have been derived under the influence of gradually diverging genomes in species that were once very similar or they may have arisen suddenly in isolated populations.   It is not possible to reach a decision on this question, but it has been demonstrated that cytoplasmically inherited changes can arise within one species as a rapid event and, so far as is known, without induction by mutation of a chromosomal gene.   Examples in yeast and in *Neurospora* are illustrative.

In contrast to situations in *Epilobium* and *Oenothera* in which cyto-

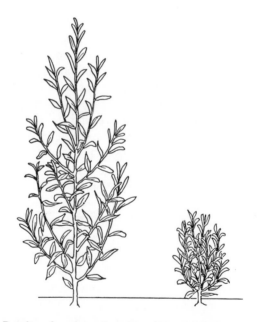

Figure 11.1.   Results of reciprocal crosses between closely related species of *Epilobium* showing cytoplasmic inheritance of plant size.   After Michaelis (672).

Table 50.   Cytoplasmic Inheritance Shown by Reciprocal Crosses in Oenothera [*]

Chromo-some Complex	Appearance of Hybrid	Appearance of Hybrid
Bv	Nonviable	Normal
Bl	Normal	Weak and yellow
lv	Mostly normal, lower leaves sometimes yellow	Weak and yellow
lI	Mostly normal, lower leaves sometimes yellow	Nonviable

Cross: ♀ *O. Berteriana* ✕ ♂ *O. odorata*   Cross: ♀ *O. odorata* ✕ ♂ *O. Berteriana*
        Bl   *vI*   *vI*   *Bl*

[*] Weier and Stocking (1058).

plasmic differences were derived from related species, the iojap character in maize was derived from chromosomal mutation.   The iojap gene in corn is a recessive located on the seventh chromosome, and in the homozygous condition it gives rise to green and white, or occasionally yellow striped plants.   In reciprocal crosses with a normal strain it yields a variety of progeny when iojap (*ij*) is used as the maternal parent but only green plants when the maternal parent is normal (*Ij*). These results are diagrammed in Figure 11.2 which also includes a summary of crosses from F_1 striped plants.   All F_1 plants from the first reciprocal crosses have the same constitution with respect to chromosomal genes as shown in the figure.   The white F_1 seedlings do not mature, but the striped seedlings do and can be used in further crosses. As shown they give rise to either *ij* or *Ij* mother cells, and when these are fertilized with normal pollen both types produce green, striped, and white plants.   Thus it is established that, although the recessive *ij* gene is necessary to establish the inherited cytoplasmic condition, this state will continue after replacing the gene with its dominant allele. It is, of course, possible that many cases of cytoplasmic inheritance may have arisen in this fashion.

Another cytoplasmically inherited and gene-dependent character in maize is that of male sterility (495, 832).   This has been put to a very important practical use in commercial production of hybrid seed.

Of the two principal examples that have been classed as cytoplasmic

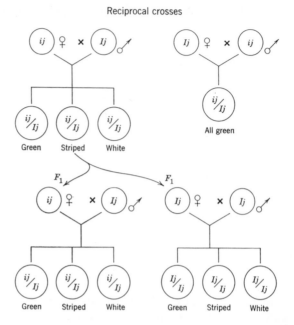

Figure 11.2. Cytoplasmic inheritance in maize showing the influence of the recessive gene *ij*. *ij/ij* gives green and white striped plants and some with yellow stripes.

inheritance in *Drosophila,* the most extensively investigated is CO_2-sensitivity discovered by L'Heritier and Teissier (582). An extensive review by L'Heritier (581) and a more recent presentation by Seecof (892) describe the evidence that the CO_2-insensitivity character in *Drosophila* is due to a virus-like agent (sigma, σ) which itself is mutable and is inherited as a cytoplasmic unit. Resistant flies are narcotized quickly in an excess of CO_2, but they recover quickly and completely even after hours of exposure. Sensitive flies are also narcotized, but even after a half-minute exposure they remain paralyzed and die. Sensitives occur in natural populations and are apparently completely normal except in the presence of high concentrations of CO_2.

In addition to transmission by crossing, the σ factor can be transferred from fly extracts by injection into σ-free flies, a procedure which permits measurement of the number of units on the assumption that one is sufficient to yield CO_2-sensitivity after replication of σ. On this basis it has been observed that sensitive strains may contain more than 10^4 σ units per fly while stabilized strains produce less and at a much slower rate. The factor has been characterized to some extent with purification by centrifugation (792, 892). It will pass a 300 mμ pore filter to some extent but is retained by a 100 mμ filter, and it bands sharply in a cesium-chloride density gradient in the analytical centrifuge. These properties place the factor in the range of characteristic of a number of viruses.

Furthermore, after transfer to a new host, there is an eclipse phase (a period when no active units are detected) before extensive replication occurs. This and the evidence for recombination between different σ characters (751) constitute strong evidence for a virus-like character of the factor. The known mutant characters of σ are based on temperature sensitivity, and different manifestations of host compatibility.

Of special interest is the CO^2-sensitivity system in the ρ strain of flies (581). This is a stabilized line in which σ character inheritance is strictly maternal. Most progeny are CO_2-resistant but on aging a few show sensitivity symptoms. Flies of this line contain very few σ units, and, if injected, the factor replicates very slowly. A related line of flies (PO) studied by Seecof (892) is highly stabilized and sensitive but carries very few infectious units. This apparent immunity to σ by ρ and PO lines has been compared to lysogeny in the bacteria-phage system with the proposal of the existence of an integrated provirus class, vegetative virus giving immunity to superinfection in stabilized germ lines and σ only in nonstabilized lines (892). It is of interest to note that the virus-like nature of σ was recognized early in the work by L'Heritier and collaborators, but, as pointed out by L'Heritier, if the stabilized σ situation had been encountered first, the case likely would have been classified as one of cytoplasmic inheritance without recognition of the virus-like component.

The "sex ratio" character in *Drosophila prosaltans* has similarities to the foregoing CO_2-sensitivity example and also to the "killer" character in paramecium (149, 650). In this case the "sex ratio" females carry the cytoplasmic factor O (omicron) and, for continued maintenance of O, the chromosomal gene combinations Sr/Sr or Sr/sr. Progeny from such females are all females due to death of XY eggs when O is present. The O factor is not transmitted through male gametes and sr/sr individuals lose the factor and behave normally. The genes Sr and sr have no phenotypic effects on the flies other than that associated with the maintenance of O. Natural populations of the fly carry the genes in both forms and the cytoplasmic factor O. Poulson and Sakaguchi (806) have presented evidence that the factor O is actually a filamentous parasite or symbiont having all the characteristics of a small spirochete resembling the genus *Treponema*. It can be readily transferred by injection into many *Drosophila* species with the production of the "sex ratio" character and microscopically visible filaments to about 5 μ in length.

Lints (620), in an extensive study of reciprocal crosses among *Drosophila,* demonstrated cytoplasmic differences influencing viability, size, and other quantitative characters, and came to the conclusion that these environmentally influenced rate effects are not to be considered as hereditary.

The sexual cycle of the fungus *Neurospora crassa* also is characterized by a zygote with a major portion of the cytoplasm originating in the female parent, and four distinguishable cytoplasmic characters affecting growth rate and the respiratory system have received attention (686, 695, 788). These are designated *W* (wild type), *po* (poky or *mi-1,* for maternal inheritance), *mi-3,* and *mi-4.* Growth characteristics of the first three mutants are shown in Figure 11.3, and the behavior of *mi-4* is very similar to that of *po.* A second aspect of the phenotypes, that of abnormalities in cytochrome composition, is shown in Figure 11.10, p. 492, and again *mi-4* is similar to *po.* Inheritance patterns among these strains are illustrated in Figure 11.4. Crosses among *W, po,* and *mi-3* are fertile in all combinations and yield essentially 100% maternal inheritance through many generations. Mutant *mi-4*

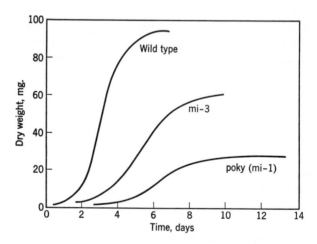

Figure 11.3. Growth characteristics of wild-type *Neurospora* and strains that exhibit cytoplasmic inheritance (*mi* = maternal inheritance).

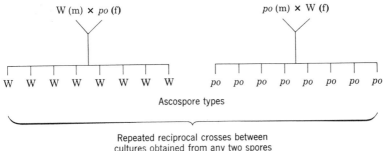

Ascospore types

Repeated reciprocal crosses between
cultures obtained from any two spores
of opposite mating types gives the
same result as in the first two crosses.

Figure 11.4. Cytoplasmic inheritance in *Neurospora*. *W* = wild-type; *po* = poky; (*m*) = maternal or protoperithecial parent; (*f*) = fertilizing parent.

is female sterile, but it behaves maternally in a cross as a heterocaryon (788). It is of interest to note that two slow-growing chromosomal mutants (*C115* and *C117*) which also have cytochrome abnormalities have been obtained and, in addition, a chromosomal suppressor of *poky* slow growth but not of *poky*'s cytochrome abnormality has been studied (693).

The genetic behavior of the *Neurospora* strains in systems of cell fusion by heterocaryosis has also been studied with several interesting results. Gowdridge (346) introduced nutritional mutant markers to limit growth into *W, po,* and *mi-3* and observed that *po-W* hetero-caryons were wild in phenotype and gave only the *W* character from hyphal tips. Heterocaryosis was assured since both parental mutation markers were recovered from the tips. Mixtures containing *mi-3* and *po* usually gave the *mi-3* phenotype but occasionally gave *po*, while the *mi-3–W* combination yielded one or the other. No intermediate types were observed in any combination nor did any heterocaryon grow faster than the most rapidly growing component. In contrast to these results a complementary action was observed with the mixture *po–mi-4* (788). The initial growth rate was essentially that of wild type for many hours on an agar surface, but eventually the rate dropped to that of the parental mutant strain. In liquid cultures growth was also more than doubled in the mixture but the cytochrome abnormalities were retained. An extensive search for a virus-like factor in the *Neurospora* mutants has not yet been successful (683).

A situation involving cytoplasmic inheritance in *Neurospora* has also been studied in interspecific crosses (945). A slow-growth maternal character *SG* obtained in *N. crassa* after acriflavin treatment was found to suppress a chromosomal gene expressed as a conidial mutant in *N. sitophila.* The gene was transferred to *SG* cytoplasm by repeated back-crossing. Thus this situation is the reverse of that mentioned earlier in which a chromosomal gene provided suppression of a cytoplasmic character. The *SG* character does not resemble the *mi* series in the cytochrome aspect of the phenotype.

Rizet and collaborators (655, 845) have investigated two interesting situations of cytoplasmic inheritance in the fungus *Podospora* which has a life cycle similar to that of *Neurospora.* One involves senescence or death in old mycelia. Reciprocal crosses using fruiting bodies from old mycelia demonstrated maternal inheritance of the characters, senescent and young. Furthermore, in heterocaryons it was shown that fusion of old and young resulted in the young phenotype throughout the culture even without nuclear migration. Rapid migration of a cytoplasmic unit from the young mycelium is indicated.

The "barrage" phenomenon in *Podospora* (844) concerns structure formations produced at the margin between two strains which are allowed to grow together on an agar surface. In normal contact hyphae of two strains interpenetrate, undergo fusion, and become pigmented. In the "barrage" situation the advancing hyphae branch and intertwine to form a dense network of mycelia with no pigmentation and apparently no hyphal fusion. "Barrages" are formed in mixtures of the normal *S* with the mutant *s,* and in crosses *S* appears in half the progeny but *s* does not appear at all. Instead, half the progeny are of the type designated s^S which does not give a "barrage" with either *S* or *s.* Strain s^S is quite stable yielding all s^S progeny in crosses. Occasionally *s S* shows spontaneous reversions to *s,* and when this occurs or when s^S is placed in contact with an *s* culture the *s* character spreads rapidly throughout the culture. As illustrated in Figure 11.5, these results have been interpreted in terms of induction of a cytoplasmic modification in the *S s* zygote under the influence of gene *S.* A strain s^S then carries gene *s* but is deficient in some way with respect to a rapidly spreading cytoplasmic factor which can be reinstated by reversion or heterocaryosis with mutant *s.* In addition to the foregoing examples a number of others of potential value have been reported in fungi (143).

Turning now to cytoplasmic inheritance in organisms which give cytoplasmic mixing through bridges or fusion in their sexual cycles, one which has been studied very extensively is that in *Paramecium* by Sonneborn and collaborators (927, 928) and Beale (61). This in-

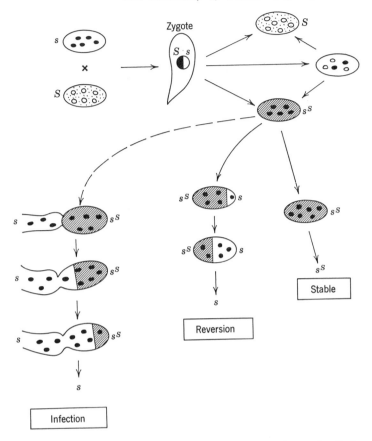

Figure 11.5. Schematic interpretation of "barrage" inheritance in *Podospora*. Changes in shaded areas indicate progressive changes of s^S to the s phenotype through infrequent reversion (center) or rapid "infection" by contact with s type (left). From Rizet (844).

volves a gene-controlled inheritance of the killer trait and the cytoplasmic unit kappa. It will be recalled from Chapter 3 that the exchange of nuclear genes in *Paramecium* takes place by a partial fusion of two animals to form a small cytoplasmic bridge between them. The haploid nuclei from the two parents fuse after migration through this bridge, and after subsequent divisions and equal distribution of the newly formed nuclei the bridge is sealed off and the two animals separate. They then reproduce by direct fission. An important aspect of the conjugation process so far as cytoplasmic inheritance is concerned is the fact that the cytoplasmic bridge sometimes persists for a longer time than necessary for nuclear exchanges and permits an exchange of relatively large amounts of cytoplasm.

As shown in Figure 11.6, cytoplasmic inheritance in *Paramecium* is clearly demonstrated on this basis by crosses and cytoplasm exchange among the three races indicated.   Race 1, the killer strain, carries the dominant gene *K* in the nucleus; it produces large cytoplasmic particles known as kappa, and releases other large particles called paramecin into the culture fluid.   Both kappa and paramecin appear to contain large quantities of deoxynucleoprotein, and paramecin is only produced when kappa is present in the cells.   Cells which do not contain kappa are sensitives and are killed by coming in contact with paramecin produced by a killer.   The sensitives are of two types (2 and 3, Figure

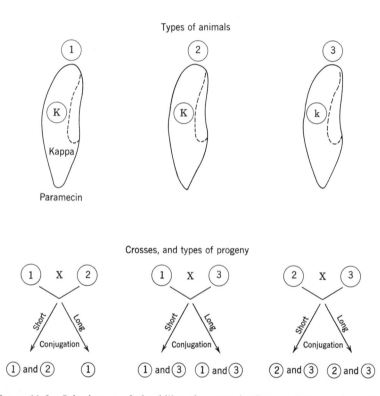

Figure 11.6.   Inheritance of the killer character in *Paramecium aurelia*.   *K* = dominant gene; *k* = recessive gene; *kappa* = cytoplasmic unit present with *K*; paramecin = toxic substance produced in the presence of *K* and *kappa*.   After Sonneborn (925, 926).

11.6), one of which carries the dominant gene *K* whereas the other carries only its recessive allele *k*. Killing does not occur during conjugation, and the results of crosses among 1, 2, and 3 are shown in Figure 11.6. Crosses of 1 $\times$ 3 and 2 $\times$ 3 show only segregation of *K* and *k,* since kappa does not persist in the absence of *K* even when it is transferred by long conjugation to type 3. In the cross of 1 $\times$ 2 long conjugation produces a cytoplasmic transfer of kappa, and, since 2 carries the gene *K,* kappa persists and is reproduced to convert 2 into a killer.

Besides this straightforward demonstration of cytoplasmic inheritance in *Paramecium,* experimental work has produced two other highly significant kinds of data. These are concerned with the various environmental factors that influence the presence of kappa and paramecin and with the fact that different kinds of kappa and paramecin have been shown to exist. Particles of kappa, which are in the order of 0.2 μ in diameter, can be observed in killer strains by Feulgen staining (for DNA). They may contain as many as 1600 of these granules per cell, whereas none has been observed in sensitive strains. When the organism is grown under conditions conducive to rapid cell multiplication, reproduction of kappa particles lags, and eventually cells are produced that are free of kappa. By nutritional limitations of cell multiplication, animals with small numbers (presumably as small as one) of kappa particles can be induced to reproduce a normal number of the granules. Killers can also be converted to sensitives by exposure of high temperatures, x-rays, or nitrogen mustard. Sensitives that carry the gene *K* can be reconverted to killers by the cytoplasmic transfer already described or by exposing them to a concentrated, cell-free suspension of ground-up killer animals.

That kappa and paramecin can exist in a number of forms with different biological specificities but still conditioned by the same gene *K* has been demonstrated (220). These different kinds of units were obtained by experimentally reducing the number of kappa particles per cell to a very low level and then permitting a regeneration period. Five types were obtained, distinguishable on the basis of the quantity or quality of paramecin produced. For example, an original strain of killer produces paramecin that causes blisters and cellular distortions of sensitives with death occurring in about 24 hours. Of two new cultures that were derived from this by the kappa-dilution and regeneration technique, one produces a limited amount of paramecin with killing characteristics like the original whereas the other produces a paramecin that causes sensitive animals to spin rapidly about the longitudinal axis and to die in about eight hours. Other killer strains were obtained that produce paramecin that will kill different killers, although in no case does paramecin act on the animals that produce it. All these killer characters that are correlated with the presence or absence of kappa are inherited by the process described (Figure 11.6).

Paramecia and also the related ciliate *Tetrahymena* display an interesting character which Nanney has termed epigenetic homeostasis (734). A simple example involving *Paramecium aurelia* (62) is presented in Figure 11.7. Here a single paramecium was allowed to multiply and the clone produced was divided into two portions. One was incubated at 15°C, and after a short time an antigenic character designated as S appeared. At the same time the 25°C culture showed the antigenic character designated as G. Both populations were then continued at 20°C and it was found that the established S and G states remained indefinitely regardless of the common environment. The potential for establishment of antigen type is gene controlled, and it is reasonably assumed that this is not a phenomenon concerned with a cytoplasmic genetic unit but with regulation of potential function into stabilized alternative systems.

Among organisms which yield zygotes through cell fusion, examples of cytoplasmic inheritance that have received extensive analysis are to

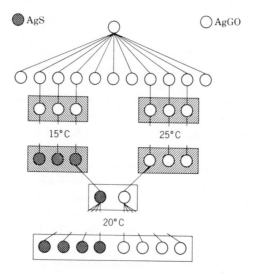

Figure 11.7. Environmental influences on antigen types in *Paramecium*. An example of "epigenetic homeostasis." From Nanney (734).

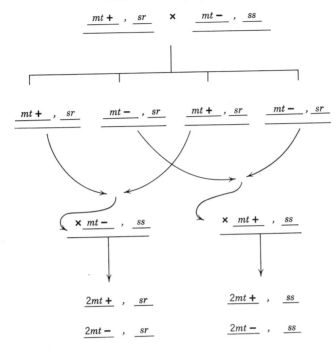

Figure 11.8. Extrachromosomal inheritance in *Chlamydomonas*. F_1 progeny from a cross of *sr* (streptomycin resistant) to *ss* (streptomycin sensitive) are all *sr* in phenotype but show segregation of mating type. Progeny of backcrosses show complete inheritance of *sr* or *ss* phenotypes associated with the *mt+* parent even though the mating types (+) or (−) segrate normally. From Sager and Ryan (873).

be found in yeast and in the green alga *Chlamydomonas*. The latter organism has two mating-type forms *mt+* and *mt−* which fuse to give a zygote and eventually four progeny, two each of the two mating types and two each of any pair of chromosomal gene alleles in a cross. The organism is normally inhibited by the antibiotic streptomycin, but treatment of a large population yielded resistant mutants in small numbers (871, 873). In some, resistance was found to be inherited in a Mendelian fashion, but in others it was not, a uniparental inheritance was demonstrated as illustrated in Figure 11.8. As shown a *mt+* (mating type) *sr* (resistant) mutant crossed to a *mt−*, *ss* (sensitive) yielded all resistant progeny but with normal segregation of mating type (and other chromosomal markers). Back crosses of the four F_1 progeny again yielded normal mating-type segregation, but four resistant or four sensitive progeny depending on parental mating types. Thus, in spite of cell fusion in the zygote, the *sr* character was transmitted only when present with the *mt*+ mating type. The result persisted through many successive backcross generations, and reversions of *sr* to *ss* have not been observed. The evidence for cytoplasmic inheritance in this case is quite straightforward, but there is a complete compatibility dependence on mating-type genes, and even though a zygote is derived by cell fusion it is not obvious how the result obtains if cytoplasmic mixing is complete.

Zygote formation in yeast (*Saccharomyces cerevisiea*) also occurs by cell fusion followed by production of four spores with two-to-two segregation of chromosomal markers. The *petite* characters in yeast which have been studied very extensively by Ephrussi and collaborators (257, 259) are slow in growth, form small colonies, and have abnormalities in mitochondrial cytochrome systems. The phenotype results either from gene mutation (segregational *petite*) or from alterations in cytoplasmic inheritance. In the latter category are the two extreme types, one which yields only normal yeast among the progeny from a cross to normal, and one which yields only *petites* in crosses to normal (suppressive petites). Inheritance patterns of these characteristics are illustrated in Figure 11.9. Suppressiveness is found to different degrees in different cell lines. *Petites* have been obtained by mass conversion of normal yeast in the presence of euflavin and by heat treatments and ultraviolet irradiation (824).

The situation in bacteria with respect to cytoplasmic inheritance is

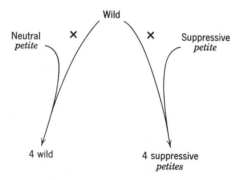

Figure 11.9. A simplified summary of the genetic behavior of the neutral and suppressive *petite* mutants of yeast. These cytoplasmic expressions repeat through successive generations but infrequent changes occur to give mixed populations in cultures. From Ephrussi, Hottinguer, and Roman (259).

actually covered quite extensively in Chapter 3 in connection with the regular patterns of inheritance through cytoplasm bridges and phage transfer. It is especially interesting to note that it is not always easy to distinguish between chromosomal and cytoplasmic inheritance in this situation. The mating-type characters F^+ and Hfr behave as cytoplasmic units as do phage in the lysogenic situation. Pieces of bacterial chromosome can be transferred by transduction from one bacterial cell to another and can evidently function as genetic units with or without association with the bacterial chromosome. However, the morphological and functional equivalence of the bacterial chromosome-cytoplasm systems to those of higher organisms is open to question. Nevertheless, the bacterial systems unquestionably present situations which can exist in more highly ordered morphological states and thus they present useful and interesting models for physical structures and metabolic actions.

Biochemical Aspects. Generally speaking, there is little direct information on which to base a biochemical interpretation of extrachromosomal inheritance. If the bacterial system is included in the heterogeneous collection of examples that have been considered, then cytoplasmic units may actually be chromosome fragments which can function in replication and control of primary protein structure in a manner similar to the functions of chromosomal units (Chapter 8). By inference the other examples of nucleic-acid-containing particles as in the CO_2-sensitivity and sex-ratio phenomena in *Drosophila* and the "killer" situation in *Paramecium* are interpretable in the same way, but direct evidence as to what primary products may be produced is not yet available. Among other examples such as chlorophyll loss in maize and mitochondrial abnormalities in yeast and *Neurospora,* the phenotypes expressed concern normal biochemical cellular units. Thus both chloroplasts and mitochondria have been described as being carriers of cytoplasmic genetic units. Since chloroplastids seem to carry a small amount of DNA, it is possible that they carry some genetic material as DNA independent of the nucleus. However this is by no means proved.

In both yeast and *Neurospora* the cytoplasmic mutants are similar in phenotypes to certain single-gene chromosomal mutants. In addition to slow-growth character, all have abnormalities in their respiratory systems. The cytochrome components are easily observed with a hand spectroscope even in living material, and a summary of patterns shown is given in Figure 11.10. Suppressive *petites* in yeast (p. 490) are similar to *petites* shown, and the mutant *mi-4* in *Neurospora* is not dis-

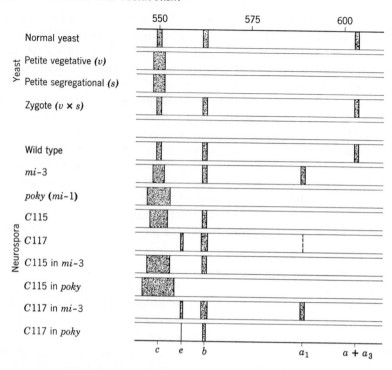

Figure 11.10. The cytochrome components of a number of strains of yeast and *Neurospora*. The α absorption bands are indicated with wider bands for higher concentrations but actual differences are much larger than indicated. Positions of some of the known cytochrome bands are shown at the bottom as c, e, b, a, and $a + a_3$.

tinguishable from *poky* by this criterion. In the zygotes from crosses with the segregational *petite,* the vegetative (neutral) *petite* is normal, indicating different biochemical causes for the chromosomal and cytoplasmic phenotype. This direct observation has not been made with *Neurospora,* but as shown the phenotypes of the chromosomal mutants in different mutant cytoplasms are more extreme than in wild, and, again, different biochemical patterns of origin are indicated. At the same time it is clear that in the cytoplasmic mutants the genetic control which determines the cytochrome abnormalities does not necessarily reside in the mitochondria which are affected. Respiratory defects in both organisms are more extensive than shown. In addition to the cytochrome anomalies the *petites* (*v*) of yeast is deficient in cytochrome c reductase, α-glycerophosphate dehydrogenase, and has much reduced (two- to fivefold) activities of aconitase, fumarase, and DPN-isocetric dehydrogenase (443). It also has a new DPN-independent malic-cytochrome c reductase activity and an increased amount (twofold) of lactic-acid dehydrogenase. Though some of these components are normally mitochondrial constituents, it is clear that these structures must be abnormal to some extent. However the *petites* do contain cytological identifiable mitochondria (1127).

In *Neurospora* some evidence has been obtained for structural abnormalities in the *poky* mutant mitochondria (677). Electron-microscope pictures of germinating conidia show a great deal of double membranous material but not morphologically distinguishable mitochondria. On the other hand this may also be a transient character of the wild-type strain and one which only shows better in the slow-growing *poky* mutant. In spite of these possible morphological differences, fractional centrifugation of *poky* homogenates yielded material corresponding to a mitochondrial fraction under the same conditions as the wild strain (containing essentially all the cellular succinic acid dehydrogenase). In terms of expected functional activities these preparations from different age cultures of *poky* behave very much like intact mycelium in respiratory activities (1012). Some properties of the latter are illustrated in Figure 11.3 where it is shown that the *poky* character is most extreme at the very early stages of growth, and in older cultures it becomes more nearly normal. Nevertheless, the cycle is repeated in new cultures using small innocula. As shown, the intact mycelium of *poky* shows as much as sixteenfold accumulation of cytochrome c, but this is less evident in the fractionated particulate material. Here, in contrast to the wild strain where the cytochrome is quite firmly bound to mitochondria, the major part of the cytochrome is not bound. Numerous other composition differences have also been observed between particulate preparations from *poky* and wild strains. The material from *poky* contains a potent proteolytic activity not exhibited by wild, approximately a twofold excess of riboflavin and about a fifteen-fold excess of long-chain unsaturated fatty acid (403). It is relatively deficient in polysaccharide and RNA. No DNA was detected in mitochondria preparations from either *poky* or wild strains. Some other undefined but interesting likenesses and differences between the two fractionated systems were obtained by separation of fractions and mixing in different combination to give anomolous respiratory characteristics. Some data are shown in Table 51. Neither particulate nor supernatant fractions showed respiration alone, and reconstituted *poky* (PP + PS) gave the least inhibition by azide as expected from the

Table 51. Respiratory Interaction in Fractions from Wild and *Poky* *Neurospora* *

	Oxygen Uptake	Per Cent Inhibition, $10^{-3}M$ Azide	Oxygen Uptake	Per Cent Inhibition, $10^{-3}M$ Azide
	(From 3-day wild and 3.5-day *poky* cultures)		(From 3-day wild and 5-day *poky* cultures)	
WP + WS	150	36	102	41
PP + PS	90	5	225	20
WP + PS	221	63	230	64
PP + WS	35	43	154	38

* From Tissieres et al. (1012).

Note: Cell debris was removed from homogenates of wild and *poky* by low-speed centrifugation and the supernatant further fractionated to give particulate (*P*, containing mitochondria) and supernatant (*S*) fractions. All reaction mixtures contained TPN and combinations of fractions and phosphate-mannitol buffer to a volume of 2.5 ml in Warburg flasks. Oxygen uptake is in μl per hour.

deficient cytochrome system of the particulate material. However both abnormal combinations of fractions (WP + PS and PP + WS) yielded unexpected oxygen uptakes and inhibitions indicating very marked differences in enzyme and substrate compositions.

Less extensive but similar information on composition and biochemical function has been presented in relation to cytoplasmic characters in *Epilobium* (855) and *Paramecium* (910), and the generalization is permitted that cytoplasmic mutants usually exhibit widespread quantitative differences from normal with respect to a variety of metabolites and catalytic activities. Such differences are also characteristic of many chromosomal mutant plietropic effects, and though useful in tracing primary effects of mutation this goal has not been achieved. The data only emphasize that the cell particulates that are most obviously affected by mutation in characteristics determined extrachromosomally are not necessarily the carriers of cytoplasmic genetic units.

General Comments

One of the prominent results of studies of extrachromosomal inheritance has been the invention of new words to describe presumed cytoplasmic genetic units. Among these are pangenes, bioplasts, plasmagenes, chondrogenes, cytogenes, genoids, plasmids, protomeres, proviruses, and episomes. Chloroplastid and mitochondrial inheritance have also been used descriptively. Undoubtedly the reason for such heterogeneity in the delineation of phenomena lies in the heterogeneity of the phenomena themselves with respect to origin, transmission, and expression. It is possible that such heterogeneity is real in terms of fundamental mechanism, but in the absence of information concerning these mechanisms, and with the conviction that order will yet prevail, the discussions in this chapter avoid both the descriptive words for cytoplasmic genes and the classification of examples based on presumed mechanisms. The one operational classification made here, that concerned with the relative amount of cytoplasm contributed by maternal and paternal parents in zygote formation, is also of dubious value as shown by the description of the genetic results. For example, similar inheritance patterns of similar phenotypes are shown in yeast and in *Neurospora,* and yet the former derives a zygote by fusion of essentially equal amounts of cytoplasm and the latter by very unequal contributions from the parents. Thus the two kinds of systems are no doubt operationally useful where distribution of small numbers of large particles is concerned but there is not necessarily a fundamental difference with respect to genetic function.

Concerning the actual and hypothetical mechanisms of cytoplasmic inheritance as a useful basis for continued study, careful analyses have been presented by Ephrussi (257), L'Heritier (581), and Nanney (733) as well as by others in earlier literature. Three different but not necessarily mutually exclusive hypotheses have been presented by these authors.

1. Extrachromosomal inheritance is derived from a symbiont situation in which virus infection is most prominent. Several of the examples discussed clearly fall in this category, but, as pointed out by L'Heritier, this makes them no less interesting from the genetic standpoint. Such virus-like units can be so firmly integrated with the host genotype and so unobtrusive as to behave like true cellular components. Thus a real distinction between a virus and an extrachromosomal genetic factor is not always possible. Pathogenicity is not adequate since it does not always obtain in virus infections, and infectivity is of little value since this can be a normal process, at least in bacteria (transformation and transduction). Further, infectivity through germ lines is an expected property of any cytoplasmically inherited genetic unit. Since it is likely that all of the examples of cytoplasmic inheritance can

be rationalized on the basis of virus infection, perhaps the critical question is that of the origin of such units. Actually this does not help since such units may be derived by evolution externally from the host where found, but they may also be evolved from normal components within the host.

2. Extrachromosomal inheritance is a phenomenon associated with the production of ordered structure from primary gene products which have been ordered only in linearity by chromosomal genes. Structural units which might be involved here are, for example, chloroplasts and mitochondria. These may contain replicable nucleic acid as genetic material guiding reproduction of three-dimensional order or other kinds of molecules may function similarly. Perhaps the best evidence here is in situations where chloroplasts are not produced unless there are some in existence (p. 474).

3. Extrachromosomal inheritance is a phenomenon of alternate steady states of metabolic systems which can exist within the potentialities conferred by the chromosomal gene complement. Changes may be induced through the environment and perhaps implemented through cytoplasmic substances entering the nucleus and influencing the primary action of genes. There is no need in this case for replicable cytoplasmic units, and such a change could come about suddenly as the result of an environmental shock or it could come about more slowly through an orderly sequence of composition changes to yield a differentiated nucleus.

These have been the principal themes in interpretations of extrachromosomal inheritance phenomena, and all are tenable, but the virus-like unit is so far the only chemical substance having to do with the problem that has been captured. It must be said that, with the present development of nucleic-acid chemistry and its established role in protein synthesis, a prime and unobtrusive candidate for a cytoplasmic genetic unit is a potentially stable molecule of messenger RNA. Virus RNA does replicate and function in cell cytoplasm, and it is quite feasible that certain chromosomal mutations might give rise to messenger RNA resistant to and replicable in its cytoplasmic environment. Details of its usual fate are not known, but without doubt a full understanding of chromosomal mechanism is essential to the cytoplasmic inheritance problem. It is not certain that the picture of DNA-base sequence → messenger RNA-base sequence → primary amino-acid sequence, is adequate to yield more than useful parts for construction of the complex machinery of the cell. It can be argued that preexisting cytoplasmic parts are necessary in or during construction of complexes or even in helping to direct primary amino-acid sequence determination (143, 899). Thus an extreme view in one direction is to consider all inheritance as bipartite, a composite of cytoplasmic function and chromosomal function. In this case mutation of cytoplasmic components would show only under special circumstances since there would always be large numbers of normal units to compete in function and transmission with mutant forms, whereas mutation in the more specialized but interdependent chromosomal system would usually show in view of the small number of units transmitted to progeny. At the other extreme the cytoplasmic inheritance phenomena can be considered as rare and relatively insignificant in genetics. Most likely neither of these extreme views will serve adequately, but there is no doubt of the existence of fundamental mechanisms of reciprocal control between chromosomal and extrachromosomal units manifested especially in processes of differentiation and development.

12 Gene Action and Development

One of the chief riddles confronting the biologist is differentiation and the other aspects of development. The problem has been recognized since at least the time of Aristotle, and was stated in its modern form in the nineteenth century by the cell biologists and embryologists who founded modern cellular biology. Many generations of biologists have devoted their attention to the analysis of development, and, in their attempts to explain it, have made many of the significant findings about cell function and structure which have been discussed in this book. The fact that they have not solved the main problem is no great surprise, as will be made evident in the following discussion. The problem is a formidable one, and its eventual solution will require the combined efforts of individuals from practically all the different subdisciplines of experimental biology and biochemistry.

When modern genetics started anew at the beginning of the present century, it was recognized by a number of embryologists, chief among them being T. H. Morgan, that here was a new tool that might be applied to the study of development. Immediate results were not forthcoming as far as embryology was concerned, but as genetics developed new light was thrown on the problems of embryology as the result of genetic fact and speculation, and the fusion of genetics with biochemistry. The result is that today we are probably close to the end of the beginning of the solution to the riddle.

The primary objective of this chapter is to consider what is known about the genic control of development. The literature relating to this topic is massive in bulk, and some of the important aspects of it date back to the middle of the nineteenth century. But only the more important recent concepts that have developed in this field will be considered. It should be recognized, however, that many of the "recent concepts" are merely old ones in newer garb.

Development cannot be understood merely by considering primary gene action at the molecular level. It is the result of the interaction of the products of many genes. To appreciate the role of genes in development it is obviously necessary to understand some of the processes that go on in the development. This will be the first matter for attention. Primary emphasis will be placed on these processes in the higher animals.

Developmental Processes

Development may be considered the resultant of three component processes which may be called differentiation, organization, and growth. These, it may be emphasized, are not independent processes, but aspects of a single total process which results in the elaboration of a complete, functional organism.

The first of these component processes, *differentiation,* is perhaps the one most often emphasized in embryological research. It involves progressive changes in cell structure and chemistry which lead to the formation of different tissues. Although the term is most often used in connection with multicellular organisms, in the broad sense, even unicellular organisms show differentiation in function and structure within the cell during the life of the individual.

In the developing organism differentiation is accompanied by *organization,* or localization of differentiated cells leading to morphogenesis, or the establishment of a definite pattern of structure. It is usually but not invariably accompanied by *growth,* a process which will be defined here as increase in mass of protoplasm with or without increase in number of cells.

If the action of genes is considered in the light of the three principal aspects of development that have been adumbrated, it becomes evident that genes have a number of functions. They must control differentiations, for the types of cells and tissues produced are unique for each type of organism; they must control the organization of parts, for structure and form are inherited; they must control growth, for this is obviously not an uncontrolled process, but an ordered one resulting in various sizes and shapes of organisms, and in organs and parts of regulated size.

Differentiation. Differentiation in the multicellular animals and plants is probably the most spectacular aspect of development. Superficially, it involves the formation of distinctly different types of cells from a zygote mother cell. Thus the nerve, muscle, epithelial, and other cells of an animal, or the cortical, phloem, and mesophyll cells of a plant are all derived from a zygote. Differentiated cells are not only morphologically different, but, as befits their different functions, show chemical, immunological, and behavioral differences as well.

These changes which occur in cells as they differentiate are quite profound, and are the result of factors which operate both within and without the cells. Ordinarily, it is considered that once a cell has differentiated to a particular functional state such as a neuron, a muscle cell, and so on, it is stable in that state and does not revert back to its preceding condition, that is, it cannot, or does not, dedifferentiate. Hence it is often stated that differentiation is a one-way process, and that during the process the cells pass a point of no return from beyond which they cannot go back to the original or even an intermediate state. This is certainly true of vertebrate erythrocytes, for example, particularly those which lose their nuclei. And as for cells such as neurons, it is quite possible that they cannot dedifferentiate. For one thing they lose their capacity to divide on arriving at the functional state. It is well established that some differentiated animal cells do change considerably in this tissue culture, but whether this can be called true dedifferentiation to an embryonic or indifferent state is debatable. Chondrocytes (cartilage-forming cells) from ten-day-old chick embryos lose the capacity to synthesize chondroitin sulfate, a characteristic of cartilage cells, and no longer are recognizable as cartilage cells. In a like manner the muscle cells of the stump of a salamander tail or limb bud seem to degenerate and form mononucleated cells which contribute to the blastema from which a new limb regenerates (453). The current point of view for vertebrate cells, at least, is that once a cell has arrived at a terminal state it cannot go back, but that under certain conditions it may lose certain of its differentiated characters which may be regained when conditions are changed (370). The issue of de-differentiation is still not settled for animal cells, but it is important to recognize that the differentiated state is, under usual conditions, a *stable* one, and that many different kinds of cells can arise from a single kind, the egg cell.

With respect to dedifferentiation in plants, it is quite possible that this process does occur. Starting with a single, nucleated carrot-phloem cell nourished *in vitro* with liquid endosperms, it is possible to grow a complete plant (972). But even here the possibility remains that differentiated cells may exist within the plant which will not demonstrate this versatility.

As a result of differentiation, cells within the same organism, and even within the same organ, develop quite different nutritional requirements, and different physiological, chemical, and behavioral characteristics. It has long been known that the culture of tissues *in vitro* requires careful attention to the nutrient medium, since tissues differ in their nutritional requirements both qualitatively and quantitatively. This is as true for adult tissues as for embryonic tissues which are in the process of differentiation. Spratt (941, 942) has shown that chick-embryo heart tissue will develop at very low concentrations of glucose, whereas nervous tissue will not. Almost twofold more glucose must be supplied to nerve tissue before development will proceed. These differential nutritional requirements are, of course, indications of basic metabolic differences among tissues—a fact more directly demonstrated by results from inhibition studies. Iodoacetate, malonate, cyanide, and azide inhibit the growth of chick brain but have little effect on the heart. Fluoride, on the other hand, inhibits the heart but not the brain (942). Duffy and Ebert (236) found that the fluoride-sensitive area of the early primitive-streak chick embryo corresponds to the heart-forming area. Hence this sensitivity is developed very early in embryogenesis. They also found that the antibiotic antimycin A inhibited the formation of the heart and other mesodermal derivatives at concentrations of 0.05 μg per ml *in vitro,* but had little or no effect on the differentiation of endodermal and ectodermal derivatives. The inhibition by fluoride and antimycin A may be interpreted to mean that heart tissue differs from other tissues such as brain in its electron-transport system (236). This may be considered an effect (or a cause) of its differentiation.

It can be assumed that metabolic differences reflect quantitative and perhaps qualitative enzymatic differences. Data are presented in Tables 1 and 2 which show the correctness of this assumption. The activities of a number of different enzymes are given for the nuclei and cytoplasms of several different tissues in the calf. Two things are quite evident: (1) that different tissues have different relative enzyme activities, and (2) that the nuclei and cytoplasm of cells of the same tissue may have quite different enzyme activities.

These differences may be qualitative as well as quantitative. A number of enzymes have been found to exist in a number of different molecular forms called isozymes as discussed in Chapter 6. All isozymes of a specific enzyme are related insofar as they appear to have the same general role to play in catalysis, but they may differ somewhat in substrate specificity, and always demonstrate certain other chemical and physical differences. Lactic dehydrogenase, which has been studied extensively, particularly in mammalian tissues, can be demonstrated to have up to five isozymes separable by electrophoresis on starch gel (1115).

Significant in this context is the fact that the isozymes of an enzyme show differences in distribution in different tissues or organs. While some tissues of an animal may contain all five forms of lactic dehydrogenase, for example, the forms may be present in quite dissimilar proportions, as shown in Figure 12.1 (1032). Since this phenomenon has been demonstrated for a number of other enzymes in different species as shown in Table 52, we may assume that it is probably of general occurrence.

The differences in synthetic capacity expected as a result of changed metabolism of differentiated cells is manifested in form, chemical constitution, physiological and immunological response, secretory ability, and so on. Although certain differences in content among cells may be ascribed to some absorbing and storing compounds synthesized by all, it is obvious that many are unique sources of certain compounds. For example, the production of hormones in an animal is the function of the endocrine glands which produce specific hormones with specific physiological properties. The capacity to respond physiologically to these hormones, at least to a recognizable degree, is again specific for certain types of tissues. Thus secretin produced by the small-intestine mucosa in the vertebrates apparently affects only the pancreas, inducing it to discharge digestive enzymes despite the fact that secretin is widely distributed throughout the body soon after food enters the small intestine from the stomach.

It has long been known that the antibodies induced in a host animal by antigenic components from an unrelated donor show considerable differences, depending on the type of tissue from which the antigens

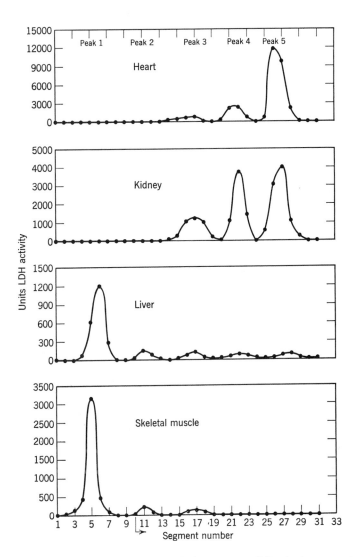

Figure 12.1. Distribution of lactic dehydrogenase activity in homogenates of human heart, kidney, liver, and skeletal muscle as determined by starch-block electrophoresis. The origin is indicated by the arrow. From Vesell (1032).

Table 52.   Enzymes with Different Distribution of Isozymes in Different Tissues of the Same Organism

Enzyme	Organism	Reference
Glucose-6-phosphate dehydrogenase	Rat	704
Lactate dehydrogenase	Pig	659
	Mouse	658
	Rat	1076
	Human	509
	Limulus	509
	Cynthia (a moth)	555
Peroxidase	Corn	643
Malic dehydrogenase	Rabbit	509
	Cynthia (a moth)	555
Esterase	Cynthia (a moth)	555
α-Glycerophosphate dehydrogenase	Cynthia (a moth)	555
Amylase	Human	647

are extracted.   Although each individual seems to have some antigens common to all tissues, organ or tissue-specific antigens are also produced (243).   As an organism develops, its tissues develop different biological specificities as determined by immunological techniques.   This is, of course, to be expected if differentiated cells differ in some of their protein molecules.   The enzymatic and immunological studies are in agreement that such differences exist.

Regional differences in morphological and physiological properties must arise as a consequence of prior chemical differentiation (1043)— presumably protein differentiation.   Thus differentiation must be looked on in part as a process of producing regional differences in protein structure which lead to secondary morphological and physiological differences.   The problem then immediately arises, how does this come about? Presumably it is a process guided by genes, but a consideration of the role of genes leads one to what seems to be a paradox.   Differentiation leads to the formation of cells of different phenotype, but yet theoretically of the same genotype, if it is assumed that regular mitosis is the only type of cell division occurring during development.   Superficially, this appears to be a contradiction to the previous affirmation that the genes determine specific proteins. Actually, however, it is not, when the further fact that the genes' action is modified by the environment is recalled, as discussed in Chapter 10. By *environment* is meant the internal environment which exists first as the egg cytoplasm and later as the whole internal environment, and, in addition, the external environment. It is in this fact that we must first seek the answer to the problem of the mechanism of differentiation.

The Role of the Nucleus in Development.   The approach to the problem of differentiation is a complex one, but a good start can and has been made by asking what role does the cell nucleus play? Obviously, it does play an important role because mutations on the chromosomes, either point mutations or chromosome aberrations, can have definite effects on the course of development. Furthermore, loss of even a single chromosome or part of a chromosome from the normal complement may cause a drastic breakdown in development and death before a significant degree of differentiation is attained.

The assumption is generally made that the cell divisions occurring during development are true mitoses giving rise to daughter cells with identical chromosome constitutions. The fact that the somatic chromosome number is almost always the same as that found in the zygote and double the number found in the gametes is in itself significant evidence for the regularity of true mitosis during development. Some variations in this general rule exist, however, and it is difficult to determine which are significant and which insignificant deviations. Many workers have demonstrated that somatic cells of mammalian tissues may have more or less chromosomes than the diploid number usually assumed for the species (471). But since the variations are generally not regular, as far as can be determined, but appear to be at random with the average number for all tissue types being the assumed diploid number, the variations may not be significant.

Studies of mitosis in plants by Huskins (473) and Wilson, Tsou, and Hyypio (1086), have shown that variations in mitosis occur in *Allium root* tips. These workers suggest the possibility that segregation of chromosomes similar to that occurring in meiosis, and the production of polyteny and polyploidy during different stages of development, are behind the process of development. Similar observations have been made in insects and other animals in which diminution of chromosome number, by the loss of whole chromosomes, and increase and decrease in ploidy and polyteny may occur regularly during the course of development at very definite stages characteristic of the species. However, none of these phenomena has yet been shown to be general or of importance in differentiation. The question of their general significance remains to be clarified.

It follows from the available evidence, then, that the nucleus of a highly differentiated cell should be no different in chromosome number from the zygote from which it is derived. Accidental occurrences of nondisjunction during mitosis may produce exceptions, but presumably after differentiation has occurred, and hence are probably of no significance as the cause of differentiation.

The fact that the chromosome number does not change is no evidence, however, that the chromosomes themselves do not undergo change. It is quite evident, in fact, to any cytologist that the chromosomes of different cells in the same organism may show quite significant differences. Indeed, it has been demonstrated that quite spectacular changes occur during development in the chromosomes of certain species of Diptera that have been studied (67, 68, 778). The polytene chromosomes of the Diptera show a definite banded structure as has been discussed earlier, with a definite order and arrangement of identifiable bands which remains intact throughout the body of the individual so far as can be determined. However, the appearance of the bands is extremely variable during larva development. The main changes are related to the condensation or puffing of certain areas of the chromosomes resulting in great enlargement of some bands and almost complete obliteration of others. A significant observation is that the appearance of the puffs is accompanied by changes in RNA and DNA concentration (see p. 71). It can be assumed that changes in these important substances during development is not without significance for they would be expected to be indicators of changes in protein synthesis which in turn would amount to changes in enzyme activity. Thus puffs may be visible signs in changes in gene activity during development (778).

In Chapter 4 the role of heterochromatin in the manifestation of the variegated position effect was pointed out, and, following the convention accepted by many geneticists, heterochromatin was contrasted to euchromatin as a rather distinctly different type of genetic material. This distinction may not be real. It has been argued that the evidence for heterochromatin and euchromatin being "alternative chromosomal states that . . . portray within one nucleus and chromosome different conditions or behavior on the part of different regions within one chromosome" (183), is just as good or better than for their being fundamentally different types of genetic material. The heterochromatic regions of the *Drosophila* polytene chromosomes show "heteropycno-sis" or heavily stainable regions of chromatin in interphase and early prophase chromosomes. These heteropycnotic regions may represent "puffs" similar to those seen in the polytene chromosome of *Rhynchosciara* and other Diptera (183).

If the changes in chromosomes just described are of any significance in the differentiation process, it should be possible to demonstrate actual nuclear differentiation by (1) chemical methods and (2) biological methods. Nuclei have been isolated free of cytoplasm from different types of differentiated cells and shown to be chemically different (679) (Tables 1, 2, and 3), and the application of certain biological techniques seem to demonstrate that nuclei as well as cytoplasm actually differentiate during development. This latter revolves about the demonstration of loss of totipotency during development.

Many experiments have been performed to determine by mechanical manipulation whether nuclei present in segmenting eggs are totipotent, or capable of continuing to control the process of normal development when moved from one part of the early embryo or cleavage stage to another. These experiments date back to the classical work of O. Hertwig (438) and Spemann (934) with the early stages of the amphibian embryo. By means of pressure applied unequally to the egg Hertwig altered the normal course of cleavage in the frog's egg so that the nuclei resulting in cleavage occupied different parts of the segmented egg than they would have if segmentation had been allowed to proceed undisturbed. No effect was noted on the embryo from eggs treated in this fashion. All developed normally to completion. Spemann applied a ligature to a newt egg before cleavage in such a way that one-half contained a nucleus and the other half did not. The two halves were connected by a bridge of cytoplasm. He then demonstrated (1) that cleavage did not occur in the enucleated half, but did so in the nucleated half, and (2) that nuclei from the half undergoing cleavage occasionally escaped through the cytoplasmic bridge at the 16-to-32 cell stage to the undeveloping, enucleated half. Even if these nuclei were several generations removed from the original zygote nucleus they started cleavage in the enucleated half which culminated in the production of a complete, normal embryo.

Experiments similar in intent, but applying the technique of destroying nuclei in insect eggs undergoing cleavage by ultraviolet light, have shown that up to the 128-cell stage in the damsel fly, *Platycnemis,* the intact nuclei are capable of replacing those destroyed by subsequent mitoses so that development proceeds undisturbed (895).

In all of the early experiments just described totipotency of nuclei was demonstrated, but, it is important to note, only for nuclei a few cell generations from the original zygote. A more valid test of possible nuclear loss of totipotency, and hence differentiation, would be to test nuclei after definite visible differentiation of cells. This has been accomplished by Briggs and King (110, 111, 112), who performed the difficult feat of transplanting nuclei from cells of various stages of frog embryos back into enucleated eggs which had not been previously fertilized (Figure 12.2). These eggs now had diploid nuclei, and after being properly induced started cleavage. In most of the experiments nuclei were taken from the endodermal (yolk) area of the donor embryos. Endoderm nuclei were tested from blastula on through to the tail-bud stages. In terms of development this span represents going

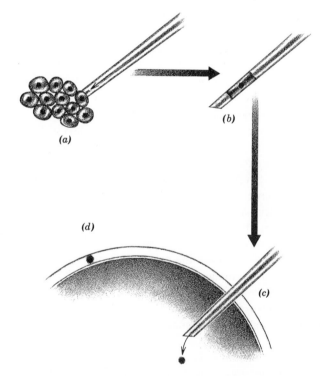

Figure 12.2. Technique for nuclear transplantation to amphibian eggs. (*a*) Diploid nucleus plus some cytoplasm taken up in micropipette as in (*b*). (*c*) Nucleus deposited in egg cytoplasm from which haploid nucleus (*d*) has been removed. After Briggs and King (109).

Table 53. Development of Embryos from Eggs Containing Transplanted Nuclei from Various Stages *

Types of Development
Per Cent of Total

Stage of Donor	Arrested Differentiation and Death	Abnormal Development and Death	Normal Larvae
Early gastrula	10	13	77
Late gastrula	27	53	20
Neural fold and tail bud	72	24	4

* From Briggs and King (109).

from a ball of cells to a highly differentiated embryo. The blastula nuclei tested came from blastulae with 8000 to 16,000 cells. These nuclei gave completely normal embryos which continued development, thus demonstrating that blastula nuclei are not irreversibly differentiated.

However, when endoderm nuclei from the gastrula and later stages were tested, it became apparent that starting with gastrulation the nuclei showed less and less totipotency, that is, more differentiation, as development of the hosts proceeded (Table 53). Endoderm nuclei from early gastrula produce a majority of normal larvae, but those from neural fold produce a majority of embryos which stop development at the blastula to early gastrula stages. Particular attention was given to the fate of those embryos arising from eggs with endoderm nuclei from late gastrula donors. Four types of embryos were obtained from these transplants with a consistent distribution of types from one experiment to the next (Figure 12.3). Type 1 embryos, which constituted about 40% of the total, stopped development after neurulation and consistently showed *normal* development of *endodermal* derivatives, but *degenerative* changes in *ectodermal* and *mesodermal* organs. This is what one would expect if the nuclei were differentiated in the direction of endoderm, and had lost the capacity for balanced mesodermal and ectodermal differentiation. Type 2 embryos continued development normally to the larval stage and constituted about 20% of the total. These would be expected on the grounds that some endodermal nuclei are still not differentiated at the late gastrula stage and are totipotent.

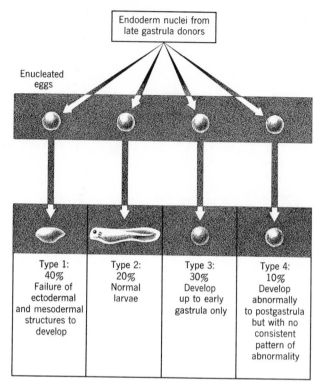

Figure 12.3. Fate of enucleated eggs into which diploid endoderm nuclei from late gastrula donors have been transplanted. After Briggs and King (110).

The other two types, 3 and 4, constituted 30% and 10% of the total, respectively. Type 3 embryos showed a very early arrest in development before or up to early gastrula. Type 4 embryos developed to the post-gastrula condition, but failed to show any consistent pattern of degeneration as in type 1.

Thus, whereas the development in types 1 and 2 can be rationally explained, in the other two it cannot. The possibility that nuclei were damaged in these latter types during transplantation was ruled out by controls. But a cytological study of these embryos showed that they were aneuploids with variable chromosome constitution. The origin of this aneuploidy is not completely understood, but it is believed to arise as a result of prior differentiation of the donor nuclei in the original donor. This differentiation confers on them the inability to divide properly in all parts of the host cytoplasm, and as a consequence nondisjunction, fragmentation, and loss of chromosomes result in aneu-ploidy. The results of aneuploidy have been discussed in Chapter 9.

In order to determine whether nuclei which seem to have lost their totipotency continue in this condition for more generations than could be tested in a single individual, descendents of transplanted differentiated nuclei were transplanted to new eggs for several successive generations as shown in Figure 12.4. The nuclei from the intermediate donor cells were transplanted early to prevent further effects of differentiation. The results clearly demonstrated that the differentiated state of the original nucleus was inherited as a stable condition through many cell genera-tions. Once acquired the differentiated state is not reversible, at least under the conditions tested. In this respect the phenomenon resembles whole-cell differentiation.

The evidence from the work of Briggs and King is strongly in favor

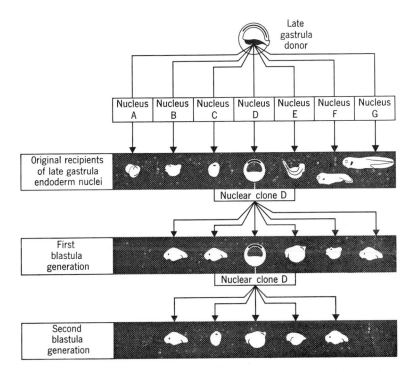

Figure 12.4. Diagram illustrating results of an experiment on serial transplanta-tion of endoderm nuclei. A nucleus from the endoderm of a late gastrula persists in producing deformed embryos despite passage through several blastulae. From King and Briggs (525).

of the stable differentiation of nuclei during development. Accepting this as a tentative conclusion, we may then ask is this nuclear differentiation a cause or an effect of the total differentiation process? This question cannot be answered decisively, but many facts known about the early stages of development tend to indicate that a tentative answer may be made. These facts relate primarily to the structure of the egg cytoplasm and the events that occur within it in animal development.

The Role of the Cytoplasm in Development. Chemical differentiation within the cytoplasm could easily lead to immediate changes in gene action by the different regions of the cytoplasm becoming separated during cleavage. In this way the same genotypes derived by mitosis from a single zygote nucleus could conceivably be injected into different environments.

Animal eggs generally show a high degree of localization of cytoplasmic material which becomes especially marked after centrifugation (Figure 12.5). In most of them mitochondria are demonstrable, as in mature cells, and lipoid material may occur as droplets. In addition, eggs of many species contain pigment and yolk granules which either may be distributed evenly throughout the cytoplasm or localized in specific areas or bands. The peripheral portions of all eggs form a gel-like cortical layer which can be distinguished from the more liquid interior by a variety of techniques. In some animals the stiff cortex may

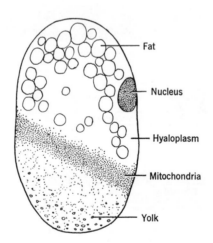

Figure 12.5. Centrifuged egg of *Limnaea stagnalis,* a snail. After Raven and Bretschneider (825).

be differentiated into morphologically or physiologically distinct areas which can usually be shown to have considerable importance in the future development of the egg. Examples of these cortical modifications are the gray crescent of the amphibia, the somewhat similar yellow crescent of the ascidians, and the polar plasm of the molluscs.

Many experiments have been performed in an attempt to demonstrate that the distribution of the visible granules in the egg have a role to play in subsequent development by moving them about by centrifugation. For the most part the results have been inconclusive or negative. It is quite possible that such significant differentiation that exists in the egg resides in the cortex, which is relatively stiff and not subject to reorganization by centrifugation as are the particles in the medullary portions of the egg.

Notwithstanding the essentially negative results obtained by reorganizing the arrangement of visible particles in the cytoplasm, other experiments of a somewhat different nature have given conclusive evidence for the significance of the cytoplasm in development. In the annelid worms and the gastropod molluscs the blastomeres formed during the early stages of development can be shown to have determined fates. The blastomeres are predestined in subsequent divisions to produce cells going into the formation of specific organs or parts of the fully formed larvae. Cleavage takes place not only with a great deal of coordination among the dividing cells but with almost perfect spatial organization of the blastomeres relative to one another. The earliest blastomeres can be numbered and their descendants followed through the subsequent divisions until tissues and organs are formed. It can then be established that specific blastomeres recognized by their position or even physical peculiarities go into formation of specific areas in the larvae, and thus the prospective fate of each cell can be mapped with a considerable degree of accuracy from early in development, as shown in Figure 12.6.

Although these observations are of interest in showing the great degree of organization achieved during the early stages of development, they do not in themselves prove that the blastomeres are irrevocably destined to fulfill certain functions in development. Their fate may be determined only by their position relative to one another and not by any intrinsic difference had by each initially. Experiments performed to test this important point have shown conclusively that in the Annelida and Gastropoda the removal or destruction of blastomeres causes drastic alterations in the course of development. The part that ordinarily develops from the absent blastomere or blastomeres does not develop,

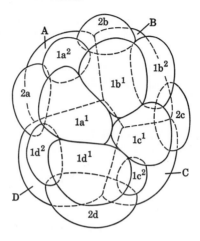

Figure 12.6.   The sixteen-cell stage of the annelid worm *Nereis*.   After Wilson (1084).   The blastomeres are numbered according to the accepted scheme which indicates their derivation.   The determined fates of some of these blastomeres are as follows: *D* will give rise to the mesodermal elements such as the coelomic lining, gonads, and longitudinal muscles as well as part of the alimentary canal. 2*d* forms the ventral nerve cord and the circular muscles.   The cells marked 1*a*², 1*b*², 1*c*², and 1*d*² form the larval ciliated belt.

and a defective larva or adult results.   Figure 12.7 illustrates the result of removing the polar lobe of the egg of *Dentalium* during its first division.   A larva completely devoid of mesoderm is produced.   Less drastic effects can be obtained by removing blastomeres in later cleavage stages.   Similar results have been observed with a large number of other gastropods, Annelida, certain ascidians (sea squirts), and ctenophores (comb jellies).   Isolated single blastomeres have also been observed to continue development, to divide and differentiate, but to produce only that portion of the embryo which would have been their fate had they remained part of the whole.

A number of experiments have been performed on the early stages of some of the animals in these groups in order to establish whether the nuclei after cleavage starts determine the fates of the blastomeres they occupy.   As described in Figure 12.6 the blastomeres of *Nereis* have determined fates.   Because of the type of highly coordinated cleavage in these animals the nuclei are always distributed in a precise fashion so that the nuclear products of mitosis can be numbered just as precisely as the blastomeres.   This leads to the possibility that nuclear segregation of some sort (not of the type due to meiosis, however) is the cause of the differentiation rather than any segregation of the cytoplasmic properties. This possibility had early been advanced by Roux (856) and Weismann (1059) to explain differentiation. Weismann, for example, hypothesized that elements which he called biophores came off of the chromosomes during development and different ones of these were distributed to different cells, thereby determining their fate. Wilson (1081) put this to a test by placing eggs under pressure between two plates causing the nuclei to divide with a different orientation than they would otherwise divide. The result was that nuclei ended up in blastomeres that they would not normally occupy. This had no significant effect on the future course of development of the blastomeres, as indicated in Figure 12.6. The conclusion seems inescapable that it is not the nuclei as such which establish this early differentiation in the early cleavage stages. It does not mean, of course, that nuclei do not become differentiated in later stages of development as has been fairly well established in the frog.

Since the blastomeres in these invertebrates possess an inherent capacity to self-differentiate in a determined direction, the role of one of them cannot be assumed by another. At the time of commencement of cleavage of the eggs these forms must achieve a degree of organization which entails localization of materials and capacities. Cleavage presumably separates these heterogenous elements and thus the course

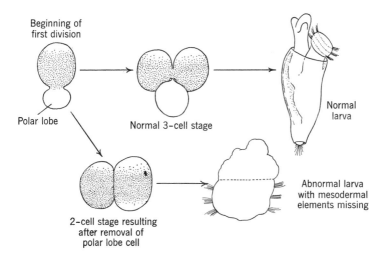

Beginning of
first division

Polar lobe

Normal 3-cell stage

Normal larva

2-cell stage resulting
after removal of
polar lobe cell

Abnormal larva
with mesodermal
elements missing

Figure 12.7. The effect of the removal of the polar lobe of the egg on the development of the larvae of the mollusc *Dentalium*. After Wilson (1084).

of development of each part is determined. This may be taken to be proof that differentiation is at least in part the result of unequal distribution of cytoplasmic metabolic capacities.

Investigations on the eggs and early developmental stages of other animals such as the Echinodermata and Amphibia give results which are subject to essentially the same interpretation provided that it is recognized that heterogeneity among blastomeres may be achieved in somewhat different ways and at different stages in some organisms than in those described in the preceding paragraphs.

If in an echinoderm, such as the sea urchin, the first two blastomeres are separated and reared independently, each will develop into a perfectly formed half-sized larva. If in a later stage of cleavage the cell mass is bisected in a plane passing through the animal and vegetal poles (Figure 12.8), two perfect larvae are again obtained. Furthermore, the removal of certain blastomeres will in no way drastically affect the development of the embryo. Other blastomeres are able to repair the loss with certain of their descendants which would otherwise carry out an entirely different role. Thus it is evident that in the sea urchin the blastomeres are capable of different types of development; their fate is not irrevocably determined. This is in direct contrast to the results obtained with the molluscan and annelidan eggs. For this reason the sea-urchin egg and its blastomeres are described as *regulative;* that is, they possess the capacity to make up for the defects incurred by destroying or removing parts during development. They possess a degree of versatility of differentiation into various channels not demonstrated by gastropods, and other invertebrates like them, which start development from eggs organized into different areas of development patterns and hence have a *mosaic* organization. Actually, the distinction between the regulative and mosaic types breaks down upon further analysis of those forms like the sea urchin and frog which exhibit the regulative type of organization.

If the sea-urchin egg is divided in two by ligature in a plane perpendicular to the animal-vegetal axis, and the half without the zygote nucleus fertilized, two perfect larvae are not formed. On the contrary, a larva with cilia covering the external surface and no endodermal derivatives results from the animal half—the so-called animalized larva (Figure 12.8). From the vegetal half there is produced generally an exogastrula (vegetalized larva) with an evaginated over-sized gut, or a more nearly normal larva with a larger gut than usual (Figure 12.8). Thus it would appear that the animal half produces primarily ectodermal derivatives, whereas the vegetal half produces primarily endo-

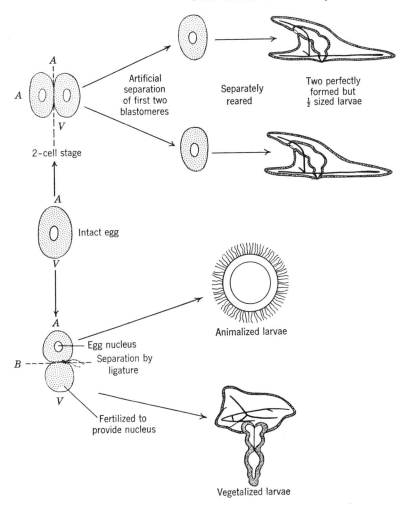

Figure 12.8. The effects of the artificial division, in two different planes, of the sea-urchin egg.

dermal material. Horstadius (463) has recombined various animal and vegetal portions of the sea-urchin egg and shown that there are two forces operating in its development—the animal and the vegetal. He has maintained with Runnström (858) that in the intact egg the two are in balance, and a larva with the normal complement of tissues and organs results within the normal range of size so long as this balance is maintained intact. However, if the balance is upset by the removal of animal or vegetal material of the egg, the early cleavage cell mass, or the blastula, the course of development is swung in favor of the material present in most abundance. Presumably any part of the egg, even a small fragment, containing a nucleus will develop into a near normal larva provided that both animal and vegetal material is represented (414).

Although the sea urchin is regulative to a large extent, it is quite clear that it also displays mosaic properties. Like the Annelida and the Gastropoda it does show a differentiation of cytoplasmic capacity from a very early stage of development in the obvious differences between the animal and the vegetal areas of the egg.

Changes in the development of the sea-urchin embryo can be brought about by treatment with chemicals. Lithium salts such as LiCl cause a normal egg to develop in the vegetal direction, producing an exo-gastrula similar to that developing from the isolated vegetal half of an egg (427). Certain amino acids such as proline and arginine also cause vegetalization (464). On the other hand, NaSCN and the amino acids, serine and lysine, produce a pronounced animalization of the same type expected in the isolated animal half of the egg (464). The effects of these salts and organic compounds have been interpreted as meaning that the animal and vegetal parts of the egg have different types of metabolisms which respond, as would be expected, differently to inhibitors and stimulators of metabolism.

Direct proof of a metabolic difference has been difficult to establish, but it now seems quite clear that there are differences, which become particularly prominent just before the onset of gastrulation. Boveri (88) had early postulated the existence of stratification along the animal-vegetal axis of the egg, in which each stratum had a different developmental significance or fate. This, he maintained, was in the nature of a gradient of reactions from one pole to the opposite one (90). Some evidence for the existence of such a gradient was obtained by Child (162), who showed that in certain developmental stages up to the late blastula there is a gradient of reduction capacity. The most rapid reducing region is at the animal pole and the least rapid at the vegetal. Lindahl and Holter (614), however, were unable to detect any evidence of this gradient by using the Cartesian diver technique of determining the respiration of isolated animal and vegetal halves of eggs. Both halves were found to have identical respiratory rates. The same was found to be true for the dipeptidase activity of the two halves (614). Undoubtedly more refined methods are needed to detect any differences in metabolism which may exist between the two parts of

the egg. One approach has been to study the distribution of mitochondria during the early stages of development. It is apparent that
during development of the sea-urchin egg a rapid increase in numbers
of mitochondria starts at the end of the blastula stage (383, 902). Some
of the workers in this field maintain that the distribution of mitochondria
in the early embryo is asymmetric so that there is a gradient from the
animal to the vegetal pole (382). This observation has not been verified in all species of sea urchins that have been studied, however, and
it is not possible at present to state that a segregation of mitochondria
is of general significance in differentiation in the sea urchin (902).
There is definite differential segregation of the mitochondria in the egg
and embryos of the ascidian *Phallusia* (831), but no evidence exists
that this segregation is a significant aspect of differentiation. However,
while recognizing the importance of mitochondria in metabolism, it is
quite possible that the segregation is of some importance.

The frog's egg, like the sea-urchin egg, is also regulative insofar as
it may be cut in half in a certain way and two normal larvae will develop.
But if the egg is cut in such a way that the gray-crescent material is
not represented in one half, an abnormal embryo will develop from
that half. The other half develops normally. The gray crescent is an
area which appears shortly after fertilization in the cortex on the dorsal
side of the egg. It is from this area that the notochord and mesoderm
of the embryo are derived, as described hereafter.

It is possible in view of our current knowledge to make the generalization that animal eggs have a differentiation manifested as an asymmetric
distribution of potentialities either before or soon after fertilization.
This can be considered a very important primary factor guiding future
differentiation in development. But these considerations raise the
question as to how these cytoplasmic localizations of potentialities arise
in the egg. Any attempt to answer this question must involve a consideration of the factors that influence the development of the egg in
the process of oögenesis.

Oögenesis. At the outset it must be admitted that very little is known
about this process. Its initial stages occur in the ovary of the animal,
and it is known that the egg nucleus and cytoplasm during oögenesis
show many singular characteristics not found in somatic cells. Eggs
start out developing from relatively undifferentiated cells within the
ovary and ordinarily arrive at a rather advanced stage of characteristic
organization before they are released from the ovary. During the
period before release their development is undoubtedly influenced by

the surrounding ovarian follicle and nurse cells, and prior to and following their release we may assume that they are under strong control from their own nuclei.

Amphibian eggs, and the eggs of many other animals (fishes, reptiles, and birds), characteristically develop *lampbrush chromosomes* during oögenesis (93). These chromosomes have a structure like that illustrated in Figure 2.18. They consist of a very long and thin axial fiber with numerous chromomeres from which loops of various size and thickness extend as shown in the figure, and described in Chapter 2. The current interpretation of the role of the loops is that they are extremely active centers of the synthesis of RNA which is transmitted via the nucleoli into the cytoplasm. The movement of RNA into the cytoplasm from the nucleus is well established as discussed at length previously, but whether in this particular instance the transfer of RNA from lampbrush chromosomes to the oöcyte cytoplasm is of any unique significance in the future organization of the egg is not established. The occurrence of lampbrush chromosomes at this particular phase of the life cycle, however, does lead to the provocative hypothesis that they represent chromosomes in an extremely active phase of synthesis and influence the egg cytoplasm in a somewhat different fashion than the chromosomes of somatic cells influence their associated cytoplasm. It may be that at the time the lampbrush chromosomes of the oöcyte are active they are organizing the cytoplasm to bring about the state of cytoplasmic localization which it has at the time it commences cleavage.

The external environment of the egg must also be considered. There is considerable evidence that the surrounding follicle or nurse cells pass materials into the developing oöcytes. Electron-microscope studies of developing frog oöcytes show that they have large numbers of microvilli projecting and interdigitating with microvilli projecting from the surrounding follicles cells (519). In insects the ovarium nurse cells come into very intimate contact with the oöcytes. In the beetle, *Dytiscus,* for example, the nurse cells develop actual cytoplasmic interconnections with the oöcytes (748). In the wasp, *Habrobracon,* the nurse cells actually pass entirely within the egg, break down, and become incorporated into the egg cytoplasm (1041). What significance the contribution of materials to the oöcytes from surrounding ovarian cells has with respect to the organization of the egg cytoplasm, we do not know, but the possibility exists that it may be considerable.

Extrinsic factors of a physical nature must also be considered. Some of the obvious ones are gravity which may cause a certain stratification of cytoplasmic elements, and light which may be of particular significance in those animals which develop outside the body of the mother. The formation of the gray crescent in the frog's egg is an extremely important event. The location of the crescent on the egg cortex determines the future symmetry of the egg. If it does not form the egg cannot develop. One factor which determines the location of the gray crescent is the point of entrance of the sperm, but another is reported to be light. The gray crescent may be induced to form before fertilization on the side of the egg most brightly illuminated (21).

No firm conclusions can be reached regarding the immediate causes of the initial organization of the egg cytoplasm, but it must be recognized that it occurs under the control of genes. The cytoplasm of the zygote, while not initially a product of the genome within it, must be considered as having been produced primarily under the control of the maternal genotype at the time the egg is formed in the ovary.

The demonstrated organization of the egg cytoplasm may be taken as the first significant step toward the later differentiation of cells derived from it, and it therefore plays an important role in the mechanism of development and the determination of the phenotype.

The Influence of One Part upon Another. Although it must be assumed on the basis of the available evidence that one factor in the origin of differentiation exists in the cytoplasm of the egg, this does not mean that all differentiation and development is solely the result of a cytoplasmic sorting out or segregation. The intrinsic factors which make the cells different during cleavage set the stage for the next phase of development in which the differentiated cells begin to affect one another. Upon the primary differentiation resulting from segregation of capacities during cleavage, there must be superimposed a second phase of differentiation arising from new environmental circumstances being created around each cell or cell layer, owing to differences in its neighbors. The extent to which the environment around each cell may be different may be readily appreciated by considering the consequences of the formation of a ball of cells (Figure 12.9).

The critical proof of the effect of one type of cell on another was advanced by Spemann as a result of his experiments on the amphibian embryo which culminated in the concept of the "organizer." The general application of this concept has since been made in many other animals with results similar to those obtained in Amphibia, and it is now a phenomenon generally recognized as being of utmost importance in the development of all vertebrate animals. It is probably also of significance in development of most invertebrate animals with the ex-

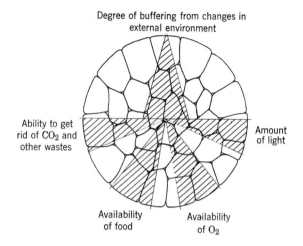

Figure 12.9.   Diagram to illustrate some of the internal environmental differences inside a ball of cells.   The shaded triangular areas are intended to show that some factors increase in intensity, and some decrease in intensity, from the surface to the center of the cellular mass.

ception of those which show strict determinate development from mosaic-type eggs.

The best examples of these influences are to be found in the Amphibia, which have a regulative type of development, and an egg which can be kept under continuous observation from the very beginning of the embryonic development following fertilization.   After the formation

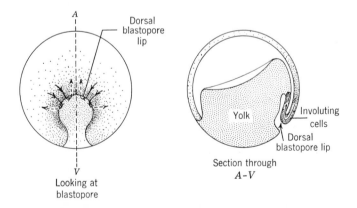

Figure 12.10.   A diagram of gastrulation in the frog.

of the gray crescent the egg starts cleavage and a blastula results which undergoes gastrulation by a process of inward turning of certain of the surface cells in the area of the original gray crescent. This establishes the location of the dorsal lip of the blastopore (Figure 12.10).

A "fate map" (Figure 12.11) can be made for a frog's egg similar to those constructed for the Mollusca and Annelida (p. 514). However, the fates of the parts of the egg are *presumptive* in the frog and not *determined*. An example will make this clear. A region of the very early gastrula is known to go into the formation of the adult eye. That this is its fate can be demonstrated by marking the area with a

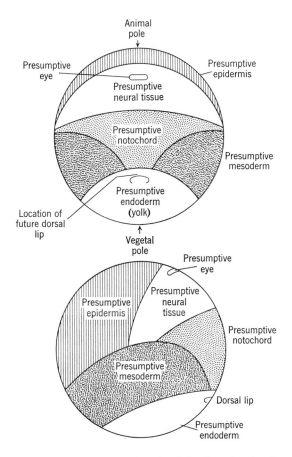

Figure 12.11. Two views of the late blastula of the frog showing the presumptive values of the various parts of the surface.

nondiffusible inert substance such as carbon and then following the marked area through normal development to its new location in the eye region. If a piece of ectoderm from this area is excised and transplanted into an older embryo, as shown in Figure 12.12, the future course of its development will depend upon where it is placed. When transplanted to the head region of the host it will form eye, brain, and mesodermal material characteristic of the head region. In other regions, however, it will form organs and tissues characteristic of those regions in normal development. It should be clear why this area in the early gastrula is called *presumptive* eye. It will develop into an eye only in the proper environment or spatial relationship to other organs and tissues. If put into a "neutral" environment, saline for example, it will continue to exist only as undifferentiated ectoderm. As presumptive

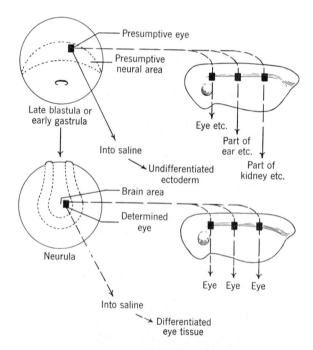

Figure 12.12. A diagram of a series of transplantation experiments in the early frog embryo. The presumptive eye area of a late blastula can develop either into an eye or into entirely unrelated parts, depending upon the region in the host to which it is transplanted. If transferred to saline it will not differentiate. If, however, the eye area is removed shortly after gastrulation is completed, its fate is determined, and it will develop into eye tissue regardless of the environment in which it is placed. After Barth (45).

eye tissue it is competent to react in a number of different ways and hence demonstrates a regulative type of development.

When the gastrula is allowed to continue to develop to the neurula stage in which the nerve cord begins to form, it is found that what formerly was merely presumptive eye now becomes *determined* eye. Transplanting this same area to an older embryo results only in the formation of an eye, no matter in what area introduced (Figure 12.12).

Two very important conclusions can be derived from these experimental observations, which incidentally have been repeated in other types of vertebrates, as well as with different presumptive areas of the early frog embryo. (1) The fates of the parts of the early embryo (for example, gastrula in frog) are determined to a large extent by their location relative to other parts. They are, in other words, competent to develop in many directions. (2) The period of competence or flexibility is not indefinite. There arrives a time during development (neurula stage in case of eye of frog) when the cells, because of their prolonged association with a specific type of surrounding cells, become fixed or determined and lose their competence to form more than one type of organ.

At the time of onset of gastrulation the various parts of the embryo show rather distinctive characteristics. The embryo may be divided into three zones: (1) *ectodermal,* comprising presumptive epidermis and neural tissue; (2) *marginal,* comprising presumptive notochord and mesoderm; and (3) *endodermal,* comprising the yolk area, much of which is covered by overlying ectoderm by this stage (Figure 12.10). Cells removed from any part of the ectoderm develop into unspecialized epidermis cells when explanted into a nutrient solution. Cells from the marginal zone (or the chordamesoderm) can develop into a variety of different tissues depending in part on the area from which they are selected within this zone. An explant from the lateral presumptive notochordal area may give rise to notochord and somites (mesoderm). If taken directly from the dorsal blastopore lip region, an explant may differentiate and organize into a partial embryo having axial organization and consisting of a notochord, neural tube with a brain-like enlargement of the anterior end, epidermis, and mesenchyme cells. For these reasons the cells from the marginal area are said to have a high degree of ability to "self-differentiate" in a number of different directions, and to organize themselves.

Explanted cells from the endoderm can and do develop only into endodermal derivatives, that is, epithelia of the gut, liver, and pancreas. Hence the endoderm appears to be determined at this time.

At the time of onset of gastrulation, then, different parts of the embryo show differing capacities to differentiate. They also show differences in competence or ability to develop into a variety of tissues. This can be demonstrated by transplantation experiments. Depending on the area into which it is transplanted, presumptive ectoderm can be induced to form any kind of epidermal, neural, or mesodermal tissue and will also form notochord. Cells from the marginal area have about the same potentialities as the ectoderm, depending again on the region to which they are transplanted. Endoderm, on the other hand, is determined to go into endodermal derivatives. It shows no other competence.

What, however, determines the other types of cells? The answer is to be found in part in the events occurring during gastrulation. Gastrulation consists of a variety of cell movements among which is an involution or folding back of cells starting at the dorsal blastoporal lip. The presumptive mesodermal and notochordal cells slip inside as shown in Figure 12.10. The dorsal lip at this time merely constitutes a sharp corner around which the cells bend in moving from the outer surface inward. Inside the gastrula these cells become notochord and somitic mesoderm; collectively the whole invading mass is referred to as the chordamesoderm. The ectoderm immediately above the notochordal material develops into a neural tube and brain parallel to the notochord. The ectodermal area over the somitic mesoderm on either side of the notochord is destined to become epidermis and organs of epidermal origin. If shortly after the invasion of the chordamesoderm to the inside, ectoderm from the medullary plate area (Figure 12.13) overlying the notochord is explanted or transplanted, it will proceed to develop into neural tissue. It is no longer competent to develop into any other kind of tissue. The reason for this is that it has been subjected to the influence of the underlying chordamesoderm which has *induced* it to differentiate in the direction of neural tissue. The chordamesoderm is referred to as the inductor, and unless it exerts its influence the undifferentiated ectoderm will not develop into the nervous system. This is true not only of the presumptive neural ectoderm, but of that part of the ectoderm which normally becomes epidermis and epidermal derivatives. If chordamesoderm is transplanted under presumptive epidermis, it too will develop into nervous tissue. From experiments of this type it has been concluded that the chordamesoderm is capable of *inducing* other cells to follow a definite course of differentiation. The nature of this induction is not understood, but its existence demonstrates the actual influence of one part of the developing animal upon another.

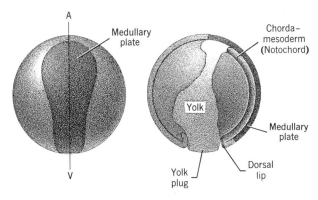

Figure 12.13. Late gastrula of frog embryo showing medullary plate on surface above chordamesoderm.

The chordamesoderm is by no means the only group of cells capable of inducing others to differentiate in a specific direction. It plays perhaps a preeminent role as an inductor, but structures induced in its presence can in turn act upon other undifferentiated tissue. Once this tissue is partially differentiated it in turn can induce still other undifferentiated cells. Hence one can visualize development as being in part a chain of inductions resulting from the activities of a succession of inductors. Figure 12.14 gives an example of such a chain.

The mechanism or mechanisms of induction remain unknown, but several things seem clear. First, it is not necessary for the inducing tissue and the reacting tissue to be in cytoplasmic contact. This has been established by interposing membranes with various-size porosities between the inducer and reactor. Cellophane prevents induction (92, 369), but membranes with pore sizes of 0.4 to 0.8 microns permit induction of metanephric kidney tubules by dorsal spinal cord. The inducing and reacting tissues may be separated by a distance of up to 80 microns (369). Data such as these make it clear that induction probably depends on macromolecules which pass by diffusion from inducer to reactor. Since induction can be rather specific, that is, only certain inducers will induce other tissues to develop in certain directions, and no tissues are competent to react with all inducers. It is also probable that rather specific types of macromolecules are produced by the different kinds of inducers.

Tremendous activity has been exerted by embryologists to identify the chemical factors involved in induction. At the present time the

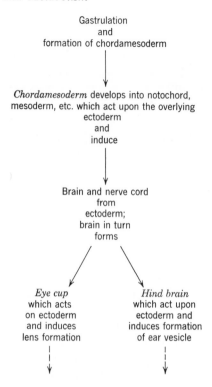

Figure 12.14.   An example of organizers acting in a chain.

most likely macromolecular types are protein and RNA or ribonucleo-protein.   Both liver and bone marrow have been shown to produce highly active inducing substances which have been identified as protein. Presumably these proteins are normally associated with RNA, but it is the proteins which are active as the inducers, since trypsin but not ribonuclease destroys the inducing activity (1117).   On the other hand, a polynucleotide of molecular weight 8000–10,000 has been isolated from embryonic spinal cord and notochord, which induces somites to differentiate vertebral cartilage (553).   RNA has also been implicated as an inducer in changes in mouse ascites tumor cells.   If ascites cells are treated with RNA from *normal* mouse or calf liver the ascites cells are apparently induced to become normal and incapable of producing new tumors.   RNA isolated from ascites tumors is incapable of bring-ing about this change (742).   It is not beyond the realm of possibility that induction is a phenomenon akin to transformation, except that RNA rather than DNA is involved.

Many different types of inductions must occur in the developing animal. Most of the early significant ones probably depend on the passage of macromolecules from one tissue type to another as just indicated. But simpler substances may also be involved. Thus Wilde (1077) has shown that pigment melanophore cells will differentiate from amphibian neural crest (see p. 428) only if phenylalanine is present. Presumably, the phenylalanine is normally produced by the archenteron roof mesoderm. In a somewhat similar vein it has also been shown that vitamin A causes cells of the chick embryo determined to become keratinized epithelium to differentiate instead as mucous-secreting, ciliated epithelium (274). Observations such as these make it quite evident that the metabolic environment of a cell exerts profound control over its differentiation.

Organization and Growth. The organization of differentiated cells in space together with regulated growth gives the form and structure to the whole which results in a completed organism. The total process may be described as morphogenesis, the molding of the whole into a definite pattern in distinction to differentiation, which is essentially a process of developing localized differences. The morphogenetic process is easily described in terms of structural change, but it is not so easily analyzed in casual terms. Even less is known about the factors involved in it than is known about differentiation. Furthermore, it is quite impossible to separate aspects of morphogenesis such as organization from differentiation, since the position of a group of cells may determine their course of differentiation. Hence to a very large extent the embryo is organized more and more as it differentiates by the influence of one part upon the other. Present conditions in the embryo are determined by antecedent ones, and so on back to the egg.

Although the fact that each developmental event depends on preceding events can only be analyzed in very vague terms, owing to the lack of knowledge as to how precisely one part affects another, a consideration of it brings to the foreground the important factor of time. Obviously the events that occur early during development are independent of those that follow, but those that follow are dependent upon preceding events. Unless there is a high degree of synchronization, orderly development will give way to chaos. For example, consider a hypothetical group of cells which by virtue of their position will, in the normal course of events, be induced to differentiate in a certain direction. In order to do so they must be competent to react to the influence of the inductor at the time the inductor is active. If they are not competent at the proper time they will either not differentiate further, or do so in another direction under the influence of a different inductor. In either event an abnormal embryo would be expected to result. It is well established that periods of competence are not indefinite in extent, and that inductors may themselves change in their capacities, so that it is evident that the timing must be precise to obtain precise end results (1044).

A second very important factor that must be considered in connection with organization is the dynamic state of the embryo in terms of internal movements of cells and cell layers. During development the whole embryo is literally on the move internally. A considerable portion of this movement is due to differential growth, but many cells, such as the neural-crest cells derived from ectoderm at the time of nerve formation, actually migrate from their site of origin to distant parts of the body. Neuroblasts send out processes which grow unerringly toward the organs they will enervate after becoming nerve fibers. In some animals the future germ cells originate in nongonadal tissue and migrate to the gonads. These are a few of the many types of cell migration which characterize every developing vertebrate. To these movements of solitary cells may be added the differential growth of tissue layers which will cover some areas but not others, or the directed growth of the pronephric tubule to the cloaca to form the Wolffian duct. What factors guide these movements? Unless each cell is endowed with intelligence and purpose, one must seek the answer in terms of affinities and internal environment.

It is well known that like cells are attracted to like, and that even different cell types show definite affinities. Holtfreter (452) has demonstrated that when ectoderm and endoderm cells from the early frog gastrula are explanted and combined *in vitro* they will at first tend to form a common mass. In time, however, each cell type tends to congregate with its own kind, and the result is a separation into ectoderm and endoderm. Thus an initial neutrality or even affinity between unlike cells gives way in time to a strong affinity of like to like and perhaps even an antagonism between unlike ones. As differentiation proceeds, the ectodermal derivatives, as an example, need not, however, show affinities because of their common origin. They may indeed show definite antagonisms toward one another, and an affinity to a tissue derived from a different germ layer. Figure 12.15, taken from Holtfreter, indicates the affinities and disaffinities of several partially differ-

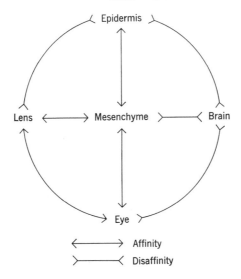

Figure 12.15. Affinities and disaffinities of some embryonic frog tissues. See text for explanation. Data from Holtfreter (452).

entiated tissues of ectodermal origin. In general, mesenchyme (a mesodermal derivative) demonstrates affinities for most cell types, which makes it sort of an internal cement of the embryo.

Muscona (729) has made intensive analysis of tissue reconstitutions from dissociated cells of chick embryos. Cells from chick-limb buds, dissociated by treatment with trypsin, produce an extracellular substance which is probably mucoprotein. It is thought to be of considerable significance in the subsequent reaggregation of such cells which, under proper conditions, will undergo organization and development. Cells from specific embryonic organs or tissues such as kidney, liver, procartilage, or skin will, after being dissociated and scrambled, reconstitute well-organized organs of their respective kinds (729, 1064). In reconstitutions of this type the cells are apparently guided by factors involved in their histogenic relationship. Chick-liver and mouse-liver cells when mixed together will form integrated chimeric liver, but chick-mesonephric and mouse-procartilage cells do not aggregate together, instead they form separate aggregates. Hence animals as widely separated as the mouse and chicken demonstrate certain similarities in the cohesive capacities of their cells. Their liver cells are not so different that they are incompatible. On the other hand, different, "unrelated" tissue cells show no affinity.

From the foregoing description of cell and tissue affinities, and capacities for self-organization, it is difficult to avoid the conclusion that the differentiated cell has a built-in capacity to "recognize" other cells. This recognition may possibly be enhanced by the mucoprotein mentioned earlier, but it is obvious that something very specific within the cell, or more probably at its surface must be responsible for the high degree of specificity of attraction which is observed.

Evidence that attractive forces between cells may be interpreted by the application of immunological theory is the basis for the Tyler-Weiss hypothesis that specific surface antigens play an important role in the selective adhesion of cells (1027, 1062). The attraction between the sea-urchin sperm and egg can, for example, be explained as the result of the complementariness of the surfaces of these cells (1027). The egg contains a protein, fertilizin, in its gelatinous coat which can be separated from the egg and shown to cause an agglutination of sperm by reacting with the protein, antifertilizin, on the sperm surface. Fertilizin and antifertilizin show the relation of antigen-antibody, as demonstrated by the fact that they can be separated from their respective cells and form a typical precipitate. It would seem clear then that they are the active factors in causing the demonstrated strong affinity between sperm and egg.

Experiments with dissociated sponge and amphibian cells in the presence of specific antisera have given results which make it quite clear that surface specificity may well explain cell affinities. It has long been known that a sponge can be dissociated by pressing it through bolting cloth, and that the isolated cells will reaggregate and form new, organized sponge within a short period.

Spiegel (937, 938) has performed experiments with two species of sponges, *Microciona prolifera* and *Cliona celata,* which have different colors of cells, the former being red and the latter yellow. When a mixture of dissociated cells of these two species is made they form an initial aggregation of both types, but then reassort themselves so that *Microciona* cells are segregated from *Cliona* cells. This segregation indicates a rather definite affinity of like for like. Antisera to *Microciona* and *Cliona* cells were produced in rabbits and tested to determine their effects on reassociation of isolated cells. The results were clear in showing that cells of both species aggregate much less readily in homologous antiserum than in the normal control serum. Furthermore, anti-*Cliona* serum had no effect on *Microciona* reaggregation. Anti-*Microciona* serum had a small effect on *Cliona* cells, however.

These observations are supported by similar ones in Amphibia (938).

Frog antigastrula serum prevents reaggregation of partially differentiated ectoderm, endoderm, and mesoderm cells which would otherwise aggregate in normal serum.

It is difficult to avoid the conclusion that specific cell surface structure governs cell adhesion. Coupled with the knowledge that the antigenic characteristics of the embryo changes as it develops, the Tyler-Weiss hypothesis gives us an interesting insight into a possible mechanism for the organization of the developing embryo. Figure 12.16 presents some models which may be used to arrive at a better understanding of how complementariness in surface may lead to affinity of similar and dissimilar cells.

On the assumption that surface specificity may account for cell affinities, it can readily be appreciated that the movements of isolated cells,

Cell surfaces which are identical and complementary. This model illustrates affinity between similar type cells as in a tissue.

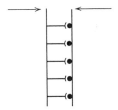

Cell surfaces which are different but complementary. Illustrates possibility of affinity between cells of different tissues as in an organ.

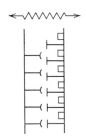

Cell surfaces which are noncomplementary. Cells with these surfaces would show disaffinity.

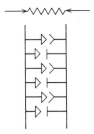

Cell surfaces which are partly complementary. Cells with these surfaces may show weak affinity.

Figure 12.16. Diagrams to illustrate possible types of complementary and noncomplementary surfaces. After the hypotheses of Weiss (1062) and Tyler (1027).

the growth of cell processes, and the growth of tubes and sheets of tissue may be guided in predetermined pathways by "tracks" of complementary surfaces. This factor may operate in addition to and in conjunction with that of following the path of least resistance.

The problem of regulated growth is an aspect of organization, since the organism regulates the growth of its own parts by some sort of internal control which is a direct result of its being organized. Perhaps the best insight into the general problem can be had by considering regeneration. If a piece of liver is removed from a rat, the remaining liver immediately begins to regenerate the missing tissues, but grows back only as much as had originally been removed. Similar phenomena are to be noted in the regeneration of other parts and organs, particularly in the invertebrate animals.

What type of control exists to determine that an organ will grow to a certain size and then cease growth? Obviously, there are many factors which may be involved. The volume-surface relationship, so often implicated as the cause of cell division, may be one factor; the competition for available substrate materials for nutrition, another. Finally, there must be considered the direct influence of one part upon another by the action of hormones, the nervous system, and so on. The net effect of these factors is a dynamic equilibrium giving the relatively constant internal milieu of the adult so much emphasized by Claude Bernard. This equilibrium, which is easily visualized as being the result of the interplay of parts manifested as self-regulation or homeostasis, must be considered as a factor in development. The growth of a part will naturally depend to a large extent on the materials available which may be in limited supply, and in part on the action of hormones and the nervous system in an animal.

Auto-inhibition has come to be recognized as an additional possible factor in the regulation of growth. The growth of a structure to a particular size can be readily explained if, as a result of its metabolic activity, it produces substances which specifically inhibit its own metabolism. The larger the structure grows, the more inhibitor is produced, with the result that when the inhibitor reaches a certain threshold concentration further increase in size is completely hindered. Rose (854) has reported experimental results which are in accord with this hypothesis. However, definitive critical experiments have yet to be done to establish the extremely important point of whether such auto-inhibitors acting in a "feedback" system actually exist naturally, and play the role they are assumed to play.

Interactions of Cytoplasm and Nucleus

In the previous section certain significant aspects of development are discussed which should enable the reader better to grasp the problems confronting us when we attempt to explain the role of genes in development. Not all pertinent aspects of development have been cited, by any means. But it should be clear from what has been described that the complement of genes in each type of cell of a developing organism must have its own unique array of activities. It is difficult to avoid this conclusion when it is recognized (1) that the nuclei of differentiated cells seem to arrive at a differentiated state themselves; (2) that certain proteins are produced in some cells, but not in others; and (3) differentiated characteristics are inherited, that is, differentiated cells remain true to type through successive mitotic cycles. If it is true that the nuclei of different kinds of cells in the same individual are not functionally equivalent, how do they arrive at this state? This is really the same question posed earlier in this chapter, but in a somewhat different form.

At this point we look upon the fertilized egg as possessing a large array of specificities which are distributed during subsequent cell divisions so as to produce cells with limited specificities, that is, cells that are differentiated. But is the process that brings this about merely segregation of cytoplasmic properties with the nuclei playing a passive role? The most reasonable answer to this question would appear to be no. The available evidence that nucleus and cytoplasm interact during development is sufficient to make this answer the most tenable one for the present.

In Hybrid Development. Evidence for nucleocytoplasmic interactions is found in a variety of different types of observations and experiments. Perhaps the most obvious, and in fact the earliest, experiments relating to this problem are to be found in the studies on the development of interspecific hybrids (705). If the egg of one species is fertilized by the sperm of another a number of alternative results may occur depending on the type of cross. (1) The sperm nucleus may not fuse with the egg nucleus, but may simply start the egg cleaving. (2) The sperm nucleus may fuse with the egg nucleus, and development may start, but end early, usually at the time of gastrulation. (3) The sperm nucleus and egg nucleus fuse and development is completed with the formation of a hybrid adult. What is of most interest to us here is result 2, in which an abnormal embryo is formed. Ordinarily these crosses are made between closely related species of animals such as frogs, sea urchins, and others, which develop outside the body of the mother. It would be expected that close relationship by morphological criteria would be reflected in close relationship genetically; the two should carry approximately the same complements of genes. Failure to complete development may conceivably be due to (1) an inability of the chromosomes of two different species to act together, or (2) the inability of the chromosomes of the male parent to act in the cytoplasm of the egg of a foreign species.

Experiments performed with amphibia have given results which enable one to decide between these alternatives. The cross *Rana pipiens* × *Rana sylvatica,* and its reciprocal, results in hybrids which develop normally to gastrulation, but normal development does not proceed beyond that point. By employing the previously described method of Briggs and King, Moore (706, 707) transferred diploid nuclei from normal *sylvatica* blastulae to enucleated *pipiens* eggs. The eggs in all cases developed abnormally; some failed to cleave normally while some did, and only proceeded to the blastula stage when further development ceased. The same type of result is obtained if an enucleated *pipiens* egg is fertilized with a *sylvatica* sperm. These results make it quite clear that an incompatibility exists between *pipiens* cytoplasm and *sylvatica* chromosomes.

To test the effect of the sojourn of the *sylvatica* chromosomes in the *pipiens* cytoplasm, nuclei from blastula containing *sylvatica* nuclei and *pipiens* cytoplasm were transferred back into enucleated *sylvatica* eggs. This is their natural environment, and one should expect normal development. This did not occur. Development ceased at the blastula stage or earlier. The reciprocal experiment of transplanting *pipiens* nuclei reared in *sylvatica* back into *pipiens* eggs gave essentially the same results. Thus, not only do the nuclei show incompatibility with the foreign cytoplasm, but they seem to be irreversibly changed by it so that they can no longer function in their own cytoplasm. In an experiment to test whether a more prolonged sojourn of *pipiens* nuclei subjected to *sylvatica* cytoplasm would cause a reversion to a normal state, Moore (708) did the following. He fertilized enucleated *sylvatica* eggs with *pipiens* sperm and transplanted the mid-blastula nuclei to enucleated eggs of *pipiens*. The nuclei resulting from blastulae from these eggs were again back-transferred to enucleated *pipiens* eggs. Back-transfers were repeated for five successive times so that the *pipiens* nuclei were subjected to prolonged exposure to *pipiens* cytoplasm.

There was no improvement in development making it quite clear that the initial replication of the *pipiens* nucleus in *sylvatica* cytoplasm results in some irreversible change.

Another type of experiment which throws some light on this question involves the transplantation of pieces of a haploid merogone (an embryo from a haploid egg containing either a sperm or haploid egg nucleus) to a normal host. Hadorn (391) produced androgenetic haploid hybrids by fertilizing enucleated *Triton palmatus* (a salamander) eggs with *Triton cristatus* sperm. The resulting merogone develops quite normally to the stage when optic cups are formed, but then dies. To determine the effect of a normal host on tissues from these merogone embryos with *cristatus* sperm and *palmatus* cytoplasm, parts of gastrulae were transplanted to gastrulae of a related species, *T. alpestris* (Figure 12.17). Transplanted pieces of presumptive brain ectoderm with underlying archenteron roof cells showed a consistent pattern of future development. The donor brain ectoderm fused and developed quite normally with the host brain, but the archenteron roof cells degenerated as they would have in the intact merogone. If presumptive epidermis is transplanted a quite striking result is obtained. The epidermis characteristic of *palmatus* develops from the transplant, not that of either *cristatus* or *alpestris*. These results lead to the interesting conclusions that (1) tissue transplants of merogones may overcome their otherwise lethal destiny in a "normal" environment, and (2) that the cytoplasm of an androgenetic merogone is not necesarily changed by the foreign nucleus. Here the *cristatus* nucleus, in fact, is overcome by the *palmatus* cytoplasm. This condition of apparent cytoplasmic control is not always obtained, however. Dalton (194) transplanted neural crest from merogones derived from enucleated eggs of the salamander *Triturus rivulario* fertilized with *Triton torosus* sperm to diploid *Triturus torosus* hosts. The pigment pattern obtained was characteristic of the nuclear donor *T. torosus,* although there was a slight indication of a cytoplasmic effect from *rivularis*. Similar nuclear control in transplanted merogone tissue of sea urchins has been demonstrated by Horstadius (462).

The apparent general lack of effect of a foreign nucleus in an egg from cleavage up to the formation of the blastula would indicate that this phase of development is under cytoplasmic control, and perhaps proceeds without much contribution from the nucleus. After this, however, it is obvious that the nucleus becomes important and that it becomes involved in the guidance of the synthesis of proteins in the cytoplasm. The net protein content of a frog embryo does not change from the unfertilized egg to the hatching of the larva from the egg (550),

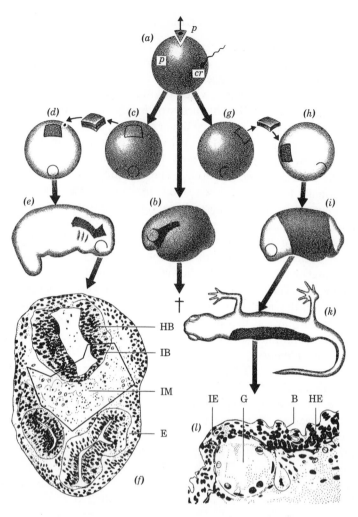

Figure 12.17. Experimental designs and results of transplantations with tissues from inviable hybrid merogonic embryos of amphibia. (*a*) Enucleation of the egg of *Triton palmatus* (*p*) and fertilization with spermatozoa of *Triton cristatus* (*cr*). (*b*) Inviable total merogone. (*c–d*) Transplantation of brain and roof of archenteron. (*e*) Host carrying implant (stippled). (*f*) Section through head region of (*e*); HB = host brain, IB = implanted brain, IM = implanted head mesenchyme (the pycnotic nuclei have been drawn a little too large for the sake of clarity), E = eye vesicle in contact with the diencephalon. (*g–h*) Transplantation of presumptive epidermis. (*i*) Position of implant (stippled) at the eye vesicle stage. (*k*) Metamorphosing host with hybrid merogonic implant, showing position of section below (*l*). (*l*) Section through skin where the implanted epidermis (IE) joins the epidermis of the host (HE); G = skin gland, B = boundary between haploid (left) and diploid (right) nuclei. From Hadorn (391).

but during this period there is a considerable change in the relative activities of the various kinds of enzymes. The evidence is also strongly in favor of a differentiation of the nuclei under, in part, the influence of the cytoplasm.

Hence, taking all lines of evidence together, we must conclude that during development there is a definite nuclear cytoplasmic interaction with the nuclei influencing the cytoplasm and vice versa. The main weakness in this analysis is the supporting evidence for cytoplasmically induced nuclear changes. Loss of total capacity of nuclei in differentiated or foreign cytoplasm for directing development of egg cytoplasm can be demonstrated as has been described in this and the previous section, but what we would like to know is whether this is just partial loss of capacity, different for each type of differentiated cell, or a similar loss for all. The results of Briggs and King with transplanted endoderm nuclei would indicate that perhaps each type of nucleus loses a specific set of potentialities but not all, since embryos developing with such nuclei are deficient in ectodermal and mesodermal structures, but not endoderm. Further information on this point which is probably the crux of the whole problem of differentiation must await the results of continuing investigations.

In Microorganisms. The best evidence for the influence of the cytoplasm on nuclear or chromosomal expression comes from studies on microorganisms, particularly the Protozoa. These organisms do not differentiate in the same sense as the multicellular organisms, but the high degree of intracellular structural development attained by some must be considered to be the result of processes similar to those involved in differentiation in higher forms.

Some of the more significant observations that have been made in the Protozoa concern the findings in *Paramecium* and *Tetrahymena* relative to the determination of mating types. Sonneborn and his co-workers have studied the inheritance of mating type in *Paramecium aurelia* over a period of more than 30 years, and have obtained data which make it quite clear that the cytoplasm has a definite role in determining nuclear differentiation in this organism.

The species category defined as *Paramecium aurelia* contains a number of varieties which are effectively sexually isolated, that is, gene exchange between them is not possible because different varieties cannot conjugate, or if they do conjugate the progeny usually die. Each variety has two mating types. Only animals of different mating types can conjugate. Mating-type inheritance is of two general types, that characteristic of four group A varieties, and that found in four group B varieties (923). These are called the group A and group B patterns, respectively, and will be considered separately.

It will be recalled from the discussion in Chapter 3 that after conjugation the exconjugants have identical genotypes. Additionally, they have no macronuclei because the parental macronuclei have degenerated during conjugation. After separation the fusion nucleus in each exconjugant divides mitotically twice, and the four products arrange themselves initially as shown in Figure 12.18. One of each of a pair of nuclei then becomes a macronucleus with the result that the exconjugants now each have two macronuclei (Figure 12.18). Each exconjugant undergoes fission and produces daughters with a single macronucleus (Figure 12.18). All fissions following this produce animals with only one macronucleus. From each pair of exconjugants, four lines or *caryonides* are produced, each exconjugant producing a pair of *sister caryonides*. The genotypes of the caryonides from a pair of conjugants are theoretically identical.

After conjugation of two animals of one of the group A varieties, animals from the same caryonide ordinarily are of the same mating type, but animals belonging to different caryonides may be of the same or different mating types. For example, from a cross between individuals of mating type I and II of variety 1 as illustrated in Figure 12.18, all four caryonides may be of mating type I or II. Or there may be all possible assortments of I and II: 3 I and 1 II, 2 I and 2 II, and 1 I and 3 II Obviously not even sister caryonides need have the same mating type. After autogamy two caryonides are produced. These may be of the same mating type or of different mating types even though derived from an original caryonide which demonstrated a single mating type. Succeeding autogamies give the same result, a continued segregation of mating type even though the individuals are homozygous.

The significant facts brought out in this analysis are (1) that the differences in mating type are due to unknown differences in macronuclei, and (2) that the differences in macronuclei arise at the time of their formation from micronuclei. Hence mating-type changes occur only at the time of nuclear reorganization, conjugation, or autogamy. Once formed a macronucleus continues to pass on the same mating type from one generation to the next.

Certain factors influence the relative proportions of mating types from a pair of conjugants or an individual undergoing autogamy. The higher the temperature shortly after conjugation, the greater the proportion of caryonides of mating type II in variety 1 (922). In addition,

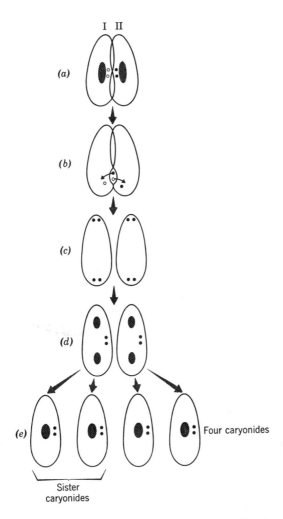

I II

(a)

(b)

(c)

(d)

(e)   Four caryonides

Sister
caryonides

Figure 12.18. Formation of caryonides in *Paramecium aurelia* after conjugation. (*a*) Pairing of conjugants. (*b*) Breakdown of macronuclei, meiosis, and exchange of haploid nuclei. (*c*) Formation of four diploid micronuclei in each exconjugant from the zygote nuclei by mitosis. (*d*) Formation of two macronuclei in each exconjugant from the micronuclei. (*e*) Binary fission and the beginnings of four caryonides. In all divisions following this the macronuclei divide.

genetic factors may be important. Some stocks of variety 1 have a single mating type, I, and never produce type II individuals after autogamy. These "one-type" stocks have been shown to differ from the "two-type" stocks, as described, which carry the potential to produce either type I or II by a single gene. Animals of the one-type are homozygous for *mt* which suppresses the development of type II. The dominant *Mt* enables animals to produce mating type II in some caryonides (923).

The group B varieties show a number of similarities in mating-type inheritance to the group A pattern. For one thing mating types do not ordinarily change during vegetative inheritance, but may at the time of nuclear reorganization. The significant difference between the A and B varieties is that in B animals the mating type of an exconjugant is usually the same as the cytoplasmic parent (930). This does not always occur, however. Occasionally, all four caryonides will demonstrate the mating type of one of the conjugants to the exclusion of the other. The reason for this seems to be cytoplasmic transmission from one conjugant to the other (923). The receptor of the cytoplasm assumes, and its progeny assume, the mating type of the cytoplasmic donor. Ordinarily, there is no change in mating type as a result of autogamy.

These observations would indicate that the inheritance of mating type in group B is cytoplasmic and has nothing to do with nuclear genes. This is far from being the case, however. Rather, the situation seems to be that the mating type expressed by a particular cytoplasm is the result of prior gene action which has caused the cytoplasm to become that way. Macronuclei determining a different mating type cannot ordinarily change the previously established mating type. In Sonneborn's (929) terminology, they become "differentiated." Evidence for this interpretation may be found in the following experiment by Sonneborn (927). If a pure type VII (genotypically and phenotypically VII) and a pure type VIII conjugate and remain connected for a long period, one or both of the pairs nearly always change mating type. This is the result of cytoplasmic exchange as stated. The effect of the exchanged cytoplasm seems to be on the new macronuclei which arise after fertilization from the new micronucleus. There are several reasons for this. Changes in mating type occur only when new macronuclei are formed from micronuclei, but not when they are formed from fragments of the old macronucleus, as occasionally happens. When the exconjugant with changed mating type divides it may produce sister caryonides of different mating type. This is interpreted to mean that one of the two macronuclei developed in the exconjugant is from old fragments which failed to disintegrate completely, whereas the other is from the newly formed fusion micronucleus as shown in Figure 12.18. The old macronucleus, having developed in a VII-type cytoplasm, cannot be changed because it is already "differentiated." The new macronucleus develops in a VIII mating-type environment from new micronuclei, and despite its genotype (heterozygous for VII and VIII) it determines only mating type VIII. This cytoplasmic condition persists even through autogamy, because the new macronuclei produced always develop in mating type VIII cytoplasm. This interpretation is supported by various other subsidiary experimental results (732).

We see from this analysis of what is outwardly a very complex system of mating-type determination that the question of whether nuclear differentiations are the result or cause of cellular differentiation may be answered by saying that they are both (929). "There is an endless series of sequential cycles of interaction between nucleus and cytoplasm in the course of successive sexual generations. Because it is endless, the choice of a starting point in describing it is entirely arbitrary" (929).

These considerations also lead to the general conclusion (which we have arrived at previously from results of a somewhat different nature) that nuclear differentiation is the restriction of gene action to different sets of genes out of the whole set present in the total genome.

Similar results, leading to essentially similar conclusions, have been obtained by Nanney (732) working with *Tetrahymena*. In addition, Lederberg and Iino (573) have described a phenomenon occurring in *Salmonella* which they have called phase variation, and which may be similar in its basic mechanism to what is found in Protozoa. Two distinct varieties or phases of *Salmonella typhimurium* may be distinguished by serological tests: those that possess the flagellar antigen i and those with flagellar antigen 1.2. These two antigens have been shown by transductional analysis to be determined by two well-separated loci, H_1^i, and $H_2^{1.2}$. However, a strain can be of only one phase or the other, not both at the same time. Hence changes in phase which occur spontaneously in culture at a frequency of about 10^{-3} to 10^{-5} per cell division are the result of alternative states of activity of these two loci. Analysis by transduction utilizing a number of different species indicate that the changes in phase are not the result of mutation of the H_1 or H_2 genes. When the H_2 gene is in its active state the cells produce antigen 1.2 and the H_1 gene is suppressed. When H_2 is inactive H_1 is expressed and only antigen i is produced. When H_1^i from cells pro-

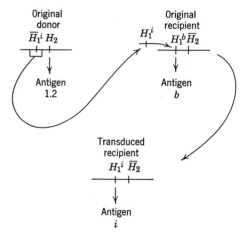

Figure 12.19.   Transduction of the capacity to produce *i* antigen in *Salmonella* from a donor which cannot produce antigen 1.2.

ducing antigen 1.2 is transduced to cells with an inactive H_2, but producing a *"b"* antigen presumably determined by an allele of $H_1{}^i$, $H_1{}^b$ the transduced cells are immediately capable of producing antigen *i* (573), Figure 12.19.   However, transduction of active H_1 or inactive H_1 to cells which produce antigen 1.2 never gives cells (except by the rare spontaneous phase change) with the active H_1 antigen *i*.   Hence the phase already maintained by the receptor determines whether the H_1 gene is to be active or not.   On the other hand, transduction of H_2 from active H_2 cells always produces progeny with antigen 1.2 whatever their previous state.   These results are summarized in Table 54.

Table 54.   Phase Specificity Inheritance in *Salmonella* [*]

Donor		Receptor		Transduced Receptor	
Gen.	Phen.	Gen.	Phen.	Gen.	Phen.
$H_1{}^i\overline{H}_2$	*i*	$H_1{}^bH_2$	*b*	$H_1{}^iH_2$	*i*
$H_1{}^i\overline{H}_2$	*i*	$H_1{}^iH_2$	1.2	$H_1{}^iH_2$	1.2
$\overline{H}_1{}^iH_2$	1.2	$H_1{}^iH_2$	*i*	$H_1{}^iH_2$	1.2
$\overline{H}_1{}^iH_2$	1.2	$H_1{}^iH_2$	1.2	$H_1{}^iH_2$	1.2

* From Lederberg and Iino (573).

Effects of Gene Change on Development

Gene mutation invariably results in some change in the development of the animal or plant within which it occurs. This results in the formation of a mutant phenotype.

One method of analyzing the role of the genotype in development is to determine the effect of a mutant gene by comparison with the effect of its normal allele. This method merely pushes back the analysis of phenotypic difference to whatever stage in development the mutant individual begins to show differences from the standard, but it enables one to come to a better understanding of the origin of phenotypic differences. Developmental studies in which mutant strains have been compared to the wild type have been made on a great number of different organisms, but primarily on chickens, mammals, and *Drosophila* (391). A few examples from among the many observations made by the workers in this field will serve to illustrate the methods of analysis and the type of results obtained.

Phase Specificity of Action. Mutant characters expressed as lethals at some stage in development have been extensively studied in *Drosophila*. By comparing the effects of a series of homozygous recessive X-chromosome deficiencies in *D. melanogaster* on the embryonic development of the larva within the egg, Poulson (805) was able to show that different lethal effects result from deficiencies of different parts of the X-chromosome. The complete absence of the X-chromosome results in disorganization of the egg during early cleavage and the death of the egg before a blastoderm is formed. Homozygous deficiencies of small segments of the X are also lethal, but, in general, embryonic development proceeds to a later stage before death ensues. A group of small deficiencies in the region of the white locus all produce essentially the same type of embryonic disturbance between the twelfth and sixteenth hour of development. Organs and parts of mesodermal and endodermal origin show highly abnormal development, whereas the ectodermal derivatives develop apparently undisturbed until death caused by the abnormal development of other parts ensues. Deficiencies including the facet locus or a region just adjacent to it cause a lethal effect in less than 12 hours. All three germ layers of the embryo are definitely disturbed in their development.

Lethality due to mutation can occur at any stage in the life of *Drosophila*, depending on the type of lethal mutation. A short list of mutant genes with lethal effects and a description of the abnormalities which presumably are directly related to the death of the organism is given in Table 55. It will be noted that for each mutant gene death occurs at a different time in development and from different causes, or at least accompanied by different symptoms.

The fact that each gene acts, or appears to act, during some limited period of development is referred to as its *phase specificity of action* (391). When dealing with lethal genes such as those described for *Drosophila,* it is well to recognize that, with respect to phase specificity, each has two periods of action which may be quite distinct. First is the *phenocritical* phase. This corresponds to the time when the first step leading to abnormal development becomes manifest. It generally

Table 55. Postembryonic Lethals of *Drosophila*

Gene and Symbol	Stage of Death	Description of Abnormality
Cy (curly)	At time of hatching from egg	Inability of larvae to continue further development after completion of embryonic development
B^{263-20}	First instar	Defective ring gland
l(2)me (meander)	Third instar	Inability to metabolize food in second half of larval life; general slowdown of growth; does not pupate; respiration extremely low in second half of larval life
l(2)gl (lethal giant larva)	End of third instar at time of pupation	"Pseudopupae" formed; larva ceases further development within puparium main imaginal discs and male germ cells degenerate before death of larva
l(3)tr (translucida)	Early or late pupal life	Larvae show abnormal signs early in first instar, small fat bodies, swollen with excessive hemolymph and quite transparent; after pupation imaginal differentiation becomes almost completely abnormal
crc (cryptocephal)	Late pupal life	Head does not evaginate but remains hidden in thorax
l(2)lgt (leg tumors)	Early imaginal life	Leg tumors

is quite separate from the *effective lethal phase* (391), which is the stage of development beyond which a lethal mutant cannot develop and consequently dies.   Lethal factors generally show a marked phase specificity of action, but the effective lethal phase may occur at different stages in different individuals.   When all individuals carrying a particular lethal gene die at the same time, the gene is described as a *monophasic lethal factor*.   All of the postembryonic lethal genes described in Table 55 are of this type, except the gene *(3)tr* which may cause death at any time during the pupal stage.   Lethal factors may also be diphasic or polyphasic in effect, that is, death may occur at either one of two specific times (diphasic), or at a number of different though specific times (polyphasic) in development.   The *(3)tr* gene may be described as polyphasic.   The Creeper *(cp)* gene in the fowl is an example of a diphasic factor.   Homozygous *Cp/Cp* fowl generally die at the end of the third day of incubation at a specific stage.   Some, however, "escape" this lethal crisis and continue to develop almost normally until the end of incubation when they die.   This escape from lethality is, of course, of some interest from the point of view of gene action, because it shows that different modifiers present in different individuals may determine the time at which lethality occurs.   Environmental factors may also be important.   What is of most interest, however, in these observations is the fact that a mutant gene may have a "specific time of action."   That is, the phenocritical phase may, and generally does, occur always at the same time whatever happens later with respect to time of death.   This is true of nonlethal as well as lethal factors.

In addition to mutant genes making themselves evident at specific stages of development, they generally demonstrate a definite specificity with respect to the type or types of cells affected.   A number of examples of mutant factors which show a definite cell specificity are given in Table 56.   The mutants of *Drosophila melanogaster*, $B^{263\text{-}20}$ and $l(2)gl$, described in Table 55 also show definite cell specificity of action.

Observations made on other animal forms have given similar results, whether they have been studies of lethal phenotypes or viable morphological mutants.   The mutant genes' effect, as compared to the normal, is to cause an upset of the normal course of events at a specific period of development which ends either in death or an abnormal phenotype. As far as can be determined from the available data, the mutation of each gene has its own characteristic effects on the pattern of development.   This effect may be so specific as to be limited to a particular kind of cell.

Table 56. Effects of Genes on Development

Gene Designation	Organism	Effect of Gene	Reference
Rodless retina, *r/r*	Mouse	Eyes never develop rods; no other abnormality evident.	516
Dwarfism, *dw/dw*	Mouse	Dwarfism. The effect of the mutant gene seems to be limited to the anterior lobe of the pituitary, and specifically to the eosinophilic cells within the organ which produces the growth hormone. These are missing.	918
Shaker, *sh/sh*	Fowl	Disturbances in coordination starting fourth week after hatching. Always accompanied by degeneration of Purkinje cells in cerebellum.	890
Brachyury, *T/T*	Mouse	Embryonic lethal. Notochord fails to develop.	160
Dominant spotting (*W*)	Mouse	*W/W* animals born anemic. Blood-forming cells in liver and bone marrow affected as well as germ cells.	80 186

Pleiotropy and Pleiotropic Patterns. A gene change generally results in more than one observable change in phenotype. Just how many changes are observed is in part a function of the acuity of the observer. This pleiotropy has long been the subject of controversy and discussion among geneticists concerned with gene action. Developmental geneticists such as Grüneberg have clearly demonstrated that complex phenotypes caused by single-gene mutations in mice or rats can frequently be accounted for as the consequence of a change in a single developmental event. For example, inherited achondroplasia in the rat (378) produces a complex syndrome in this animal which leads to early death. Among the more striking phenotypic effects are inability to suckle, high resistance in pulmonary circulation, faulty respiration and occlusion of incisors, and general arrest of development. All these abnormal conditions can be traced back to a single primary effect, an abnormal development of cartilage which first manifests itself in the early fetus. A simple "primary" cause for extremely complex end results is to be expected on the basis of our knowledge of development and metabolism. The interrelationships of the parts of the whole at all levels of organization can only be expected to have ramifying consequences resulting from a change in any part of the system. Consider for example the disease beriberi. This disease has an extremely complex set of symptoms in its acute form which include labored breathing, rapid pulse, nausea, diarrhea, decreased urine flow, palpitations, blueness of the skin, edema, neuritis, and cardiac failure—a complex phenotype produced by a deficiency of a single vitamin, thiamine.

The problem of pleiotropy is directly related to the problem of the mechanism of development. As we have seen, mutant genes do not cause uniform changes throughout an organism, but affect only specific parts. Sometimes the parts affected are easily related, at other times not. What kind of mechanisms can be invoked to explain these phenomena? Hadorn (391) has suggested some model systems which attempt to answer this question, and at the same time clarify some of the problems involved. First, he pointed out that it must be recognized that two general types of mechanisms must occur, intracellular and intercellular. The intracellular model considers the problems of how the same gene can give rise to cells of different phenotype by acting within them. This is, of course, the general problem of differentiation which has already been considered at length in the preceding sections.

Illustrated in Figure 12.20*a* and *b* are the ways in which Hadorn visualized intracellular processes giving rise to pleiotropic patterns. In Figure 12.20*a*, gene *a* becomes, as a result of cleavage, incorporated into three different cytoplasms. It produces the same primary product in all cytoplasms, but only in cells I and II are there cytoplasmic constituents ready to react. Cytoplasms I and II are different; hence the gene product acts differently in each, and two somewhat different phenotypes are formed. Cytoplasm III does not react at all, hence it is unaffected by mutation of gene *a,* whereas I and II may be drastically affected.

Also possible, and, as discussed in the previous section, probable, is the effect of the cytoplasm on the gene's action. This is illustrated in Figure 12.20*b*. In this model, gene *a* produces different kinds of products in cytoplasms I and II but remains "silent" in III. The result, as before, is a stable difference in I and II and no effect on III.

In a highly differentiated multicellular organism one must also consider the rate of intercellular interactions in which the different tissues

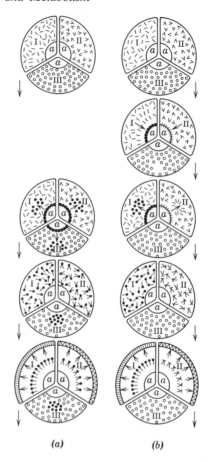

(a)   (b)

Figure 12.20.   Diagram based on an hypothesis to explain how *intracellular* processes may give rise to a pleiotropic effect.   See text for details.   From Hadorn (391).

and organs interact at many different levels, both at the height of the developmental process and later.   A model to illustrate what might be expected is presented in Figure 12.21.   In this model the gene *b* produces a particular phenotype in cell type, tissue, or organ I by *intracellular interaction*.   The differentiation of II, IV, and V is controlled indirectly by gene *b* through an *intercellular interaction* as shown in the figure.   Cells of type III are unaffected by gene *b*, and hence a mutation of *b* should affect only types I, II, IV, and V.

Two types of pleiotropy should be expected to be found according to the foregoing discussion. First, a *mosaic* pleiotropy is expected, resulting from the differential action of the same gene in different cytoplasms as in Figure 12.20*a,* and second a *relational pleiotropism* in which a gene acts directly in only one type of cell but exerts an indirect influence on other cells (391). Each of these cells reacts in its own fashion, and a pleiotropy results as shown in Figure 12.21. The intercellular influence is presumably through diffusible substances such as hormones, amino acids, and the like.

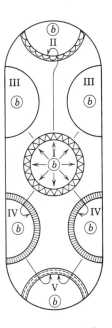

Not all complex phenotypes can be related to a single initial cause in development. The careful study of the *Sd* strain of mice by Glucksohn-Schoenheimer (331, 332) provides an interesting example of this. *Sd* (Danforth's short tail) is a dominant mutation which is lethal shortly after birth in homozygotes and semilethal in heterozygotes. Abnormalities are found in both the axial skeleton and the urogenital system; the spine and tail develop improperly; and the metanephric kidneys either fail to develop at all (homozygotes) or are quite abnormal (heterozygotes). A detailed study of the early stages of development failed to reveal any single causative factor which manifested itself morphologically to connect abnormalities in these two systems. Both systems are of mesodermal origin, but other mesodermal derivatives do not appear to be affected by the mutation. Other examples of extreme pleiotropy which are difficult to resolve with a one-primary-function hypothesis for the gene may be found among the mutations that affect coat color in the mammals. Generally, these have effects in addition to changes in

Figure 12.21. Diagram based on an hypothesis to explain how *intercellular* processes may give rise to pleiotropic results. See text for details. From Hadorn (392).

pigmentation. The gray lethal mutant of the mouse apparently blocks the production of yellow pigment in the hair and causes gross abnormalities of the skeletal system (377). In the same animal, the yellow coat-color mutation, A^y, one of the alleles of the agouti series, is lethal when homozygous but produces viable offspring with yellow coat color when heterozygous (847). It is difficult to relate the mere absence or presence of a melanin pigment with extensive internal abnormalities. The

A^y gene homozygous actually results in cessation of development in the blastocyst (blastula) stage.

The question of whether every gene has but a single primary action, and whether pleiotropy is always the result of subsequent interactions, as described before, has been debated for many years by developmental geneticists. The general tendency has been to (1) consider that each gene has but a single product, and (2) the product in turn has but a single function. These conclusions are, however, by no means proved, and, in fact, it is quite certain that the second one is not always true. In order to prove that the first conclusion is true it will be necessary to show that each DNA genetic "code" can be read in one and only one way so as always to give the same amino-acid sequence. This has not been done. The universality of the second conclusion is refuted by the recent findings that active glutamic-acid dehydrogenase from mammalian liver is an aggregate of identical subunits, and the subunits have little activity for glutamic acid, but are active as an alanine dehydrogenase (1016).

If it is established that other enzymes are also aggregates whose specific function depends on the degree of aggregation and disaggregation of subunits, then it is evident that we cannot ascribe a single catalytic function to the product of each gene. In addition to the two strictures just noted, it should also be noted that a decision regarding the nature of pleiotropy depends directly on how "gene" is defined. These considerations demonstrate the importance and significance of pleiotropy both in development and in such basic matters as the primary action of genes.

Of course, it should also be appreciated that pleiotropic effects may be expected from the accumulation of a single simple molecule resulting from a genetic block. Such a molecule, if it inhibited an enzyme or enzymes, could have widespread effects. Pleiotropy through accumulation is to be considered as an important general explanation. An actual example is given in Chapter 14 in connection with the discussion of galactosemia in man (p. 611).

Degeneration in Development. It is not unusual to find, in investigating the developmental mechanics of mutant phenotypes, that certain organs or tissues show normal development from a very early stage in embryogeny, then suddenly begin to develop abnormally. Frequently, a regression, or perhaps dedifferentiation, sets in, and the affected tissue may all but disappear. This phenomenon is illustrated in the inherited recessive rumpless condition found in chickens (1139) which results in the incomplete development of the tail axial skeleton. The tail rudiment, including both the mesodermal- and neural-tissue elements, develops normally to at least the middle of the fourth day. At this time the neural tissue forming the nerve cord of the tail begins to degenerate and disappear, accompanied to some extent by the degeneration of notochordal material. In contrast to this condition in recessive rumpless, a superficially similar condition, dominant rumpless, demonstrates abnormalities of the tail bud at the time of its first appearance in embryogeny.

A degeneration similar to recessive rumpless in the fowl is found in the *Sd* mutant of the mouse. In this mutant the tail develops normally up to the age of ten days after fertilization. At this juncture all parts of the tail begin to show signs of degeneration, and by the twelfth day the neural tube, gut notochord, and somite components of the tail have disappeared.

Degenerative phenomena are not unknown in normal development, as for example the breakdown of larval tissue in insect pupae, and probably therefore should not be thought strange in abnormal conditions. The dramatic shift, however, in the direction of development backward as the result of a single gene change, is of considerable interest, nonetheless. It may be a manifestation of the failure of the proper inductor to develop at the right time in some cases; and in others it may be due to improper metabolic conditions in the locality in which the degenerating tissue finds itself. On the other hand, the determining factor may be intrinsic and quite independent of external influences. A means of approaching this general problem has been found in transplantation experiments discussed in the following section.

Transplants between Different Genotypes. A considerable amount of attention has been given to transplanting a piece of tissue of one genotype into a host of another with the object of determining the reaction of the transplant. In general, the results have been quite simple; the transplant develops autonomously according to its own genotype as if it had not been transplanted. The few exceptions which have demonstrated heteronomous development have, however, been most instructive in demonstrating the interrelations which exist in development by showing that some parts of an embryo may be affected only indirectly by a mutant gene which presumably exists in all cells.

The heteronomous development of transplants has already been illustrated in the development of eye color in *Drosophila* (p. 326). It will be recalled that vermilion and cinnabar eye anlage when planted in a larva of normal genotype responded by producing brown pigment. The underlying cause for failure to develop pigmentation was discovered to be a metabolic deficiency in the mutants, resulting in their inability to form the intermediates for eye-pigment production. The intermediates in this example happen to be readily diffusible through the bodies of the flies, and hence are supplied by the normal host—an effect on the phenotype identical in many respects to the induction of growth in a nutritional mutant of *Neurospora* on the addition to the medium of a required compound. It is noteworthy, however, in the example of *Drosophila,* that, of 30-odd mutant eye-color anlage of different geno-type tested by transplantation, only these two genotypes have shown a heteronomous response. Precisely why vermilion and cinnabar are heteronomous and the others not is unknown, but it is reasonable to assume that factors such as permeability, diffusibility, and cell or tissue organization may well be involved.

The developmental aspects of the inherited Creeper condition in the fowl have been studied to some advantage by implanting the embryonic parts of mutant fowl into the chorioallantoic membrane (extraembryonic membrane with a respiratory and excretory function), coelom, or eye regions of normal hosts. The Creeper condition is caused by a dominant gene which is lethal homozygous. The homozygous embryos die in the third or fourth day of incubation, presumably because of an aberrant yolk-sac circulation. Occasionally, a "breakthrough" occurs, and a few individuals escape to develop almost until the end of the incubation period (see Figure 12.22). These individuals show extreme malformation of bone and cartilage structures (399). The eyes are also affected. These are smaller than normal and afflicted with colomba (546). The heterozygotes are viable but have extremely short and bent legs resulting from the abnormal development of cartilage preceding bone formation (546).

By means of growing tissue explants of 60-hour homozygous Creeper embryos in tissue culture and in the chorioallantoic membranes of normal hosts, David (200) was able to demonstrate that the lethal action of the Creeper gene is superficially nonautonomous in its effect. All tissues tested, such as heart, presumptive cartilage, and so on, survived and grew for periods long past the time they would have been expected to cease growth had they been left *in situ*. Presumptive Creeper bud materials from embryos as young as 30 hours old, however, do not develop into normal limb buds but clearly retain the Creeper charac-teristics (399, 857). This is true for limbs derived from both homo-zygous and heterozygous donors. Although the development of the

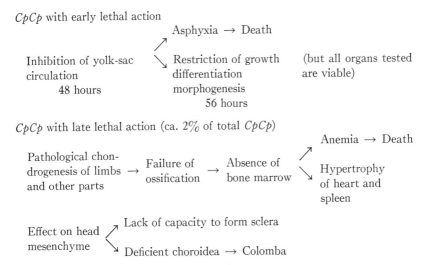

$CpCp$ with early lethal action

$CpCp$ with late lethal action (ca. 2% of total $CpCp$)

Figure 12.22. Action of homozygous Creeper in the fowl. From Hamburger (399).

limb bud is clearly autonomous, that of the eye in the homozygous mutants can clearly be shown to be heteronomous. Homozygous Creeper eye primordia planted into normal host sites after removal of the host eyes develop normally despite their genotypic constitution (399).

The Creeper gene has a definite pleiotropic effect, as shown in Figure 12.22, which is accentuated if the embryos succeed in surmounting the original difficulty of poor yolk-sac circulation. Of the two major manifestations of the mutant gene in the embryos which succeed in developing to the end of incubation, the effect on limb formation appears to be definitely more direct than that on the eyes, because the limbs develop autonomously. Thus it may be assumed that the mutation has primary effects on limb formation presumably by interfering with the normal course of cartilage and bone formation, and secondary effects on the eyes. Hamburger (399) interpreted the aberrant eye development in the mutants as being the result of primary gene effect on the mesoderm of the head since when placed in contact with normal head mesoderm the eye reacts with normal development. One interpretation of these observations is that the Creeper gene acts *directly* only in the formation of bone, etc., and head mesoderm, and *indirectly,* or *is inactive,* in eye tissue.

The existence of gross differences in the ability of the tissues from a lethal-containing animal to survive and continue growth is particularly well illustrated by the results of transplanting various organs or other anlage from the lethal giant larvae (*lgl*) of *D. melanogaster* to a normal host. Larvae homozygous for this lethal do not metamorphose (Table

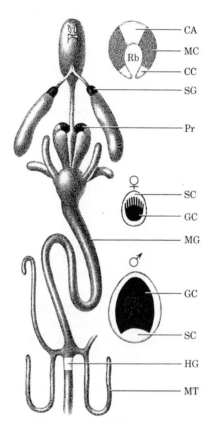

Figure 12.23. Diagrammatic representation of the anatomy of a *Drosophila* larva and of the development of transplanted imaginal primordia of the mutant lethal giant larvae (*lgl*). Groups of cells which are incapable of development are shown in black, i.e., imaginal rings of the salivary glands (SG), the proventricles (Pr), the male germ cells (♂GC). Limited imaginal differentiation, i.e., female germ cells (♀GC) and imaginal aggregates of cells of the midgut (MG). Complete imaginal differentiation, i.e., corpus allatum (CA) and corpus cardiacum (CC) of the ring gland, (RG) imaginal ring of the hindgut (HG), and somatic cells of the gonads (SC). Malpighian tubules (MT); main cells of the ring gland (MC). From Hadorn (391).

55). They show a definite pleiotropic effect of the *lgl* gene, for the ring glands, imaginal discs, salivary glands, fat bodies, and gonads of the larvae are retarded or degenerate in development while the other tissues seem quite normal. The further development of all tissues ceases, of course, at the time of onset of pupation. Various parts of *lgl/lgl* larvae have been transplanted to normal host larvae in order to observe the effect of the normal environment. Results are summarized in Figure 12.23. Some of the transplanted rudiments develop normally in the new environment, others do not. Thus the *lgl* gene is expressed in the former but not in the latter. Here again is another example of specificity of action of a mutant gene which can be adequately analyzed only by the use of transplantation techniques.

General Considerations and Conclusions

Perhaps the most significant conclusion to emerge from the foregoing discussion is that different arrays of active proteins are to be found in different types of differentiated cells. The main evidence supporting this generalization may be summarized as follows: (1) differentiated cells differ quantitatively and qualitatively in their enzyme content; (2) they show different immunological characteristics; and (3) they show different nutritive requirements and capacities for synthesis. All these facts indicate that differentiated cells differ in their capacities to produce specific proteins. Since the specificity of proteins is apparently dictated by genes, it follows that in one type of differentiated cell different sets of genes should be inactive than those in another type of cell. Of course, presumably, both types have many active genes in common.

In harmony with the foregoing conclusion are the numerous facts considered in this chapter concerning the more superficial differences among differentiated cells such as behavioral differences, affinities, ability or inability to react, inducing cells or substances, and so on. Also in agreement are the observations described on the effects of mutation. Mutation of a gene may apparently affect some cells in an organism but not others. Presumably a gene's mutation will cause changes only in those cells in which it is active.

If differentiated cells differ in their arrays of active genes, the next question to be considered is what makes some genes inactive and allows others to be active? If we define gene activity as ability to produce a specific protein, the problem becomes one of the regulation of specific protein biosynthesis in the presence of the gene specific for that bio-

558   *Genetics and Metabolism*

synthesis.   Reference to Figure 12.24 makes it clear that protein biosynthesis is subject to control at many points in the chain of reactions leading from the production of messenger RNA to the completion of the folding of the polypeptide chain that presumably comes off the ribosomes.   The control mechanism or mechanisms may act to prevent completion of synthesis of a functional protein at any one of these points, and at the present time it can only be surmised which are the most likely.

It is very possible that the controlling genes discussed in Chapter 9 play a role in differentiation.   McClintock, in fact, advanced this hypothesis soon after she discovered the controlling elements in maize (638).   More recently Monod and Jacob (703) have revived this suggestion and have presented a series of models based on their explanation of the action of controlling genes in *Escherichia coli* (see Figure 9.8,

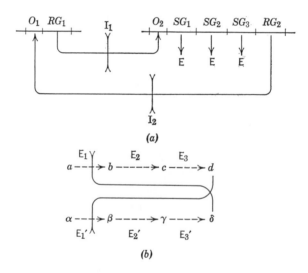

Figure 12.24.   (*a*) The regulator gene RG_1 controls the activity of an operon containing three structural genes (SG_1, SG_2, SG_3) and another regulator gene RG_2. The regulator gene RG_1 itself belongs to another operon sensitive to the repressor synthesized by RG_2.   The action of RG_1 can be antagonized by an inducer I_1, which activates SG_1, SG_2, SG_3, and RG_2 (and therefore inactivates RG_1).   The action of RG_2 can be antagonized by an inducer I_2 which activates RG_1 (and therefore inactivates the systems SG_1, SG_2, SG_3, and RG_2).   (*b*) The reactions along the two pathways $a \to b \to c \to d$, and $\alpha \to \beta \to \gamma \to \delta$, are catalyzed by enzymes E_1, E_2, E_3, and E_1', E_2', E_3'.   Enzyme E_1 is inhibited by δ, the product of the other pathway.   Conversely, enzyme E_1' is inhibited by metabolite d, produced by the first pathway.   From Monod and Jacob (703).

p. 424) which purport to relate to the problem of differentiation. An example of one type of model proposed by them is given in Figure 12.24a. In this system a regulator gene RG_1 controls an operon consisting of an operator, O_2, the structural genes SG_1, SG_2, and SG_3, and another regulator, RG_2. The regulator gene RG_1 belongs to an operon with operator O_1 controlled by RG_2. The inducer I_1 inhibits RG_1. As a consequence the enzymes E_1, E_2, and E_3 are synthesized, and RG_2 is activated and represses the operon containing RG_1. Once the system gets going in the presence of initial amounts of I_1, it should produce enzymes E_1, E_2, and E_3 even in the absence of I_1. In a like fashion, in the presence of I_2, RG_1 should be activated and the synthesis of E_1, E_2, and E_3 and the repressor produced by RG_2 would be repressed. Each of these alternatives would be virtually irreversible. Thus the fate of a cell containing this system would be determined by the inducer with which it comes in contact.

A simpler model, proposed originally by Delbrück would have the end product of one system inhibiting an enzyme involved in another (Figure 12.24b). Since the inhibition is mutual, it would mean that the first reaction to get started would permanently inhibit the other. This system could, of course, work through the proposed repressor mechanism of Jacob and Monod (485) as shown in Figure 12.24b. Instead of the end products acting on regulator-gene products controlling their own pathways to activate repressors, they act on the regulators of other pathways than their own. Such systems have not yet been discovered, but it is quite feasible that they do exist.

Considerations such as the foregoing regarding interactions between different biosynthetic pathways serve to bring to the foreground a second main conclusion, namely, that an organism is a highly balanced entity. It is impossible to study development and not to recognize that an organism consists of many different interacting systems. We arrived at the same conclusion from the study of gene interaction, and it should now be recognized that the study of development is also in part a study of gene interaction and vice versa. The interacting systems must be acting at the intracellular as well as at the intercellular level as pointed out in the examples given from Hadorn (391). In connection with the interactions between the different systems of an organism, it should be recognized that two very different organisms may have exactly the same potential for producing the same types of enzymes and yet be quite different. The differences could quite easily be in the control systems. If these are different, the organisms might be expected to develop differently in quite separate pathways although starting out with essentially the same structural genes.   Therefore we arrive at a third tentative conclusion, namely, that the important determinants of what an egg is going to develop into are the system(s) which control the production of its enzymes.

It should be clear that even if this conclusion is correct the problem of explaining differentiation in a developing animal or plant still remains a formidable enterprise.   It involves an understanding first of all the possible reactions that go on in the cell and their interrelationships.   Furthermore, it involves a better understanding of structure as related to function in cells than we have at present.   In short, we probably will not understand development until we know a great deal more about how cells function and interact than we do at the present.

13 Plant Pigments

The study of the inheritance of plant pigments has contributed a great deal to the understanding of the chemical basis of heredity, primarily by making it clear that inherited differences are basically chemical differences. To the early workers in this area must go the credit for establishing beyond doubt that the substitution of one gene by another may result in a specific chemical change in a compound or group of compounds. It was from findings like these that the one gene-one enzyme hypothesis and its successors developed, once organisms and chemical systems were found that lent themselves more easily to protein and enzyme analysis than the higher plants and their pigments-synthesizing pathways.

In this chapter two pigmentary systems will be considered: the flavonoid and the carotenoid. These two systems have very little in common. Not only are they quite different chemically, but the flavonoids when glycosylated are water soluble pigments found in solution in the cell sap of plants, whereas the carotenoids are strictly water insoluble pigments confined primarily to the chloroplastids. Color characteristics of both classes of compound have provided the basis for extensive investigations which have been and continue to be important in genetics.

Chemistry of Flavonoids and Related Compounds

Many of the plant pigments which give such great variety in color to flowers and other plant parts are derived from a chemical structure containing two aromatic C_6 rings connected together by a C_3 unit. For the flavonoids the ring system and numbering conventions are as follows:

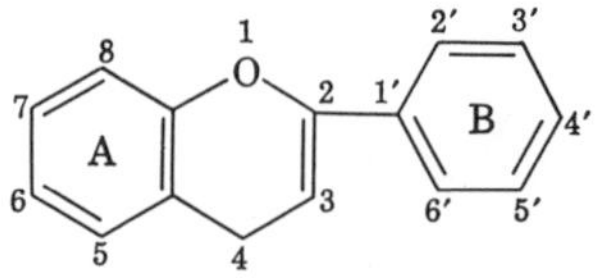

More specifically, five of the major classes of pigments have ring systems with substitutions as represented below where R can be —H, —OH, and, sometimes, —OCH$_3$.

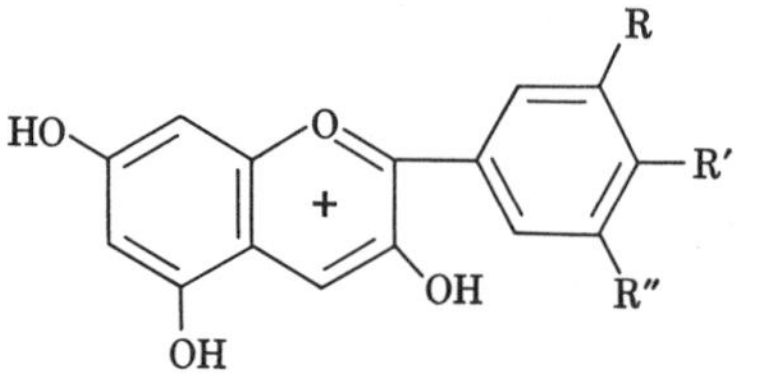

Anthocyanidins

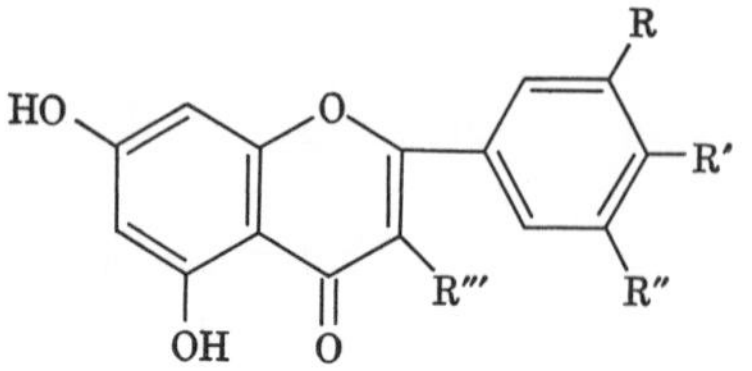

Flavones, R''' = H
Flavonols, R''' = OH
(Anthoxanthins)

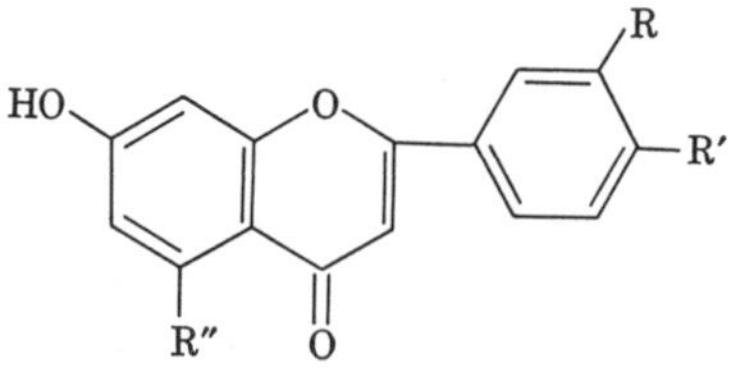

Flavonones

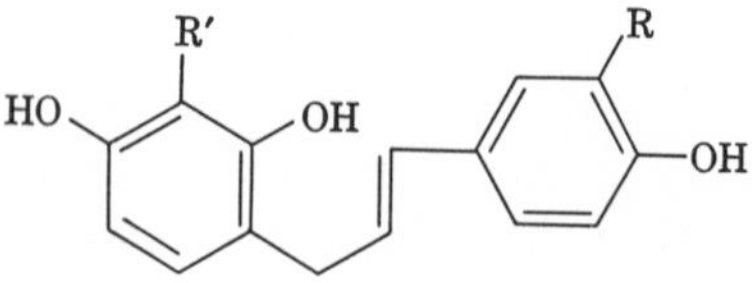

Chalcones

Aurones

The structures shown for the anthocyanidins with the + charge in the middle ring is a representation of the resonant allylic structures invoked by Shriner and Moffett (905) to account for the ionic properties of the 2-phenlybenzopyrilium nucleus. The R groups show positions of ring substitutions of the kinds indicated, but the very common further substitutions by sugars are not shown. The chemical structures of some of the more common pigments to be dealt with in this chapter are listed in Table 57. For details on other compounds reference should be made to the compilation by Karrer (512). That the chemical problem in this field is not a simple one is clear from the fact that Karrer's list includes 59 different anthocyanidins and 238 flavones and flavonols.

Anthocyanidins occur only rarely in nature as the aglycones. When free of sugar, they are insoluble in water. Usually they are associated with the hexose sugars, glucose and galactose, or with disaccharides,

Table 57. Common Anthocyans, Flavones and Flavonols of Natural Occurrence

Anthocyanidins
Pelargonidin	3,5,7,4-Tetrahydroxy-2-phenylbenzopyrilium
Cyanidin	3,5,7,3,4-Pentahydroxy-2-phenylbenzopyrilium
Delphinidin	3,5,7,3,4,5-Hexahydroxy-2-phenylbenzopyrilium
Peonidin	3,5,7,4-Tetrahydroxy-3-methoxy-2-phenylbenzopyrilium
Malvidin	3,5,7,4-Tetrahydroxy-3,5-dimethoxy-2-phenylbenzopyrilium
Petunidin	3,5,7,4,5-Pentahydroxy-3-methoxy-2-phenylbenzopyrilium

Anthocyans
Pelargonin	Pelargonidin-3,5-diglucoside
Cyanin	Cyanidin-3,5-diglucoside
Callistrephin	Pelargonidin-3-monoglucoside
Chrysanthemin	Cyanidin-3-monoglucoside
Primulin	Malvidin-3-monogalactoside
Peonin	Peonidin-3,5-diglycoside

Flavones, Flavonols
Quercitin	3,5,7,3,4-Pentahydroxy-2-phenylbenzopyrone
Luteolin	5,7,3,4-Tetrahydroxy-2-phenylbenzopyrone
Kaempferol	3,5,7,4-Tetrahydroxy-2-phenylbenzopyrone
Apeginin	5,7,4-Trihydroxy-2-phenylbenzopyrone
Myricetin	3,5,7,3,4,5-Hexahydroxy-2-phenylbenzopyrone

such as gentiobiose.   The sugar molecule or molecules are bound to the anthocyanidin in glycosidic linkage.   The methyl-pentose sugar, rhamnose, is occasionally found involved, as are compounds formed by acylation with organic acids such as *p*-hydroxybenzoic acid, malonic acid, *p*-coumaric acid, and *p*-hydroxycinnamic acid.   Either one or two sugar molecules may be attached to the ring with one of the sugars, nearly always at the 3-position; the other, if present, will be at the 5-position.

The chemical linkage of the sugar or sugars and the anthocyanidin produces the water-soluble *anthocyans* or *anthocyanins,* which range in color from red through violet and blue.   The degree of hydroxylation and methoxylation of the 2-phenylbenzopyrilium nucleus is an important factor in determining the color of the anthocyan—the more hydroxyl groups the bluer the color, the more methoxyl groups the redder the color.   The addition of another sugar molecule to a monoglucoside anthocyan increases the blueness.   The pH of the cell sap is also a contributing factor, for the anthocyanidins are color indicators.   For example, cyanin is red at pH 3.0, violet at pH 8.5, and blue at pH 11. In addition to these factors, the final color of flower petals containing anthocyans is modified by the presence of anthoxanthins, or, occasionally, tannins.   This is not caused alone by the blending of the anthocyan with the anthoxanthin, but by a weak chemical linkage between the molecules of the two types of pigments which results in the phenomenon of copigmentation.   In general, copigmentation results in a bluer color than would be the case if the anthocyan were alone.

The flavones and flavonols also occur as glycosides in nature.   In general, these pigments range in color from ivory white to orange, and they may occur, like the anthocyans, in the stems, leaves, and flowers of the plant, although their occurrence in the flowers is more apparent since they are not masked by chlorophyll and carotenoids as they are in other parts.   A list of some commonly occurring flavones and flavonols is given in Table 57.   Many more are known, but little or nothing is known of their inheritance.

Most plants have been found to contain colorless compounds which on treatment with mineral acids results in their conversion to anthocyanidins.   These are known as the leuco-anthocyanins.   The evidence is good that the leuco compounds are precursors of the colored anthocyanins and flavonoids (13, 523, 962, 963).   Leuco-anthocyanins are present in high concentration in unpigmented flower buds and gradually become converted to pigmented compounds as the flowers mature.

Inheritance of Flavonoid Pigments

Starting with the work of Onslow and Bassett in 1913, a considerable amount of information has been obtained concerning the inheritance of flower pigments in a number of cultivated plants. Practically all of the earlier work, most of which was done by a group of English geneticists at the John Innes Horticultural Institute, has been adequately reviewed by Lawrence and Price (562) and Lawrence (560). Here the discussion will be primarily concerned with work done since 1950, because newer chemical techniques, particularly chromatography, have made it possible to obtain more meaningful data than had been possible to obtain earlier.

The cultivated snapdragon, *Antirrhinum majus,* provides some of the best illustrative material for an introductory consideration of inheritance of plant pigments because it has been studied for over a period of 50 years by workers in the United States, England, and Germany. Since the independent efforts of these workers has yielded results which are in quite close agreement, we shall use them as the basis of our discussion. In this discussion the system of gene symbolism introduced by Geissman, Jorgensen, and Johnson (317) will be used for the main genes. Table 58 gives the equivalent symbols used by other recent workers; keys to the symbols used by earlier workers may be found in Böhme and Schutte (82).

Antirrhinum plants of the genotype *nn* are always albino, or dead white. Furthermore, they contain no compounds that have a flavonoid

Table 58. Symbols Used for Genes Affecting Flavonoid Pigments in *Antirrhinum* *

Geissman et al. (317)	Haney (400)	Stubbe (983)	Dayton (206)
N	*W*	*Niv*	*Y*
P	*lv*	*Inc*	*R*
Y	*Y*	*Sulf*	*l*
M	*Dil*	*Eos*	*B*

* *Note:* Symbols in horizontal rows are equivalent.

(C_6—C_3—C_6) type of structure (314), but they do contain C_6—C_3 derivatives which are probably esters of *p*-coumaric and caffeic acids.

HO—⟨ ⟩—CH=CH·COOR   HO—⟨ ⟩—CH=CH·COOR

p-coumaric acid   caffeic acid

It is possible that these compounds actually represent precursor material for flavonoids, which react with other phenyl-ring-containing compounds to give different flavonoids.   Therefore *p*-coumaric acid might be expected to be the precursor for pelargonidin, apeginin, and kampferol, all of which have a single hydroxyl group in the B ring.   Similarly, caffeic acid would be expected to contribute the B ring and the C_3 for luteolin, quercitin, and cyanidin (314).   When the dominant *N* allele is present, pigment is always produced in the flower, but the kind and quantity are dependent on other genes.

Data presented in Table 59 relate to the flavonoid-aglycone content of flowers of different genotypes which are otherwise all *NN* or *Nn* in genotype.   The three genes represented are the major factors which have been found to segregate in cultivated snapdragons to give the different kinds of colors of flowers listed in the second column.   Many other shades are known, but these are the result of modifying factors which will not be considered here.   It will be noted that when dominant *P* is present, anthocyanidins are formed, but in the presence of homozygous *p*, anthocyanidins are absent.   In addition, the two flavonols, quercitin and kampferol, are also formed only in the presence of *P*. If the effects of substitution at the *M* locus are considered, it will be seen that in the presence of *M* and *P*, cyanidin, luteolin, and quercitin are formed, but when *m* is present with *P*, no luteolir is produced, and the anthocyanidin and flavonol formed are pelargonidin and kampferol, respectively.   Substitutions at the *Y* locus have no qualitative effect, but as can be seen from Figure 13.1 they do have a definite quantitative effect both on the concentration of anthocyans and the yellow pigment aureusidin.   When *Y* is substituted for *y* there is a slight elevation in the amount of anthocyans, but a drastic reduction in the amount of aureusidin produced in the flower.   This latter effect is particularly notable in *pp mm*—plants.   Although not shown in Table 59, substitutions at the *M* locus have a definite quantitative effect on the concentrations of the two flavones apeginin and luteolin.   Apeginin is present

Table 59. Qualitative Differences in Flavonoid Aglycones and Aureusidin in *Antirrhinum majus* of Different Genotypes *

Genotype	Phenotype	Cyanidin	Pelargonidin	Luteolin	Apeginin	Quercitin	Kampferol	Aureusidin
PPMMYY	Magenta	+		+	+	+		+
PPMMyy	Orange red	+		+	+	+		+
PPmmYY	Pink		+		+		+	+
PPmmyy	Yellow orange		+		+		+	+
ppMMYY	Ivory			+	+			+
ppMMyy	Yellow			+	+			+
ppmmYY	Ivory				+			+
ppmmyy	Ivory							

* From Geissman, Jorgensen, and Johnson (317) and Sherratt (904).

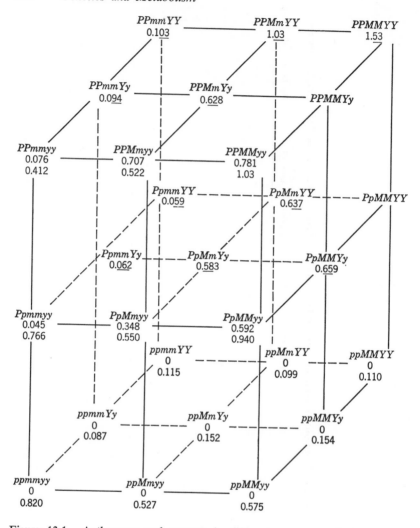

Figure 13.1. Anthocyans and aurones in different genotypes of *Antirrhinum majus.* The top figures are the relative concentration of anthocyans, the bottom relative concentrations of aurones. From Jorgensen and Geissman (497).

whether *M* is present or not, but there is ⅓ to ⅙ *less* apeginin in *M* plants than in *mm* plants. In other words, when luteolin is present there is less apeginin.

The foregoing facts have been used to formulate an hypothetical scheme for the biosynthesis of flavonoids in *Antirrhinum majus* (Figure 13.2). The first thing to be noticed about this proposed pathway is

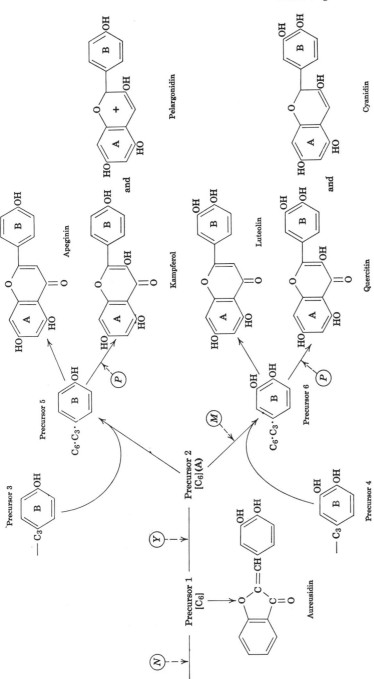

Figure 13.2. Proposed pathway for the biosynthesis of the flavonoids in *Antirrhinum majus*. After Geissman et al. (317) and Sherratt (904).

that it involves a number of substrates which are common to two or more pathways and that all of the pigments have a common precursor (precursor 1). The gene N is assumed to act toward the production of precursor 1, since in the presence of its recessive allele no pigments are formed. This interpretation is supported by the accumulation of *p*-coumaric and caffeic acid in *nn* plants. This would be expected if these compounds served as precursors 3 and 4, respectively, and were not used. The reason for assuming that Y acts to convert the hypothetical C_6-precursor 1 to 2 is simply that in its presence the amount of flavonoids is increased and that of aureusidin decreased. Maximum aureusidin production is found in *yy* plants, because most of the available precursor 1 is converted to aureusidin in the presence of the block between 1 and 2. This block must be only a partial block, however, since all the flavonoids are formed in the presence of *yy*. Hence *y* should be considered a hypomorph in this scheme. The precursor 2 is common to all flavonoids as the "A" part of the flavonoid molecule. If it combines with precursor 3, flavonoids hydroxylated only at the 4′ position are formed, whereas if it combines with precursor 4, 3′, 4′ hydroxylated flavonoids are formed. The scheme presupposes that in the presence of M most of precursor 2 combines with 4, leaving only a small amount to be converted to apeginin. When *mm* is present, the pathway to the 3′, 4′ compounds is blocked and only the 3′ compounds formed. More apeginin would be expected to be formed in *mm* plants according to this scheme. This is in agreement with the data already discussed. The P gene, it is assumed, acts to make flavonols and anthocyanidins out of precursors 5 and 6, presumably by the same type of reaction in each case. In the presence of *pp MM* only luteolin and apeginin are formed among the flavonoids, as would be expected.

The scheme given in Figure 13.2 is in general agreement with the data now available on flower color for the snapdragon and other plants. Experiments employing C^{14}-labeled acetate show that acetate is selectively incorporated into the A ring of the flavonoids (1053). This would be expected if the flavonoids were synthesized by the pathway shown in Figure 13.2.

In general, most flowers containing both flavonols and anthocyanidins have both types of pigments present in the same degree of hydroxylation, that is, cyanidin is accompanied by quercitin and pelargonidin by kampferol (402). Thus, in the carnation, *Dianthus caryophyllus,* plants with the dominant factor R contain both cyanidin and quercitin with some kampferol but no pelargonidin in the flowers. But *rr* plants contain only kampferol and pelargonidin (316). No traces of cyanidin or quercitin are present.   Again in the *Cyclamen, C. persicum,* and the French bean, *Phaseolus vulgaris,* it is apparent that anthocyanidins and flavonols come from the same precursors (275, 900).   These observations should not be taken to mean that those flavonoids oxidized at the 3′ position never occur simultaneously with those oxidized at the 3′, 4′ or 3′, 4′, 5′ positions, however.   In the diploid potato, petunidin, which is oxidized at the 3′, 4′ and 5′ position on the B ring, is present in one strain in both flowers and tubers simultaneously along with cyanidin and peonidin, which are oxidized at the 3′, 4′ positions (224) (see Table 57).   Furthermore, different parts of a plant may have anthocyanidins in different stages of oxidation.   Thus, also in the potato, the flower may contain cyanidin and the tuber pelargonidin.   An extreme case has been reported in the nasturtium, *Tropaeolum majus,* variety, Empress of India.   This strain has pelargonidin in the petals, cyanidin in the sepals, and delphinidin in the leaves (848).

   The inheritance of qualitative differences among the anthocyans has been extensively studied in a number of species.   Three significant types of changes in the anthocyan molecule have been found to be inherited in a simple Mendelian fashion: (1) hydroxylation; (2) methoxylation; and (3) number of sugar molecules attached to the anthocyanidin nucleus.   The Cape primrose (*Streptocarpus* sp.) hybrids, investigated by Lawrence, Scott-Moncrieff, and Sturgess (561, 564), show in their inheritance all three of these differences, with three major genes, *O, R,* and *D,* involved.   Figure 13.3 illustrates the effect of substituting the dominant and recessive alleles *R* and *r, O* and *o,* and *D* and *d* on flower pigments in *Streptocarpus* hybrids derived from a cross between *S. Rexii* and *S. Dunni,* which are two pure-breeding wild strains with blue and red flowers, respectively.   Since the genes *R, D,* and *O* appear to be completely dominant to their respective recessive alleles, the genotype designations in the figure are shortened to the minimum necessary to describe the genotypic state.   Thus *rod* is equivalent to *rroodd,* and *RoD* is equivalent to *RrooDd, RRooDD, RrooDD,* or *RRooDd,* and so on for the others.

   When *r* is substituted for by *R,* it will be noted that a higher degree of oxidation is obtained, both in those plants with a mixture of mono- and diglycosides (*rod*) and those with diglycosides only (*roD*).   The oxidation may result simply in hydroxylation, in which case cyanidin derivatives are produced, or in methoxylation, with the production of peonidin derivatives.   Further oxidation at both the 3′ and 5′ position caused by the substitution of *O* for *o* results in methoxylation only, with the formation of malvidin derivatives.   The state of *R* in *O* plants is

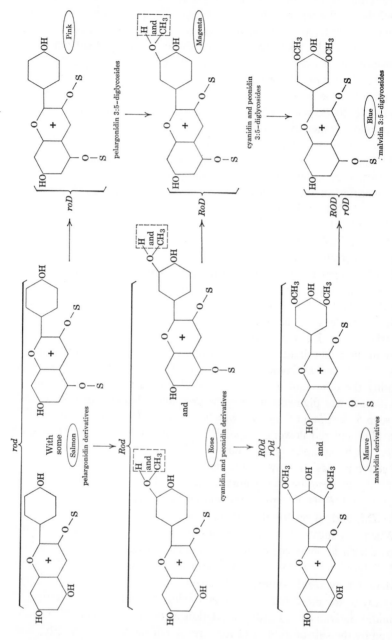

Figure 13.3.   The effect of gene substitutions in *Streptocarpus* sp. on the degree of oxidation and the number of sugars in the flower anthocyans.   (S = sugar.)   After Lawrence, Scott-Moncrieff, and Sturgess, and Lawrence (564, 561).

unimportant, since both *OR* and *Or* plants are phenotypically identical with respect to the degree of oxidation. *O,* therefore, "masks" the effect of *R,* or is *epistatic* to *R.* The term *epistasis* is used frequently to describe the "dominance" of one gene over a nonallelic gene, as is exhibited here.

The effect of gene *D* is to cause the appearance of almost pure 3,5-diglycosides in place of a mixture of mono- and diglycosides. It does this whatever the states of the other two genes *R* and *O.*

To summarize, it is clear that the action of *R* and *O* is to determine the degree of oxidation of the anthocyanidin. *R* results in oxidation, either hydroxylation or methoxylation, at the 3′ position, and *O* at both the 3′ and 4′ positions. *D* has nothing to do with the state of oxidation but determines an increase in the amount of 3,5-diglycosides.

Although it is quite clear from the foregoing that definite effects on the chemical structure of the flavonoid pigments are brought about by gene changes, it should be also recognized that there are many different patterns of effects and that frequently the shifts from one pattern to another may be more quantitative than qualitative. In these cases it is difficult to state just precisely what chemical state or change is controlled by a given gene. The results of the investigations on the balsam, *Impatiens balsamina,* illustrates some of the complexities that are found to exist when a careful chemical analysis is made to accompany the genetic analysis (14, 172, 395). The effects of three genes, *H, L,* and *P,* segregating in this plant have been investigated. *P* has three identifiable alleles, *P, P^g,* and *P^r,* which control anthocyan production but have no effect on flavonol production. The quantitative effect of this locus depends also on the states of the *L* and *H* genes. In addition, the effects are different depending on whether one is considering differences in petals or sepals. In the following discussion the remarks apply primarily to petal pigments. The gene *L* determines the presence of malvidin glycosides and myricetin, a flavonol which, like malvidin, is oxidized at the 3′, 4′, 5′ positions of the B ring. It has no effect on another flavonol, kampferol, also present in the petals.

H, like *P,* determines the amount of anthocyanidin in the petals, but in a rather complex fashion as will be seen from Table 60. In *hhpp* plants only malvidin, among the anthocyanidins, is present in the petals, but in *Hpp* plants both pelargonidin and malvidin are present. Pelargonidin *is* produced in *hh* plants, however, when *L* is homozygous recessive and either *P^g* or *P^r* are present. *H* is no simple determiner of pelargonidin production, therefore, but neither is *P* because no pelargonidin is produced in *Lhh* plants whatever the state of *P.* It is also

Table 60.   The Anthocyanidins and Flavonols Present in Different Genotypes of *Impatiens balsamina* *

Genotype	Phenotype	Pelargonidin	Malvidin	Myrecitin
lhp	White	−	−	−
Lhp	Pale lavender	−	+	+
LHp	Rose lavender	+	+	+
LhP^g	Lavender	−	+	+
LHP^g	Pink lavender	+	+	+
LhP^r	Purple	−	+	+
LHP^r	Magenta	+	+	+
lHP^r	Red	+	−	−
lhP^g	Pale pink	+	−	−
lhP^r	Pink	+	−	−

* From Alston and Hagen (14) and Clevenger (172).

evident from the data in Table 60 that *P*, *P^g*, and *P^r* cause an increase in the amount of anthocyanidins formed in the petals in the order indicated.   This has been verified spectrophotometrically by Hagen (395).   Substitutions at this locus do not seem to have an effect on flavonol production, however.   On the other hand, the presence of *H* causes a reduction in kampferol.

A multiple allelic series with effects on the seed-coat color of the French bean, *Phaseolus vulgaris,* also gives rather peculiar results with respect to anthocyan and flavonol production.   In the series *C*, *C^r*, and *C^u*, the homozygous recessive *c^{u}c^u* produces no anthocyanins or flavonol glycosides.   In the presence of *C* or *c^n* both types of compounds are produced, but in *C* plants anthocyanins are found only in those cases in which a flavonol oxidized at the 3′, 4′, 5′ positions occur.   In these the anthocyanin is also oxidized at the same three positions on the B ring.   In *c^r* plants anthocyans with all three levels of oxidation of the B ring are found (mono-, di-, and trihydroxy) but only mono- and dihydroxy flavonols (275).

Observations on the inherited differences of glycosidal types of anthocyans have been confined primarily to determining inheritance of mono- and diglycosides.   The carnation, *Dianthus caryophyllus* (315), Chinese aster *Cheiranthus cheiri* (1091), and *Cyclamen* (900) have strains in which either the 3-monoglycosides or 3,5-diglycosides are represented in the flower anthocyans.   In general, the presence of two sugars is dominant to one. *Verbena* (63) is an exception; in some strains the diglycosides are dominant and in others recessive to the monoglycosides. Many modifiers are involved because a mixture of different kinds of sugars in the glycosides is frequently found rather than a single type.

As mentioned previously, the pH of the medium in which the anthocyan is in solution may have considerable effect on the color. In many of the plant species which have been investigated, pH differences in the flower-petal cell sap have been found among the color varieties. In at least three of these species, *Primula sinensis, Papaver rhoeas,* and *Primula acaulis,* pH differences are inherited in a simple Mendelian fashion, with the lower pH dominant over the higher. The pH difference is about 0.5 pH unit, which is a wide enough range to cause some color change in certain anthocyans. The difference in pH seems to be confined to the cell sap of the petals; the rest of the plant parts are unaffected, unless they, too, contain anthocyans. For this reason it is suspected that the more acid pH of the petals in some strains is associated in some way with the type of anthocyan or precursor formed in the petals.

Since, in general, flavonol and flavone differences are correlated with changes in anthocyans, the inheritance of the two types cannot properly be considered apart. The clearest example of this relationship is to be found in the work of Lawrence and Scott-Moncrieff (563) on the inheritance of flower color in *Dahlia variabilis,* the cultivated *Dahlia.* This plant is an alloöctoploid (an octoploid with inheritance similar to that of a tetraploid with tetrasomic inheritance), which makes it possible to have five different combinations of a pair of alleles, that is, *AAAA, AAAa, AAaa, Aaaa,* and *aaaa.*

Four dominant genes are known to affect flower color. Each has a recessive allele which seems to be completely ineffective in pigment production. The dominant genes *A* and *B* are directly concerned with the production of anthocyans containing either cyanidin or pelargonidin. Gene *Y* is accompanied by the presence of a chalkone, discovered in *Dahlia* by Price (1017). The fourth gene, *I,* is related to the production of a flavone, apeginin.

Plants which are homozygous recessive for *a, b, y,* and *i,* that is, *aaaa, bbbb, yyyy,* and *iiii,* are devoid of flower pigment, but the presence of dominant alleles of any one or more of the four result in some pigment. However, the production of the three pigments together in one flower because of a dominant *A* or *B, Y* and *I* is not a simple matter (see Figure 13.4). The amount of one pigment produced may be very

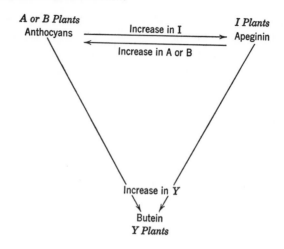

Figure 13.4. The relationship of anthocyans, butein, and apeginin in *Dahlia variabilis* as determined by the genes *A, B, Y,* and *I*.

much dependent on the production of another. For example, in plants homozygous recessive for *y, a,* and *b,* but with dominant *I* present, apeginin appears in the flowers in an amount dependent on the number of *I* alleles present. *Iiii* plants have a small amount, and the genotypes *IIii, IIIi,* and *IIII* have equivalent and the maximum attainable amount. *I* is therefore completely dominant to *i* when it is present twice, provided that the other genes are in the recessive state.

When *A* is present as the dominant in homozygous recessive *y, i, b* plants, anthocyan appears in the flowers and increases with increase in the dosage of *A* from *Aaaa* to *AAAA*. If now both *A* and *I* are present together in the same plant, an interesting effect is noted in the production of both apeginin and anthocyan. No anthocyan is present in plants in which *A* is represented once and *I* three or four times, but apeginin *is* present. In the other possible dosage combinations of *A* and *I*, anthocyans are present with apeginin. The relative concentrations of the two pigments are dependent on the relative dosages of *A* and *I*. The introduction of *I* in increasing dosages into *A* plants causes a reduction in anthocyan correlated with an increase in apeginin, and, on the other hand, an increase in *A* is accompanied by an increase in anthocyan and a concomitant decrease in apeginin. Thus there exists an inverse correlation between anthocyan and apeginin production which has been interpreted as meaning that the two types of pigments are competing in metabolism for a common, limited precursor substance.

Further discussion is given this interpretation in Chapter 9. For the present it is sufficient to point out that a relationship in the inheritance of the two pigments is proved in *Dahlia variabilis*.

The production of chalkone in plants recessive for *a* and *b* shows a pattern similar to the one just described. Only a trace of apeginin but considerable chalkone is found in *Yyyy IIii* plants, whereas in *YYyy IIii* individuals apeginin seems to be completely replaced by chalkone. Furthermore, an examination of the effects of the possible dosage combinations between *Y* and *B,* and *Y* and *A* reveals that chalkone is produced at the expense of anthocyan. Thus it is apparent that all three pigments, anthocyans, apeginin, and chalkone, are related in their inheritance and perhaps are therefore associated in biogenesis.

In addition to having provided some of the early significant studies contributing to the development of biochemical genetics, the continued study of plant flavonoid pigments has brought to light some extremely interesting facts about gene interaction. There is no doubt that far more remains to be learned about gene interaction by the study of these pigments and their inheritance, particularly with respect to its role in the development of plants. The turning on and off of pigment production in various parts of the plant as development proceeds may well be under the control of controlling elements in general. A number of cases have already been cited in which this is certain in maize. A study of the enzymes involved in the biosynthesis of the flavonoids and their changes in apparent activity might well make important contributions to an understanding of the physiology of development in plants and animals.

Carotenoids and Derivatives

Carotenoids are water-insoluble pigments produced by and generally retained within the cell plastids. They range in color from yellow through orange and red. A few are known to be colorless and are presumably precursors to the colored forms. They consist of repeating isoprene units to give highly conjugated systems as shown in the examples depicted in Figure 13.5. The differences among the various carotenoids are those of stereoisomerism including *cis-trans* isomerism, the presence or absence of allylcylic rings at the ends of the chains, the number and arrangement of the double bonds, and the degree of oxidation. Carotenoids are generally divided into two groups: the carotenes, which are simple polyene hydrocarbons; and xanthophylls, the oxidized

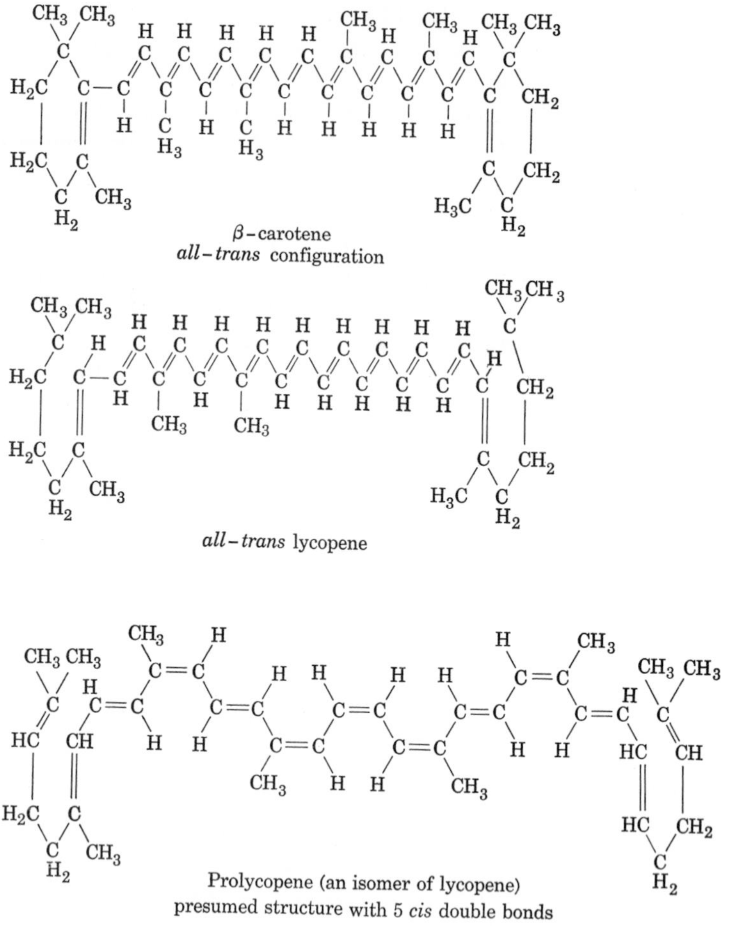

β-carotene
all-trans configuration

all-trans lycopene

Prolycopene (an isomer of lycopene)
presumed structure with 5 *cis* double bonds

Figure 13.5. Structures of the principal types of carotenoids known to occur in the fruit of the tomato. The structure of *all-trans* lycopene is identical to β-carotene except for the absence of the rings at the ends of the chain.

forms of the carotenes. In the green parts of plants both the carotenoid and the chlorophyll derivatives are present together in the chloroplasts, but the green of the chlorophyll masks the red and yellow of the carotenoids. The carotenoid colors, however, are readily recognized in the chlorophyll-free structures, such as the fruits of the tomato and the roots of the carrot, and in mutant forms in which the chlorophyll is absent but the carotenoids present.

The pigmentation of the tomato fruit is produced by carotenes. *All-*

trans lycopene, *cis* isomers of lycopene, and β-carotene are the predominant types (Figure 13.5). The familiar red tomato contains primarily *all-trans* lycopene and β-carotene, but there are other color varieties with qualitative and quantitative differences. Genetic analysis reveals that at least four nonallelic genes are involved in the inheritance of these differences. Fruit color is related to two of these, *R* and *T*, as follows (488, 630, 632):

$$\left.\begin{array}{l} RRTT \\ RrTt \\ RrTT \\ RRTt \end{array}\right\} \text{Red}$$

$$\left.\begin{array}{l} RRtt \\ Rrtt \end{array}\right\} \text{Orange (tangerine variety)}$$

$$\left.\begin{array}{l} rrTT \\ rrTt \end{array}\right\} \text{Yellow}$$

$$rrtt \quad \left.\begin{array}{l} \text{Intermediate between yellow and} \\ \text{orange (yellow tangerine)} \end{array}\right.$$

Some data of Mackinney and Jenkins are given in Table 61 which show the quantitative and qualitative differences between the four

Table 61. The Carotene Content of Four Strains of the Tomato, *Lycopersicon esculentum* *

	Microgreams of Carotenoid per Gram of Fruit			
Carotene Type	RT Red	Rt Orange	rT Yellow	rt Intermediate
all-trans Lycopene	70–130	. . .	0–0.5	. . .
cis Isomers of lycopene	. . .	40–55	. . .	10–15
β-Carotene (*all-trans*)	5–10	3–12	1–3	0.5–1.0
poly-cis-Carotene	. . .	8–15	. . .	. . .
all-trans-ζ-Carotene	0–0.1	20–15	. . .	0.01
Phytofluene	3–5	4–7	ca. 0.1	0.7–1.0
Total carotenes	80–150	75–150	3–7	15–20

* From Tomes et al. (1014) and Mackinney and Jenkins (632).

phenotypes. It will be noted that yellow (rT) fruits are characterized by a very low total carotene content while the red (RT) and orange (Rt) contain the most carotene. The double recessive $rrtt$ contains intermediate amounts.

Thus, total carotenes considered, the following quantitative order is found: $rT < rt < Rt = RT$, showing that the major effect of substituting R for r is to increase the total amount of carotene. If now the qualitative differences are analyzed, it is seen that *cis* isomers occur in detectible amounts only in those plants in which t is homozygous, namely those bearing orange—and intermediate—colored fruit. In these it is also evident that *all-trans* lycopene is absent, although β-carotene and ζ-carotene are present to represent the *trans* isomers.

A third gene, discovered in a Mexican strain of tomatoes by Jenkins and Mackinney (489), apricot (at), gives a phenotype somewhat similar to yellow ($rrTT$) when homozygous recessive. Accordingly, it has a similar content of carotenoids (Table 62). The major difference between apricot and yellow seems to lie in the much higher β-carotene content of the apricot. The main effect of the recessive gene, at, seems to be the suppression of lycopene pigment in RT plants.

R appears to act early in the biosynthesis of carotenoids, probably controlling the production of an early common precursor. However, the difficulty with this interpretation of the data is that $rrtt$ tomatoes have considerably more carotenes than those of genotype $rrTT$.

Table 62.   Carotene Content of Apricot Tomatoes *

	atat Apricot	*rr at at* Yellow apricot	*tt at at* Tangerine apricot
Lycopene isomers	2.5	. . .	6.5–21
ζ-Carotene	. . .	. . .	0.6–1.0
β-Carotene	6–10	1.2	0.7–2.1
Phytofluene	0.2	Trace	0.5–1.1
Total	8–15	1–2	9–25

* From Jenkins and Mackinney (489).

Lincoln and Porter (613) have described a fourth gene, *B,* which in its recessive form *b* reduces the amount of β-carotene. Concomitant with the reduction of β-carotene, the concentration of lycopene is increased (1015). Thus *bb* tomatoes contain about 10–20% β-carotene and 80–90% lycopene, while *Bb* or *BB* tomatoes have 70–90% β-

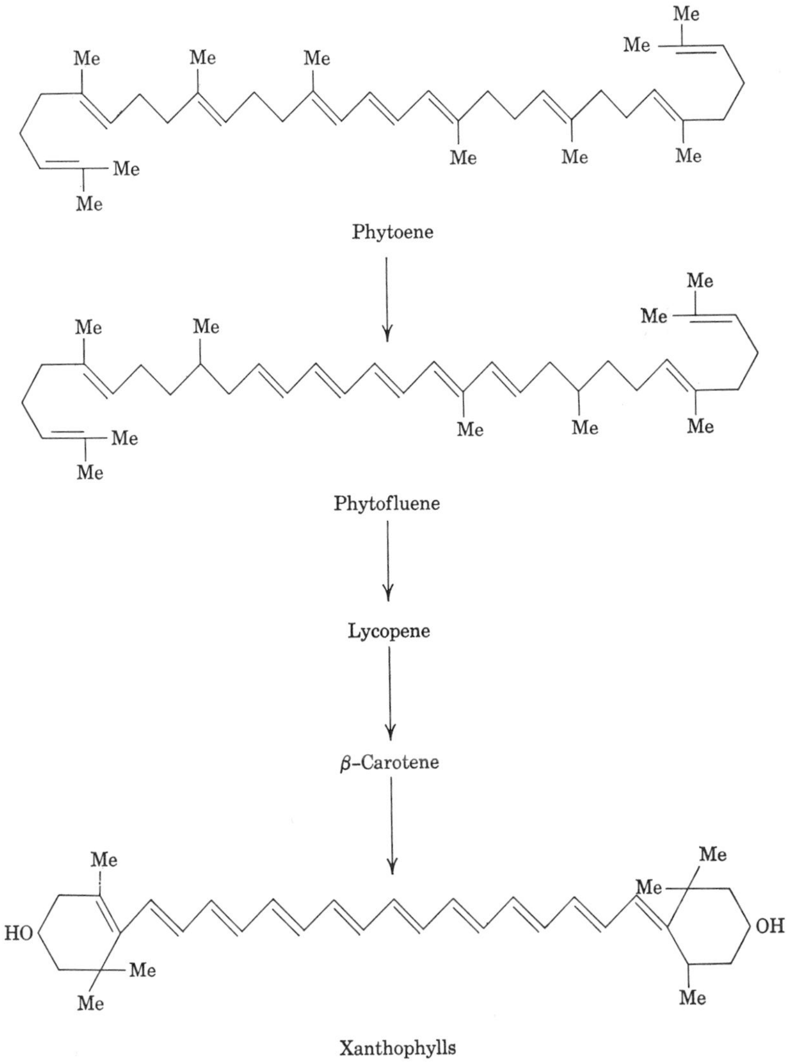

Phytoene

Phytofluene

Lycopene

β-Carotene

Xanthophylls

Figure 13.6. Suggested biosynthetic pathway for carotenoids.

carotene and 1–30% lycopene. There is also an increase in γ-carotene in the *B* strains. The reciprocal relation between lycopene and β-carotene exhibited in these strains indicates that these two pigments are derived from the same precursor which is limited in amount during the period of active carotene synthesis.

A definite quantitative effect of the dominant gene *Y* has been shown to exist in maize on the amount of total carotenoid in the endosperm (654, 820). As the number of *Y* genes increases starting from *yyy* endosperms, which contain only small amounts of carotenoids, to *YYY* endosperms, the amount of carotenoids increases linearly. In a tetraploid maize plant, the *YYYYYY* endosperm contains about 40% more carotene than the triploid endosperm of the diploid plant.

Little work has been done with derived mutants of microorganisms in order to attempt to elucidate the pathway or pathways of biosynthesis leading to the carotenes. The most thorough analysis has been made by Haxo (418), who analyzed a number of *Neurospora crassa* mutants with carotenoid deficiencies. These he was able to divide into four groups: (1) those which did not produce acidic xanthophylls; (2) those which produced no xanthophylls and few carotenes; (3) those which produced phytoenes (colorless, presumed saturated, precursors of carotenes); and (4) those which produced no detectible polyenes or traces. From these results he arrived at the tentative conclusion that the formation of carotenoids in *Neurospora* starts from partially saturated phytoenes and proceeds through the various steps shown in Figure 13.6 to the xanthophylls. Similar schemes have been advanced to explain the data from tomatoes (341), but the observed actions of the genes *B, T, R,* and *At* do not easily lend themselves to any simple analysis, and none of the schemes presented so far is satisfactory in explaining all the data.

General Considerations

Although practically no enzyme work has been done with plant pigments, the results obtained by the study of the pigments alone have been most valuable in the general elucidation of the problems which beset us in the attempt to analyze inherited metabolic differences. As we have already seen by the study of melanin inheritance in mammals such as the guinea pig, the production of a single type or related group of pigments is dependent on many inherited factors. The results obtained from the study of the inheritance of plant pigments present us with parallel complexities, but at the same time make it evident that probably one of the controlling factors in metabolism is the intracellular concentration of intermediate metabolites. Precursors must be present in limited concentrations, and the rates of production of end products with a common precursor must be dependent not only on the concentration of the precursor, but the relative efficiencies of the different enzymes acting on it. Thus it is easy to see why, when one end product is not formed due to a gene change, other products dependent on the same substrate increase in amount. At the same time it should be apparent that any attempt to increase the production of a given end product by gene substitution will, for the same reasons, decrease the amount of other products.

These considerations have general significance in interpreting certain practical breeding problems in plants and animals. Frequently, when a breeder attempts to develop a strain with certain characteristics of economic value, he finds that he may succeed, but only at the expense of other equally valuable characteristics which may disappear or be modified concomitantly. Metabolic systems may be changed drastically through gene substitution, but there are definite limitations to the number of changes that may be incorporated in any given strain. At the present time, the breeder can only operate empirically to determine what is possible, but as more information is obtained about the interrelations existing in metabolism, and the genetic control of these, breeding will be placed on a more predictable basis.

14 Biochemical Genetics in Humans

Man is not a particularly suitable object for the study of genetics, because he neither breeds according to a plan nor keeps, except in rare instances, very good pedigrees. Furthermore, from the point of view of biochemical studies, experimentation upon him is severely restricted to moderate changes in nutrition and physiological conditions, to blood, saliva, and urine analyses, and to the occasional use of small tissue slices from biopsies and autopsies.

Despite these difficulties, it is a matter of self-interest that man's biochemical genetics be studied as extensively as possible. Of all organisms, man has the greatest control over his environment. In this resides great danger, because he is capable of doing both very remarkable and very foolish things. Man changes his environment through alteration of foods, the use of drugs, and the pollution of air, water, and land. These changes provide both favorable and unfavorable conditions for gene expression, and some of them provide for a high rate of gene mutation and a maximum of preservation of genes with bad effects. It is frequently difficult to know which conditions are favorable and which unfavorable, unless the subject material, man, is understood. This is a prime reason why the study of human biochemical genetics is important. Its study is not merely the investigation of peculiar and interesting, rare human abnormalities, but the investigation of the "normal" individual and how he differs biochemically from other individuals. This knowledge is necessary if we are to understand how man's changing environment affects man favorably and unfavorably. It is more than a matter of curiosity; the lives of our descendants will depend on it.

The Human System

Thus far it has been profitable to carry over principles of function, both genetic and biochemical, from one organism to another, even though there are many obvious quantitative and qualitative differences to contend with. Therefore it is reasonable to proceed with a consideration of human genetics on the assumption that all of the principles discussed in earlier chapters and derived from studies of various organisms are applicable here, too. For example, there are obvious differences in the mechanisms of gene transmission and in recombination in bacteriophages and humans, but it is likely that the principles of specific pairing of nucleic acids in replication, and in function in the designation of order in polypeptide chains are quite similar. It is likely, also, that the kinds of molecular changes that constitute mutation and the ranges of variations in gene products obtain in humans and other organisms, even though ultimate expression may be quite different. These are important assumptions which demand continued reexamination, for we cannot do the same kinds of experiments with humans as with other organisms and still remain human.

Humans as Genetic Subjects. It is not at all a simple thing to maintain a genetically homogeneous population in any organism. With the assumption of about 10^4 genes in *Drosophila* (and of the same order in humans), and a reasonable average spontaneous mutation rate per gene of 10^{-5} per generation, one gamete in ten will carry a new mutation. Thus, about 20% of the individuals in a new diploid generation may be expected to carry different new mutations. In experimental organisms where controlled breeding and selection can be carried out, such new mutations can be eliminated at will when detectible, but in the human population they are retained, multiplied, and eliminated only at some rate which is a function of their degree of reduction of reproductivity. Obviously, then, the human population is extremely heterogeneous genetically, and, with increasing population and mixing, it will likely become more so. Genetic analyses in humans are then confronted with serious difficulties. It is essentially never possible to study the effects of mutation of one gene against reasonably homogeneous backgrounds, since different individuals are rarely very similar except for identical siblings from the same egg. That this is true is quite evident from the fact that in the gametes of an individual who is heterozygous at only one locus in each of the 23 pairs of chromosomes there are over eight million possible and equally probable combinations (968) that can be passed on to progeny. This is a low degree of heterozygosity, and it is obvious that two individuals from the same family will not be really alike. Furthermore, in view of the numerous possible bad combinations from heterozygous parents, it is sufficient for great joy when offspring are normal individuals, whether or not there are strong family resemblances.

It is clear, then, that only extreme genetic expression which shows through variable heterozygosity is likely to be observed and adequately recorded over a sufficient number of generations to permit the kind of analysis of inheritance patterns which is routine, for example, in *Drosophila*. Even when records of families are adequate, numbers of individuals are small. Still, useful information has been gained in this fashion, as in hemophelia in the descendants of Queen Victoria and in facial characteristics, especially the protruding jaw, in the Hapsburg dynasty. In the latter case, part of the record is in family portraits, and the character was sufficiently expressive as to survive the interpretations of the various artists who painted the portraits. Details of these and many other familial-trait pedigrees are available—see, for example, Stern (968) and Vogel (1037). Now, many more satisfactory family records are being kept, especially with relation to inherited pathological conditions, but it would be helpful if keepers of family trees could include objective vital statistics and noticeable physical characteristics.

Although kindred studies in humans are profitable and should become more so in the future, much of the present understanding of human genetics comes from consideration of statistical frequencies of specific genes in large populations. The fundamental basis for this approach was derived by Hardy and by Weinberg in 1908. For an understanding of the results of mixing and random breeding among two equal populations, one homozygous for gene A and the other for an allele a, the following may be considered. Here, four combinations of marriages are to be expected with equal frequency:

	Female		Male	Frequency
(1)	AA	$\times$	AA	$\frac{1}{2} \times \frac{1}{2}$
(2)	AA	$\times$	aa	$\frac{1}{2} \times \frac{1}{2}$
(3)	aa	$\times$	AA	$\frac{1}{2} \times \frac{1}{2}$
(4)	aa	$\times$	aa	$\frac{1}{2} \times \frac{1}{2}$

But types 2 and 3 are identical as far as offspring are concerned, and if each marriage yields the same average number of children the genotypes of progeny will occur in this same ratio: $AA = \frac{1}{2} \times \frac{1}{2}$; $Aa = 2 \times \frac{1}{2} \times \frac{1}{2}$; and aa $\frac{1}{2} \times \frac{1}{2}$ or $\frac{1}{4} : \frac{1}{2} : \frac{1}{4}$. Treating the next generation in the same way with all possible combinations of the three genotypes in males and females, $4(\frac{1}{16})AA$, $4(\frac{1}{8})Aa$ and $4(\frac{1}{16})aa$, or exactly the same ratio as in the parental population, results. This constancy holds for subsequent generations and other initial frequency ratios will also be maintained in the same way. It is clear, then, that in such a population the relative frequencies of the alleles A and a will remain constant, and if the frequency of A is represented by p and a by q, then the proportions of AA, Aa, and aa are present in the ratio $p^2 : 2pq : q^2$. This is the Hardy-Weinberg Law, and it represents the equilibrium frequencies of A and a in a population after one generation of random mating. A temporary disturbance such as nonrandom mating will be succeeded by a return to the same equilibrium since this does not change the total frequencies of p and q.

The foregoing considerations are based on assumptions that often are not valid. For example, mating is not random as a rule, for tall people tend to marry tall people and redheads tend not to marry redheads, and so on, but, nevertheless, for a number of less visible and emotionally neutral characters such as MN blood groups, the Hardy-Weinberg equilibrium holds remarkably well in each of several different populations. Mutation and selection also produce deviations from the rule, as discussed hereafter, but, nevertheless, it provides a valuable baseline to serve as a reference in the analysis of deviations. One obvious valuable application is in evaluating the total numbers of heterozygotes in a population. As shown in Table 63, for rare mutant genes, heterozygotes are enormously more frequent than homozygotes, and if the mutant is recessive and deleterious when homozygous, there are relatively large numbers of carriers which will produce one in four affected offspring when two of them marry. From the standpoint of affected individuals, this is undesirable, and it can be avoided by nonrandom mating when the heterozygous condition can be recognized. On the other hand, lowered reproductivity in unfavorable homozygotes

Table 63. Incidence of Heterozygotes and Homozygotes at Different Gene Frequencies *

Frequency of Gene a q	Incidence of Heterozygotes Aa $2pq$	Incidence of Homozygotes aa q^2	Ratio Heterozygotes to Homozygotes $2pq:q^2$
0.5	0.5	0.25	2:1
0.2	0.32	0.04	8:1
0.1	0.18	0.01	18:1
0.01	0.0198	0.0001	198:1
0.001	0.001998	0.000001	1998:1
0.0001	0.0001998	0.00000001	19998:1

* From Harris (410).

provides the mechanism for the elimination of deleterious genes from the gene pool of the population, and selective breeding is not necessarily a good thing for the population as a whole. Thus the conflict between what is favorable for the individual and for the population poses an extremely difficult problem in many aspects of human genetics, as it does in social and political matters. Hardly anyone likes to be eliminated, even for the "good" of the race, and in our social and medical practices we continue to make it less probable that "slow rabbits get eaten."

Equilibria in human populations are, in general, only approached, even under random mating, since mutations occur spontaneously and are likely to occur at ever-increasing rates with our modifications of our chemical and physical environments. When a new mutation does occur, it is expected to approach a frequency level in a population dependent on mutation rate and the rate of elimination through lowered reproduction fitness. An example comparing two mutation rates to equal fitness is shown in Figure 14.1. These are realistic expectations, as shown in Table 9, where estimated rates to mutations which give rise to a number of human inherited diseases are presented. In the case of dominant abnormalities which result in a reproductive fitness of zero (as by early death or sterility), it is clear that the frequency in a population is directly related to the mutation frequency; however, in other situations the mutations accumulate, and equilibria and elimination must be taken into account. Very approximate estimates suggest about

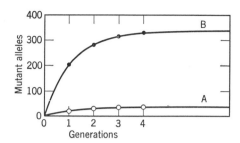

Figure 14.1. Frequency of an abnormal dominant *A* allele in the course of generations after the onset of mutations from *a* to *A*. (Size of population: 1,000,000. Reproduction fitness of *A*: 0.4). Curve A: Mutation rate 1 in 100,000. Curve B: 1 in 10,000. From Stern (968).

a fifty-fold greater frequency of detrimental mutants in a population than would be expected from new mutations.

With this brief outline of some basic concepts concerned with humans as subjects for genetic studies, it is clear that much can and must be done to simplify analysis with this system, for people can only be taken as they exist and are not easily manipulated. With its enormous heterozygosity the human population presents a very large pool of alternative alleles that yields uniqueness to nearly all individuals and an avalanche of characters for study. Serious limitations exist in precision of genetic analyses; for example, in addition to those already discussed, it is well known in controlled breeding systems that the same phenotype can result from mutation of several different genetic loci, and the population gene frequency method can be, and at times has been, misleading. Thus a very important adjunct to purely genetic studies of humans is the analysis of phenotypes in biochemical rather than morphological terms. Invaluable information can be obtained at all levels—those of composition and metabolic patterns, which represent integrated results of genetic constitution, and those of macromolecular structure, which more nearly define the characteristics of specific genes.

Humans as Biochemical Subjects. It is not an acceptable practice to grind up people for the purpose of extracting enzymes or other interesting substances, and hence there are significant limitations to the most direct approaches to studies in biochemical genetics in humans. However, three primary materials available in quantity from humans are blood, urine, and saliva, and under special circumstances small amounts of various tissues are available through operations and by autopsy. But a major indirect source, which has a great potential for the future, is material from tissue culture. It is possible now to culture many different kinds of human tissue cells in quantity and for indefinite periods of time (603). This provides means for studies of biochemical genetics in cell lines from different hosts as well as in each cell line alone. Without question this is a useful approach, and, although changes in cell characteristics may occur in culture, genetically regulated primary structure and deficiencies of macromolecules should be maintained. This has been demonstrated (539) for human cell cultures from a galactosemic patient, a condition now well characterized as an inherited specific enzyme deficiency in man (see p. 611).

Among the three more readily available biological materials from man—blood, urine, and saliva—it is clear that urinary components are, in general, the least direct products of an individual's gene functions. Normally, urine contains only small amounts of macromolecular substances, and, though its composition can be markedly influenced by diet under reasonably constant conditions, its composition is quite constant and characteristic for one individual. Williams and collaborators (1080) have carried out extensive analyses of urine from individuals to derive the characteristic patterns illustrated in Figure 14.2. (Estimates of some taste sensitivities and saliva compositions are also included.) Patterns 11 and 12, the only two that are similar, are those of identical twins. These characteristics of all individuals were analyzed repeatedly over a considerable period of time, and it is reasonable to consider each pattern as a partial phenotype of each individual. Considering the great heterogeneity of the gene pool in a human population, such variations in composition in materials from individuals within the range of "normal," in terms of concentrations, is broad indeed. This general approach is applicable, especially by simple chromatographic and electrophoretic methods, to analyses of any kind of tissue, and, though it would be extremely difficult to explain any of these pattern differences in terms of actions and interactions of any specific genes, the principle is illustrative of a growing trend. In essence, this trend is toward leaving the genetics out of biochemical genetics— for the moment at least, and especially in studies of man where genetic analysis is difficult. Now, as discussed earlier (p. 268) and subsequently in connection with human hemoglobins (p. 614), the trend is to isolate pure gene products and deduce from their structures the structures of genes. This is clearly a profitable trend in the direction of gene structure but not gene behavior.

Imposed upon the background of "normal" individual variations in amounts of urinary components are numerous situations in which abnormal components appear, or quantities of normal constituents change in great excess. These may result from genetic influences on metabolism of special dietary components, on abnormalities in tissue metabolism, or on renal function. A variety of examples of such characters are described briefly in Table 64 (for further details see 310, 410, 552, 968). The data presented are far from complete, but a number of points of interest are illustrated. Such information shows clearly the general value of urine-composition studies in the delineation of specific genetic disorders. Although complex situations are often encountered in abnormal balances of normal components, studies of unusual urinary components can point quite directly to specific enzyme deficiencies in tissues, as in alcaptonuria. Further, it is likely that many other interesting and useful situations can be found from studies of urinary constituents in relation to dietary components, as in the examples given (asparagus and beets). Such cases, in which mutant gene frequencies are quite high, have a great potential usefulness as genetic markers and as source materials for gene-protein structure studies.

A second point of interest in the data of Table 64 concerns the fact that many of the disorders, so far as they have been defined, resemble very closely many of the metabolic enzyme-deficiency situations observed in organisms other than man (Chapter 6). It may be that total deficiencies such as in some of the nutritional mutants of microorganisms do not survive at all in man, but many are obviously sufficient to result in accumulations of metabolic intermediates. For example, conditions equivalent to galactosuria in bacteria (508) and to orotic-acid aciduria in *Neurospora* (step 4, Figure 7.12, p. 316) are well known, and there are marked similarities between xanthosis and the *ma-1* and *ry* mutant conditions of *Drosophila* (p. 326). Many other parallels can be drawn which indicate similar fundamental bases for disorders found in man and lower forms. It is especially interesting to note, however, the characteristics of mutant phenotypes in a complex animal like man. As noted in the table, several apparently simple situations of enzyme deficiency with accumulations result in mental disturbances. The accumulations of abnormal quantities of normal metabolites may be as important in such cases as the enzyme deficiency itself, as indicated by the interpretations of the various aspects of the human galactosemia phenotype.

A further important point in the data of Table 64 is the uncertainty in the column "Gene." As shown, a majority of the characters listed are probably recessive (R) rather than dominant (D), but in a number of cases this is far from clear-cut, with some intermediate pheno-

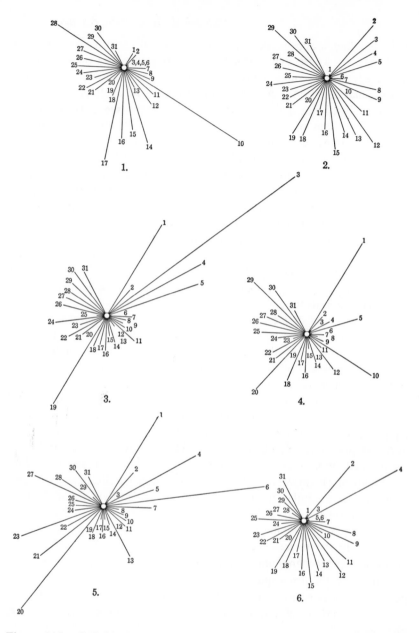

Figure 14.2. Individual patterns for humans showing characteristics of taste sensitivities and concentrations of salivary and urinary constituents. The lengths of the polar coordinates indicate the relative amount of the various constituents for each individual. Notice the similarity of the patterns for individual numbers 11 and 12, who are identical twins.

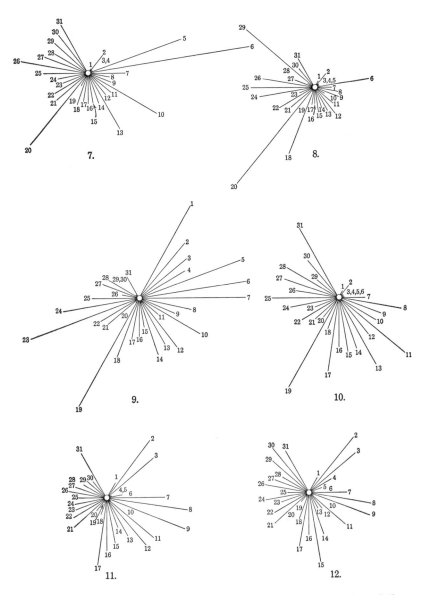

Taste sensitivity: 1, creatinine; 2, sucrose; 3, KCl; 4, NaCl; 5, HCL. Salivary constituents: 6, uric acid, 7, glucose; 8, leucine; 9, valine; 10, citrulline; 11, alanine; 12, lysine; 13, taurine; 14, glycine; 15, serine; 16, glutamic acid; 17, aspartic acid. Urinary constituents: 18, citrate; 19, base Rf 0.28; 20, acid Rf 0.32; 21, gonadotropin; 22, pH; 23, pigment/creatinine; 24, chloride/creatinine; 25, hippuric acid/creatinine; 26, creatinine; 27, taurine; 28, glycine; 29, serine; 30, citrulline 31, alanine. From Williams (1080).

Table 64. Deviations from "Normal" of Certain Urine Components in Inherited Human Disorders or Conditions

Character	Gene	Disorder	Excretion	Phenotypic Effects
Dietary				
Asparagus	R*	Metabolism of component	Methylmercaptan (odor)	Urine odor
Beets	D (exc.)	Metabolism pigment	Pigment (red)	Urine odor
Tissue metabolism				
Phenylketonuria	R	Phenylalanine oxidation	Phenylpyruvic acid and others	Mental deficiency Early death
Alcaptonuria	R, D?	Homogentisic acid degradation	Homogentisic acid	Blackening urine Little effect
Goitrus cretinism	R	Dehalogenase	Iodotyrosines	Physical and mental retardation
Hartnip disease	R	Tryptophan metabolism?	Many amino acids	Pellagra-like neurological disorder
Cystathioninuria	?	Methionine synthesis	Cystathionine	Mental deficiency
α-Aminobutyric aciduria	?	Pyrimidine degradation	α-Aminoisobutyric acid	Little effect
Arginosuccinic aciduria	R?	Urea cycle	Arginosuccinic acid	Mental deficiency
Galactosemia	R	Transferase	Galactose	Growth and mental retardation

Disease	Inheritance[*]	Process/Enzyme	Substance	Effect
Xyloketosuria	R	Sugar metabolism	Xyloketose	Little effect
Maple-sugar disease	R?	Keto acid decarboxylase?	Branched-chain amino acids, odors	Neurological disorders, Early death
Hypophosphatasia	R	Alkaline phosphatase	Ethanolamine phosphate	Bone abnormalities
Hyperoxaluria	R	Glycine oxidation?	Oxalic acid	Oxalic-acid kidney stones, early death
Congenital porphyria	R	Heme synthesis	Porphyrins	Photosensitivity, Skin ulcers
Xanthosis	?	Xanthine oxidase	Low uric acid, high xanthine, and hypoxanthine	Little effect
Orotic aciduria	?	Pyrimidine synthesis	Orotic acid	Anemia
Marfan's syndrome		?	Mucopolysaccharide	Mental deficiency
Renal				
Cystinuria	R?	?	Cystine, lysine, arginine, ornithine	Cystine kidney stones
Cystinosis	R?	?	Many amino acids, sugar	Cystine in tissues, Deformities, early death
Renal glycosuria	?	?	Glucose	Little effect
Diabetis insipidus	D?	?	Water	Little effect, high excretion

Note: R = recessive condition; D = dominant.

typic aspects in apparent heterozygotes. Furthermore, in a number of cases there is already good evidence for the involvement, in different kindreds, of different genetic loci which results in very similar phenotypes. This is, of course, expected from studies of other organisms, but in man resolution by genetic analysis is frequently not feasible, and distinctions must come from thorough and careful analyses of biochemical phenotypes. Urine analyses may be helpful, but studies of activities and structures of macromolecular components of blood and other tissues are perhaps more directly to the point.

In man, blood is a tissue available in quantity from single individuals, and it contains a great variety of components more directly related to immediate gene products than does urine. It is, of course, used, as are secretions such as saliva and gastric juice, in the same general fashion as urine as a source of information on genetically determined individual differences in composition with respect to innumerable low-molecular-weight metabolites. However, of greatest interest in blood are the protein components, many of which are present in sufficient quantities to permit extensive studies on single individuals. In view of the great genetic heterogeneity of man, this is especially important, and it may be expected that deductions concerning gene structure from structures of proteins of individuals will become increasingly important in human genetics. Work with hemoglobin has already contributed a great deal in this direction, and this is discussed in detail in a later section (p. 614).

Besides hemoglobin and specific enzymes, blood contains several major components of genetic interest. Included here are the γ globulins, α_1, α_2, and β globulins, albumins, and fibrinogen. An important method for study of such components is electrophoresis, and an example of a direct separation of these fractions is shown in Figure 14.3. Such a picture can be obtained easily from serum by free boundary or paper electrophoresis. As is well demonstrated, this means only that serum contains groups of similar proteins, and not that each of these is a pure substance. Even so, certain mutations in man result in near deletion of one or another of these major groups. For example, a number of cases have been studied in which the γ globulin fraction is reduced to less than 25% of the normal level (congenital agammaglobulinaemia (410). Since this fraction contains most antibodies, such individuals are very susceptible to bacterial infections (but surprisingly not to viral infections) by the end of the first half year of life when maternal γ globulins have disappeared. Most cases have been in males, and early death from infection is frequent. It is interesting that skin grafts from

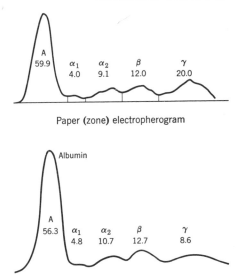

Paper (zone) electropherogram

Standard (boundary) electropherogram

Figure 14.3. Comparison of paper and free boundary electrophoresis of human serum. From Cooper and Mandel (182).

unrelated individuals are accepted by affected individuals, presumably due to lack of immunological response, but grafts of γ globulin-forming tissues (lymph nodes) would be more helpful, if it is possible.

Several cases of albumin deficiency (analbuminaemia) have been studied, and, surprisingly, the clinical consequences of a 2000-fold reduction in albumin are quite mild. Affected individuals have up to twice the normal levels of α, β, and γ globulins.

In Wilson's disease, which is apparently due to a rare recessive gene, a 75% reduction of the serum copper protein, caeruloplasmin, has been noted. This substance migrates electrophoretically as an α_2 globulin and is concerned some way in copper utilization. Progressive degeneration in brain tissue and cirrhosis of the liver occur in this disease. It has not been established whether the genetic defect acts directly on caeruloplasmin synthesis or on other copper-binding proteins.

Fibrinogen is a major protein component of blood plasma which is responsible for clotting. This process of clotting is a complex one involving at least eight different proteins in unknown obligatory sequences of reactions. Primarily, fibrinogen (mol. wt. 330,000) is acted upon by thrombin to split off peptides of a size in the range of 4 to 8000. The residual protein then polymerizes to the insoluble fibrin (mol. wt. 5×10^6), which forms the matrix of a clot. A genetically determined abnormality in any of the components of the system can result in a hemorrhagic disease, and several of these are listed in Table 65. Each of these coagulation disorders is derived from malfunction of a different gene and can be the cause of serious bleeding and death from minor cuts in affected individuals. Fibrinogen is the major protein component, comparable in quantity to those shown in Figure 14.3, and it is removed in the preparation of serum. The condition afibrinogenemia is evidently due to a rare recessive gene, but there is some evidence for reduction of the amount of fibrinogen in heterozygotes. The reduction is very great in the disease. The two types of hemophilia, A and B, come from rare sex-linked genes, and recognition of the B type as distinct from A is relatively recent. Approximately 12% of the cases previously classed just as hemophilia are of type B. The specific protein deficiencies here and those of Owren's and Hageman's diseases, which come from abnormal functions of different autosomal genes, concern blood components normally present in small quantities, and the deficiencies would not be noted in the electrophoresis pattern illustrated in Figure 14.3. However, in all of these cases of clotting deficiency, the genetic defects appear to be directly concerned with synthesis of specific functional proteins. Defectiveness is relative in several kindreds, and altered proteins with partial function are likely to be found.

The foregoing examples represent several kinds of gene-determined alterations in protein components of human blood and illustrate the general usefulness of this approach in both genetical and biochemical studies of man. However, as pointed out previously, many of these

Table 65. Blood Diseases Associated with Protein Abnormalities

Protein Abnormality	Disease
Fibrinogen	Afibrinogenemia
Antigemophilic globulin	Hemophilia A
Christmas factor	Hemophilia B
Factor V	Owren's disease
Factor VII	Hageman's disease

major components of blood are heterogeneous mixtures, and, for a more detailed understanding of gene-protein relations, much better resolution and purification is essential. Many methods for obtaining pure proteins are available, but for purposes of surveying large numbers of individual blood samples, refinements of electrophoresis, such as the use of starch gel (919) and acrylamide polymers (156), and immunological methods have proven very productive. At the same time, it should be kept in mind that both of these techniques are limited. Differences may be significant, but similarities are not. For example, in electrophoresis primary polypeptide chains which differ in only one amino-acid residue will usually separate if the alternatives have different charges (such as leucine and glutamic acid, or leucine and lysine), but they probably will not separate if the alternatives have essentially the same charge contribution (such as leucine and valine, or glutamic acid and aspartic acid). In immunological methods only alternative substitutions which change the conformation of antigen-antibody attachment sites are detectible, and thus specificity can be remarkably good or very poor. In any case, application of these methods to studies of the blood components, the haptoglobins, the blood-group substances, β globulins, and hemoglobins has yielded important information, and some details and interpretations are presented in the following section.

Haptoglobins and Blood-Group Substances. All tissues contain mucoid substances made up of conjugates of polypeptides and carbohydrates, and, though it is well established that these are of outstanding biological importance, they are not well defined chemically. This is primarily due to difficulties in purification as well as to their complex structures. Among the many kinds of such substances that exist, the haptoglobins of human blood and the blood-group substances of blood and other body fluids have received considerable attention in genetics. The haptoglobins are mucopolysaccharides which migrate electrophoretically in the α_2 region (Figure 14.3), and they form stable complexes with hemoglobin. These complexes migrate without dissociation, and they will function as peroxidases. The blood-group substances are glycoproteins which exist attached in red-cell membranes and also in the free form in various body fluids such as saliva. The richest source for isolation is the fluid of ovarian cysts. As discussed later, there exist many genetically determined blood-group substances (as in the familiar ABO system), but, so far as the information goes, these are all closely related in chemical composition. There is an interesting relation, also, to haptoglobin composition, as shown in the generalized data in Figure 14.4.   Here a major difference is in the proportion of carbohydrate to polypeptide, but the blood-group substances are much larger, having molecular weights in the range 3×10^5 to 3×10^6 compared to the haptoglobins at 8×10^4 to 1.6×10^5.   Furthermore, the neuraminic-acid portion of the haptoglobin apparently exists in a side chain to the central structure, and it is an interesting point that one blood-group substance with a similar side chain has been isolated (817). The polypeptide portion of blood-group substances contains some 17 amino acids, with a very large excess of threonine and serine (about ten- and fivefold, respectively).   Based on the large number of carbohydrate end groups per molecule and partial hydrolysis studies, Morgan (711) proposed a general structure for blood-group substances consisting of polypeptide chains linked together by carbohydrate chains. Carbohydrate chains linked to peptide at one end only are also proposed to provide specific end groups which confer immunological specificity. This general picture is reproduced in Figure 14.5.   Although supporting evidence is lacking, it is possible that haptoglobins and other mucopolysaccharides possess some similar structures, with fewer or shorter cross-linking carbohydrate chains in the haptoglobins.

The foregoing presentation of somewhat limited facts and rather excessive speculation is designed in an attempt at unification of the uniquely and fundamentally important problem of genetic designation of structure that exists in relation to these classes of compounds.   As will be shown, genetic differences may appear to be reflected solely in

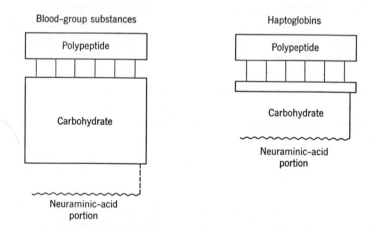

Figure 14.4.   A comparison of haptoglobins and blood-group substances with respect to carbohydrate components.

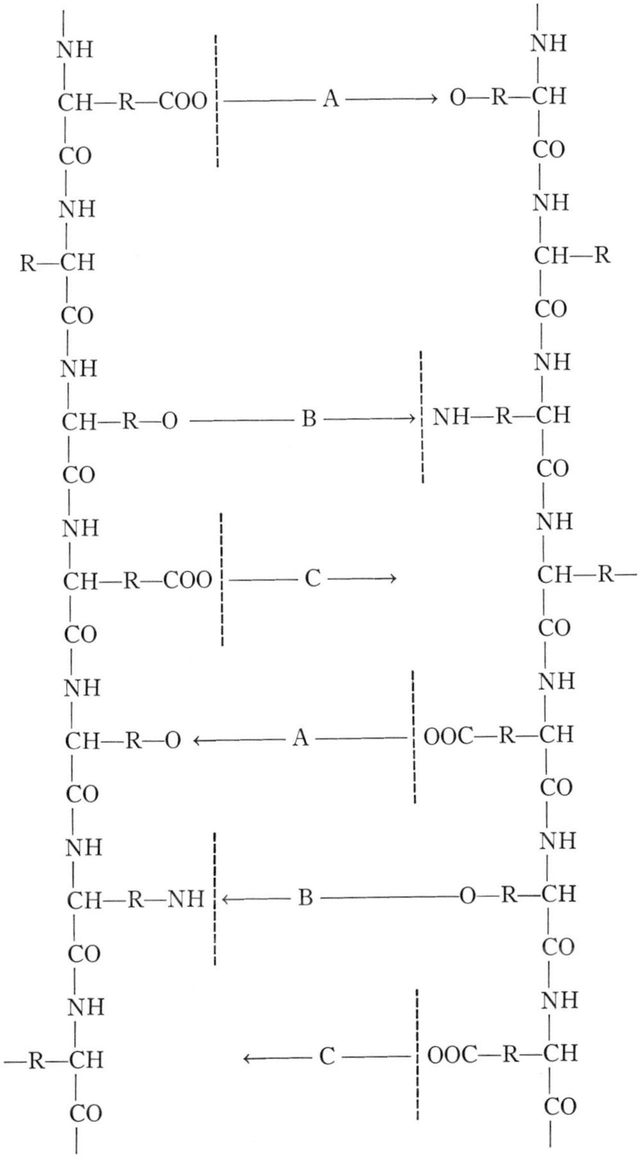

```
    |                                        |
    NH                                       NH
    |                                        |
    CH—R—COO ⌐———— A ———→  O—R—CH
    |         ⌐                              |
    CO        ⌐                              CO
    |                                        |
    NH                                       NH
    |                                        |
R—CH                                         CH—R
    |                                        |
    CO                                       CO
    |                                        |
    NH                         ⌐             NH
    |                          ⌐             |
    CH—R—O ———— B ———→ ⌐ NH—R—CH
    |                          ⌐             |
    CO                         ⌐             CO
    |                                        |
    NH           ⌐                           NH
    |            ⌐                           |
    CH—R—COO ⌐——— C ———→        CH—R—
    |            ⌐                           |
    CO           ⌐                           CO
    |                                        |
    NH                         ⌐             NH
    |                          ⌐             |
    CH—R—O ←——— A ——— ⌐ OOC—R—CH
    |                          ⌐             |
    CO                         ⌐             CO
    |                                        |
    NH           ⌐                           NH
    |            ⌐                           |
    CH—R—NH ⌐←——— B ————— O—R—CH
    |            ⌐                           |
    CO           ⌐                           CO
    |                                        |
    NH                         ⌐             NH
    |                          ⌐             |
—R—CH           ←——— C ——— ⌐ OOC—R—CH
    |                          ⌐             |
    CO                         ⌐             CO
    |                                        |
```

Figure 14.5. A possible macromolecular structure for the blood-group substances. A, B, and C represent different polysaccharide chains. From Harris (410).

carbohydrate portions of these molecules, and yet we have, so far, evidence only for direct genetic designation of primary sequences of amino acids in polypeptide chains. This should be kept in mind during the following discussion of the genetical and related biochemical aspects of the problem.

Haptoglobin inheritance in man provides a good illustration of the complexities of the situation in gene-glycoprotein relations. Early work described two alleles, Hp^1 and Hp^2, which give rise to different hapto-globin patterns on electrophoresis (920, 181, 968). Hp^1 yields a single component by starch-gel electrophoresis but Hp^2 yields at least eleven components. Material from a heterozygous individual (Hp^1/Hp^2) yields at least six components most of which are different from those from Hp^1 or Hp^2. Evidence that the haptoglobins are made up of unlike subunits was obtained by electrophoresis in urea (to break H bonds) and mercaptoethanol (to break S-S bonds), and it was estab-lished that heterozygote haptoglobin which contains hybrid molecules can be dissociated to yield two polypeptide components, one like that from Hp^1 and one like that from Hp^2. This situation is well illustrated, as shown in Figure 14.6, by comparison of haptoglobins from two suballele types, Hp^{1F} and Hp^{1S} (180a). As shown, α-chain compo-nents from F and S are unlike and the F/S heterozygote material yields both. The more similar common zone material contains the carbo-

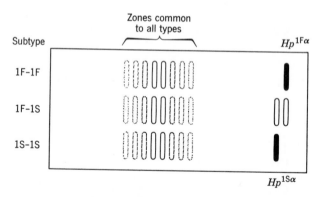

Figure 14.6. Starch-gel electrophoresis of haptoglobins types Hp^{1F} and Hp^{1S} in the presence of urea and mercaptoethanol. Differences in α-polypeptide chains are shown at the right, and similarities in other products derived from the hapto-globin dissociation are shown near the center. (The relative darkness of the common zones is reduced photographically to permit reproduction of the photo-graph. From Connell, Dixon, and Smithies (180a).

hydrate material and probably also polypeptide but no data are yet available on these substances. Some further classification has come from amino-acid composition and sequence studies of α-chains (920a). Those of Hp^{1F} and Hp^{1S} differ only by a single residue, a substitution of an acidic amino acid by a lysine. The Hp^2 α-chain has a great deal of amino-acid sequence in common with the Hp^1 allele types but it has a much higher molecular weight. Smithies and collaborators (920a) have suggested that the Hp^2-type α-chain is derived from duplication at the gene level. Thus the haptoglobin inheritance patterns present a very interesting situation for more extended analyses and suggest that the genetic differences result in α-chain alterations. These changes may account for hybrid molecule formation but it is by no means clear how they can control carbohydrate composition or other alterations that may exist in undissociated haptoglobins.

The genetic picture with respect to the blood-group substances is far more complex than that for haptoglobins, perhaps due only to application of immunological methods of detection of differences. The major known genetically controlled chemical changes rest on carbohydrate-composition differences. There are at least nine blood-group systems, each represented by at least two alleles and some by many. The first, and perhaps best known, is that which carries the alleles A^1, A^2, A^3, B, and O. These were first defined (with A^1, A^2, and A^3 as A) on the basis of agglutination of blood cells, which can result from mixing of blood cells and serum from different individuals. Results from A, B, and O combinations are illustrated in Figure 14.7, and, as shown, when blood-group substances A, B, or both are present, corresponding antigens are not present in sera of the same individuals. These, of course, are the observations which provided a basis for practical limitations on transfusions in medical practice.

In addition to delineation of blood groups by the foregoing method, definition by more general means is possible, since the group substances elicit specific antibody formation when injected into other animals. On this basis, as well as by agglutination and the use of refined techniques, the A type has been subdivided into three subgroups, A_1, A_2, and A_3, and the O type has been found to carry a group substance H, particularly in the saliva in the presence of the secretor gene Se. The latter determines the presence or absence of the ABO and Lewis (L_e^a, L_e^b) group substances in body fluids other than blood. Thus there is an intimate relation among the three genetic loci represented by the ABO series, the secretor gene, and the L_e gene. In addition to these, the L series (Landsteiner) has six known alleles including the M and N

Group	Antigens in red blood cells	Antibody present in serum	Reaction to serum (listed to left) of red blood cells from group			
			O	A	B	AB
O	O	Anti-A Anti-B				
A	A	Anti-B				
B	B	Anti-A				
AB	AB	——				

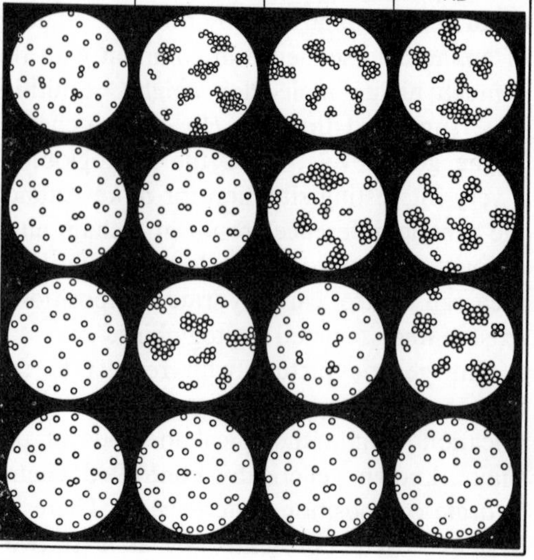

Figure 14.7.   Reactions of red blood cells of O, A, B, and AB individuals to antibodies, anti-A, and anti-B.   After *Biology, Its Principles and Duplications,* by Garrett Hardin.   W. H. Freeman Co., San Francisco, 1961.

antigens, while the *R* series has at least eight alleles.   The latter, the familiar Rh group, is considered by some to involve three closely linked genes.   Other less well-known series are *P, K, Lu, Fy,* and *Jk* (968), all of which have been at least partly defined on the basis of induction of specific and distinguishable antibodies.   It has been mentioned previously that man is genetically very heterogeneous, and the blood-group diversity provides a very convincing proof of the point. As an example, 475 Londoners were blood-group tested, using only 15 of the more than 25 known blood-group substance antibodies.   Of these, 211 were found to be unique individuals, and at the other extreme, 10 individuals were of one composite type.   Clearly, complete unique genetic identification of each individual from blood-type antigens alone should be possible with only a relatively few more test reagents.

To return now to the gene-antigen structure problem, studies of the A, B, H, and L_e antigen specificities have yielded a good deal of important information.   Morgan (712) and Kabat (502) have made use of the facts that specific carbohydrate groups are responsible for antigenic specificity and that simple sugars will interfere with antigen-antibody reactions. In a like manner, simple sugars will inhibit the action of specific glycolytic enzymes obtainable from various microorganisms. Some pertinent data obtained from the enzyme inhibition are shown in Table 66. From this it is clear that especially significant units in specificities are N-acetyl-D-galactosamine for A substance, galactose for B substance, and fucose for H substance. Furthermore, the appropriate enzymes release galactose and fucose, respectively, from B and H substances. Additional information gained from studies of effects of oligosaccharides from various sources, including those from partial hydrolysis of the blood-group substances themselves, indicates that A

Table 66. Percentage Substrate Remaining Unchanged When the *T. foetus* Enzyme Acts on the A, B, and H Blood-Group Substances in the Presence of Different Sugars *

	Percentage Unchanged Substrate		
Sugar Added	A Substance	B Substance	H Substance
Control (no sugar added)	3	1	1
L-Fucose	3	1	50–100
D-Fucose	3	1	1
L-Galactose	...	1	3
D-Galactose	6	50–100	1
N-Acetyl D-glucosamine	1	1	1
N-Acetyl D-galactosamine	50–100	6	1
D-Glucosamine	3	1	1
D-Galactosamine	3	6	50–100
α-Methyl L-fucopyranoside	3	1	3
β-Methyl L-fucopyranoside	...	...	1
α-Methyl L-fucofuranoside	3	1	1
β-Methyl L-fucofuranoside	3	1	1
α-Methyl D-galactopyranoside	3	50–100	1
β-Methyl D-galactopyranoside	3	50–100	1
β-Methyl D-galactofuranoside	...	3	1
Lactose	...	50–100	1
Melibiose	...	50–100	1
Methyl-N-acetyl-galactosaminide	50–100	...	...

* (410).

specificity comes from the terminal structure O-(N-acetyl-α-D-galac-tosaminoyl) $(1 \rightarrow 3)$ galactose ––. B specificity is derived by virtue of an O-α-D-galactosyl-$(1 \rightarrow 3)$-D galactose end group and H by a terminal O-α-L-fucosyl group. Further details on H substance are not available, but L_e^a substance evidently carries a branched-chain end group with fucose and galactose attached to acetylglucosamine. L_e^b substance is similar but has an additional fucose linked to the next monosaccharide unit beyond the acetylglucosamine. One other point of importance in connection with the end-group specificities is the virtual demonstration that in the AB blood type the A and B specificities are both part of the same macromolecule. This conclusion is derived from the fact that A and B substances are separable from mixtures by antibody precipitations but not when derived from an AB genotype. In AH combinations both AH and H substances were observed. This evidence supports the conclusion that the heterozygotes yield hybrid macromolecules (see Chapter 8).

A second valuable approach to the problem of specific structures of blood-group substances described by Watkins (1054) represents an extension of the specific enzyme-function method. B substance was treated with an enzyme to yield galactose and a material with H substance specificity. Successive treatments with two other specific enzymes first yield fucose + L_e^a active substance, and then a product with a strong pneumococcus type XIV activity. All ABO blood-group substances share the last characteristic to some extent, but it is increased with partial acid hydrolysis. These data are in agreement with other findings discussed, but they also demonstrate the potential for enzymatic interconversion of the different type substances.

It should be clear from the foregoing discussion concerning mucopolysaccharide structure and its genetic designation that carbohydrate units, rather than amino-acid residues, occupy the center of attention. At present there are no grounds for expecting direct genetic designation of carbohydrate sequences, and it is of interest to inquire whether the existing information can be accounted for on the basis of designation of amino-acid sequences in the polypeptide portions of the mucoid substances. This, by rank speculation, can be done as indicated in Figure 14.8. Here it is assumed that the determining alleles produce polypeptides of relatively small size, and many (ten in the diagram) are needed to yield the ultimate macromolecule. From these subunits, attached polysaccharide chains for bridging and end groups are produced, and it is assumed that the altered amino-acid sequence in B polypeptide influences the nature of the product either through spe-

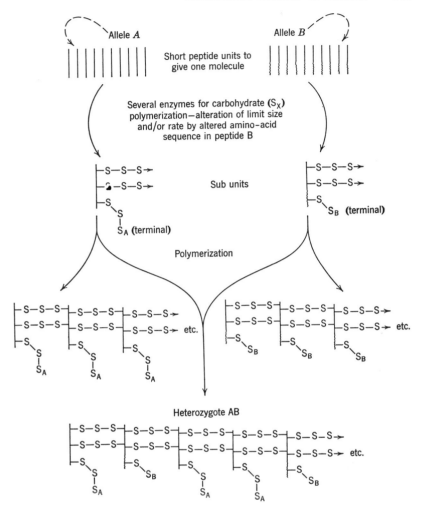

Figure 14.8. A speculation concerning the genetic control of mucopolysaccharide synthesis.

cificity or rate influences. Several enzymes must be involved here, and the situation has similarities to that of specific glucosylation of phage DNA (p. 116). In the next stage, polymerization to some limiting size is assumed, and in a heterozygote both types of subunits (and resulting end groups) are available for production of a random mixture containing both kinds of chains. Many variations on the scheme are possible but are not worth further discussion here, since the whole postulate can be easily evaluated by some careful work on size and sequences of polypeptide chains. The point here is that it is possible to account for different sugar end-group and hybrid-molecule formation on the basis of primary sequence in polypeptide chains.

Enzyme Deficiencies

Many examples of enzyme deficiency or alteration resulting from mutation in microorganisms have been discussed in earlier chapters, and all the same principles should be applicable in man. In man, however, total enzyme deficiency more often may result in lethality, and we must study what exists rather than select for extreme cases as can be done in experimental organisms. Nevertheless, there are now a good many examples of gene-determined enzyme deficiencies in humans. A number of these are noted in Table 64, page 594, and some further details of some of these and other cases are of special interest in human biochemical genetics (410, 552).

Aromatic Amino Acids. In contrast to microorganisms and plants which are able to produce aromatic amino acids, man requires dietary tryptophan and phenylalanine. He is able to convert tryptophan to niacin (Figure 7.10c, p. 310), and mutations affecting this system in man have not been described. Several mutations, however, are known which result in enzyme deficiencies concerned in the metabolism of phenylalanine. A generalized summary is shown in Figure 14.9. The reaction block shown at 1 indicates an enzyme deficient in liver and associated with the disease phenylketonuria. The oxidative reaction is a complex one involving at least two enzymes, one of which is quite unstable. Unfortunately, this is the one deficient in the disease, and it has not been obtained in a purified form, nor can it be proven that it is totally deficient in affected individuals. Phenylalanine accumulates in the blood and phenylpyruvic acid in the cerebral spinal fluid. The latter compound is also excreted in urine in large quantities along with the further reaction products indicated, and o-hydroxyphenyllactic, indolelactic, and indolacetic acids. Heterozygous individuals also exhibit these characteristics, but to a much lesser degree, and they do not show the extreme mental deficiencies characteristic of homozygotes.

Since a marked improvement of the disease can be achieved by dietary restriction of phenylalanine intake, it is probable that the patho-

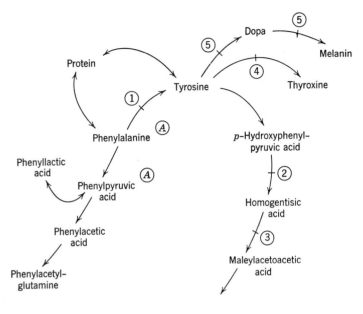

CITRIC–ACID CYCLE

Figure 14.9. Part of the pattern of the metabolism of phenylalanine and tyrosine in man.

logical effects are due to the accumulations, or other metabolic imbalances, rather than the enzyme deficiency itself.

Reaction 2 in Figure 14.9 presumably is defective in the disease tyrosinosis, but only one case has been described. The patient excreted in the urine large amounts of tyrosine and *p*-hydroxyphenylpyruvic acid. Individuals with a deficiency at step 3 are, however, more common, and affected individuals are alcaptonurics. There is evidence for cause by a recessive (more common) and a dominant gene. Individuals with alcaptonuria excrete large quantities (several grams/day) of homogentisic acid in the urine, but the level in blood is low. The urine, on exposure to air, turns black due to oxidation and polymerization of the diphenol, but affected individuals are not usually seriously affected. With advancing age, darkening of cartilagenous tissues occurs and arthritic symptoms may appear. It was in connection with this disease that Garrod (313) first proposed a direct gene-enzyme relationship. Gross (372) in 1914 provided experimental support for the relationship. Le Du and collaborators (544) more recently demonstrated that liver tissue from alcaptonurics is either totally deficient or extremely low in the enzyme homogentisic-acid oxidase but is essentially normal with respect to other enzymes of this reaction series.

The other specific reaction deficiencies indicated in Figure 14.9 concern several probable enzyme deficiencies (at least five) in iodine metabolism (step 4) and the deficiency of tyrosinase in human albinism (step 5).

Carbohydrate Metabolism. There are now several well-established inherited conditions in man which result in gross disturbances in glycogen deposition (410, 552). Reaction patterns involved are shown in Figure 14.10. In McArdle's disease (step 1), an extreme deficiency of phosphorylase has been noted. The disease is biochemically complex, with a large accumulation of glycogen in the liver evidently due to irreversible synthesis through an alternate uridine-diphosphoglucose system. Accumulation in muscle is relatively small, but muscle dysfunction is marked. Muscle degeneration is also a result of low levels of the debranching enzyme in liver and muscle at step 2. Organs become excessively large and contain large amounts of an abnormal glycogen with short side chains.

General organ enlargement and glycogen deposition is also characteristic of an apparent deficiency of brancher enzyme (step 3). In

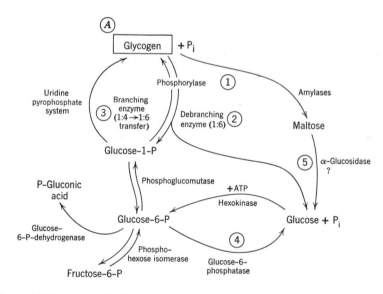

Figure 14.10. A metabolic pattern concerned with glycogen formation in man.

this case the glycogen is abnormal with long chains and few branch points. Though it seems likely that there is not a total enzyme deficiency, the condition is serious, resulting in liver cirrhosis and early death. The inherited condition represented by an extreme enzyme deficiency in glucose-6-phosphatase (step 4) has received considerable attention, and extreme and moderate forms likely represent different alleles. In this disease there is an excessive enlargement of the liver and kidneys with a large accumulation of normal glycogen. It is interesting to note that glucose-6-phosphatase is normally low in muscle, and thus muscle tissue is essentially normal with this enzyme deficiency. The enzymatic dysfunction represented at step 5 is tentatively associated with a shortage of α-glucosidase (428), though the situation is complex. In this case there is a great accumulation of normal glycogen with great enlargement and early failure of the heart.

In all of the foregoing conditions it appears likely that specific enzyme deficiencies are the fundamental bases for pathological, inherited disorders, even though phenotypic effects are complex. Generalized mechanisms for such secondary complications are discussed in earlier chapters, and an interesting example in man has been discussed by Sidbury (906). This concerns the disease galactosemia which is characterized by hepatomegaly, neonatal jaundice, cirrhosis, cataracts, mental retardation, growth failure, albuminuria, galactosuria, aminoaciduria, a bleeding tendency, and an abnormal galactose-tolerance curve, all resulting from a rare recessive mutation.

Though there is a considerable variation in severity of the symptoms, they all require explanation, and the question here is concerned with whether it is possible to do so on the basis of the deficiency of a single enzyme. Probably it is, as is evident in the following discussion.

An essential portion of the metabolic pattern of normal galactose metabolism is shown in Figure 14.11, and, as indicated, there is a deficiency of the enzyme uridyl transferase in galactosemia (24, 505, 506, 587). These reactions are probably most prominent in liver but occur in other tissues, and studies of the enzyme deficiency in erythrocytes have been especially useful. The disease is severe with a normal dietary intake of galactose, and thus it appears early in infancy where nearly half of the caloric intake is from the lactose of milk. At the same time, the fact that very early limitation of dietary galactose greatly alleviates disease symptoms suggests that it is not the enzyme deficiency itself which causes greatest difficulties. Rather, it is the result of metabolic imbalance resulting from the genetic block, and a large part of it may be due to accumulation of galactose-1-phosphate.

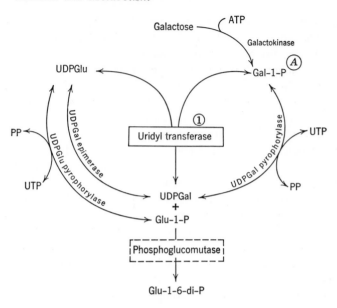

Figure 14.11.   Reactions related to galactose metabolism in man.

This accumulation in various tissues has been observed repeatedly, and, as shown in Figure 14.11, this is a relative rate-balance situation, since at least one alternative metabolic pathway (to step 1) exists normally. As to the mechanisms by which galactose-1-phosphate may cause the extensive pleiotropy observed, Sidbury (906) has presented some interesting evidence and suggestions.   First, based on the mechanism of action of phosphoglucomutase as evaluated by Najjar (731), it is clear that Gal-1-P is a multiple substrate inhibitor acting in the metabolism of Glu-1-P.   This is shown at the top in Figure 14.12, and though the reaction equilibrium is much in favor of Gal-1-P, the mechanisms for the galactose and glucose-phosphate interconversions are the same, with diphosphate sugars and dephosphorylated enzyme as intermediates.   This provides a basis for inhibition.   That is, continuously accumulated Gal-1-P reacts with enzyme phosphate (E-P) to yield enzyme (E), which in turn is reconverted to E-P by reaction with glucose diphosphate.   With depletion of the diphosphate, the Glu-1-P → Glu-6-P interconversion is limited.   As shown at the bottom in Figure 14.12, phosphoglucomutase participates in a number of other reactions of this type, and many of the galactosemia symptoms can be rationalized on this general basis—the inhibition of function of phosphoglucomutase by Gal-1-phosphate.

Competitive inhibition of phosphoglucomutase

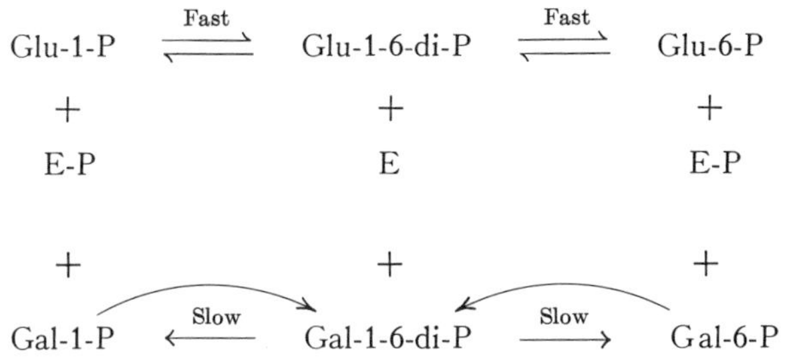

$$\text{Glu-1-P} \;\underset{\longleftarrow}{\overset{\text{Fast}}{\longrightarrow}}\; \text{Glu-1-6-di-P} \;\underset{\longleftarrow}{\overset{\text{Fast}}{\longrightarrow}}\; \text{Glu-6-P}$$

$$+ \qquad\qquad + \qquad\qquad +$$

$$\text{E-P} \qquad\qquad \text{E} \qquad\qquad \text{E-P}$$

$$+ \qquad\qquad + \qquad\qquad +$$

$$\text{Gal-1-P} \;\xleftarrow{\;\;\text{Slow}\;\;}\; \text{Gal-1-6-di-P} \;\xrightarrow{\;\;\text{Slow}\;\;}\; \text{Gal-6-P}$$

Some reactions involving phosphoglucomutase

1. Glucose-1-P → Glucose-6-P
2. Glucosamine-1-P → Glucosamine-6-P
3. Acetylglucosamine-6-P → Acetylglucosamine-1-P
4. Mannose-1-P → Mannose-6-P
5. Ribose-1-P → Ribose-5-P

Figure 14.12. Reaction diagrams indicating some expected effects of the accumulation of galactose-1-phosphate in man.

Jaundice, an early symptom, is caused by excess free bilirubin in serum, and it is normally removed as a glucuronide. It is postulated that phosphoglucomutase inhibition limits production of uridine-PP-glucuronic acid which should normally come from glucose through phosphorylation and the reverse of reaction 1 in Figure 14.12.

In the similar vein, it is expected that mental retardation may result from limitation of carbohydrate components needed for production of cerebrosides in nervous tissue (reactions 2 and 3, and again a limitation of available Glu-1-P). Similarly, cataract may be due to limitations in substrates (reactions 2, 3, and 4) for formation of normal, soluble and properly transparent, glyco-proteins in the lense.

Hypoglycemia can be accounted for in several ways, among them a limitation in reaction 1 (Figure 14.12), which would normally be followed by glucose production by action of glucosephosphatase. Reaction 5 indicates a situation in which inhibition would be expected to cause accumulation of another substance (ribose-1-phosphate) which would augment the effects of galactose-1-phosphate on phosphoglucomutase.

It is quite clear that there are many more possibilities than those outlined for metabolic interaction responsible for the pleiotropic effects of galactosemia. Nevertheless, this represents a type of analysis that is likely to become increasingly important. In human genetic diseases the prospects for restoration of production of lost enzymes are, at best, poor, but prospects are reasonably good for evolvement of corrective measures at the metabolic level in the near future. Thus a thorough understanding of metabolic results of mutation can become very beneficial to affected individuals and helpful for detection of recessive heterozygotes who can potentially produce more affected individuals.

Protein Structure and Genes in Man

Frequently it is difficult to obtain in man a sufficient amount of data to establish conclusively the mode of inheritance of some particular characteristic, and fine structure analysis such as has been described for microorganisms is quite out of the question. At the same time, man represents an enormous pool of mutants, a large proportion of which have phenotypes that are within the range of "normality." These and many inherited pathological conditions tend to be preserved in man, through medical attention to individuals and mutation, and they all provide valuable material for study even with the limited prospects for genetic analysis. If it is assumed, from evidence obtained using other organisms, that primary amino-acid sequences in polypeptides are a direct reflection of gene structure, then it is reasonable to use the mutant pool of man in reverse and see what can be deduced about gene structure from protein structure. Many examples discussed in this chapter present excellent material for work in this direction, but the one which has been exploited the most extensively is that concerned with hemoglobin, and some pertinent observations are presented in the following paragraphs.

The red blood cells of a "normal" man characteristically contain one of two kinds of hemoglobin: hemoglobin F, which is the major constituent of fetal blood, and hemoglobin A, which is the major constituent from early infancy through adulthood. These are easily distinguishable from each other by electrophoresis and from several other components that are present in very small quantities. Considered in general, hemoglobin has a molecular weight of 66,000 and contains four heme groups which are exchangeable at neutral pH values and removable at acidic pH values. Reconstitution from globin and protohemin can be effected to yield native hemoglobin. The protein itself is also dissociable into four subunits representing two different kinds of primary polypeptide chains.

Hemoglobin A yields the two designated as α and β chains and thus by present conventional nomenclature it has the constitution $\alpha_2{}^A\beta_2{}^A$ (1033). Hemoglobin F has the same α chain, but it is associated with a different polypeptide designated γ, and thus it has the constitution $\alpha_2{}^A\gamma_2{}^F$. Dissociation can be effected at pH 3 to 6 or 11 by treatment with IN NaCl or high concentrations of urea (656), and, at high pH values at least, it is stepwise from four units to two unlike units to single units (1034).

The different kinds of polypeptide primary chains are separable by ion-exchange chromatography, and sequence analyses being carried out in several laboratories are approaching completion (100, 529, 883). Existing information is summarized in Table 67 (517). As shown, α chains carry 141 amino-acid residues, β chains 146, and γ chains are of similar size. Though α and β chains have the same residues at 66 positions as they are aligned, and γ chains also have sequence similarities, they all have major differences, and it may reasonably be considered that each is derived from the action of a different genetic locus. This conclusion is supported by mutant studies as discussed later. At the same time, there exists the possibility that a functional hemoglobin might be produced from a single-chain type (e.g., α_4, β_4, γ_4) if mutational changes can provide compatible structures for polymerization in this way. This situation can exist, as again shown by studies of mutants. Thus the full potential for combinations in hemoglobin structure in an individual is represented by all possible combinations of products, from two alleles each, of at least three genetic loci.

Normally, genes giving rise to β (adult) and γ (fetal) chains do not function at the same time, and, so far, subunits appear in pairs for the most part, but these are not necessary limitations. Actual limitations must lie in specific sequences which permit appropriate secondary and tertiary structures compatible with formation of a functional (quaternary) four-unit system. As shown in Figure 14.13 (781), the four basic units must have the proper shape, size, and mutual combining sites for each other and for heme in order to form a functional hemoglobin molecule.

From the foregoing discussion of hemoglobin inheritance alone, from the standpoint of primary amino-acid sequences in polypeptide chains, it is evident that the mutant pool in man should contain many variant primary structures, and this is indeed the case. The method most gen-

Table 67. Amino-Acid Sequences of Hemoglobin Chains (α, β, γ) *

α	β	γ	α	β	γ
Val	Val	Gly	40 Lys	Glu	?
His	His	His	Thr	40 Arg	Arg
Leu	Leu	Phe	Tyr	Phe	Phe
Ser	Thr	Thr	Phe	Phe	
Pro	5 Pro	Glu	Pro	Glu	
5 Ala	Glu	Glu	45 His	Ser	
Asp	Glu	Asp	Phe	45 Phe	
Lys	Lys	Lys		Gly	
Thr	Ser	Ala	Asp	Asp	
Asp	10 Ala	Thr	Leu	Leu	
10 Val	Val	Ileu	Ser	Ser	
Lys	Thr	Thr	50 His	50 Thr	
Ala	Ala	Ser		Pro	
Ala	Leu	Ileu		Asp	
Try	15 Try	Try		Ala	
15 Gly	Gly	Gly		Val	
Lys	Lys	Lys		55 Met	
Val	Val	Val	Gly	Gly	Gly
Gly	Asp	Asp (NH₂)	Ser	Asp	Asp (NH₂)
Ala			Ala	Pro	Pro
20 His			Glu	Lys	Lys
Ala	20 Val	Val	55 Val	60 Val	Val
Gly	Asp	Glu	Lys	Lys	Lys
Glu	Glu	Asp	Gly	Ala	Ala
Tyr	Val	Ala	His	His	His
25 Gly	Gly	Gly	Gly	Gly	Gly
Ala	25 Gly	Gly	60 Lys	65 Lys	Lys
Glu	Glu	Glu	Lys	Lys	Lys
Ala	Ala	Thr	Val	Val	Val
Leu	Leu	Leu	Ala	Leu	Leu
30 Glu	Gly	Gly	Asp	Gly	Thr
Arg	30 Arg	Arg	65 Ala	70 Ala	Ser
Met	Leu	Leu	Leu	Phe	Leu
Phe	Leu	Leu	Thr	Ser	Gly
Leu	Val	Val	Asp	Asp	Asp
35 Ser	Val	Val	Ala	Gly	Ala
Phe	35 Tyr	Tyr	70 Val	75 Leu	Ileu
Pro	Pro	Pro	Ala	Ala	
Thr	Try	Try	His	His	
Thr	Thr	?	Val	Leu	

α	β	γ	α	β	γ
Asp	Asp		Thr	Val	
75 Asp	80 Asp		Leu	Leu	
Met	Leu		110 Ala	115 Ala	
Pro	Lys		Ala	His	
Asp	Gly		His	His	
Ala	Thr		Leu	Phe	
80 Leu	85 Phe		Pro	Gly	
Ser	Ala		115 Ala	120 Lys	
Ala	Thr		Glu	Glu	
Leu	Leu		Phe	Phe	
Ser	Ser		Thr	Thr	
85 Asp	90 Glu		Pro	Pro	
Leu	Leu		120 Ala	125 Pro	
His	His		Val	Val	
Ala	Cys		His	Glu	
His	Asp		Ala	Ala	
90 Lys	95 Lys		Ser	Ala	
Leu	Leu		125 Leu	130 Tyr	
Arg	His		Asp	Glu	
Val	Val		Lys	Lys	
Asp	Asp		Phe	Val	Met
95 Pro	100 Pro		Leu	Val	Val
Val	Glu		130 Ala	135 Ala	Thr
Asp	Asp		Ser	Gly	Gly
Phe	Phe		Val	Val	Val
Lys	Arg		Ser	Ala	
100 Leu	105 Leu		Thr	Asp	
Leu	Leu		135 Val	140 Ala	
Ser	Gly		Leu	Leu	
His	Asp		Thr	Ala	
Cys	Val		Ser	His	
105 Leu	110 Leu		Lys	Lys	Arg
Leu	Val		140 Tyr	145 Tyr	Tyr
Val	Cys		Arg	His	His

* From Keil (517).

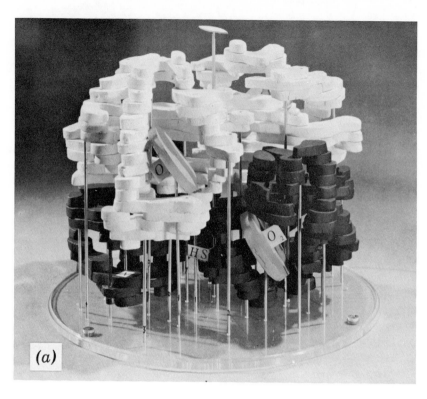

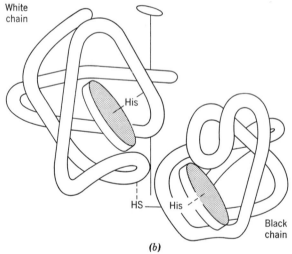

Figure 14.13. Models of the α and β polypeptide chains of human hemoglobin. From Perutz et al. (781). (a) Model with four units. Heme groups are indicated by the gray disks (two visible). (b) Chain configurations of the two subunits.

erally applied as a primary criterion for hemoglobin difference has been that of electrophoresis. Although this is convenient for handling large numbers of samples, it should be emphasized at the outset that this will not usually reveal mutational amino-acid substitutions involving residues that have similar charges. Other rapid methods for detection of differences are needed. In any case, many altered hemoglobins have already been found, and an example of the electrophoretic behavior of several is illustrated in Figure 14.14. This group of twelve represents variants found from examinations of samples from many populations and families, and a total of more than 30 is now known.

Detection of differences in this way has provided material for more elaborate analyses. A very useful analysis is that developed by Ingram (476). This method depends on specific hydrolysis of protein by proteolytic enzymes to yield a mixture of peptide fragments. Trypsin, for example, splits polypeptides adjacent to lysine and arginine, and mixed products from hemoglobin A yield, following electrophoresis in one direction and chromatography on paper in the other, a characteristic pattern (fingerprint) of ninhydrin-reacting peptides, as shown at the left in Figure 14.15. Corresponding patterns for three different hemoglobins are shown at the right in the figure, and, as indicated, peptide four of A is not present in S or C. A new peptide was obtained from S and two from C, and subsequent analyses demonstrated specific single amino-acid substitutions at the same position as shown in Figure 14.16. Arrows indicate points of enzyme splitting of the original chain, and the reason for production of two peptides from C

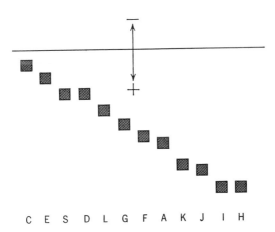

C E S D L G F A K J I H

Figure 14.14. Diagrammatic representation of electrophoretic separation of different hemoglobins on filter paper at pH 8.6. From Harris (410).

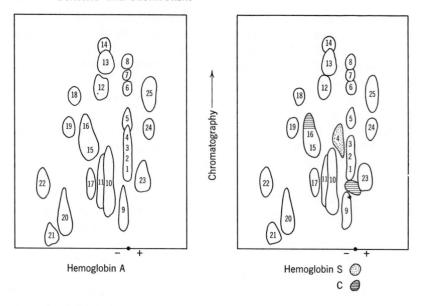

Hemoglobin A

Hemoglobin S ◉
C ◉

Figure 14.15. Summary tracing of fingerprints of tryptic digests prepared from hemoglobins A, S, and C. Peptides carrying alterations are shaded. From Ingram (477).

is clearly due to the substitution of arginine for glutamic acid. S and C are therefore clearly derived from similar mutations affecting the β chain. The fact that the S and C mutant genes appear, from the available genetic data, to be allelic is of considerable interest. As pointed out previously for electrophoresis alone, this method will not necessarily reveal all changes, but more extensive analysis has not demonstrated other substitutions in S and C hemoglobins.

More elaborate methods of analysis, based on column chromatog-

$$\text{HbA}...\text{Val}-\text{His}-\text{Leu}-\text{Thr}-\text{Pro}-\underline{\text{Glu}}-\text{Glu}-\text{Lys}...$$

$$\text{HbS}...\text{Val}-\text{His}-\text{Leu}-\text{Thr}-\text{Pro}-\underline{\text{Val}}-\text{Glu}-\text{Lys}...$$

$$\text{HbC}...\text{Val}-\text{His}-\text{Leu}-\text{Thr}-\text{Pro}-\underline{\text{Lys}}-\text{Glu}-\text{Lys}...$$

Figure 14.16. Amino-acid substitutions in a peptide portion of hemoglobins A, S, and C. From Ingram (477).

raphy for separation of hemoglobin subunits, peptide fragments from partial hydrolysis and ultimate hydrolytic products (the amino acids), have been applied extensively, and specific amino-acid substitutions and sequences for quite a number of different hemoglobins are now known. Some of these data are summarized in Table 68. It usually appears that the new hemoglobins found by these methods differ by a single amino-acid substitution, and in one case (C versus S) the substitution is at the same position. As expected, most changes observed involve acidic or basic amino acids, since these are the changes most easily detectible by electrophoretic methods. It is interesting to note, however, that, as would be expected, position by electrophoresis does not establish identity (see Figure 14.14), since the two G and three M types listed in the table are all different and changes are not even in the same chain. However, both G types change in the direction of

Table 68.   **Amino-Acid Substitutions in Hemoglobins** *

	Residue No.	Hb A Residue	Substitution
α-Chain			
I	16	Lys ⟶ Asp	
Norfolk	57	Gly ⟶ Asp	
M-Boston	58	His ⟶ Tyr	
G-Philadelphia	68	Asp ⟶ Lys	
D α	...	...	...
Hopkins 2	...	...	...
Q	...	...	...
P	...	...	...
β-Chain			
M-Emory	63	His ⟶ Tyr	
M-Milwaukee	67	Val ⟶ Glu	
G-San Jose	7	Glu ⟶ Gly	
C, X	6	Glu ⟶ Lys	
S	6	Glu ⟶ Val	
E	26	Glu ⟶ Lys	
Zürich	63	Ala ⟶ Arg	
D	121	Glu ⟶ Glu (NH$_2$)	
δ	9	Ser ⟶ Thr
	12	Thr ⟶ Asp
	22	Glu ⟶ Ala
	50	Thr ⟶ Ser

* From Keil (517).

less acidity and all three M types in the direction of more acidity. In addition to the kinds of alterations indicated here, it has been reported that H hemoglobin has the constitution β_4^H, and "Bart's" hemoglobin the constitution γ_4^F.

Now, by the proposition that a gene determines the amino-acid sequence in one polypeptide chain, it may be considered that the α, β, and γ chains represent products of three genetic loci, and that the products represented in Table 68 correspond to allelic series, one for the α locus and one for the β locus. Thus, in an adult heterozygous for one of these loci, at least two types of hemoglobin are to be expected unless there is complete dominance. This is the case in A/S heterozygotes (see p. 623). Furthermore, given sufficient data of the kind in Table 68, it should be possible to evaluate the proposition that the amino-acid sequence is determined by specific sequences of triplets of nucleotides in messenger RNA and DNA from which this is presumed to be copied. So far, there is reasonable agreement between substitutions found and code combinations (Figure 6.10, p. 277). For example, the C and S types which show the changes Glu → Lys and Glu → Val can be represented in nucleotide triplets as AUG → AUA and AUG → UUG, respectively. Mutation would thus require the substitution of only one but a different base to yield these two different substitutions at the same position in the polypeptide sequence. It remains to be seen how far such an analysis can be carried, but, in any case, a firm foundation for the true peptide-nucleotide relation in man should come out of these investigations.

So far, the genetic and physiological aspects of the hemoglobin problem in man have been almost completely ignored in the discussion. This was done deliberately in an attempt to focus attention on the expectations of the problem at the biochemical level. That is, a great deal of very valuable information can be expected by structure studies on hemoglobins from individuals taken at random without reference to, or knowledge of, their genetic constitutions. Mutational differences are, perhaps, more frequent among different races and in mild or severe pathological disorders, but it is unlikely that there is such a thing as a standard "normal" hemoglobin A. More likely, there are many which function normally. In any case, the whole hemoglobin problem started from consideration of a familial disease known as sickle-cell anemia. This inherited disease was shown by Neel (737) to be derived as a rare single-gene recessive mutation. Blood cells of affected individuals assume a sickle shape in the presence of low-oxygen tension, and the individuals frequently suffer a severe hemolytic anemia which may result in death in early childhood.

Sickle-cell anemia is quite common among Negro populations in Africa and the United States, and from studies of some 75 families it was established that sickling is also characteristic of cells from individuals heterozygous for the mutant gene. Such individuals are relatively normal and are referred to as carriers of the sickle-cell trait. In 1949 Pauling and collaborators (777) observed that the hemoglobins from sickle-cell anemia, sickle-cell trait (sicklemia), and normal individuals behaved differently on electrophoresis. As shown by the patterns for electrophoretic migrations of carbonmonoxy-hemoglobins in Figure 14.17, normal and sickle-cell anemia hemoglobins are quite different and the heterozygote has both types. These correspond to hemoglobins A, S, and a mixture of A and S (Table 68). As discussed earlier, and as established by Ingram, the difference in A and S is in a single amino-acid substitution, and this is in position 6 of the β chain. It is evident, then, that the mutation is in the gene corresponding to the β chain, and in the heterozygotes both normal and mutant alleles produce hemoglobin. These two products would be $\alpha_2{}^A\beta_2{}^A$ and $\alpha_2{}^A\beta_2{}^S$, and it is interesting to note that a mixed molecule such as $\alpha_2{}^A\beta_1{}^A\beta_1{}^S$ evidently is not produced. Sickle-cell-trait individuals do show considera-

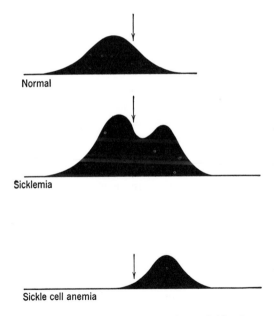

Normal

Sicklemia

Sickle cell anemia

Figure 14.17. Characteristics of carbon monoxyhemoglobins from normal, sicklemia, and sickle-cell anemia red-blood cells. Scanning patterns from electrophoresis. From Pauling et al. (777).

ble variation in proportions of the two types of hemoglobin, however. Different families show values grouped around 41 and 35 and 26% of the S type. Gene interaction may be involved. Physiologically, the sickle-cell anemia appears to be due to the physical properties of the S hemoglobin in relation to packing into red cells. In the reduced form the S hemoglobin is much less soluble than A, and the cells which carry it are much more fragile. This is true to some extent in sicklemia, and life expectancy is lowered in these individuals.

After the general characterization of hemoglobin S, other situations involving anemia were examined, and the C and D hemoglobins were thus discovered by Itano and Neel (682) and Itano (481). These, however, are derived from individuals with less severe sickle-cell characteristics, even though both C and D are substituted in the β chain, and C at the same site as in S. The properties of the resulting protein are evidently not so unfavorable as in sickle-cell anemia. Actually, three types of D have been found (476, 477).

The observations on hemoglobin S and sickle-cell anemia more or less set the pattern for quite a number of other altered hemoglobins that have been found. Several are associated with anemias, but relatively little is known of the mode of inheritance in most cases. Usually, individuals carrying two types are considered to be heterozygotes. It has been pointed out that no abnormalities have been found which are associated with substitution involving only uncharged amino acids, but this is not surprising since charge alterations would be expected to have greater influences on hemoglobin properties, and neutral substitutions may usually yield normal phenotypes. Further, in hemoglobin E, involving the substitution Glu $\rightarrow$ Lys, an extreme charge change does not appear to cause any particular disease.

Another variant in the human hemoglobin picture is thalassemia major, or Cooley's anemia. This is associated with a genetic locus probably different from those determining primary structure, and in homozygotes it results in half or more of the hemoglobin in adults of the F type. This and the adult type present appear normal, and a regulation defect appears likely. Either production of adult hemoglobin is restricted, or production of F (γ chain) is not restricted at the proper time. The disease is frequently fatal at an early age, but there is little disturbance in heterozygotes.

A point of some interest with respect to the altered hemoglobin carriers and other similar manifestations in man is distribution in different populations. Sickle-cell trait is found among different tribes in central Africa to the extent of some 2–25% with an average of 20%. It is less frequent among the Negro population in the United States

(9%) and almost nonexistent in other parts of the world. The trait associated with hemoglobin C occurs in localized areas of the African Gold Coast to the extent of 3–20%, and it is very infrequent elsewhere. Types D and E are also localized, D in northwest India (1%), and E in southwest Asia (10%). Similarly, thalassemia is most prominent in northern Italy, Sicily, and Sardinia (410). Most of these frequencies are much too high to be derived from usual mutation rates, and studies of the sicklemia situation are quite suggestive of accumulation of heterozygotes through environmental selective advantage. Allison (8) noted that sicklemia is especially high in areas where malaria is prominent and presented evidence to show that the trait provides resistance to the malaria parasite. Other data are conflicting, and Raper (821) found confirmation only for cerebral malaria. As pointed out by Neel (738), a demonstration of selective advantage on such a basis must also include information on reproductive advantage, and this has not been demonstrated. However, it does seem quite likely that such localized high-gene frequencies must usually depend on selective advantage factors, and much is to be gained at both genetic and biochemical levels by studies of populations in localized environments.

General Comments

The genetic diversity of man is immediately obvious to the most casual observer just from looking about. As fingerprints can be used to identify an individual, so can innumerable other physical characteristics and combinations of characteristics of shape, size, and color. This is true even in situations of striking family resemblance if one looks for differences as carefully as for likenesses. But most of these characteristics have a complex genetics, and involve genes with highly pleiotropic effects. Consequently, a decisively important aspect of human genetics is to reach more and more into the root of things by studies of the most basic gene products that it is possible to investigate. This is at the level of protein biochemistry, and, as exemplified by the developments in the hemoglobin-structure problem, both genetic analysis and physiological results are becoming more meaningful and useful. Without information on different types of hemoglobin which give rise to red-cell sickling, the different types would likely be classed together, making the picture much more complicated than it is. Thus the purely biochemical findings can aid greatly in delineation of the geneticist's problem. In the other direction, evaluations of a genetic character at the molecular level can be of great assistance in understanding pleiotropy, as has been illustrated in the studies of galactosemia. Thus directed and even random examinations of structures of human protein will likely contribute greatly to furthering understanding of human inheritance. For the future it is to be expected that similar principles will hold when methods become available for isolation and sequence determinations in different molecular species of nucleic acids. Perhaps this lies in the immediate future for soluble RNA's at least.

It has been pointed out that man, as a subject for studies in biochemical genetics, is not very favorable in some respects, but in others he is unique and especially interesting. Fundamental principles of many kinds are more easily obtained by studies of other organisms, but the unique and subtle facets of the problem must be examined using man himself. These involve medical and social problems related to the frequently conflicting interests of the individual and the population. Through animal and plant breeding and control of disease, we carefully select stock lines most favorable for our own use, among organisms upon which we depend for existence, and at the same time we exert no such controls on ourselves. Furthermore, we deliberately preserve clearly unfavorable genes through medical practice, and we deliberately produce environmental conditions, as by increasing radiation in our environment, which increase the probability of accumulating more unfavorable genes. Our own genetic destruction could be in progress. But this is not a necessary outcome of the situation that exists; it is only a statement of the probable outcome if the problem is ignored.

At the same time this is not a plea for any mass control of breeding combinations in the human populations. There are favorable and unfavorable genes in all populations of the human race, but no one knows what combinations are most desirable or how to meet the ever-changing effects of continued mutation. The prospects, then, for our general benefit, which lie within the acceptable limits of individual rights and responsibilities, are in the direction of voluntary reduction in known unfavorable combinations. As repeatedly emphasized by Stern (968), we already know of, and can detect, many dominant and recessive genes which should not be perpetuated, and, with more and more specific knowledge, voluntary genetic marriage counseling can become an effective factor in our destiny. Surely we can individually accept that much responsibility. For the future it should be possible to extend present capacities for prediction from obvious gene-controlled pathological conditions to the greater subtleties of chronic susceptibilities to various diseases and to such things as variations in mental disorders and capacities. For all of these things, we need to know man much better, but without doubt we will gain most rapidly by action on all fronts to obtain more understanding of biochemical genetics in many organisms.

References

1. Abrams, R. (1961), *Ann. Rev. Biochem.*, 30, 165.
2. Afzelius, B. A. (1955), *Expl. Cell Res.*, 8, 147.
3. Agol, H. (1931), *Genetics*, 16, 254.
4. Alberty, R. A. (1956), *Advan. Enzymol.*, 17, 1.
5. Alberty, R. A. (1959), in *The Enzymes*, vol. I, p. 143, edited by P. D. Boyer, H. Lardy, and K. Myrbäck, Academic Press, New York.
6. Alderson, T. (1961), *Nature*, 191, 251.
7. Alexander, M. L. (1960), *Genetics*, 45, 1019.
8. Allison, A. C. (1954), *Brit. Med. J.*, 1, 290.
9. Allfrey, V. (1961), in *The Cell*, vol. I, p. 193, edited by J. Brachet and A. E. Mirsky, Academic Press, New York.
10. Allfrey, V., A. E. Mirsky, and S. Osawa (1957), *J. Gen. Physiol.*, 40, 451.
11. Allfrey, V. G., A. E. Mirsky, and H. Stern (1955), *Advan. Enzymol.*, 16, 411.
12. Allfrey, V. G., J. W. Hopkins, J. H. Frenster, and A. E. Mirsky (1960), *Ann. N.Y. Acad. Sci.*, 88, 722.
13. Alston, R. E. (1958), *Am. J. Botany*, 45, 289.
14. Alston, R. E., and C. W. Hagen (1958), *Genetics*, 43, 35.
15. Altman, R. (1889), *Arch. Anat. Physiol., Physiol. Abstr.*, 524.
16. Ames, B. N. (1957), *J. Biol. Chem.*, 228, 131.
17. Ames, B. N., and Garry, B. (1959), *Proc. Nat. Acad. Sci.*, 45, 1453.
18. Ames, B. N., and P. E. Hartman (1961), *Symp. Fundamental Cancer Res.*, 15th, Houston, Tex. 1961, 322.

19. Ames, B. N., B. Garry, and L. A. Herzenberg (1960), *J. Gen. Microbiol.*, 22, 369.
20. Anagnostopoulos, C., and I. P. Crawford (1961), *Proc. Nat. Acad. Sci.*, 47, 378.
21. Ancel, P., and P. Vintemberger (1948), *Bull. Biol. France*, Suppl. 31.
22. Anderer, F. A., H. Uhlig, E. Weber, and G. Schramm (1960), *Nature*, 186, 922.
23. Anderson, T. F. (1958), *Cold Spring Harbor Symp. Quant. Biol.*, 23, 59.
24. Anderson, E. P., H. M. Kalckar, K. Kurahashi, and K. J. Isselbacher (1957), *J. Lab. Clin. Med.*, 50, 469.
25. Anderson, T. F., E. L. Wollman, and F. Jacob (1957), *Ann. Inst. Pasteur*, 93, 450.
26. Anfinsen, C. B. (1959), *The Molecular Basis of Evolution*, John Wiley and Sons, New York.
27. Archer, R. (1960), *Ann. Rev. Biochem.*, 29, 547.
28. Astbury, W. L., and F. O. Bell (1939), *Cold Spring Harbor Symp. Quant. Biol.*, 6, 109.
29. Astrachan, L., and E. Volkin, *Biochem. Biophys. Acta*, 29, 536.
30. Atwood, K. C., and T. H. Pittinger (1955), *Am. J. Botany*, 42, 496.
31. Auerbach, C. (1951), *Hereditas*, 37, 1.
32. Auerbach, C. (1961), *Nat. Acad. Sci. Nat. Res. Council Publ. 891.*
33. Auerbach, C., and H. Moser (1953), *Z. Induktive Abstammungs-Vererbungslehre*, 85, 479.
34. Auerbach, C., and J. M. Robson (1947), *Proc. Roy. Soc. Edinburgh*, B 62, 271.
35. Auerbach, C., and J. M. Robson (1947), *Proc. Roy. Soc. Edinburgh*, B 62, 284.
36. Auerbach, C., and B. Woolf (1960), *Genetics*, 45, 1691.
37. Austin, M., L. D. Widmayer, and L. Walker (1956), *Physiol. Zool.*, 29, 261.
38. Austrian, R., and H. P. Bernheimer (1955), *J. Clin. Invest.*, 34, 920.
39. Avery, O. T., C. M. MacLeod, and M. McCarty (1944), *J. Exp. Biol. Med.*, 79, 137.
40. Avery, A. G., S. Sativa, and J. Rietsema (1959), *Blakeslee: The Genus Datura*, The Ronald Press Co., New York.
41. Bacq, Z. M. (1951), *Experientia*, 7, 11.
42. Baker, C. L. (1933), *Arch. Protistenk.*, 80, 434.
43. Baker, W. K. (1949), *Genetics*, 34, 167.
44. Barron, E. S. G. (1954), *In Radiation Biology*, vol. 1, p. 283, McGraw-Hill Book Co., New York.
45. Barth, L. G. (1953), *Embryology*, rev. ed., Dryden Press, New York.
46. Bateson, W. (1909), *Mendel's Principles of Heredity*, Cambridge University Press, New York.
47. Baur, E. (1909), *Z. Induktive Abstammungs-Vererbungslehre*, 1, 330.
48. Bauer, H. (1939), *Chromosoma*, 1, 343.
49. Bauer, H. (1942), *Chromosoma*, 2, 407.
50. Bautz, E., and E. Freese (1960), *Proc. Nat. Acad. Sci.*, 46, 1585.
51. Baylor, M. B., D. Hurst, S. Allen, and E. Bertani (1957), *Genetics*, 42, 104.

52. Beadle, G. W. (1944), *J. Biol. Chem.,* **156,** 683.
53. Beadle, G. W. (1945), *Chem. Rev.,* **37,** 15.
54. Beadle, G. W., and V. L. Coonradt (1944), *Genetics,* **29,** 291.
55. Beadle, G. W., and B. Ephrussi (1937), *Genetics,* **22,** 76.
56. Beadle, G. W., and E. L. Tatum (1941), *Proc. Nat. Acad. Sci.,* **27,** 499.
57. Beadle, G. W., and E. L. Tatum (1941), *Am. Naturalist,* **75,** 107.
58. Beadle, G. W., and E. L. Tatum (1945), *Am. J. Botany,* **32,** 678.
59. Beadle, G. W., E. L. Tatum, and C. W. Clancy (1938), *Biol. Bull.,* **75,** 447.
60. Beale, J. H. (1952), *Genetics,* **37,** 62.
61. Beale, G. H. (1954), *The Genetics of Paramecium aurelia,* Cambridge University Press, New York.
62. Beale, G. H. (1957), *Intern. Rev. Cytol.,* **6,** 1.
63. Beale, G. H., J. R. Price, and R. Scott-Moncrieff (1940), *J. Genet.,* **41,** 65.
64. Bearn, A. G., and E. C. Franklin (1958), *Science,* **128,** 596.
65. Becker, H. J. (1959), *Chromosoma,* **10,** 654.
66. Becker, H. J. (1960), *Genetics,* **45,** 519.
67. Beermann, W. (1956), *Cold Spring Harbor Symp. Quant. Biol.,* **21,** 217.
68. Beermann, W. (1959), *Developmental Cytology,* The Ronald Press Co., New York.
69. Beermann, W. (1961), *Chromosoma,* **12,** 1.
70. Bendich, A., H. B. Pahl, H. S. Rosenkranz, and M. Rosoff (1958), *Symp. Soc. Exp. Biol.,* **12,** 31.
71. Benedict, S. R. (1916), *J. Lab. Clin. Med.,* **2,** 1.
72. Bennett, D., L. C. Dunn, and S. Badenhausen (1959), *Genetics,* **44,** 795.
73. Benzer, S. (1955), *Proc. Nat. Acad. Sci.,* **41,** 344.
74. Benzer, S. (1959), *Proc. Nat. Acad. Sci.,* **45,** 1607.
75. Benzer, S. (1957), *The Chemical Basis of Heredity,* edited by W. D. McElroy and B. Glass, The Johns Hopkins Press, Baltimore.
76. Benzer, S. (1961), *Proc. Nat. Acad. Sci.,* **47,** 403.
77. Berg, P. (1961), *Ann. Rev. Biochem.,* **30,** 293.
78. Bernstein, H. (1961), *J. Gen. Microbiol.,* **25,** 41.
79. Berthet, J., and C. deRuve (1951), *Biochem. J.,* **50,** 174.
80. Bianchi, A., and C. Manera (1953), *Bibliotheca Haematol.,* **13,** 1.
81. Birkana, B. N. (1938), *Biol. Zh.,* **7,** 653.
82. Böhme, H., and H. R. Schütte (1956), *Biol. Zentralblatt,* **75,** 597.
83. Boivin, A., R. Vendrely, and C. Vendrely (1948), *Compt. Rend.,* **226,** 1061.
84. Bonner, D. (1946), *Am. J. Botany,* **33,** 788.
85. Bonner, D. M., Y. Suyama, and J. A. DeMoss (1960), *Federation Proc.,* **19,** 926.
86. Bonner, D. M., C. Yanofsky, and C. W. H. Partridge (1952), *Proc. Nat. Acad. Sci.,* **38,** 25.
87. Borsook, H., and G. L. Keighley (1935), *Proc. Roy. Soc. (London), Ser.* B 118, 488.
88. Boveri, T. (1901), *Verhandl. Physiol. Med. Ges. Würzburg,* N.F. **34,** 145.
89. Boveri, T. (1902), *Verhandl. Physiol. Med. Ges. Würzburg,* N.F. **35,** 165.
90. Boveri, T. (1910), *Festschrift R. Hertwig,* **3,** 131.

91. Boyer, P. B., H. Lardy, and K. Myrbäck, editors (1959), *The Enzymes,* vol. I, Academic Press, New York.
92. Brachet, J. (1950), *Experientia,* 6, 56.
93. Brachet, J. (1957), *Biochemical Cytology,* Academic Press, New York.
94. Brachet, J. (1961), in *The Cell,* vol. II, edited by J. Brachet and A. E. Mirsky, Academic Press, New York.
95. Brachet, J., and A. E. Mirsky (1959), in *The Cell,* vols. I–V, Academic Press, New York.
96. Brachet, J., and A. E. Mirsky, editors (1961), in *The Cell,* vol. I, Academic Press, New York.
97. Bradley, S. G., and J. Lederberg (1956), *J. Bacteriol.,* 72, 235.
98. Bradley, S. G., D. L. Anderson, and L. A. Jones (1959), *Ann. N.Y. Acad. Sci.,* 81, 811.
99. Braendle, D. H., and W. Szybalski (1959), *Ann. N.Y. Acad. Sci.,* 81, 824.
100. Braunitzer, G., N. Hilschmann, K. Hilse, B. Liebold, and R. Müller (1960), *Z. Physiol. Chem.,* 322, 96.
101. Brenner, S. (1959), The Mechanism of Gene Action, in *Ciba Found. Symp. Biochem. Human Genet.,* Little, Brown and Co., Boston.
102. Bresch, C. (1955), *Z. Naturforsch.,* 106, 545.
103. Bresch, C. (1959), *Ann. Rev. Microbiol.,* 13, 313.
104. Breuer, M. E., and C. Pavan (1955), *Chromosoma,* 1, 371.
105. Bridges, C. B. (1913), *J. Exp. Zool.,* 15, 587.
106. Bridges, C. B. (1916), *Genetics,* 1, 1.
107. Bridges, C. B. (1939), Cytological and Genetic Basis of Sex, in *Sex and Internal Secretions,* 2nd ed., chap. 2, p. 15, Williams and Wilkins, Baltimore.
108. Bridges, C. B., and K. S. Brehme (1944), *Carnegie Inst. Wash. Publ.,* 552.
109. Briggs, R., and T. J. King (1955), in Biological Specificity and Growth, *12th Growth Symposium* (E. G. Butler, editor), p. 207, Princeton University Press, Princeton, N.J.
110. Briggs, R., and T. J. King (1959), in *The Cell,* vol. I, edited by J. Brachet and A. E. Mirsky, Academic Press, New York.
111. Briggs, R., and T. J. King (1960), *Develop. Biol.,* 2, 252.
112. Briggs, R., T. J. King, and M. A. DiBerardino (1960), in *Symposium on Germ Cells and Development,* p. 441; Institut International d'Embryologie and Fondazione A. Baselli.
113. Brink, R. A. (1954), *Genetics,* 39, 729.
114. Brink, R. A. (1958), *Cold Spring Harbor Symp. Quant. Biol.,* 23, 379.
115. Brink, R. A., and R. A. Nilan (1952), *Genetics,* 37, 519.
116. Brink, R. A., and D. F. Brown, J. Kermicle, and W. H. Weyers (1960), *Genetics,* 45, 1297.
117. Buchmann, W., and N. W. Timofeeff-Ressovsky (1935), *Z. Induktive Abstammungs-Vererbungslehre,* 70, 130.
118. Buchmann, W., and N. W. Timofeeff-Ressovsky (1936), *Z. Induktive Abstammungs-Vererbungslehre,* 71, 335.
119. Buller, A. H. R. (1941), *Botan. Rev.,* 7, 335.
120. Burdette, W. J. (1940), *Univ. Texas Publ.,* 4032, 157.
121. Burgeff, H. (1912), *Ber. Deuts. Botan. Ges.,* 30, 679.
122. Burns, J. J., and A. H. Conney (1960), *Ann. Rev. Biochem.,* 29, 413.
123. Butenandt, A., and G. Neubert (1958), *Ann.,* 618, 167.

124. Buttin, G. (1961), *Cold Spring Harbor Symp. Quant. Biol.,* **26,** 213.
125. Buxton, E. W. (1956), *J. Gen. Microbiol.,* **15,** 133.
126. Byers, H. L. (1954), *Proc. Intern. Congr. Genet., 9th Bellagio, Italy,* **2,** 694.
127. Cahn, R. D., N. D. Kaplan, L. Levine, and E. Zwilling (1962), *Science,* **136,** 962.
128. Calvin, M. (1958), *Brookhaven Symp. Biol.,* **11,** 160.
129. Canellakis, E. S. (1962), *Ann. Rev. Biochem.,* **31,** 271.
130. Carlson, E. A. (1959), *Quart. Rev. Biol.,* **34,** 33.
131. Carlson, E. A. (1959), *Genetics,* **44,** 347.
132. Carlson, J. G. (1941), *Proc. Nat. Acad. Sci.,* **27,** 42.
133. Case, M. E., and N. H. Giles (1958), *Proc. Nat. Acad. Sci.,* **44,** 378.
134. Case, M. E., and N. H. Giles (1960), *Proc. Nat. Acad. Sci.,* **46,** 659.
135. Caspari, E. (1949), *Quart. Rev. Biol.,* **24,** 185.
136. Caspersson, T. (1936), *Skand. Arch. Physiol.,* **73,** (Suppl. No. 8).
137. Caspersson, T. (1962), *The Molecular Control of Cellular Activity,* p. 127, edited by J. M. Allen, McGraw-Hill Book Co., New York.
138. Castle, W. E. (1948), *Genetics,* **33,** 22.
139. Castle, W. E. (1954), *Genetics,* **39,** 35.
140. Catcheside, D. G. (1947), *J. Genet.,* **48,** 31.
141. Catcheside, D. G. (1947), *J. Genet.,* **48,** 99.
142. Catcheside, D. G. (1948), *Advan. Genet.,* **2,** 271.
143. Catcheside, D. G. (1959), *Nature,* **184,** 1012.
144. Catcheside, D. G. (1960), in Microbial Genetics, *10th Symp. Soc. Gen. Microbiol.,* Cambridge University Press, New York.
145. Catcheside, D. G., and D. E. Lea (1945), *J. Genet.,* **47,** 25.
146. Catcheside, D. G., and M. J. Mathieson (1955), *J. Gen. Microbiol.,* **13,** 72.
147. Catcheside, D. G., and A. Overton (1958), *Cold Spring Harbor Symp. Quant. Biol.,* **23,** 137.
148. Catsch, A. (1948), *Z. Induktive Abstammungs-Vererbungslehre,* **82,** 155.
149. Cavalcanti, A. G. L., D. N. Falcão, and L. E. Castro (1957), *Am. Naturalist,* **91,** 327.
150. Cavalieri, L. F., and B. H. Rosenberg (1962), *Ann. Rev. Biochem.,* **31,** 247.
151. Cavalieri, L. F., B. H. Rosenberg, and J. F. Deutsch (1959), *Biochem. Biophys. Res. Commun.,* **1,** 124.
152. Chamberlain, M., and P. Berg (1962), *Proc. Nat. Acad. Sci.,* **48,** 81.
153. Chance, B. (1956), *Advan. Enzymol.,* **17,** 90.
154. Chance, B. (1960), *J. Biol. Chem.,* **235,** 2426.
155. Chance, B. (1960), *J. Biol. Chem.,* **235,** 2440.
156. Chang, L. O., A. M. Srb, and F. C. Steward (1962), *Nature,* **193,** 756.
157. Chapeville, F., F. Lipman, G. von Ehrenstein, B. Weisblum, W. J. Ray, Jr., and S. Benzer (1962), *Proc. Nat. Acad. Sci.,* **48,** 1086.
158. Chargaff, E. (1955), in vol. 1, *The Nucleic Acids,* edited by E. Chargaff and J. N. Davidson, Academic Press, New York.
159. Chase, M., and A. H. Doermann (1958), *Genetics,* **43,** 333.
160. Chesley, P. (1935), *J. Exp. Zool.,* **70,** 429.

161. Chichester, C. O., H. Yokoyama, T. O. M. Nakayama, A. Lukton, and G. MacKinney (1959), *J. Biol. Chem.*, **234**, 598.
162. Child, C. M. (1941), *Problems and Patterns of Development*, University of Chicago Press, Chicago.
163. Chovnick, A. (1958), *Proc. Nat. Acad. Sci.*, **44**, 333.
164. Chovnick, A. (1961), *Genetics*, **46**, 493.
165. Chovnick, A., and A. S. Fox (1953), *Proc. Nat. Acad. Sci.*, **39**, 1035.
166. Chovnick, A. R., J. Lefkowitz, and A. S. Fox (1956), *Genetics*, **41**, 589.
167. Christensen, A. K., and D. W. Fawcett (1960), *Anat. Record*, **136**, 333.
168. Church, G. (1955), *J. Morphol.*, **96**, 565.
169. Clausen, R. E., and D. R. Cameron (1950), *Genetics*, **35**, 4.
170. Clayton, F. (1954), *Univ. Texas Publ.*, **5422**, 189.
171. Clayton, F. (1958), *Genetics*, **43**, 261.
172. Clevenger, S. (1958), *Arch. Biochem. Biophys.*, **76**, 131.
173. Cohen-Bazire, G., and M. Jolit (1953), *Ann. Inst. Pasteur*, **84**, 1.
174. Cohen, S. S. (1961), *Federation Proc.*, **20**, 641.
175. Cohn, C. (1958), *J. Immunol.*, **80**, 73.
176. Cohn, M. (1957), *Bacteriol. Rev.*, **21**, 140.
177. Cohn, M., and A-M. Torriani (1953), *Biochem. et Biophys. Acta*, **10**, 280.
178. Cole, R. M., and J. J. Hahn (1962), *Science*, **135**, 722.
179. Colowick, S. P., and N. O. Kaplan, editors, *Methods in Enzymology*, vol. I, Academic Press, New York.
180. Conger, A. D., and N. H. Giles (1950), *Genetics*, **35**, 397.
180a. Connell, G. E., G. H. Dixon, and O. Smithies (1962), *Nature*, **193**, 505.
181. Connell, G. E., and O. Smithies (1959), *Biochem. J.* **72**, 115.
182. Cooper, G. R., and E. E. Mandel (1954), *J. Lab. Clin. Med.*, **44**, 636.
183. Cooper, K. W. (1959), *Chromosoma*, **10**, 535.
184. Correns, C. (1900), *Berichte Deutsch. Bot. Gesellsch.*, **18**, 158.
185. Correns, C. (1909), *Z. Induktive Abstammungs-Vererbungslehre*, **1**, 291.
186. Coulombre, J. L., and E. S. Russell (1954), *J. Exp. Zool.*, **126**, 277.
187. Cox, R. A., and A. R. Peacocke (1956), *J. Chem. Soc.*, 2499.
188. Crawford, I. P., and C. Yanofsky (1958), *Proc. Nat. Acad. Sci.*, **44**, 1161.
189. Crick, F. H. C., J. S. Griffeth, L. E. Orgel (1957), *Proc. Nat. Acad. Sci.*, **43**, 416.
190. Crow, J. F. (1946), *Am. Naturalist*, **80**, 663.
191. Crow, J. F. (1948), *Genetics*, **33**, 477.
192. Crow, J. F. (1952), in *Heterosis*, edited by J. W. Gowen, Iowa State College Press, Ames, Iowa.
193. Dalton, A. J. (1961), in *The Cell*, vol. II, p. 603, edited by J. Brachet and A. E. Mirsky, Academic Press, New York.
194. Dalton, H. C. (1946), *J. Exp. Zool.*, **103**, 169.
195. Dalton, H. C. (1950), *J. Exp. Zool.*, **115**, 17.
196. Dalton, H. C. (1953), in *Pigment Cell Growth*, p. 17, edited by M. Gordon, Academic Press, New York.
197. Dalton, H. C., and Z. P. Krassner (1958), in *Pigment Cell Biology*, p. 51, edited by M. Gordon, Academic Press, New York.
198. D'Amato, F., and O. Hoffmann-Ostenhof (1956), *Advan. Genet.*, **8**, 1.
199. Danneel, R. (1941), *Ergeb. Biol.*, **18**, 55.
200. David, P. R. (1936), *Arch. Entwicklungsmech.*, **135**, 521.

201. Davis, B. D. (1948), *J. Am. Chem. Soc.*, 70, 4267.
202. Davis, B. D. (1950), *Experientia*, 6, 41.
203. Davis, B. D. (1955), in *Amino Acid Metabolism*, p. 799, edited by W. McElroy and B. Glass, The Johns Hopkins University Press, Baltimore.
204. Davis, R. H. (1961), *Science*, 134, 470.
205. Davis, R. H. (1962) *Genetics*, 47, 351.
206. Dayton, T. O. (1956), *J. Genet.*, 54, 249.
207. d'Herelle, F. (1917), *Compt. Rend. Acad. Sci.*, Paris, 165, 373.
208. Delbrück, M. (1946), *Biol. Rev.*, 21, 30.
209. Delbrück, M., and C. E. Luria (1942), *Arch. Biochem.*, 1, 111.
210. Demerec, M. (1932), *Proc. Nat. Acad. Sci.*, 18, 430.
211. Demerec, M. (1937), *Genetics*, 22, 469.
212. Demerec, M., and Z. Hartman (1956), *Carnegie Inst. Wash. Publ.*, 612, 5.
213. Demerec, M., and H. Ozeki (1959), *Genetics*, 44, 269.
214. Demerec, M., I. Goldman, and E. L. Lahr (1958), *Cold Spring Harbor Symp. Quant. Biol.*, 23, 59.
215. deSerres, F. J. (1956), *Genetics*, 41, 668.
216. deSerres, F. J. (1960), *Genetics*, 45, 555.
217. deSerres, F. J. (1962), *Nat. Acad. Sci. Nat. Res. Council Publ.*, No. 950.
218. DeVries, H. (1889), *Intracelluläre Panagenesis*, Jena.
219. DeVries, H. (1900), *Compt. Rend. Acad. Sci.*, Paris, 130, 845.
220. Dippell, R. V. (1950), *Heredity*, 4, 165.
221. Djordjevic, B., and W. Szybalski (1960), *J. Exp. Med.*, 112, 509.
222. Dobzhansky, Th. (1951), *Genetics and the Origin of Species*, 3rd ed., Columbia University Press, New York.
223. Dobzhansky, Th., and J. Schultz (1934), *J. Genet.*, 34, 135.
224. Dodds, K. S., and D. H. Long (1956), *J. Genet.*, 54, 27.
225. Dodge, B. O. (1942), *Bull. Torrey Botan. Club*, 69, 75.
226. Doermann, A. H. (1953), *Cold Spring Harbor Symp. Quant. Biol.*, 18, 3.
227. Doerman, A. H., M. Chase, and F. W. Stahl (1955), *J. Cellular Comp. Physiol.*, 45 (Suppl. 2), 51.
228. Donohue, J., and K. Trueblood (1960), *J. Mol. Biol.*, 2, 363.
229. Doty, P., J. Marmur, J. Eigner, and C. Schildkraut (1960), *Proc. Nat. Acad. Sci.*, 46, 461.
230. Doty, P., H. Boedtker, J. R. Fresco, R. Haselkorn, and M. Litt (1959), *Proc. Nat. Acad. Sci.*, 45, 482.
231. Doudney, C. O., and F. L. Haas (1960), *Genetics*, 45, 1481.
232. Doy, C. H., and F. Gibson (1959), *Biochem. J.*, 72, 586.
233. Dulbecco, R. (1952), *J. Bacteriol.*, 63, 199.
234. Dubinin, N. P., and B. N. Sidorov (1934), *Biol. Zh. Mosk.*, 2, 132.
235. Dubinin, N. P., and B. N. Sidorov (1934), *Am. Naturalist*, 68, 377.
235a. Dubinin, N. P., and B. N. Sidorov (1935), *Biol. Zh., Mosk.*, 4, 555.
236. Duffey, L. M., and J. D. Ebert (1957), *J. Embryol. Exp. Morphol.*, 5, 324.
237. Dunn, L. C. (1956), *Cold Spring Harbor Symp. Quant. Biol.*, 21, 187.
238. Eagle, H., K. A. Piez, R. Fleischman, and V. I. Oyama (1959), *J. Biol. Chem.*, 234, 592.
239. East, E. M. (1935), *Genetics*, 20, 443.
240. Ebersold, W. T., R. P. Levine, E. E. Levine, and M. A. Olmsted (1962), *Genetics*, 47, 531.

241. Eberhardt, K. (1939), *Chromosoma,* 1, 317.
242. Eberhart, B. M., and E. L. Tatum (1959), *J. Gen. Microbiol.,* 20, 43.
243. Ebert, J. D. (1959), in *The Cell,* vol. I, p. 14, edited by J. Brachet and A. Mirsky, Academic Press, New York.
244. Edgar, R. S. (1958), *Genetics,* 43, 235.
245. Edgar, R. S. (1962), personal communication.
247.* Edgar, R. S., R. P. Feynman, S. Klein, I. Lielansis, and C. M. Steinberg (1962), *Genetics,* 47, 179.
249.* Edström, J. E. (1960), *J. Biophys. Biochem. Cytol.,* 8, 47.
250. Emerson, R. A. (1921), *Cornell Univ. Agr. Exp. Sta. Mem.,* 39, 1.
251. Emerson, R. A., G. W. Beadle, and A. C. Fraser (1935), *Cornell Univ. Agr. Exp. Sta. Mem.,* 180, 83.
252. Emerson, S. (1948), *Proc. Nat. Acad. Sci.,* 34, 72.
253. Emerson, S. (1952), in *Heterosis,* Iowa State College Press, Ames, Iowa.
254. Emerson, S. (1955), *Handbuch d. physiologisch- und pathologisch-chemischen Analyse,* 10 Aufl. Bd. II, 443.
255. Emmens, C. W. (1937), *J. Genet.,* 34, 191.
256. Ephrussi, B. (1953), *Nucleo-cytoplasmic Relations in Microorganisms,* Clarendon Press, Oxford.
257. Ephrussi, B. (1958), *J. Cellular Comp. Physiol.,* 52, 35.
258. Ephrussi, B., and J. L. Herold (1945), *Genetics,* 30, 62.
259. Ephrussi, B., H. Hottingeur, and H. Roman (1955), *Proc. Nat. Acad. Sci.,* 41, 1065.
260. Ephrussi-Taylor, H. (1951), *Cold Spring Harbor Symp. Quant. Biol.,* 16, 445.
261. Ephrussi-Taylor, H. (1960), in *10th Symp. Soc. Gen. Microbiol.,* p. 132.
262. Epstein, R. H. (1957), Multiplicity Reaction in Bacteriophage, dissertation, University of Rochester.
263. Eyster, H. C. (1950), *Plant Physiol.,* 25, 630.
264. Fabergé, A. C., and G. H. Beale (1942), *J. Genet.,* 43, 173.
265. Fahmy, O. G., and M. J. Fahmy (1957), *J. Heredity,* 55, 28.
266. Fahmy, O. G., and M. J. Fahmy (1960), *Heredity,* 15, 115.
267. Fahmy, O. G., and M. J. Fahmy (1960), *Genet. Res.,* 1, 173.
268. Fahmy, O. G., and M. J. Fahmy (1960), *Genetics,* 45, 1191.
269. Fahmy, O. G., and M. J. Fahmy (1961), *Genetics,* 46, 361.
270. Fahmy, O. G., and M. J. Fahmy (1961), *Genetics,* 46, 447.
271. Fano, U. (1942), *Quart. Rev. Biol.,* 17, 244.
272. Fawcett, D. W., and S. Ito (1958), *J. Biophys. Biochem. Cytol.,* 4, 135.
273. Feldherr, C. M., and A. B. Feldherr (1960), *Nature,* 185, 250.
274. Fell, H. B., and E. Mellanby (1953), *J. Physiol.,* 119, 470.
275. Feenstra, W. J. (1960), *Mede del. Landbouwhogeschool,* 60, 1.
276. Feughelman, M., R. Langridge, W. E. Seeds, A. R. Stokes, H. R. Wilson, C. W. Hooper, M. H. F. Wilkins, R. K. Barclay, and L. D. Hamilton (1955), *Nature,* 75, 834.
277. Feulgen, R., and H. Rossenbeck (1924), *Z. Physiol. Chem.,* 135, 203.
278. Fincham, J. R. S. (1959), *Ann. Rev. Biochem.,* 28, 343.
279. Fincham, J. R. S. (1959), *J. Gen. Microbiol.,* 21, 600.

* References 246 and 248 have been eliminated.

280. Fincham, J. R. S. (1960), *Advan. Enzymol.,* 22, 1.
281. Fincham, J. R. S., and P. R. Bond (1960), *Biochem. J.,* 77, 96.
282. Fincham, J. R. S., and J. A. Pateman (1957), *Nature,* 179, 741.
283. Flaks, J. G., J. Lichtenstein, and S. S. Cohen (1959), *J. Biol. Chem.,* 234, 1507.
284. Flemming, W. (1882), *Zellsubstanz, Kern und Zelltheilung,* Leipzig.
285. Forrest, H. S., E. Glassman, and H. K. Mitchell (1956), *Science,* 124, 725.
286. Forrest, H. S., E. W. Hanley, and J. M. Lagowski (1961), *Genetics,* 46, 1455.
287. Forrest, H. S., R. Hatfield, and C. van Baalen (1959), *Nature,* 183, 1269.
288. Foster, M. (1956), *Genetics,* 41, 396.
289. Foster, R. J., and C. Niemann (1951), *J. Am. Chem. Soc.,* 73, 1552.
290. Foster, R. J., and C. Niemann (1953), *Proc. Nat. Acad. Sci.,* 39, 371.
291. Foure-Fremiet, E. (1910), *Anat. Anz.,* 36, 186.
292. Fraenkel-Conrat, H. (1959), in *The Enzymes,* vol. I, p. 589, edited by D. B. Boyer, H. Lardy, and K. Myrbäck, Academic Press, New York.
293. Franklin, R. E., and Gosling, R. C. (1953), *Nature,* 171, 740.
294. Freese, E. (1957), *Genetics,* 42, 671.
295. Freese, E. (1959), *Brookhaven Symp. Biol.,* 12, 63.
296. Freese, E. (1959), *Proc. Nat. Acad. Sci.,* 45, 622.
297. Freese, E. (1959), *J. Mol. Biol.,* 1, 87.
298. Freese, E. (1961), *Proc. Intern. Congr. Biochem.,* 5th, Moscow.
299. Friedman, M., and S. O. Byers (1948), *J. Biol. Chem.,* 175, 727.
300. Fries, N. (1947), *Nature,* 159, 199.
301. Fries, N., and B. Kihlman (1948), *Nature,* 162, 573.
302. Fresco, J. R., and D. B. Straus (1962), *Am. Scientist,* 50, 158.
303. Fritz-Niggli, H. (1959), *Strahlenbiologie,* Georg Thieme, Stuttgart.
304. Fruton, J. S., and S. Simmonds (1958), *General Biochemistry,* 2nd ed., John Wiley and Sons, New York.
305. Furst, S. S., P. M. Roll, and G. B. Brown (1950), *J. Biol. Chem.,* 186, 251.
306. Furberg, S. (1949), *Nature,* 164, 22.
307. Gall, J. G. (1958), in *Chemical Basis of Development,* edited by W. D. McElroy and B. Glass, The Johns Hopkins University Press, Baltimore.
308. Gall, J. G. (1959), *J. Biophys. Biochem. Cytol.,* 6, 115.
309. Gans, M. (1953), *Bull. Biol. France et Belgique,* Suppl. 38.
310. Gardner, L. T. (1959), *Molecular Genetics and Human Disease,* Charles C Thomas Publisher, Springfield, Ill.
311. Garnjobst, L. (1953), *Am. J. Botany,* 40, 607.
312. Garnjobst, L., and J. F. Wilson (1956), *Proc. Nat. Acad. Sci.,* 42, 613.
313. Garrod, A. E. (1923), *Inborn Errors in Metabolism,* Oxford University Press, Oxford.
314. Geissman, T. A., and J. B. Harbone (1955), *Arch. Biochem. Biophys.,* 55, 447.
315. Geissman, T. A., and A. L. Mehlquist (1947), *Genetics,* 32, 410.
316. Geissman, T. A., E. H. Hinreiner, and E. C. Jorgensen (1956), *Genetics,* 41, 93.
317. Geissman, T. A., E. C. Jorgensen, and B. L. Johnson (1954), *Arch. Biochem. Biophys.,* 49, 368.
318. Gibson, Q. H., and E. Antonini (1960), *Biochem. J.,* 77, 328.

636   *Genetics and Metabolism*

319. Giles, N. H. (1951), *Cold Spring Harbor Symp. Quant. Biol.,* 16, 283.
320. Giles, N. H. (1958), *Proc. Intern. Congr. Genet., 10th, Montreal,* 1, 261, University of Toronto Press, Toronto.
321. Giles, N. H., and C. W. H. Partridge (1953), *Proc. Nat. Acad. Sci.,* 39, 479.
322. Giles, N. H., F. J. deSerres, and E. Barbour (1957), *Genetics,* 42, 609.
323. Giles, N. H., C. W. H. Partridge, and N. J. Nelson (1957), *Proc. Nat. Acad. Sci.,* 43, 305.
324. Gilvarg, C. (1960), *Federation Proc.,* 19, 948.
325. Ginsburg, B. (1944), *Genetics,* 29, 176.
326. Glanville, E. V., and M. Demerec (1960), *Genetics,* 45, 1359.
327. Glass, B., and R. K. Ritterhoff (1956), *Science,* 124, 314.
328. Glassman, E., and H. K. Mitchell (1959), *Genetics,* 44, 153.
329. Glassman, E., and H. K. Mitchell (1959), *Genetics,* 44, 547.
330. Gloor, H. (1943), *Rev. Suisse Zool.,* 54, 637.
331. Gluecksohn-Schoenheimer, S. (1945), *Genetics,* 30, 29.
332. Gluecksohn-Schoenheimer, S. (1949), *Growth,* 9, 163.
333. Goldschmidt, R. (1920), *Die Quantitative Grundlagen von Vererbung und Artbildung,* Springer, Berlin.
334. Goldschmidt, R. (1934), *Bibliog. Genet.,* 11, 1.
335. Goldschmidt, R. (1937), *Univ. Cal. (Berkeley) Publ. Zool.,* 41, 277.
336. Goldschmidt, R. (1938), *Physiological Genetics,* McGraw-Hill Book Co., New York.
337. Goldstein, A. (1944), *J. Gen. Physiol.,* 27, 529.
338. Good, N., R. Heilbrunner, and H. K. Mitchell (1950), *Arch. Biochem.,* 28, 464.
339. Goodgal, S. H., and R. M. Herriot (1957), in *Symposium on the Chemical Basis of Heredity,* edited by W. D. McElroy, and B. Glass, The Johns Hopkins University Press, Baltimore.
340. Goodspeed, T. H., and P. Avery (1939), *J. Genet.,* 38, 381.
341. Goodwin, T. W. (1958), *Encyclopedia Plant Physiology,* 10, 186.
342. Gorini, L., and W. Gundersen (1961), *Proc. Nat. Acad. Sci.,* 47, 961.
343. Gorini, L., W. Gundersen, and M. Burger (1961), *Cold Spring Harbor Symp. Quant. Biol.,* 26, 173.
344. Gorz, H. J., and F. A. Haskins (1960), *J. Heredity,* 51, 74.
345. Goto, M., and H. S. Forrest (1961), *Biochem. Biophys. Res. Commun.,* 6, 180.
346. Gowdridge, B. (1956), *Genetics,* 41, 470.
347. Gowen, J. W., and E. H. Gay (1933), *Science,* 77, 312.
348. Gowen, J. W. (1946), *Anat. Record,* 94, 344.
349. Graf, G. E., E. Hadorn, and H. Ursprung (1959), *J. Insect Physiol.,* 2, 120.
350. Granick, S. (1950), *J. Biol. Chem.,* 183, 713.
351. Granick, S. (1961), in *The Cell,* vol. II, edited by J. Brachet and A. E. Mirsky, Academic Press, New York.
352. Gray, C. H., and E. L. Tatum (1944), *Proc. Nat. Acad. Sci.,* 30, 404.
353. Gray, L. H. (1951), *J. Cellular Comp. Physiol.,* 39, Suppl. 1, 57.
354. Green, D. E., and Y. Hatifi (1961), *Science,* 133, 13.
355. Green, D. E., and R. L. Lester (1959), *Federation Proc.,* 18, 987.

356. Green, D. McD. (1959), *Exp. Cell Res.,* 18, 466.
357. Green, M. M. (1946), *Genetics,* 31, 1.
358. Green, M. M. (1954), *Proc. Nat. Acad. Sci.,* 40, 92.
359. Green, M. M. (1955), *Proc. Nat. Acad. Sci.,* 41, 375.
360. Green, M. M. (1957), *Genetics,* 29, 1.
361. Green, M. M. (1959), *Heredity,* 13, 302.
362. Green, M. M. (1959), *Proc. Nat. Acad. Sci.,* 45, 549.
363. Green, M. M. (1960), *Heredity,* 13, 302.
364. Green, M. M. (1961), *Genetics,* 46, 671.
365. Green, M. M., and K. C. Green (1949), *Proc. Nat. Acad. Sci.,* 35, 586.
366. Green, M. M., and K. C. Green (1956), *Z. Indukt. Abst. Vererbungsl.,* 87, 708.
367. Grell, R. F. (1960), *Genetics,* 44, 911.
368. Griffen, A. B., and W. S. Stone (1940), *Univ. Texas Publ.,* 4032, 190.
369. Grobstein, C. (1957), *Exp. Cell Research,* 13, 575.
370. Grobstein, C. (1959), in *The Cell,* vol. I, edited by J. Brachet and A. E. Mirsky, Academic Press, New York.
371. Gros, F., W. Gilbert, H. H. Hiatt, G. Attardi, P. F. Spahr, and J. D. Watson (1961), *Cold Spring Harbor Symp. Quant. Biol.,* 26, 111.
372. Gross, A. (1914), *Biochem. Z.,* 61, 165.
373. Gross, S. R., and A. Fein (1960), *Genetics,* 45, 885.
374. Grunberg-Manago, M. (1962), *Ann. Rev. Biochem.,* 31, 301.
375. Grüneberg, H. (1937), *Nature,* 140, 932.
376. Grüneberg, H. (1938), *Proc. Roy. Soc., Ser. B,* 125, 123.
377. Grüneberg, H. (1943), *J. Genetics,* 45, 22.
378. Grüneberg, H. (1947), *Animal Genetics and Medicine,* Paul R. Hoeber, New York.
379. Grumbach, M. M., A. Morishima, and E. H. Y. Chu (1960), *Acta Endocrinol.,* 35, Suppl. 51, 633.
380. Gulland, J. M., and D. O. Jordon (1947), *Symp. Soc. Exp. Biol.,* 1, 56.
381. Gulland, J. M. (1947), *Cold Spring Harbor Symp. Quant. Biol.,* 12, 95.
382. Gustafson, T., and P. Lenicque (1952), *Exp. Cell Research,* 3, 251.
383. Gustafson, T., and P. Lenicque (1955), *Exp. Cell Research,* 8, 114.
384. Haas, F. L., and C. O. Doudney (1957), *Proc. Nat. Acad. Sci.,* 43, 871.
385. Haas, F. L., C. O. Doudney, and T. Kada (1961), *Nat. Acad. Sci. Nat. Res. Council Publ.* No. 891.
386. Haas, F. L., E. Dudgeon, F. E. Clayton, and W. S. Stone (1954), *Genetics,* 39, 453.
387. Haas, F. L., M. B. Mitchell, B. N. Ames, and H. K. Mitchell (1952), *Genetics,* 37, 217.
388. Hadorn, E. (1955), *Lethalfaktoren,* Georg Thieme Verlag, Stuttgart.
389. Hadorn, E. (1956), *Cold Spring Harbor Symp. Quant. Biol.,* 21, 363.
390. Hadorn, E. (1958), in *Chemical Basis of Development,* p. 779, edited by W. D. McElroy and B. Glass, The Johns Hopkins Press, Baltimore.
391. Hadorn, E. (1961), *Developmental Genetics and Lethal Factors,* John Wiley and Sons, New York.
392. Hadorn, E., and G. E. Graf (1958), *Zool. Anz.,* 160, 231.
393. Hadorn, E., and H. K. Mitchell (1951), *Proc. Nat. Acad. Sci.,* 37, 650.
394. Hadorn, E., and I. Ziegler (1958), *Z. Vererbungslehre,* 89, 221.

395. Hagen, C. W. (1959), *Genetics,* 44, 787.
396. Haldane, J. B. S. (1927), *Biol. Rev.,* 2, 199.
397. Hall, B. D., and P. Doty (1959), *J. Molecular Biol.,* 1, 111.
398. Hall, C. E., and M. Litt (1958), *J. Biophys. Biochem. Cytol.,* 4, 1.
399. Hamburger, V. (1942), *Biol. Symp.,* 6, 311.
400. Haney, W. J. (1954), *J. Heredity,* 45, 146.
401. Hannah, A. (1951), *Advan. Genet.,* 4, 87.
402. Harbone, J. B. (1958), *Biochem. J.,* 68, 129.
403. Hardesty, B., and H. K. Mitchell (1963), *Arch. Biochem. Biophys.,* 100, 330.
404. Harm, W. (1956), *Virology,* 2, 559.
405. Harnly, M. H. (1936), *Genetics,* 21, 84.
406. Harnly, M. H. (1942), *Biol. Bull.,* 82, 215.
407. Harnly, M. H., and M. L. Harnly (1936), *J. Exp. Zool.,* 74, 41.
408. Harold, F. M. (1960), *Biochem. Biophys. Acta,* 45, 172.
409. Harrap, B. S., W. P. Gratzer, and P. Doty (1961), *Ann. Rev. Biochem.,* 30, 269.
410. Harris, H. (1959), *Human Biochemical Genetics,* Cambridge Press, England.
411. Hartman, P. E., Z. Hartman, and D. Serman (1960), *J. Gen. Microbiol.,* 22, 354.
412. Hartman, P. E., J. C. Loper, and D. Serman (1960), *J. Gen. Microbiol.,* 22, 323.
413. Hartman, S. C., and J. M. Buchanan (1959), *Ann. Rev. Biochem.,* 28, 365.
414. Harvey, E. B. (1951), *Ann. N.Y. Acad. Sci.,* 51, 1336.
415. Haskins, F. A., and H. K. Mitchell (1952), *Am. Naturalist,* 86, 231.
416. Hawthorne, D. C., and L. K. Mortimer (1960), *Genetics,* 45, 1085.
417. Haxo, F. (1949), *Arch. Biochem. Biophys.,* 28, 450.
418. Haxo, F. (1952), *Biol. Bull.,* 103, 286.
419. Hayes, W. (1952), *Nature,* 169, 118.
420. Hayes, W. (1953), *J. Gen. Microbiol.,* 8, 72.
421. Hayward, M. D., and B. D. Bower (1960), *Lancet,* 2, 844.
422. Hearon, J. Z., S. A. Bernhard, S. L. Friess, D. J. Botts, and N. F. Morales (1959), in *The Enzymes,* vol. I, p. 49, edited by P. D. Boyer, H. Lardy, and K. Myrbäck, Academic Press, New York.
423. Heidenthal, G. (1940), *Genetics,* 25, 197.
424. Hein, G., and C. Niemann (1961), *Proc. Nat. Acad. Sci.,* 47, 1341.
425. Helinski, D. R., and C. Yanofsky (1962), *Proc. Nat. Acad. Sci.,* 48, 173.
426. Henning, V., and C. Yanofsky (1962), *Proc. Nat. Acad. Sci.,* 48, 183.
427. Herbst, C. (1892), *Z. Wiss. Zool.,* 55, 446.
428. Hers, H. G. (1961), *Chem. Weekblad,* 57, 437.
429. Hersh, A. H. (1930), *J. Exp. Zool.,* 57, 283.
430. Hershey, A. D. (1953), *J. Gen. Physiol.,* 37, 1.
431. Hershey, A. D. (1958), *Cold Spring Harbor Symp. Quant. Biol.,* 23, 19.
432. Hershey, A. D., and E. Burgi (1956), *Cold Spring Harbor Symp. Quant. Biol.,* 21, 91.
433. Hershey, A. D., and M. Chase (1951), *Cold Spring Harbor Symp. Quant. Biol.,* 16, 471.
434. Hershey, A. D., and M. Chase (1952), *J. Gen. Physiol.,* 36, 39.

435. Hershey, A. D., and R. Rotman (1949), *Genetics,* 34, 44.
436. Hershey, A. D., A. Garen, D. Fraser, and J. D. Hudis (1954), *Carnegie Inst. Wash. Year Book,* 53, 210.
437. Hertwig, O. (1875), *Morphologisches Jahrbuch.*
438. Hertwig, O. (1893), *Arch. Mikrobiol. Anat.,* 42, 662.
439. Hexter, W. M. (1957), *Genetics,* 42, 376.
440. Hexter, W. M. (1958), *Proc. Nat. Acad. Sci.,* 44, 768.
441. Hirota, Y., and P. H. A. Sneath (1961), *Japan J. Genetics,* 36, 307.
442. Hirs, C. H. W., S. Moore, and W. H. Stein (1960), *J. Biol. Chem.,* 235, 633.
443. Hirsch, H. M. (1952), *Biochim. Biophys. Acta,* 9, 674.
444. Hogeboom, G. H., and W. C. Schneider (1952), *J. Biol. Chem.,* 204, 233.
445. Hogness, D. S. (1962), in *Molecular Control Cellular Activity,* edited by J. M. Allen, McGraw-Hill Book Co., New York.
446. Hogness, D., and H. K. Mitchell (1954), *J. Gen. Microbiol.,* 11, 401.
447. Hogness, D. S., M. Cohn, and J. Monod (1955), *Biochim. Biophys. Acta,* 16, 99.
448. Holden, J. (1962), in *Symposium on Free Amino Acids,* edited by J. Holden and E. Roberts, Elsevier Press, Amsterdam.
449. Hollaender, A., and C. W. Emmens (1941), *Cold Spring Harbor Symp. Quant. Biol.,* 9, 179.
450. Hollander, W. F., J. H. D. Bryan, and J. W. Gowen (1960), *Genetics,* 45, 412.
451. Holloway, B. W. (1955), *Genetics,* 40, 117.
452. Holtfreter, J. (1939), *Arch. Exp. Zellforsch.,* 23, 169.
453. Holtzer, H. (1961), in *Synthesis of Molecular and Cellular Structure,* edited by D. Rudnick, p. 13, The Ronald Press Co., New York.
454. Hopwood, D. A. (1959), *Ann. N.Y. Acad. Sci.,* 81, 887.
455. Horowitz, N. H. (1946), *J. Biol. Chem.,* 162, 413.
456. Horowitz, N. H., and H. K. Mitchell (1951), *Ann. Rev. Biochem.,* 20, 465.
457. Horowitz, N. H., and R. D. Owen (1954), *Ann. Rev. Physiol.,* 16, 81.
458. Hurwitz, I., A. Bresler, and R. Diringer (1960), *Biochim. Biophys. Res. Commun.,* 3, 15.
459. Horowitz, N. H., M. Fling, H. Macleod, and N. Sueoka (1960), *J. Mol. Biol.,* 2, 96.
460. Horowitz, N. H., M. Fling, H. Macleod, and N. Sueoka (1961), *Genetics,* 46, 1015.
461. Horowitz, N. H., M. Fling, H. Macleod, and Y. Watanabe (1961), *Cold Spring Harbor Symp. Quant. Biol.,* 26, 233.
462. Horstadius, S. (1936), *Mem. Musée Roy. Hist. Nat. Belg.,* 3, 801.
463. Horstadius, S. (1939), *Biol. Rev.,* 14, 132.
464. Horstadius, S., and T. Gustafson (1948), *Symp. Soc. Exp. Biol.,* 2, 50.
465. Hotchkiss, R. D. (1951), *Cold Spring Harbor Symp. Quant. Biol.,* 16, 457.
466. Hotchkiss, R. D., and A. H. Evans (1958), *Cold Spring Harbor Symp. Quant. Biol.,* 23, 85.
467. Hotchkiss, R. D., and A. H. Evans (1960), *Federation Proc.,* 19, 912.
468. Hotchkiss, R. D., and J. Marmur (1954), *Proc. Nat. Acad. Sci.,* 40, 55.
469. Houlahan, M. B., and H. K. Mitchell (1947), *Proc. Nat. Acad. Sci.,* 33, 223.

470. Houlahan, M. B., and H. K. Mitchell (1948), *Proc. Nat. Acad. Sci.*, 34, 465.
471. Hsu, T. C., and C. M. Pomerat (1953), *J. Morph.*, 93, 301.
472. Hsu, T. C., and C. E. Somers (1961), *Proc. Nat. Acad. Sci.*, 47, 396.
473. Huskins, C. L. (1948), *J. Heredity*, 39, 311.
473a. Hurwitz, J., J. J. Furth, M. Anders, P. J. Ortiz, and J. T. August (1961), *Cold Spring Harbor Symp. Quant. Biol.*, 26, 91.
474. Hutt, F. B. (1949), *Genetics of the Fowl*, McGraw-Hill Book Co., New York.
475. Ingram, V. M. (1956), *Nature*, 178, 792.
476. Ingram, V. M. (1959), in *Molecular Genetics and Human Disease*, edited by L. I. Gardner, Charles C Thomas Publisher, Springfield, Ill.
477. Ingram, V. M. (1961), *Hemoglobin and its Abnormalities*, Charles C Thomas, Springfield, Ill.
478. Ingram, V. M. (1962), in *Molecular Control of Cell Activity*, p. 179, edited by J. M. Allen, McGraw-Hill Book Co., New York.
479. Isbell, E. R., H. K. Mitchell, A. Taylor, and R. J. Williams (1942), *Univ. Texas Publ.*, 4237, 81.
480. Ishitani, C., Y. Ikeda, and K. Sagaguchi (1957), *J. Gen. Appl. Microbiol.*, Tokyo, 2, 401.
481. Itano, H. A. (1951), *Proc. Nat. Acad. Sci.*, 37, 775.
482. Itano, H. A., and J. V. Neal (1950), *Proc. Nat. Acad. Sci.*, 36, 613.
483. Ives, P. T. (1950), *Evolution*, 4, 236.
484. Jacob, F., and J. Monod (1961), *J. Mol. Biol.*, 3, 318.
485. Jacob, F., and J. Monod (1961), *Cold Spring Harbor Symp. Quant. Biol.*, 26, 193.
486. Jacob, F., and E. L. Wollman (1958), *Symp. Soc. Exp. Biol.*, 12, 75.
487. Jacob, F., and E. L. Wollman (1961), *Sexuality and the Genetics of Bacteria*, Academic Press, New York.
488. Jenkins, J. A., and G. Mackinney (1953), *Genetics*, 38, 107.
489. Jenkins, J. A., and G. Mackinney (1955), *Genetics*, 40, 715.
490. Jensen, K. A., I. Kirk, G. Kolmark, and M. Westergaard (1951), *Cold Spring Harbor Symp. Quant. Biol.*, 16, 245.
491. Jinks, J. L. (1952), *Proc. Roy. Soc. Ser. B*, 140, 83.
492. Jinks, J. L. (1961), *Heredity*, 16, 241.
493. Johannsen, W. (1911), *Am. Naturalist*, 45, 129.
494. Johnson, M. J. (1950), *Respiratory Enzymes*, p. 255, edited by H. J. Lardy, Burgess Publishing Co., Minneapolis, Minn.
495. Jones, D. F. (1950), *Genetics*, 35, 507.
496. Jordon, D. O. (1960), *The Chemistry of Nucleic Acids*, Butterworths, London.
497. Jorgensen, E. C., and T. A. Geissman (1955), *Arch. Biochem. Biophys.*, 55, 389.
498. Josse, J., and A. Kornberg (1960), *Federation Proc.*, 19, 305.
499. Judd, B. H. (1955), *Genetics*, 40, 739.
500. Judd, B. H. (1959), *Genetics*, 44, 34.
501. Judd, B. H. (1961), *Proc. Nat. Acad. Sci.*, 47, 545.
502. Kabat, E. (1956), *Blood Group Substances*, Academic Press, New York.
503. Käfer, E. (1958), *Advan. Genet.*, 9, 105.

504. Kaiser, A. D. (1955), *Virology,* 1, 424.
505. Kalckar, H. M. (1958), *Advan. Enzymol.,* 20, 111.
506. Kalckar, H. M. (1961), *Am. J. Clin. Nutr.,* 9, 676.
507. Kalckar, H. M., and T. A. Sundararajan (1961), *Cold Spring Harbor Symp. Quant. Biol.,* 26, 227.
508. Kalckar, H., K. Kurahashi, and E. Jordan (1959), *Proc. Nat. Acad. Sci.,* 45, 1776.
509. Kaplan, N. O., and M. M. Ciotti (1961), *Ann. N.Y. Acad. Sci.,* 94, 701.
510. Kaplan, R. W. (1947), *Z. Naturforschung,* 26, 308.
511. Karlson, P., and A. Schweiger (1961), *Z. Physiol. Chem.,* 323, 199.
512. Karrer, W. (1958), *Konstitution und Vorkommen der Organischen Pflanzenstoffe,* Birkhauser Verlag, Basel.
513. Kaudewitz, F. (1959), *Z. Naturforschung,* 14b, 528.
514. Kaufmann, B. P. (1947), *Am. Naturalist,* 81, 77.
515. Kaufmann, B. P., H. Gay, and M. R. McDonald (1960), *Intern. Rev. Cytol.,* 9, 77.
516. Keeler, C. E. (1928), *Z. Vergleich. Physiol.,* 7, 736.
517. Keil, B. (1962), *Ann. Rev. Biochem.,* 31, 139.
518. Kellenberger, G., M. L. Zichichi, and J. J. Weigle (1961), *Proc. Nat. Acad. Sci.,* 47, 869.
519. Kemp, N. E. (1956), *J. Biophys. Biochem. Cytol.,* 2, 281.
520. Kendrew, J. C., G. Bodo, H. M. Dintzes, R. G. Parrish, and H. Wyckoff (1958), *Nature,* 181, 662.
521. Kendrew, J. C. (1959), *Proc. Roy. Soc. (London),* Ser. A, 253, 70.
522. Kendrew, J. C., R. E. Dickerson, B. E. Strandberg, R. G. Hart, and D. R. Davies (1960), *Nature,* 185, 422.
523. King, F. E., and W. Bottomley (1954), *J. Chem. Soc. (London),* 1399.
524. King, R. C. (1962), *Genetics,* Oxford University Press, London.
525. King, T. J., and R. Briggs (1956), *Cold Spring Harbor Symp. Quant. Biol.,* 21, 271.
526. Kiritani, K. (1962), *Japan J. Genetics,* 37, 42.
527. Kit, S., C. Beck, O. L. Graham, and A. Gross (1958), *Cancer Res.,* 18, 598.
528. Kogut, M., M. Pollock, and E. J. Tridgell (1956), *Biochem. J.,* 62, 391.
529. Konigsberg, W., and R. J. Hill (1961), *Intern. Congr. Biochem., 5th, Moscow,* Symp. No. 1.
530. Kornberg, A. (1962), in *Molecular Control of Cellular Activity,* p. 245, edited by J. M. Allen, McGraw-Hill Book Co., New York.
530a. Kornberg, A. (1962), *Enzymatic Synthesis of DNA,* John Wiley and Sons, New York.
531. Kornberg, A., S. B. Zimmerman, S. R. Kornberg, and J. Josse (1959), *Proc. Nat. Acad. Sci.,* 45, 772.
532. Kornberg, S. R., S. B. Zimmerman, and A. Kornberg (1961), *J. Biol. Chem.,* 236, 1487.
533. Koshland, D. E., Jr. (1959), in *The Enzymes,* vol. I, p. 305, edited by P. D. Boyer, H. Lardy, K. Myrbäck, Academic Press, New York.
534. Koshland, D. E., Jr. (1960), *Advan. Enzymol.,* 22, 45.
535. Kotval, J. P., and L. H. Gray (1947), *J. Genet.,* 48, 135.
536. Kozloff, L. N. (1953), *Cold Spring Harbor Symp. Quant. Biol.,* 18, 209.

537. Krebs, H. A. (1962), in *The Molecular Control of Cellular Activity*, p. 279, edited by J. M. Allen, McGraw-Hill Book Co., New York.
538. Kroeger, H. (1960), *Chromosoma*, 11, 129.
539. Krooth, R. S., and A. Weinberg (1960), *Biochem. Biophys. Res. Commun.*, 3, 518.
540. Kruis, K., and J. Satava (1918), *Nakl. C., Akar. Praha*, 67 pp.
541. Kubitschek, H. E., and H. E. Bundigkeit (1961), *Genetics*, 46, 105.
542. Kunitz, M., and J. H. Northrup (1935), *J. Gen. Physiol.*, 18, 433.
543. Lacy, A. M., and D. M. Bonner (1958), *Proc. Intern. Congr. Genet., 10th, Montreal*, 2, 157.
544. LaDu, B. N., V. G. Zannoni, L. Lester, and J. E. Seegmiller (1958), *J. Biol. Chem.*, 230, 251.
545. Lamy, R. (1947), *J. Genet.*, 48, 223.
546. Landauer, W. (1935), *J. Genet.*, 30, 303.
547. Landauer, W. (1948), *Growth Symp.*, 12, 171.
548. Landauer, W. (1954), *J. Cell. Comp. Physiol.*, 43, Suppl. 1, 261.
549. Landauer, W. (1956), *J. Exp. Zool.*, 132, 25.
550. Lang, C. A., and P. Grant (1961), *Proc. Nat. Acad. Sci.*, 47, 1236.
551. Langridge, R., H. R. Wilson, C. W. Hooper, M. H. F. Wilkins, and L. D. Hamilton (1960), *J. Mol. Biol.*, 2, 19
552. Larner, J. (1962), *Ann. Rev. Biochem.*, 31, 569.
553. Lash, J., F. Zilliken, and F. A. Hommes (1961), *Am. Zool.*, 1, 367.
554. Laskowski, M., and M. Laskowski (1954), *Advan. Protein Chem.*, 9, 203.
555. Laufer, H. (1961), *Ann. N.Y. Acad. Sci.*, 94, 825.
556. Laughnan, J. R. (1948), *Genetics*, 33, 488.
557. Laughnan, J. R. (1952), *Genetics*, 37, 375.
558. Laughnan, J. R. (1955), *Proc. Nat. Acad. Sci.*, 41, 78.
559. Laughnan, J. R. (1955), *Am. Naturalist*, 89, 91.
560. Lawrence, W. J. C. (1950), *Biochem. Soc. Symp.*, No. 4, 3.
561. Lawrence, W. J. C. (1957), *Heredity*, 11, 337.
562. Lawrence, W. J. C., and J. R. Price (1940), *Biol. Rev.*, 15, 35.
563. Lawrence, W. J. C., and R. Scott-Moncrieff (1935), *J. Genet.*, 30, 155.
564. Lawrence, W. J. C., R. Scott-Moncrieff, and V. C. Sturgess (1939), *J. Genet.*, 38, 299.
565. Lea, D. E. (1947), *Actions of Radiations on Living Cells*, Cambridge University Press, New York.
566. Lea, D. E., and D. G. Catcheside (1942), *J. Genet.*, 44, 216.
567. Lebedeff, G. A. (1939), *Genetics*, 24, 553.
568. Lederberg, E. M. (1960), in Microbial Genetics, *10th Symp. Soc. Gen. Microbiol.*, Cambridge University Press, New York.
569. Lederberg, J. (1947), *Genetics*, 32, 505.
570. Lederberg, J. (1949), *Proc. Nat. Acad. Sci.*, 35, 178.
571. Lederberg, J. (1955), *J. Cellular Comp. Physiol.*, 45, Suppl. 2, 75.
572. Lederberg, J. (1957), *Proc. Nat. Acad. Sci.*, 43, 1060.
573. Lederberg, J., and T. Iino (1956), *Genetics*, 41, 743.
574. Lederberg, J., and E. L. Tatum (1946), *Nature*, 158, 558.
575. Lederberg, J., and N. Zinder (1948), *J. Am. Chem. Soc.*, 70, 4267.
576. Lederberg, J., L. L. Cavalli, and E. M. Lederberg (1952), *Genetics*, 37, 720.

577. Lefevre, G., Jr. (1955), *Genetics,* 40, 374.
578. Lehman, I. R., and E. A. Pratt (1960), *J. Biol. Chem.,* 235, 3254.
579. Lehninger, A. L. (1960), *Federation Proc.,* 19, 952.
580. Leidy, G., E. Hahn, and H. E. Alexander (1956), *J. Exp. Med.,* 104, 305.
581. L'Heritier, P. (1958), *Advan. Virus Res.,* 5, 195.
582. L'Heritier, P., and G. Teissier (1937), *Compt. Rend.,* 205, 1099.
583. Lein, J., and P. S. Lein (1950), *J. Bacteriol.,* 60, 185.
584. Lein, J., and P. S. Lein (1952), *Proc. Nat. Acad. Sci.,* 38, 44.
585. Lein, J., H. K. Mitchell, and M. B. Houlahan (1948), *Proc. Nat. Acad. Sci.,* 34, 435.
586. LeJeune, J., R. Turpin, and F. Gautier (1959), *Ann. Genet.,* 1, 41.
587. Leloir, L. F. (1951), in *Phosphorus Metabolism,* 1, 67, edited by W. D. McElroy and B. Glass, The Johns Hopkins Press, Baltimore.
588. Leloir, L. F., J. M. Olavarria, S. H. Goldemberg, and H. Carmenatti (1959), *Arch. Biochem. Biophys.,* 81, 508.
589. Lengyel, P., J. F. Speyer, and S. Ochoa (1961), *Proc. Nat. Acad. Sci.,* 47, 1936.
590. Lerman, L. S., and L. J. Tolmach (1959), *Biochim. Biophys. Acta,* 33, 371.
591. Lester, H. E., and S. R. Gross (1959), *Science,* 129, 572.
592. Leupold, U. (1956), *J. Genetics,* 54, 411.
593. Leupold, U. (1958), *Cold Spring Harbor Symp. Quant. Biol.,* 23, 161.
594. Leupold, U., and N. H. Horowitz (1951), *Cold Spring Harbor Symp. Quant. Biol.,* 16, 65.
595. Levene, P. A., and L. W. Bass (1931), *Nucleic Acids,* Chemical Catalog Co., New York.
596. Levine, R. P., and W. T. Ebersold (1958), *Cold Spring Harbor Symp. Quant. Biol.,* 23, 101.
597. Levine, R. P., and W. T. Ebersold (1960), *Ann. Rev. Microbiol.,* 14, 197.
598. Levine, R. P., and P. T. Ives (1953), *Proc. Nat. Acad. Sci.,* 39, 817.
599. Levinthal, C. (1954), *Genetics,* 39, 169.
600. Levinthal, C. (1956), *Proc. Nat. Acad. Sci.,* 42, 394.
601. Levinthal, C. (1958), *Proc. Intern. Congr. Genet., 10th, Montreal,* University of Toronto Press, Toronto.
602. Levinthal, C., and P. F. Davidson (1961), *Ann. Rev. Biochem.,* 30, 641.
603. Levintow, L., and H. Eagle (1961), *Ann. Rev. Biochem.,* 30, 605.
604. Lewin, R. A. (1953), *J. Genetics,* 51, 543.
605. Lewin, R. A. (1954), in *Sex in Microorganisms,* edited by D. H. Wenrich, AAAS Publication, Washington, D.C.
606. Lewis, E. B. (1945), *Genetics,* 30, 137.
607. Lewis, E. B. (1949), personal communication.
608. Lewis, E. B. (1950), *Advan. Genet.,* 3, 73.
609. Lewis, E. B. (1951), *Cold Spring Harbor Symp. Quant. Biol.,* 16, 159.
610. Lewis, E. B. (1952), *Proc. Nat. Acad. Sci.,* 38, 953.
611. Lewis, E. B. (1954), *Am. Naturalist,* 88, 225.
612. Lewis, E. B. (1955), *Am. Naturalist,* 89, 73.
613. Lincoln, R. E., and J. W. Porter (1950), *Genetics,* 35, 211.
614. Lindahl, P. E., and H. Holter (1940), *Compt. Rend. Trav. Lab. Carlsberg,* 23, 249.
615. Lindegren, C. C. (1955), *Science,* 121, 605.

616. Lindegren, C. C., and G. Lindegren (1947), *Ann. Missouri Botan. Garden,* 34, 95.
617. Lindegren, C. C., and G. Lindegren (1951), *J. Gen. Microbiol.,* 5, 885.
618. Lindstrom, E. W. (1941), *Genetics,* 26, 387.
619. Linderstrom-Lang, K., and J. A. Schellman (1959), in *The Enzymes,* vol. I, p. 443, edited by P. D. Boyer, H. Lardy, K. Myrbäck, Academic Press, New York.
620. Lints, F. A. (1960), *Genetica,* 31, 188.
621. Litman, R. M., and A. B. Pardee (1956), *Nature,* 178, 529.
622. Loper, J .C. (1961), *Proc. Nat. Acad. Sci.,* 47, 1440.
623. Lwoff, A. (1950), *Problems of Morphogenesis in Ciliates,* John Wiley and Sons, New York.
624. Lwoff, A., and H. Dusi (1935), *Compt. Rend. Soc. Biol.,* 119, 1092.
625. Luce, W. M., H. Quastler, and L. S. Skaggs (1949), *Am. J. Roentgenol. Radium Therapy,* 62, 555.
626. Lumry, R. (1959), in *The Enzymes,* vol. I, p. 157, edited by P. D. Boyer, H. Lardy, and K. Myrbäck, Academic Press, New York.
627. Luria, S. E., and M. Delbrück (1943), *Genetics,* 28, 491.
628. Luria, S. E., and R. Dulbecco (1949), *Genetics,* 34, 93.
629. Luria, S. E., D. K. Fraser, J. N. Adams, and J. W. Burrous (1958), *Cold Spring Harbor Symp. Quant. Biol.,* 23, 71.
630. Mackinney, G., and J. A. Jenkins (1949), *Proc. Nat. Acad. Sci.,* 35, 284.
631. MacKendrick, M. E., and G. Pontecorvo (1952), *Experientia,* 8, 390.
632. Mackinney, G., and J. A. Jenkins (1952), *Proc. Nat. Acad. Sci.,* 38, 48.
633. McCalla, D. R. (1962), *Science,* 137, 225.
634. McCarty, M., and O. T. Avery (1946), *J. Exp. Med.,* 83, 89.
635. McCauley, A. L., and J. M. Ford (1947), *Heredity,* 1, 247.
636. McClintock, B. (1934), *Z. Zellforsch. Mikroskop. Anat.,* 21, 294.
637. McClintock, B. (1944), *Genetics,* 29, 478.
638. McClintock, B. (1951), *Cold Spring Harbor Symp. Quant. Biol.,* 16, 13.
639. McClintock, B. (1956), *Cold Spring Harbor Symp. Quant. Biol.,* 21, 197.
640. McClintock, B. (1956), *Brookhaven Symp. Biol.,* 8, 58.
641. McClintock, B. (1961), *Am. Naturalist,* 95, 265.
642. McClung, C. E. (1902), *Biol. Bull.,* 3, Nos. 1 and 2.
643. McCune, D. C. (1961), *Ann. N.Y. Acad. Sci.,* 94, 723.
644. McElroy, W. D. (1947), *Quart. Rev. Biol.,* 22, 25.
645. McElroy, W. D., and H. H. Siliger (1961), in *Light and Life,* p. 219, edited by W. D. McElroy and B. Glass, The Johns Hopkins University Press, Baltimore.
646. McElroy, W. D., and C. P. Swanson (1951), *Quart. Rev. Biol.,* 26, 348.
647. McGeachin, R. L., and J. M. Reynolds (1961), *Ann. N.Y. Acad. Sci.,* 94, 996.
648. Maas, W. K., and B. D. Davis (1952), *Proc. Nat. Acad. Sci.,* 38, 785.
649. Macklin, M. T. (1960), *Am. J. Human Genet.,* 12, 1.
650. Magni, G. E. (1952), *1st Lombardi Sci. Lettere,* 75, 3.
651. Maguire, M. P. (1960), *Genet. Res.,* 1, 487.
652. Mampell, K. (1946), *Genetics,* 31, 589.
653. Mandelstam, J. (1960), *Bacteriol. Rev.,* 24, 269.
654. Mangelsdorf, P. C., and G. S. Fraps (1931), *Science,* 73, 241.

655. Marcou, D., and J. Schecroun (1959), *Compt. Rend. Acad. Sci.,* Paris, 248, 280.

656. Margoliash, E. (1961), *Ann. Rev. Biochem.,* 30, 549.

657. Margolin, P. (1956), *Genetics,* 41, 685.

658. Markert, C. L., and E. Appella (1961), *Ann. N.Y. Acad. Sci.,* 94, 678.

659. Markert, C. L., and F. Møller (1959), *Proc. Nat. Acad. Sci.,* 45, 753.

660. Markert, C. L., and W. K. Silvers (1956), *Genetics,* 41, 429.

661. Marmur, J., and D. Lane (1960), *Proc. Nat. Acad. Sci.,* 46, 453.

662. Mason, H. S. (1959), in *Pigment Cell Biology,* edited by M. Gordon, Academic Press, New York.

663. Matthaei, J. H., O. W. Jones, R. G. Martin, and M. W. Nirenberg (1962), *Proc. Nat. Acad. Sci.,* 48, 673.

664. Mazia, D. (1961), *Ann. Rev. Biochem.,* 30, 669.

665. Mendel, G. (1866), *Verhandl. Naturforsch. Vereins in Brünn,* 4.

666. Meselson, M., and F. W. Stahl (1958), *Proc. Nat. Acad. Sci.,* 44, 671.

667. Meselson, M., and J. J. Weigle (1961), *Proc. Nat. Acad. Sci.,* 47, 857.

668. Metzenberg, R. L., and H. K. Mitchell (1958), *Biochem. J.,* 68, 168.

669. Meyer, K. H. (1943), *Advan. Enzymol.,* 3, 126.

670. Michaelis, L., and M. L. Menton (1913), *Biochem. Z.,* 49, 333.

671. Michaelis, P. (1940), *Z. Induktive Abstammungs-Vererbungslehre,* 78, 187.

672. Michaelis, P. (1951), *Cold Spring Harbor Symp. Quant. Biol.,* 16, 121.

673. Michelson, A. M. (1961), *Ann. Rev. Biochem.,* 30, 133.

674. Michelson, A. M., W. Drell, and H. K. Mitchell (1951), *Proc. Nat. Acad. Sci.,* 37, 396.

675. Miescher, F. (1871), *Hoppe Selye Mediz. Chem. Unters. Labor. Angew. Chem.,* 4, 441.

676. Miescher, F. (1897), *Histochemischen und Physiologischen Arbeiten,* Vogel, Leipzig.

677. Miller, A. Unpublished.

678. Miller, W. J. (1954), *Genetics,* 39, 938.

679. Mirsky, A. E., and V. Allfrey (1958), in *A Symposium on the Chemical Basis of Development,* edited by W. D. McElroy and B. Glass, The Johns Hopkins Press, Baltimore.

680. Mirsky, A. E., and S. Osawa (1961), in *The Cell,* vol. II, edited by J. Brachet and A. E. Mirsky, Academic Press, New York.

681. Mirsky, A. E., and H. Ris (1949), *Nature,* 163, 666.

682. Mirsky, A. E., and H. Ris (1951), *J. Gen. Physiology,* 34, 475.

683. Mitchell, H. K. Unpublished.

684. Mitchell, H. K., and M. B. Houlahan (1946), *Am. J. Botan.,* 33, 31.

685. Mitchell, H. K., and J. Lein (1948), *J. Biol. Chem.,* 175, 481.

686. Mitchell, M. B., and H. K. Mitchell (1952), *Proc. Nat. Acad. Sci.,* 38, 442.

687. Mitchell, H. K., and J. R. Simmons (1962), in *Free Amino Acids,* Elsevier Press, Amsterdam.

688. Mitchell, H. K., M. B. Houlahan, and J. F. Nyc (1948), *J. Biol. Chem.,* 172, 525.

689. Mitchell, M. B. (1955), *Proc. Nat. Acad. Sci.,* 41, 215.

690. Mitchell, M. B. (1955), *Proc. Nat. Acad. Sci.,* 41, 935.

691. Mitchell, M. B. (1957), *Compt. Rend. Lab. Carlsberg Ser. Physiol.,* 26, 285.

692. Mitchell, M. B., and H. K. Mitchell (1952), *Proc. Nat. Acad. Sci.*, 38, 205.
693. Mitchell, M. B., and H. K. Mitchell (1956), *J. Gen. Microbiol.*, 14, 84.
694. Mitchell, M. B., and H. K. Mitchell (1956), *Genetics*, 41, 319.
695. Mitchell, M. B., H. K. Mitchell, and A. Tissieres (1953), *Proc. Nat. Acad. Sci.*, 39, 606.
696. Mitchell, M. B., T. H. Pittenger, and H. K. Mitchell (1952), *Proc. Nat. Acad. Sci.*, 38, 569.
697. Mitchell, P. (1959), *Ann. Rev. Microbiol.*, 13, 407.
698. Miyake, T. (1960), *Genetics*, 45, 11.
699. Moat, A. C., N. Peters, Jr., and A. M. Srb (1959), *J. Bacteriol.*, 77, 673.
700. Moewus, F. (1940), *Z. Induktive Abstammungs-Vererbungslehre*, 78, 418.
701. Moewus, F. (1941), *Ergeb. Biol.*, 18, 257.
702. Monod, J. (1959), *Angew. Chem.*, 71, 685.
703. Monod, J., and F. Jacob (1961), *Cold Spring Harbor Symp. Quant. Biol.*, 26, 389.
704. Moore, B. S., and P. U. Angeletti (1961), *Ann. N.Y. Acad. Sci.*, 94, 659.
705. Moore, J. A. (1955), *Advan. Genet.*, 7, 139.
706. Moore, J. A. (1958), *Exp. Cell Res.*, 14, 532.
707. Moore, J. A. (1958), *Exp. Cell Res.*, Suppl. 6, 179.
708. Moore, J. A. (1960), *Develop. Biol.*, 2, 535.
709. Morgan, T. H. (1928), *Theory of the Gene*, Yale Univ. Press, New Haven.
710. Morgan, T. H., C. B. Bridges, and J. Schultz (1937), *Carnegie Year Book*, 36, 301.
711. Morgan, W. T. J. (1954), *Lectures Sci. Basis Med.*, 4, 92.
712. Morgan, W. T. J. (1960), *Bull. Soc. Chim. Biol.*, 42, 1591.
713. Morrison, M. (1961), *Ann. Rev. Biochem.*, 30, 11.
714. Moyed, H. S., and B. Magasanik (1960), *J. Biol. Chem.*, 235, 149.
715. Mudd, S., L. C. Winterscheid, E. D. Delamater, and H. J. Henderson (1951), *J. Bacteriol.*, 62, 459.
716. Muller, H. J. (1927), *Science*, 66, 84.
717. Muller, H. J. (1928), *Genetics*, 13, 279.
718. Muller, H. J. (1932), *Proc. Intern. Congr. Genet.*, 6th, Ithaca, N.Y., 1, 213.
719. Muller, H. J. (1935), *J. Heredity*, 26, 469.
720. Muller, H. J. (1946), *Genetics*, 31, 55.
721. Muller, H. J. (1947), *Proc. Roy. Soc.*, Sect. B, 134, 1.
722. Muller, H. J. (1950), *J. Cellular Comp. Physiol.*, 35 (Suppl. 1), 9.
723. Muller, H. J. (1950), *Evidence of the Precision of Genetic Adaptation*, Harvey Lecture Series 43, 1947–1948, Charles C Thomas, Springfield, Ill.
724. Muller, H. J., and E. Altenburg (1919), *Proc. Soc. Exp. Biol. Med.*, 17, 10.
725. Muller, H. J., E. Carlson, and A. Schalet (1961), *Genetics*, 46, 213.
726. Mundry, K. W. (1959), *Virology*, 9, 722.
727. Mundry, K. W., and A. Z. Gierer (1958), *Nature*, 182, 1457.
728. Murray, N. E. (1960), *Heredity*, 15, 207.
729. Muscona, A. A. (1959), in Developing Cell Systems and Their Control, *Symp. Soc. Study of Develop. Growth*, edited by D. Rudnick, p. 45, The Ronald Press Co., New York.
730. Nägeli, C. (1884), *Mechanisch-physiologische Theorie der Abstammungslehre*, Munich and Leipzig.
731. Najjar, V. A. (1954), *Science*, 119, 631.

732. Nanney, D. L. (1956), *Am. Naturalist*, 90, 291.
733. Nanney, D. L. (1958), *Proc. Nat. Acad. Sci.*, 44, 712.
734. Nanney, D. L. (1960), *Am. Naturalist*, 94, 167.
735. Nason, A., and H. Takahashi (1958), *Ann. Rev. Microbiol.*, 12, 203.
736. Neel, J. V. (1942), *Genetics*, 27, 519.
737. Neel, J. V. (1951), *Blood*, 6, 389.
738. Neel, J. V. (1956), *Ann. Human Genet.*, 21, 1.
739. Nelson, T. C., and J. Lederberg (1954), *Proc. Nat. Acad. Sci.*, 40, 415.
740. Nirenberg, M. W., and J. H. Matthaei (1962), *Proc. Nat. Acad. Sci.*, 48, 108.
741. Newmeyer, D. (1954), *Genetics*, 39, 604.
742. Niu, M. C., C. C. Cordova, and L. C. Niu (1961), *Proc. Nat. Acad. Sci.*, 47, 1689.
743. Northrup, J. H., M. Kunitz, and R. M. Herriott (1948), in *Crystalline Enzymes, 2nd ed.*, Columbia University Press, New York.
744. Novick, A., and L. Szilard (1950), *Proc. Nat. Acad. Sci.*, 36, 708.
745. Novikoff, A. B. (1960), in *Biology of Pyelonephritis*, p. 113, edited by E. Quinn and K. Kass, Little, Brown and Co., Boston.
746. Novikoff, A. B. (1961), in *The Cell*, vol. II, edited by J. Brachet and A. G. Mirsky, Academic Press, New York.
747. Nuffer (1961), *Genetics*, 46, 625.
748. Nussbaum-Hilarowicz, J. (1917), *Z. Wiss. Zool.*, 117, 554.
749. Ochoa, S., and L. Heppel (1957), in *Chemical Basis of Heredity*, p. 615, edited by W. D. McElroy and B. Glass, The Johns Hopkins University Press, Baltimore.
750. Ofengand, E. J., M. Dieckman, and P. Berg (1961), *J. Biol. Chem.*, 236, 1741.
751. Ohanessian-Guilleman, A. (1959), *Ann. Genet.*, 1, 59.
752. Ohnishi, E., H. Macleod, and N. H. Horowitz (1962), *J. Biol. Chem.*, 237, 138.
753. Olenov, J. M. (1941), *Am. Naturalist*, 75, 580.
754. Oliver, C. P. (1940), *Proc. Nat. Acad. Sci.*, 26, 452.
755. Oliver, C. P. (1941), *Genetics*, 26, 163.
756. Oliver, C. P. (1947), *Univ. Texas Publ.*, 4720, 167.
757. Oliver, C. P., and M. M. Green (1944), *Genetics*, 29, 331.
758. Palade, G. E., and F. Sikevitz (1956), *J. Biophys. Biochem. Cytol.*, 2, 671.
759. Palay, S. L. (1958), *Frontiers in Cytology*, Yale University Press, New Haven, Conn.
760. Panshin, I. B. (1938), *Nature*, 142, 837.
761. Pardee, A. B. (1962), in *Molecular Control of Cell Activity*, p. 265, edited by J. M. Allen, McGraw-Hill Book Co., New York.
762. Pardee, A. B., F. Jacob, and J. Monod (1959), *J. Mol. Biol.*, 1, 165.
763. Pardee, A. B., and L. S. Prestidge (1959), *Biochim. Biophys. Acta*, 36, 545.
764. Pardee, A. B., and L. S. Prestidge (1961), *Biochim. Biophys. Acta*, 49, 77.
765. Pardee, A. B., and I. Williams (1953), *Ann. Inst. Pasteur*, 84, 117.
766. Partridge, C. W. H. (1960), *Biochem. Biophys. Commun.*, 3, 613.
767. Pascher, A. (1917), *Arch. Protistenk.*, 38, 1.
768. Pateman, J. A. (1960), *Genetics*, 45, 839.
769. Patterson, J. T. (1931), *Biol. Bull.*, 61, 133.

770. Patterson, J. T., and W. S. Stone (1952), *Evolution in the genus Drosophila*, The Macmillan Co., New York.
771. Patterson, J. T., M. S. Brown, and W. S. Stone (1940), *Univ. Texas Publ.*, 4032, 167.
772. Patterson, J. T., W. S. Stone, and S. Bedichek (1937), *Genetics*, 22, 407.
773. Pauling, L. (1945), *Nature of the Chemical Bond*, Cornell University Press, Ithaca, New York.
774. Pauling, L., and R. B. Corey (1953), *Proc. Nat. Acad. Sci.*, 39, 84.
775. Pauling, L., and R. B. Corey (1956), *Arch. Biochem. Biophys.*, 65, 164.
776. Pauling, L., R. B. Corey, and H. R. Branson (1951), *Proc. Nat. Acad. Sci.*, 37, 205.
777. Pauling, L., H. A. Itano, S. J. Singer, and I. C. Wells (1949), *Science*, 110, 543.
778. Pavan, C. (1958), *Proc. Intern. Congr. Genetics, 10th, Montreal*, 1, 321, University of Toronto Press, Toronto.
779. Perkins, D. D. (1949), *Genetics*, 34, 607.
780. Perrin, D., F. Jacob, and J. Monod (1960), *Compt. Rend. Acad. Sci.*, Paris, 250, 155.
781. Perutz, M. F., M. G. Rossmann, A. F. Cullis, H. Muirhead, G. Will, and A. C. T. North (1960), *Nature*, 185, 416.
782. Peterson, P. A. (1960), *Genetics*, 45, 115.
783. Phinney, B. O., C. A. West, M. Ritzel, and P. M. Neely (1957), *Proc. Nat. Acad. Sci.*, 43, 398.
784. Pimenlal, G. C., and A. L. McClellan (1960), *The Hydrogen Bond*, W. H. Freeman and Co., San Francisco.
785. Pipkin, S. (1940), *Univ. Texas Publ. No. 4032*, 126.
786. Pipkin, S. (1960), *Genetics*, 45, 1205.
787. Pittenger, T. H. (1954), *Genetics*, 39, 326.
788. Pittenger, T. H. (1956), *Proc. Nat. Acad. Sci.*, 42, 747.
789. Plaut, G. W. E. (1960), *J. Biol. Chem.*, 235, PC41.
790. Plaut, G. W. E. (1961), *Ann. Rev. Biochem.*, 30, 409.
791. Plough, H. H. (1941), *Cold Spring Harbor Symp. Quant. Biol.*, 9, 127.
792. Plus, N. (1960), *Compt. Rend. Acad. Sci.*, Paris, 251, 1685.
793. Pollister, A. W. (1930), *J. Morphol.*, 49, 455.
794. Pomper, S., and P. R. Burkholder (1949), *Proc. Nat. Acad. Sci.*, 35, 456.
795. Ponder, E. (1961), in *The Cell*, vol. II, p. 1, edited by J. Brachet and A. E. Mirsky, Academic Press, New York.
796. Pontecorvo, G. (1952), *Nature*, 170, 204.
797. Pontecorvo, G. (1953), *Advan. Genet.*, 5, 141.
798. Pontecorvo, G. (1956), *Cold Spring Harbor Symp. Quant. Biol.*, 21, 171.
799. Pontecorvo, G. (1958), *Trends in Genetic Analysis*, Columbia University Press, New York.
800. Pontecorvo, G., and E. Kafer (1958), *Advan. Genet.*, 9, 71.
801. Pontecorvo, G., and G. Sermonti (1954), *J. Gen. Microbiol.*, 11, 94.
802. Pontecorvo, G., J. A. Roper, and E. Forbes (1953), *J. Gen. Microbiol.*, 8, 198.
803. Porter, K. R. (1961), in *The Cell*, vol. II, p. 621, edited by J. Brachet and A. E. Mirsky, Academic Press, New York.

804. Porter, K. R., and R. Machado (1962), European Regional Conference on Electron Microscopy.
805. Poulson, D. F. (1945), *Am. Naturalist,* 79, 340.
806. Poulson, D. F., and B. Sakaguchi (1961), *Science,* 133, 1489.
807. Pritchard, R. H. (1955), *Heredity,* 9, 343.
808. Pritchard, R. H. (1960), *Genet. Res.,* 1, 1.
809. Pritchard, R. H. (1960), in Microbial Genetics, *Tenth Symp. Soc. Gen. Microbiol.,* Cambridge University Press, New York.
810. Prout, T. C., H. Huebschman, H. Levine, and F. J. Ryan (1953), *Genetics,* 38, 518.
811. Provasoli, L., S. H. Hutner, and I. J. Pintner (1951), *Cold Spring Harbor Symp. Quant. Biol.,* 16, 113.
812. Preer, J. R. (1959), *J. Immunol.,* 83, 378.
813. Preer, J. R. (1959), *J. Immunol.,* 83, 385.
814. Preer, J. R. (1959), *Genetics,* 44, 803.
815. Preer, J. R., and L. B. Preer (1959), *J. Protozool.,* 6, 88.
816. Price, J. R. (1939), *J. Chem. Soc.,* 1939, 1017.
817. Pusztai, A., and W. T. J. Morgan (1961), *Biochem. J.,* 78, 135.
818. Raffel, D., and H. J. Muller (1940), *Genetics,* 25, 541.
819. Rajewski, B. N., and N. W. Timofeeff-Ressovsky (1939), *Z. Induktive Abstammungs-Vererbungslehre,* 77, 488.
820. Randolph, L. F., and D. B. Hand (1940), *J. Agric. Res.,* 60, 51.
821. Raper, A. B. (1956), *Brit. Med. J.,* 1, 965.
822. Raper, J. R. (1954), in *Sex in Microorganisms,* Publication of American Association for the Advancement of Science.
822a. Rapoport, J. A. (1947), *Am. Naturalist,* 81, 30.
823. Raut, C. (1950), *Genetics,* 35, 381.
824. Raut, C. (1954), *J. Cellular Comp. Physiol.,* 44, 463.
825. Raven, C. P., and L. H. Bretschneider (1942), *Arch. Neerl. Zool.,* 6, 255.
826. Ravin, A. W. (1961), *Advan. Genet.,* 10, 61.
827. Regehr, H., T. J. Arnasow, and H. E. Johns (1950), *Nature,* 166, 228.
828. Regnery, D. C. (1944), *J. Biol. Chem.,* 154, 151.
829. Rendel, J. M. (1953), *Am. Naturalist,* 87, 129.
830. Renner, O. (1936), *Flora,* 130, 218.
831. Reverberi, G. (1957), *Publ. Am. Assoc. Advan. Sci.,* 48, 319.
832. Rhoades, M. M. (1933), *J. Genet.,* 27, 71.
833. Rhoades, M. M. (1941), *Cold Spring Harbor Symp. Quant. Biol.,* 9, 138.
834. Rhoades, M. M. (1945), *Proc. Nat. Acad. Sci.,* 31, 91.
835. Rhoades, M. M. (1946), *Cold Spring Harbor Symp. Quant. Biol.,* 11, 202.
836. Rhuland, L. E., *Nature,* 185, 224.
837. Rhyne, C. (1960), *Genetics,* 45, 59.
838. Rich, A., and D. W. Green (1961), *Ann. Rev. Biochem.,* 30, 93.
839. Richards, F. M. (1958), *Proc. Nat. Acad. Sci.,* 44, 162.
840. Rimington, C. (1950), *Biochem. Soc. Symp., Cambridge, Engl.,* 4, 16.
841. Ris, H. (1957), in *Chemical Basis of Heredity,* p. 23, edited by W. D. McElroy and B. Glass, The Johns Hopkins University Press, Baltimore.
842. Ris, H., and A. E. Mirsky (1949), *J. Gen. Physiol.,* 32, 489.
843. Rivera, A. Jr., and P. R. Srinivason (1962), *Proc. Nat. Acad. Sci.,* 48, 864.
844. Rizet, G. (1952), *Rev. Cytol. Biol. Vegetales,* 13, 51.

845. Rizet, G., D. Marcou, and J. Schecroun (1958), *Bull. Soc. Franc. Physiol. Vegetale*, 4, 136.
846. Robbins, P. W. (1960), *Federation Proc.*, 19, 193.
847. Robertson, J. G. (1942), *J. Exp. Zool.*, 89, 197.
848. Robinson, G. M., and R. Robinson (1932), *Biochem. J.*, 26, 1647.
849. Rogers, P., and W. D. McElroy (1955), *Proc. Nat. Acad. Sci.*, 41, 67.
850. Roman, H. (1956), *Cold Spring Harbor Symp. Quant. Biol.*, 21, 175.
851. Roman, H., D. C. Hawthorne, and H. C. Douglas (1951), *Proc. Nat. Acad. Sci.*, 37, 79.
852. Roper, J. A. (1952), *Experientia*, 8, 14.
853. Roper, J. A., and R. H. Pritchard (1955), *Nature*, 175, 639.
854. Rose, S. M. (1952), *Am. Naturalist*, 86, 337.
855. Ross, H. (1948), *Z. Induktive Abstammungs-Vererbungslehre*, 82, 187.
856. Roux, W. (1883), *Über die Bedeutung der Kerntheilungsfiguren*, Engelmann, Leipzig.
857. Rudnick, D. (1945), *J. Exp. Zool.*, 100, 1.
858. Runnstrom, J. (1933), *Roux Arch. Entw. Mech. Organ.*, 129, 442.
859. Russell, E. S. (1939), *Genetics*, 24, 332.
860. Russell, E. S. (1948), *Genetics*, 33, 228.
861. Russell, E. S. (1949), *Genetics*, 34, 133.
862. Russell, E. S. (1949), *Genetics*, 34, 146.
863. Russell, L. B., and J. W. Baugham (1961), *Genetics*, 46, 509.
864. Russell, L. B., and E. H. Y. Chu (1961), *Proc. Nat. Acad. Sci.*, 47, 571.
865. Russell, L. B., and W. L. Russell (1948), *Genetics*, 33, 237.
866. Russell, W. L. (1939), *Genetics*, 24, 645.
867. Russell, W. L. (1952), *Cold Spring Harbor Symp. Quant. Biol.*, 17, 327.
868. Ryan, F. J., and K. Kiritani (1959), *J. Gen. Microbiol.*, 20, 644.
869. Ryan, F. J., and L. K. Schneider (1948), *J. Bacteriol.*, 56, 699.
870. Ryan, F. J., and L. K. Schneider (1949), *J. Bacteriol.*, 58, 201.
871. Sager, R. (1954), *Proc. Nat. Acad. Sci.*, 40, 356.
872. Sager, R. (1960), *Science*, 132, 1459.
873. Sager, R., and F. J. Ryan (1961), *Cell Heredity*, John Wiley and Sons, New York.
874. Sagisaka, S., and K. Shimura (1960), *Nature*, 184, 1709.
875. Saito, H., and Y. Ikedo (1959), *Ann. N.Y. Acad. Sci.*, 81, 862.
876. Sanger, F. (1952), *Advan. Protein Chem.*, 7, 1.
877. Satina, S., A. F. Blakeslee, and A. G. Avery (1937), *J. Heredity*, 28, 193.
878. Sax, K. (1941), *Cold Spring Harbor Symp. Quant. Biol.*, 9, 93.
879. Sax, K. (1950), *J. Cellular Comp. Physiol.*, Suppl. 1, 35, 71.
880. Schmid, W. (1949), *Z. Vererbungslehre*, 83, 220.
881. Schmid, W. (1961), *Am. Naturalist*, 95, 103.
882. Schoenheimer, R. (1942), *Edward K. Dunham Lectures, 1941*, Harvard University Press, Cambridge, Mass.
883. Schroeder, W., J. T. Jones, J. R. Shelton, J. Cormick, and K. McCalla (1961), *Proc. Nat. Acad. Sci.*, 47, 811.
884. Schrödinger, E. (1945), *What is Life?*, Cambridge University Press, New York.
885. Schultz, J. (1935), *Am. Naturalist*, 49, 30.
886. Schultz, J., and C. B. Bridges (1932), *Am. Naturalist*, 66, 323.

887. Schulman, H. M., and D. M. Bonner (1962), *Proc. Nat. Acad. Sci.*, 48, 53.
888. Schwartz, D. (1962), *Proc. Nat. Acad. Sci.*, 48, 750.
889. Schwartz, D. (1960), *Proc. Nat. Acad. Sci.*, 46, 1210.
890. Scott, H. M., C. C. Morrill, J. O. Alberts, and E. Roberts (1950), *J. Heredity*, 41, 255.
891. Sears, E. R. (1948), *Advan. Genet.*, 2, 239.
892. Seecof, R. L. (1962), *Cold Spring Harbor Symp. Quant. Biol.*, 27, 501.
893. Seegers, W. H. (1955), *Advan. Enzymol.*, 16, 23.
894. Segal, H. L. (1959), in *The Enzymes*, vol. I, p. 1, edited by P. D. Boyer, H. Lardy, and K. Myrbäck, Academic Press, New York.
895. Seidel, L. (1932), *Arch. Entwicklungsmech. Organ.*, 123, 213.
896. Serebrovsky, A. S. (1927), *J. Genet.*, 18, 137.
897. Sermonti, G., and I. Spada-Sermonti (1956), *J. Gen. Microbiol.*, 15, 609.
898. Sermonti, G., and I. Spada-Sermonti (1959), *Ann. N.Y. Acad. Sci.*, 81, 854.
899. Serra, J. A. (1958), *Port. Acta Biol. Ser. A*, 5, 126.
900. Seyffert, W. (1955), *Z. Induktive Abstammungs-Vererbungslehre*, 87, 311.
901. Shapiro, H. S., and E. Chargaff (1960), *Nature*, 188, 62.
902. Shaver, J. R. (1957), *Publ. Am. Assoc. Advan. Sci.*, 48, 263.
903. Shaw, M. W., H. F. Falls, and J. V. Neel (1960), *Am. J. Human Genet.*, 12, 389.
904. Sherratt, H. S. A. (1958), *J. Genet.*, 56, 28.
905. Shriner, R. L., and R. B. Moffett (1941), *J. Am. Chem. Soc.*, 63, 1694.
906. Sidbury, J. B. (1961), in *Molecular Genetics and Human Disease*, p. 61, edited by L. I. Gardner, Charles C Thomas, Publisher, Springfield, Ill.
907. Silagi, S. (1962), *Develop. Biol.*, 5, 35.
908. Silow, R. A. (1939), *J. Genet.*, 38, 229.
909. Silver, W. S., and W. D. McElroy (1954), *Arch. Biochim. Biophys.*, 51, 379.
910. Simonsen, D. H., and W. J. Wagtendonk (1952), *Biochim. Biophys. Acta*, 9, 515.
911. Simpson, M. V. (1962), *Ann. Rev. Biochem.*, 31, 333.
912. Sinnott, E. W., L. C. Dunn, and Th. Dobzhansky (1958), *Principles of Genetics*, 5th Edition, McGraw-Hill Book Co., New York.
913. Sinsheimer, R. L. (1959), *J. Mol. Biol.*, 1, 43.
914. Sinsheimer, R. L. (1960), *Ann. Rev. Biochem.*, 29, 503.
915. Sinsheimer, R. L. (1962), in *Molecular Control of Cellular Activity*, p. 221, edited by J. M. Allen, McGraw-Hill Book Co., New York.
916. Smith, J. D., and D. B. Dunn (1959), *Biochim. Biophys. Acta*, 31, 573.
917. Smith, O. H., and C. Yanofsky (1960), *J. Biol. Chem.*, 235, 2051.
918. Smith, P. E., and E. C. MacDowell (1930), *Anat. Record*, 46, 249.
919. Smithies, O. (1955), *Biochem. J.*, 61, 629.
920. Smithies, O., and G. E. Connell (1959), in *Ciba Found. Symp. Biochem. Human Genet.*, 178, Little, Brown and Co., Boston.
920a. Smithies, O., G. E. Connell, and G. H. Dixon (1962), *Nature*, 196, 232.
920b. Smyth, D. G., W. H. Stein, and S. Moore (1963), *J. Biol. Chem.*, 238, 228.
921. Sohval, A. R. (1961), *Am. J. Med.*, 31, 397.
922. Sonneborn, T. M. (1939), *Am. Naturalist*, 73, 390.
923. Sonneborn, T. M. (1947), *Advan. Genet.*, 1, 264.
924. Sonneborn, T. M. (1948), *Proc. Nat. Acad. Sci.*, 34, 413.

925. Sonneborn, T. M. (1950), *Heredity*, 4, 11.
926. Sonneborn, T. M. (1951), in *Genetics in the 20th Century*, edited by L. C. Dunn, The Macmillan Co., New York.
927. Sonneborn, T. M. (1954), *Caryologia*, Suppl. No. 6, *Proc. Intern. Cong. Genetics, 9th, Bellagio, Italy*, 307.
928. Sonneborn, T. M. (1957), in *The Species Problem*, p. 115, edited by E. Mayr, *Am. Assn. Advan. Sci.*, Washington, D.C.
929. Sonneborn, T. M. (1960), *Proc. Nat. Acad. Sci.*, 46, 149.
930. Sonneborn, T. M., and R. V. Dippell (1946), *Physiol. Zool.*, 19, 1.
931. Spackman, D. H., W. H. Stein, and S. Moore (1960), *J. Biol. Chem.*, 235, 648.
932. Spahr, P. F., and A. Tissieres (1959), *J. Mol. Biol.*, 1, 237.
933. Sparrow, A. H. (1961), Symposium on Mutation and Plant Breeding, *Nat. Acad. Sci., Nat. Res. Council Publ.*, 891.
934. Spemann, H. (1938), *Embryonic Development and Induction*, Yale University Press, New Haven, Conn.
935. Spencer, W. P. (1944), *Genetics*, 29, 520.
936. Spencer, W. P., and C. Stern (1948), *Genetics*, 33, 43.
937. Spiegel, M. (1954), *Biol. Bull.*, 107, 130.
938. Spiegel, M. (1955), *Ann. N.Y. Acad. Sci.*, 60, 1056.
939. Spiegelman, S., B. O. Hall, and R. Storck (1961), *Proc. Nat. Acad. Sci.*, 47, 1135.
940. Spizizen, J. (1959), *Federation Proc.*, 18, 957.
941. Spratt, N. (1950), *J. Exp. Zool.*, 114, 375.
942. Spratt, N. (1950), *Biol. Bull.*, 99, 120.
943. Sprinson, D. B., and D. Rittenberg (1950), *J. Biol. Chem.*, 184, 405.
944. Spurway, H., and J. B. S. Haldane (1954), *J. Genet.*, 52, 208.
945. Srb, A. M. (1959), *Cold Spring Harbor Symp. Quant. Biol.*, 23, 269.
946. Srb, A. M., and N. H. Horowitz (1944), *J. Biol. Chem.*, 154, 129.
947. Srb, A. M., and R. D. Owen (1952), *General Genetics*, W. H. Freeman and Co., San Francisco.
948. Stadler, L. J. (1942), *Some Observations on Gene Variability and Spontaneous Mutation*, Spragg Memorial Lectures (3rd Series), Michigan State College, East Lansing, Mich.
949. Stadler, D. R. (1959), *Genetics*, 44, 648.
950. Stadler, L. J. (1946), *Genetics*, 31, 377.
951. Stadler, L. J. (1948), *Am. Naturalist*, 82, 289.
952. Stadler, L. J. (1951), *Cold Spring Harbor Symp. Quant. Biol.*, 16, 49.
953. Stadler, L. J., and F. M. Uber (1942), *Genetics*, 27, 84.
954. Stanier, R. Y. (1951), *Ann. Rev. Microbiol.*, 5, 35.
955. Stanier, R. Y. (1954), in *Aspects of Synthesis and Order in Growth*, p. 43, edited by D. Rudnik, Princeton University Press, Princeton, N.J.
956. Steer, E. (1961), *Science*, 133, 2010.
957. Steinberg, C. M., and R. S. Edgar (1962), *Genetics*, 47, 187.
958. Steinmann, E. (1952), *Exp. Cell Res.*, 3, 367.
959. Steinmann, E., and F. S. Sjostrand (1955), *Exp. Cell Res.*, 8, 15.
960. Stent, G. (1958), *Advan. in Virus Res.*, 5, 95.
961. Stephens, S. G. (1947), *Advan. Gen.*, 1, 431.
962. Stephens, S. G. (1948), *Genetics*, 33, 191.

963. Stephens, S. G. (1948), *Arch. Biochem.*, 18, 449.
964. Stephens, S. G. (1951), *Advan. Genet.*, 4, 247.
965. Stephens, S. G. (1951), *Cold Spring Harbor Symp. Quant. Biol.*, 16, 131.
966. Stern, C. (1943), *Genetics*, 28, 441.
967. Stern, C. (1948), *Science*, 108, 615.
968. Stern, C. (1960), *Principles of Human Genetics*, W. H. Freeman and Co., San Francisco.
969. Stern, C., and E. W. Schaeffer (1943), *Proc. Nat. Acad. Sci.*, 29, 351.
970. Stern, H., and A. E. Mirsky (1952), *J. Gen. Physiol.*, 37, 177.
970a. Stevens, N. M. (1905), Carnegie Institute, Washington, Publ. 36.
971. Stevenson, A. C. (1957), *Am. J. Human Genet.*, 9, 81.
972. Steward, F. C., E. M. Schantz, J. K. Pollard, M. O. Mapes, and J. Mitra (1961), from *Synthesis of Molecular and Cellular Structures*, edited by D. Rudnick, p. 13, The Ronald Press Co., New York.
973. Stone, W. S. (1942), *Univ. Texas Publ.*, 4228, 146.
974. Stone, W. S., O. Wyss, and F. Haas (1947), *Proc. Nat. Acad. Sci.*, 33, 59.
975. Stone, W. S., F. Haas, J. B. Clark, and O. Wyss (1948), *Proc. Nat. Acad. Sci.*, 34, 142.
976. Stormont, C. (1955), *Am. Naturalist*, 89, 105.
977. Strasburger, E. (1884), *Neue Untersuchungen über den Befruchtungsvorgang*, Jena.
978. Straus, O. H., and Goldstein, A. (1943), *J. Gen. Physiol.*, 26, 559.
979. Straus, B. S. (1960), *An Outline of Chemical Genetics*, W. B. Saunders, Philadelphia.
980. Strauss, B. S., and S. Pierog (1954), *J. Gen. Microbiol.*, 10, 221.
981. Streisinger, G., and V. Bruce (1960), *Genetics*, 45, 1290.
982. Stubbe, H. (1935), *Z. Induktive Abstammungs-Vererbungslehre*, 70, 533.
983. Stubbe, H. (1941), *Z. Induktive Abstammungs-Vererbungslehre*, 79, 401.
984. Sturtevant, A. H. (1913), *J. Exp. Zool.*, 14, 43.
985. Sturtevant, A. H. (1925), *Genetics*, 10, 117.
986. Sturtevant, A. H. (1945), *Genetics*, 30, 297.
987. Sturtevant, A. H. (1951), *Proc. Nat. Acad. Sci.*, 37, 405.
988. Sturtevant, A. H., and Th. Dobzhansky (1936), *Am. Naturalist*, 70, 574.
989. Sueoka, N., J. Marmur, and P. Doty (1959), *Nature*, 183, 1429.
990. Suskind, S. R. (1957), in *Chemical Basis of Heredity*, edited by W. D. McElroy and B. Glass, The Johns Hopkins University Press, Baltimore.
991. Suskind, S. R., and L. I. Kurek (1959), *Proc. Nat. Acad. Sci.*, 45, 193.
992. Suskind, S. R., and C. Yanofsky (1961), in *Control Mechanisms in Cellular Processes*, p. 3, edited by D. M. Bonner, The Ronald Press Co., New York.
993. Sutton, W. S. (1902), *Biol. Bull.*, 4, 1.
994. Sutton, W. S. (1903), *Biol. Bull*, 4, 231.
995. Swain, C. G., and J. F. Brown (1959), in *The Enzymes*, edited by P. D. Boyer, H. Lardy, and K. Myrbäck, Academic Press, New York.
996. Swift, H. (1962), in *Molecular Control of Cell Activity*, p. 73, edited by J. M. Allen, McGraw-Hill Book Co., New York.
997. Tabor, H., C. W. Tabor, and S. M. Rosenthal (1961), *Ann. Rev. Biochem.*, 30, 579.
998. Tatum, E. L., and J. Lederberg (1947), *J. Bact.*, 53, 673.

999. Tatum, E. L., R. W. Baratt, and V. M. Cutler, Jr. (1949), *Science,* **109,** 509.
1000. Taylor, A. L., and E. A. Adelberg (1960), *Genetics,* **45,** 1233.
1001. Taylor, J. H. (1958), *Genetics,* **43,** 515.
1002. Taylor, J. H. (1960), *J. Biophys. Biochem. Cytol.,* **7,** 455.
1003. Taylor, J. H., P. S. Woods, and W. L. Hughes (1957), *Proc. Nat. Acad. Sci.,* **43,** 122.
1004. Thoday, J. M. (1942), *J. Genetics,* **43,** 189.
1005. Thoday, J. M. (1953), *Heredity,* **6** (Suppl.), 299.
1006. Thom, C., and R. A. Steinberg (1939), *Proc. Nat. Acad. Sci.,* **25,** 329.
1007. Thompson, P. E. (1960), *Genetics,* **45,** 1567.
1008. Timofeeff-Ressovsky, N. W. (1932), *Biol. Zentralblatt,* **52,** 468.
1009. Timofeeff-Ressovsky, N. W. (1935), *Z. Induktive Abstammungs-Vererbungslehre,* **70,** 125.
1010. Timofeeff-Ressovsky, N. W., and K. G. Zimmer (1935), *Strahlentherapie,* **53,** 134.
1011. Timofeeff-Ressovsky, N. W., and K. G. Zimmer (1938), *Naturwissenschaften,* **26,** 362.
1012. Tissieres, A., H. K. Mitchell, and F. A. Haskins (1953), *J. Biol. Chem.,* **205,** 423.
1013. Tissieres, A., J. D. Watson, D. Schlessinger, and B. R. Hollingsworth (1959), *J. Mol. Biol.,* **1,** 221.
1014. Tomes, M. L., F. W. Quackenbush, O. E. Nelson, and B. North (1953), *Genetics,* **38,** 117.
1015. Tomes, W. L., F. W. Quackenbush, and M. McQuiston (1954), *Genetics,* **39,** 810.
1016. Tomkins, G. M., and K. L. Yielding (1961), *Cold Spring Harbor Symp. Quant. Biol.,* **26,** 331.
1017. Touster, O. (1962), *Ann. Rev. Biochem.,* **31,** 407.
1018. Trautner, T. A., M. M. Schwarz, and A. Kornberg (1962), *Proc. Nat. Acad. Sci.,* **48,** 449.
1019. Treffers, H. P., V. Spinelli, and N. O. Belser (1954), *Proc. Nat. Acad. Sci.,* **40,** 1064.
1020. Trimble, H. C., and C. E. Keeler (1938), *J. Heredity,* **29,** 281.
1021. Tschermak, E. (1900), *Berichte Deutsch. Bot. Gesellsch.,* **18,** 232.
1022. Tso, P. O. P., J. Bonner, and H. Dentzis (1958), *Arch. Biochem. Biophys.,* **76,** 225.
1023. Tsugita, A. (1962), *J. Mol. Biol.,* **5,** 284.
1024. Tsugita, A., and H. Fraenkel-Conrat (1960), *Proc. Nat. Acad. Sci.,* **46,** 636.
1025. Tsugita, A., D. T. Gish, J. Young, H. Fraenkel-Conrat, C. A. Knight, and W. M. Stanley (1960), *Proc. Nat. Acad. Sci.,* **46,** 1463.
1026, Twort, F. W. (1915), *Lancet,* **2,** 124.
1027. Tyler, A. (1947), *Growth* (10 Suppl.), **7,** 19.
1028. Van Beneden, E. (1884), *Arch. Biol.,* Liege and Paris, **5,** 111.
1029. Van Beneden, E., and A. Neyt (1887), *Bull. Acad. Roy. Med. Belg. Ser. 3,* **14,** 215.
1030. Van Schaik, N. W., and R. A. Brink (1959), *Genetics,* **44,** 725.
1031. Virchow, R. (1858), *Die Cellularpathologie in ihrer Begründung auf physiologische und pathologische Geweblehre,* Berlin.
1032. Vesell, E. (1961), *Ann. N.Y. Acad. Sci.,* **94,** 877.

1033. Vinograd, J. R., W. P. Hutchinson, and W. A. Schroeder (1959), *J. Am. Chem. Soc.,* 81, 3168.

1034. Vinograd, J. R., and W. D. Hutchinson (1960), *Nature,* 187, 216.

1035. Visconti, N., and M. Delbrück (1953), *Genetics,* 38, 5.

1036. Viscontini, M. E., and E. Möhlmann (1959), *Helv. Chim. Acta,* 42, 836.

1037. Vogel, F. (1961), *Lehrbuch der Allgemeinen Human genetik,* Springer-Verlag.

1038. Vogel, H. J. (1961), *Cold Spring Harbor Symp. Quant. Biol.,* 26, 163.

1039. Vogel, R. H., and M. J. Kopac (1959), *Biochim. Biophys. Acta,* 36, 505.

1040. Volkin, E., and L. Astrachan (1957), *The Chemical Basis of Heredity,* edited by W. D. McElroy and B. Glass, Baltimore.

1041. von Borstel, R. C. (1957), *Publ. Am. Assoc. Advan. Sci.,* 48, 175.

1042. von Euler, H. (1946), *Arkiv Kemi Mineral. Geol.,* 24A, 13.

1043. Waddington, C. H. (1948), *Symp. Soc. Exp. Biol.,* 2, 145.

1044. Waddington, C. H. (1952), *Epigenetics of Birds,* Cambridge University Press, New York.

1045. Wadkins, C. L., and A. L. Lehninger (1960), *Proc. Nat. Acad. Sci.,* 46, 1582.

1046. Wagner, R. P. (1949), *Proc. Nat. Acad. Sci.,* 35, 185.

1047. Wagner, R. P. (1959), *Biological Contributions, Univ. Texas Publ.,* 5914, 261.

1048. Wagner, R. P., and A. Bergquist (1960), *Genetics,* 45, 1375.

1049. Wagner, R. P., and C. H. Haddox (1951), *Am. Naturalist,* 85, 319.

1050. Wagner, R. P., K. Kiritani, and A. Bergquist (1962), in *The Molecular Basis of Neoplasia,* University of Texas Press, Austin.

1051. Wagner, R. P., C. H. Haddox, R. Fuerst, and W. S. Stone (1950), *Genetics,* 35, 237.

1052. Ward, F. D. (1935), *Genetics,* 20, 230.

1053. Watkin, J. E., E. W. Underhill, and A. C. Neish (1957), *Canad. J. Biochem. Physiol.,* 35, 229.

1054. Watkins, W. M. (1960), *Bull. Soc. Chim. Biol.,* 42, 1599.

1055. Watson, J. D., and F. H. C. Crick (1953), *Nature,* 171, 737.

1056. Watson, J. D., and F. H. C. Crick (1953), *Cold Spring Harbor Symp. Quant. Biol.,* 18, 123.

1057. Webber, B. B., and M. E. Case (1960), *Genetics,* 45, 1605.

1058. Weier, T. E., and C. R. Stocking (1952), *Botan. Rev.,* 18, 75.

1059. Weismann, A. (1892), *Das Keimplasma, Eine Theorie der Vererbung,* Jena.

1060. Weiss, S. B. (1960), *Proc. Nat. Acad. Sci.,* 46, 1020.

1061. Weiss, J. (1944), *Nature,* 153, 748.

1062. Weiss, P. (1947), *Yale J. Biol. Med.,* 19, 235.

1063. Weiss, P. (1962), in *Molecular Control of Cellular Activity,* p. 1, edited by J. M. Allen, McGraw-Hill Book Co., New York.

1064. Weiss, P., and A. C. Taylor (1960), *Proc. Nat. Acad. Sci.,* 46, 1177.

1065. Welshons, W. J. (1958), *Cold Spring Harbor Symp. Quant. Biol.,* 23, 171.

1066. Welshons, W. J., and E. S. von Halle (1962), *Genetics,* 47, 743.

1067. Westergaard, M. (1957), *Experienta,* 13, 224.

1068. Westergaard, M. (1958), *Advan. Genet.,* 9, 217.

1069. Westergaard, M. (1960), *Abhandl. Deut. Akad. Wiss. Berlin Kl. Med.,* 3, 30.

1070. Westergaard, M., and H. Hirsch (1954), *Proc. Seventh Symp. Colston Res. Soc.,* **7**, 171.

1071. Westergaard, M., and H. K. Mitchell (1947), *Am. J. Botany,* **34**, 573.

1072. Westheimer, F. H. (1959), in *The Enzymes,* vol. I, p. 259, edited by P. R. Boyer, H. Lardy, and K. Myrbäck, Academic Press, New York.

1073. White, M. J. D. (1954), *Animal Cytology and Evolution,* 2nd Ed., Cambridge University Press, New York.

1074. Whiting, P. W. (1935), *J. Heredity,* **26**, 263.

1075. Whitman, C. O. (1894), *Biological Lectures,* Marine Biological Laboratory, Ginn and Co., Boston.

1076. Wieland, T., and G. Pfeiderer (1961), *Ann. N.Y. Acad. Sci.,* **94**, 691.

1077. Wilde, C. E. (1956), *J. Exp. Zool.,* **133**, 409.

1078. Wilken, D. R., and R. G. Hansen (1961), *J. Biol. Chem.,* **236**, 1051.

1079. Wilkins, M. H. F., A. R. Stokes, and H. R. Wilson (1953), *Nature,* **171**, 738.

1080. Williams, R. J. (1951), *Univ. Texas Publ.,* **5109**, 7.

1081. Wilson, E. B. (1900), *The Cell in Development and Inheritance,* The Macmillan Co., New York.

1082. Wilson, E. B. (1902), *Science,* **16**, 991.

1083. Wilson, E. B. (1925), *The Cell in Development and Heredity,* The Macmillan Co., New York.

1084. Wilson, E. B. (1928), *The Cell in Development and Heredity,* The Macmillan Co., New York.

1085. Wilson, E. B. (1931), *J. Morphol.,* **52**, 429.

1086. Wilson, G. B., T. Tsou, and P. Hyypio (1952), *J. Heredity,* **43**, 211.

1087. Wilson, P. W. (1950), *Respiratory Enzymes,* edited by H. J. Lardy, Burgess Publishing Co., Minneapolis, Minn.

1088. Winge, O. (1934), *Compt. Rend. Trav. Lab. Carlsberg,* **21**, 1.

1089. Winge, O. (1935), *Compt. Rend. Trav. Lab. Carlsberg,* **21**, 77.

1090. Winitz, M. (1962), *Symposium on Free Amino Acids,* edited by J. Holden and E. Roberts, Elsevier Press, Amsterdam.

1091. Wit, F. (1937), *Genetica,* **19**, 1.

1092. Witkin, E. M. (1953), *Proc. Nat. Acad. Sci.,* **39**, 427.

1093. Witkin, E. M. (1956), *Cold Spring Harbor Symp. Quant. Biol.,* **21**, 123.

1094. Witkin, E. M., and E. C. Thiel (1960), *Proc. Nat. Acad. Sci.,* **46**, 226.

1095. Wittman, H. G. (1960), *Virology,* **12**, 609.

1096. Wittman, H. G. (1961), *Naturwissenschaften,* **48**, 729.

1097. Wolf, B., and J. F. Nyc (1959), *Biophys. Biochem. Acta,* **31**, 208.

1098. Wollman, E. L., F. Jacob, and W. Hayes (1956), *Cold Spring Harbor Symp. Quant. Biol.,* **21**, 141.

1099. Woltereck, R. (1911), *Verhandl. Deut. Zool. Ges.,* 1911.

1100. Woods, D. D. (1940), *Brit. J. Exp. Path.,* **21**, 74.

1101. Woods, D. D., and P. Fildes (1940), *J. Soc. Chem. Ind.,* **59**, 133.

1102. Woodward, D. O. (1959), *Proc. Nat. Acad. Sci.,* **45**, 846.

1103. Woodward, D. O., C. W. H. Partridge, and N. H. Giles (1958), *Proc. Nat. Acad. Sci.,* **44**, 1237.

1104. Woodward, D. O., C. W. H. Partridge, and N. H. Giles (1960), *Genetics,* **45**, 535.

1105. Woodward, V. W. (1962), *Proc. Nat. Acad. Sci.,* **48**, 348.

1106. Woodward, V. W., J. R. DeZeeuw, and A. M. Srb (1954), *Proc. Nat. Acad. Sci.,* **40**, 192.
1107. Wright, S. (1916), *Carnegie Inst. Wash. Publ. No. 241.*
1108. Wright, S. (1941), *Proc. Intern. Genet. Congr., 7th* (1939), 319.
1109. Wright, S. (1941), *Physiol. Rev.,* **21**, 487.
1110. Wright, S. (1945), *Am. Naturalist,* **79**, 289.
1111. Wright, S. (1949), *Genetics,* **34**, 245.
1112. Wright, S. (1959), *Genetics,* **44**, 1001.
1113. Wright, S. (1960), *Genetics,* **45**, 1503.
1114. Wright, S., and Z. I. Braddock (1949), *Genetics,* **34**, 223.
1115. Wroblewski, F., and K. F. Gregory (1961), *Ann. N.Y. Acad. Sci.,* **94**, 912.
1116. Wyss, O., J. B. Clark, F. Haas, and W. S. Stone (1948), *J. Bacteriol.,* **54**, 767.
1117. Yamada, T. (1958), in *A Symposium on the Chemical Basis of Development,* p. 217, edited by W. D. McElroy and B. Glass, The Johns Hopkins University Press, Baltimore.
1118. Yamada, E., and K. R. Porter (1961), *J. Biophys. Biochem. Cytol.,* **4**, 329.
1119. Yanofsky, C. (1952), *Proc. Nat. Acad. Sci.,* **38**, 215.
1120. Yanofsky, C. (1956), in *Enzymes: Units of Biological Structure and Function,* edited by O. H. Gaebler, Academic Press, New York.
1121. Yanofsky, C. (1960), *Bacteriol. Rev.,* **24**, 221.
1122. Yanofsky, C., and D. M. Bonner (1955), *Genetics,* **40**, 761.
1123. Yanofsky, C., and I. P. Crawford (1959), *Proc. Nat. Acad. Sci.,* **45**, 1016.
1124. Yanofsky, C., and J. Stadler (1958), *Proc. Nat. Acad. Sci.,* **44**, 245.
1125. Yanofsky, C., and P. St. Lawrence (1960), *Ann. Rev. Microbiol.,* **14**, 311.
1126. Yanofsky, C., D. R. Helinski, and B. D. Maling (1961), *Cold Spring Harbor Symp. Quant. Biol.,* **26**, 11.
1127. Yotsuyanagi, V. (1955), *Nature,* **176**, 1208.
1128. Zamecnik, P. C. (1962), in *Molecular Control of Cellular Activity,* p. 259, edited by J. M. Allen, McGraw-Hill Book Co., New York.
1129. Zamenhof, S. (1957), in *Symposium on Chemical Basis of Heredity,* edited by. W. D. McElroy and B. Glass, The Johns Hopkins University Press, Baltimore.
1130. Zamenhof, S. (1960), *Proc. Nat. Acad. Sci.,* **46**, 101.
1131. Zamenhof, S., R. de Giovanni, and S. Greer (1958), *Nature,* **181**, 827.
1132. Zeigler, I. (1961), *Advan. Genetics,* **10**, 349.
1132a. Zelle, M. R. (1955), *Bacteriol. Rev.,* **19**, 32.
1133. Zelle, M. R., and A. Hollaender (1954), *J. Bacteriol.,* **68**, 210.
1134. Zelle, M. R., and J. Lederberg (1951), *J. Bacteriol.,* **61**, 351.
1135. Zimmer, J. K., and N. W. Timofeeff-Ressovsky (1942), *Z. Induktive Abstammungs-Vererbungslehre,* **80**, 353.
1136. Zinder, N. D., and J. Lederberg (1952), *J. Bacteriol.,* **64**, 679.
1137. Zubay, G. (1962), *Proc. Nat. Acad. Sci.,* **48**, 894.
1138. Zuitan, A. J. (1941), *Compt. Rend. Acad. Sci.,* URSS, **30**, 61.
1139. Zwilling, E. (1956), *Cold Spring Harbor Symp. Quant. Biol.,* **21**, 349.

Index

Disomics, 442
DNA, *see* Deoxyribonucleic acids
Dominance, 18, 341–346
Dopa, 309, 609
Dopaquinone, 309
Dosage, gene, 338
  radiation, 178–195
Doves (*Columba* sp.), 385
Drosophila, attached X, 173
  complementation, 382
  CO₂-sensitivity, 480
  eye-pigment mutants, 329
  eye pigments, 324
  phenocopies, 455, 465
  polytene chromosomes, 69
  radiation effects, 187–190
  sex determination, 438–440
  sex ratio character, 481
  suppressor genes, 398
  temperature-sensitive mutants, 459
  tissue transplant, 326
Drosophila mutants, bar, 347
  bithorax, 349
  bobbed, 345
  brown, 334
  cinnabar, 328
  cryptocephal, 546
  cubitus interruptus, 339, 344, 447
  dumpy, 382
  lozenge, 347
  meander, 546
  scarlet, 334
  shaven, 338
  stubble, 351
  translucida, 546
  vermilion, 328
  vestigial, 446
  white, 333, 354
  zeste, 355
Duplicate genes, 395
Duplication, 356
Dynamic state, 234–239
Dwarf mice, 548

Eclipse period, 114, 121
Ectodermal zone, 525
Electron transport, 41
  cytochromes, 230
Electrophoresis, blood proteins, 596

Electrophoresis, lactic dehydrogenase, 202, 256, 502
  peptides, 81, 620
Endodermal zone, 525
Endomitosis, 77
Endoplasmic reticulum, 30
Endosperm, 156, 175
Energy requirements, reactions, 241
Enhancer gene, 183
Environment, adaptation, 451
  phenocopies, 455
  temperature effects, 456
Enzyme complexes, 229–234
Enzyme induction, galactosidase, 412–419
Enzyme regulation model, 424
Enzymes, action on macromolecules, 258–263
  active sites, 84
  bifunctional, 275
  chemical modifications, 258
  cytochrome, 255
  dilution effects, 225
  energy barriers, 246
  general properties, 217
  hybrid, 257
  induced fit model, 86
  influence, 219
  inhibition, 225–229
  isozymes, 504
  kinetics, 220–229
  mechanisms, 243–247
  model systems, 245
  precursors, 261
  species differences, 255
  subunits, 257
  suppressor genes, 405
  temperature sensitive, 458
Ephestia mutants, 329, 336
Epigenesis, 2
Epigenetic factors, 476
Epilobium, cytoplasmic inheritance, 476
Epinephrine, 309
Episome, 130
Equilibrium constants, 241
Erymothecium, riboflavin, 319
Escherichia coli, 288
  chromosome transfer, 123
  complementation, 375